Physiology and Physiopathology of Adipose Tissue

Jean-Philippe Bastard · Bruno Fève
Editors

Physiology and Physiopathology of Adipose Tissue

Springer

Editors

Jean-Philippe Bastard
Institut des maladies cardiométaboliques
et de la nutrition (IHU ICAN)
Paris
France

and

Faculté de Médecine Pierre et Marie
Curie site Saint-Antoine
INSERM, UPMC, UMR_S938
Paris
France

and

Service de Biochimie et Hormonologie
Hôpital Tenon
Paris
France

Bruno Fève
Institut des maladies cardiométaboliques
et de la nutrition (IHU ICAN)
Paris
France

and

Faculté de Médecine Pierre et Marie
Curie site Saint-Antoine
INSERM, UPMC, UMR_S938
Paris
France

and

Service d'Endocrinologie-Métabolisme
Hôpital Saint-Antoine
Paris
France

ISBN 978-2-8178-0342-5 ISBN 978-2-8178-0343-2 (eBook)
DOI 10.1007/978-2-8178-0343-2
Springer Paris Heidelberg New York Dordrecht London

Library of Congress Control Number: 2012949073

Printed on acid-free paper

Springer is part of Springer Science+Business Media (www.springer.com)

Foreword

Created in 1983, the French Association of Study and Research on Obesity (AFERO) will soon be 30 years old. Since the 3rd "European Congress on Obesity" (ECO) in Nice in 1991 until the 19th ECO in Lyon in 2012, what a lot of road gone through! Twenty years later, our "musketeers at that time", Gérard Ailhaud, Bernard Guy-Grand, Max Lafontan, and Daniel Ricquier knew how to find the relief team and to pass on their expertise and their enthusiasm!

The French school of obesity has robust traditions. It always wanted to weave narrow links between researchers and clinicians, between basic research and clinical research. The presidents followed one another, by respecting at best this duality, Jacques Le Magnen (1983–1986), Stelio Nicolaïdis (1987–1990), Gérard Ailhaud (1991–1994), Bernard Guy-Grand (1995–1998), Max Lafontan (1999–2002), Martine Laville (2003–2006), Yannick Le Marchand Brustel (2007–2010), and Olivier Ziegler (since 2011).

We already had the journal Obesity (Springer) and an excellent textbook on "The Medicine and Surgery for Obesity" (A. Basdevant, on 2011), which is the second version of the textbook Medicine for Obesity, (A. Basdevant and B. Guy-Grand, on 2004), and proposes a global medical approach, by opening interdisciplinary perspectives, from biology to human and social sciences. But we missed a reference publication on the adipose tissue. We warmly thank Jean-Philippe Bastard and Bruno Fève for having taken up the challenge to design and coordinate this work, under the aegis of the AFERO both in French and in English.

The objective of this joint publication is to propose to the reader the analysis and the synthesis of a large number of experimental or clinical facts, keys, hypotheses, or leads for a better understanding of the multiple facets of this "organ's pathology with systemic impact" that is obesity. New functions and new concepts were described leading to new physiopathological paradigms, which are going to develop the clinical practices.

The medicine for obesity has to be a medicine based on the proofs. The clinical reasoning and the care of the patients must be supported by a scientific quality evidence-based approach. The plan for Obesity (2010–2013, piloted by

A. Basdevant) concerns the multiple determinants or the consequences of obesity but it also gives the possibility of reforming in-depth the French health system (notion of network of care, gradation of the levels of alternative…). It is thus crucial that obesity is understood as a progressive chronic disease, the therapeutic indications of which vary according to the natural history, the co-morbidities, and the benefit/risk balance of various treatments. Research showed us the key role of the adipose tissue in the "dialogue between organs", which in certain circumstances, can lead to the development of the metabolic and cardiovascular diseases but also to cancer. Millions of persons are concerned.

This didactic work gathers a large number of researchers and clinicians, among whom the educational qualities and the scientific relevance will help the reader to better understand the physiology and pathophysiology of adipose tissue. We gratefully thank them for the contribution to the success of this essential approach to the quality of care.

Olivier Ziegler
Service de Diabétologie
Maladies Métaboliques
Nutrition CHRU de Nancy-Hôpital Brabois Adultes
Université de Lorraine
Vandoeuvre-Les-Nancy
France

Foreword

Obesity is affecting more than 100 million citizens across Europe and is recognized as a major cause of morbidity, disability, and premature death. Importantly, it is recognized that obesity disproportionately affects socioeconomically disadvantaged populations and is a heavy burden not only on individuals but also on health care systems. A major concern in the development of obesity research is that data provided by basic and clinical investigators will be found relevant for policy makers and ultimately will have an impact at societal level.

It is one of the aims of the European Association for the Study of Obesity (EASO) to promote research into obesity and to facilitate contacts between individuals and organizations across Europe, to help tackle the epidemic of obesity in our region. EASO, founded in 1986, is an umbrella organization for national associations for the study of obesity that now exist in almost all European countries. EASO organizes every year, in collaboration with the national obesity association of the host country, the European Congress on Obesity (ECO). In 2012, the ECO will be held in Lyon, France.

At the occasion of the 2012 edition of ECO, J.-P. Bastard and B. Fève have assembled the contributions of an impressive group of French researchers to provide an overview of current knowledge on the physiology and pathophysiology of adipose tissue. France has been one of the leading country in obesity research for the last 30 years with a wide array of expertise in various aspects ranging from basic to more applied research. Both editors have to be commended for this timely initiative demonstrating the strength and dynamism of this obesity research community. The various chapters of this book provide both an introductory material to the field and an account of cutting edge research. This book will be a guide for readers in their discovery of one of the most fascinating tissues of the body, which has profound effects on health. It also helps organizations like EASO in disseminating findings from European obesity research to the largest possible audience.

Jean-Michel Oppert
Department of Nutrition
Pitié-Salpêtrière Hospital (AP-HP)
Human Nutrition Research Centre Ile-de-France (CRNH IdF)
Institute of Cardiometabolism and Nutrition (ICAN)
University Pierre et Marie Curie
Paris, France

Contents

Contributors

Maud Alligier Unité INSERM U1060 - INRA 1235 - INSA-Lyon - Université-Lyon1, Laboratoire CarMeN et Centre Européen Nutrition Santé. Centre de Recherche en Nutrition Humaine Rhône-Alpes, Lyon, France; Faculté de Médecine Lyon Sud-BP 12, 165 Chemin du Grand Revoyet, 69921 Oullins Cedex, France

Jean-Sébastien Annicotte Métabolisme et Cancer, IGMM Institut de Génétique Moléculaire de Montpellier CNRS-UMR 5535, 1919 Route de Mende, 34293 Montpellier-Cedex 5, France

Judith Aron-Wisnewsky Institut des maladies cardiométaboliques et de la nutrition (IHU ICAN), Paris, France; Nutriomique U872 équipe 7, Centre de recherche des Cordeliers, INSERM, UPMC, 15, rue de l'école de médecine, 75006 Paris, France; Centre de Recherche en Nutrition Humaine Ile-de-France (CRNH IdF), Paris, France; Assistance Publique Hôpitaux de Paris, Service de Nutrition, Bâtiment Benjamin Delessert, Hôpital Pitié-Salpêtrière, 91 boulevard de L'hôpital, 75013 Paris, France

Robert Barouki INSERM UMR-S 747, Paris, France; Centre Universitaire des Saints-Pères, Université Paris Descartes, 45 Rue des Saints-Pères, 75270 Paris Cedex 06, France; Assistance Publique-Hôpitaux de Paris, Hôpital Necker-Enfants Malades, 75015 Paris, France

Jean-Philippe Bastard Institut des maladies cardiométaboliques et de la nutrition (IHU ICAN), Paris, France; Faculté de Médecine Pierre et Marie Curie site Saint-Antoine, INSERM, UPMC, UMR_S938, 27 rue Chaligny, 75012 Paris, France; Service de Biochimie et Hormonologie, Hôpital Tenon, 4 rue de la Chine, 75970 Paris cedex 20, France

Arnaud Basdevant Institut des maladies cardiométaboliques et de la nutrition (IHU ICAN), Paris, France; Nutriomique U872 équipe 7, Centre de recherche des Cordeliers, INSERM, UPMC, 15, rue de l'école de médecine, 75006 Paris, France; Centre de Recherche en Nutrition Humaine Ile-de-France (CRNH IdF), Paris,

France; Assistance Publique Hôpitaux de Paris, Service de Nutrition, Bâtiment Benjamin Delessert, Hôpital Pitié-Salpêtrière, 91 boulevard de L'hôpital, 75013 Paris, France

Véronique Béréziat Institut des maladies cardiométaboliques et de la nutrition (IHU ICAN), Paris, France; Faculté de Médecine Pierre et Marie Curie site Saint-Antoine, INSERM, UPMC, UMR_S938, 27 rue Chaligny, 75012 Paris, France

Nathalie Billon Institute of Developmental Biology and Cancer, University of Nice Ophia-Antipolis, CNRS, UMR6543, 06107 Nice, France

Anne Bouloumié INSERM 1048, Team 1, I2MC, Batiment L4, CHU Rangueil, 1 Avenue Jean poulhès, BP 84225, 31432 Toulouse Cedex 4, France

Martine Caron-Debarle Institut des maladies cardiométaboliques et de la nutrition (IHU ICAN), Paris, France; Faculté de Médecine Pierre et Marie Curie site Saint-Antoine, INSERM, UPMC, UMR_S938, 27 rue Chaligny, 75012 Paris, France

Christian Carpéné Département Métabolisme et Obésité, Institut des maladies métaboliques et cardiovasculaires (I2MC), INSERM U1048/Université Paul Sabatier Toulouse Equipe n°3, BP 84225, 31432 Toulouse Cedex 4, France

Jacqueline Capeau Institut des maladies cardiométaboliques et de la nutrition (IHU ICAN), Paris, France; Faculté de Médecine Pierre et Marie Curie site Saint-Antoine, INSERM, UPMC, UMR_S938, 27 rue Chaligny, 75012 Paris, France; Service de Biochimie et Hormonologie, Hôpital Tenon, 4 rue de la Chine, 75970 Paris cedex 20, France

Louis Casteilla Equipe 1 "Plasticité des tissus adipeux", UMR UPS/CNRS/EFS 5273, U1031 INSERM, BP 84 225, 31432 Toulouse, France; Université de Toulouse, UPS; UMR 5241, BP 84 225, 31432 Toulouse, France

Isabelle Castan-laurell INSERM U1048 I2MC, 1 Ave J. Poulhès BP 84 225, 31432 Toulouse Cedex 4, France

Carine Chavey Métabolisme et Cancer, IGMM Institut de Génétique Moléculaire de Montpellier CNRS-UMR 5535, 1919 Route de Mende, 34293 Montpellier-Cedex 5, France

Karine Clément Institut des maladies cardiométaboliques et de la nutrition (IHU ICAN), Paris, France; Nutriomique U872 équipe 7, Centre de recherche des Cordeliers, INSERM, UPMC, 15, rue de l'école de médecine, 75006 Paris, France; Centre de Recherche en Nutrition Humaine Ile-de-France (CRNH IdF), Paris, France; Assistance Publique Hôpitaux de Paris, Service de Nutrition, Bâtiment Benjamin Delessert, Hôpital Pitié-Salpêtrière, 91 boulevard de L'hôpital, 75013 Paris, France

Mireille Cormont INSERM U895, Mediterranean Center of Molecular Medicine, Team "Cellular and Molecular Physiopathology of obesity and diabetes",

Bâtiment Archimed. 151, route de St Antoine de Ginestière, BP2 3194, 06204 Nice Cedex 3, France

Béatrice Cousin Equipe 1 "Plasticité des tissus adipeux", UMR UPS/CNRS/EFS 5273, U1031 INSERM, BP 84 225, 31432 Toulouse, France; Université de Toulouse, UPS; UMR 5241, BP 84 225, 31432 Toulouse, France

Elise Dalmas Institut des maladies cardiométaboliques et de la nutrition (IHU ICAN), Paris, France; Nutriomique U872 équipe 7, Centre de recherche des Cordeliers, INSERM, UPMC, 15, rue de l'école de médecine, 75006 Paris, France

Christian Dani Institute of Developmental Biology and Cancer, University of Nice Ophia-Antipolis, CNRS, UMR6543, 06107 Nice, France

Emmanuel Disse Unité INSERM U1060 - INRA 1235 - INSA-Lyon - Université-Lyon1, Laboratoire CarMeN et Centre Européen Nutrition Santé. Centre de Recherche en Nutrition Humaine Rhône-Alpes, Lyon, France; Faculté de Médecine Lyon Sud-BP 12, 165 Chemin du Grand Revoyet, 69921 Oullins Cedex, France

Cédric Dray INSERM U1048 I2MC, 1 Ave J. Poulhès BP 84 225, 31432 Toulouse Cedex 4, France

Isabelle Dugail Centre de Recherche des Cordeliers, INSERM, U872—Équipe 8 Pathologies métaboliques et nutritionnelles, Diabète, Obésité, 15 rue de l'école de médecine, 75006 Paris, France

Lluis Fajas Métabolisme et Cancer, IGMM Institut de Génétique Moléculaire de Montpellier CNRS-UMR 5535, 1919 Route de Mende, 34293 Montpellier-Cedex 5, France

Pascal Ferré Team 8 UMRS-872, Centre de Recherches des Cordeliers, 15 rue de l'Ecole de Médecine, 75270 Paris cedex 06, France

Bruno Fève Institut des maladies cardiométaboliques et de la nutrition (IHU ICAN), Paris, France; Faculté de Médecine Pierre et Marie Curie site Saint-Antoine, INSERM, UPMC, UMR_S938, Paris, France; Service d'Endocrinologie-Métabolisme, Hôpital Saint-Antoine, 27 rue Chaligny, 75571 Paris cedex 12, France

Fabienne Foufelle Team 8 UMRS-872, Centre de Recherches des Cordeliers, 15 rue de l'Ecole de Médecine, 75270 Paris cedex 06, France

Frédéric Fumeron INSERM U695 Déterminants génétiques du diabète de type 2 et de ses complications vasculaires UFR de Médecine Paris Diderot—Paris 7, site Bichat, 16, rue Henri Huchard, BP 416, 75870 Paris Cedex 18, France

Jean Galitzky INSERM 1048, Team 1, I2MC, Batiment L4, CHU Rangueil, 1 Avenue Jean poulhès, BP 84225, 31432 Toulouse Cedex 4, France

Michèle Guerre-Millo Institut des maladies cardiométaboliques et de la nutrition (IHU ICAN), Paris, France; Nutriomique U872 équipe 7, Centre de recherche des Cordeliers, INSERM, UPMC, 15, rue de l'école de médecine, 75006 Paris, France

Anne-Laure Hafner Institute of Developmental Biology and Cancer, University of Nice Ophia-Antipolis, CNRS, UMR6543, 06107 Nice, France

Jennifer Jager School of Medicine, Division of Endocrinology Diabetes and Metabolism, University of Pennsylvania, 735 Clinical Research Building, 415 Curie Boulevard, Philadelphia, PA 19104, USA

Vincent Kaddai Department of Pathology and Immunology, Geneva University, Centre Médical universitaire. 1, rue Michelet-Servet, 1211 Geneve 4, Switzerland

Claude Knauf INSERM U1048 I2MC, 1 Ave J. Poulhès BP 84 225, 31432 Toulouse Cedex 4, France

Michel Krempf Institut du Thorax, Endocrinologie, Maladies métaboliques et Nutrition Hôpital Laennec, 44093 Nantes CX, France

Nadège Lay Institute of Developmental Biology and Cancer, University of Nice Ophia-Antipolis, CNRS, UMR6543, 06107 Nice, France

Annie Ladoux Institute of Developmental Biology and Cancer, University of Nice Ophia-Antipolis, CNRS, UMR6543, 06107 Nice, France

Max Lafontan INSERM 1048, Team 1, I2 MC, Batiment L4, CHU Rangueil, 1 Avenue Jean poulhès, BP 84225, 31432 Toulouse cedex 4, France

Sylviane Lagarrigue Métabolisme et Cancer, IGMM Institut de Génétique Moléculaire de Montpellier CNRS-UMR 5535, 1919 Route de Mende, 34293 Montpellier-Cedex 5, France

Patrick Laharrague Equipe 1 "Plasticité des tissus adipeux", UMR UPS/CNRS/EFS 5273, U1031 INSERM, BP 84 225, 31432 Toulouse, France; Université de Toulouse, UPS; UMR 5241, BP 84 225, 31432 Toulouse, France; Laboratoire d'Hématologie, CHU Toulouse, TSA 50032, 31 059 Toulouse, France

Dominique Langin I2MC, INSERM UPS U1048, Obesity Research Laboratory, Team 4, CHU Rangueil, Institute of Metabolic and Cardiovascular Diseases, 1 avenue Jean Poulhès, BP 84225, 31432 Toulouse Cedex 4, France; UMR1048, University of Toulouse, Paul Sabatier University, Toulouse, France; Clinical Biochemistry Department, Toulouse University Hospitals, Toulouse, France

Martine Laville Unité INSERM U1060 - INRA 1235 - INSA-Lyon - Université-Lyon1, Laboratoire CarMeN et Centre Européen Nutrition Santé. Centre de Recherche en Nutrition Humaine Rhône-Alpes, Lyon, France; Faculté de

Médecine Lyon Sud-BP 12, 165 Chemin du Grand Revoyet, 69921 Oullins Cedex, France

Soizic Le Lay INSERM U1063, «Stress Oxydatif et pathologies métaboliques»—IBS-IRIS, Rue des Capucins, 49100 Angers, France

Anne Lorsignol Equipe 1 "Plasticité des tissus adipeux", UMR UPS/CNRS/EFS 5273, U1031 INSERM, BP 84 225, 31432 Toulouse, France; Université de Toulouse, UPS; UMR 5241, BP 84 225, 31432 Toulouse, France

Pauline Martin Institute of Developmental Biology and Cancer, University of Nice Ophia-Antipolis, CNRS, UMR6543, 06107 Nice, France

Yannick Le Marchand-Brustel INSERM U895, Centre Méditerranéen de Médecine Moléculaire, Université de Nice Sophia Antipolis, 151 route de St Antoine de Ginestière, BP 2 3194, 06204 Nice cedex 3, France

Tala Mohsen-Kanson Institute of Developmental Biology and Cancer, University of Nice Ophia-Antipolis, CNRS, UMR6543, 06107 Nice, France

Etienne Mouisel I2MC, INSERM UPS U1048, Obesity Research Laboratory, Team 4, CHU Rangueil, Institute of Metabolic and Cardiovascular Diseases, 1 avenue Jean Poulhès, BP 84225, 31432 Toulouse Cedex 4, France; UMR1048, University of Toulouse, Paul Sabatier University, Toulouse, France

Jean-Michel Oppert Institut des maladies cardiométaboliques et de la nutrition (IHU ICAN), Paris, France; Université Pierre et Marie Curie Paris 6, Paris, France; Centre de Recherche en Nutrition Humaine Ile-de-France (CRNH IdF), Paris, France; Assistance Publique Hôpitaux de Paris, Service de Nutrition, Bâtiment Benjamin Delessert, Hôpital Pitié-Salpêtrière, 91 boulevard de L'hôpital, 75013 Paris, France

Luc Pénicaud Centre des Sciences du Goût et de l'Alimentation, UMR 6265 CNRS, 1324 INRA-Université de Bourgogne, 15 rue Hugues Picardet, 21000 Dijon, France

Pascal Peraldi Institute of Developmental Biology and Cancer, University of Nice Ophia-Antipolis, CNRS, UMR6543, 06107 Nice, France

Magali Plaisant Institute of Developmental Biology and Cancer, University of Nice Ophia-Antipolis, CNRS, UMR6543, 06107 Nice, France

Valérie Planat-Benard Equipe 1 "Plasticité des tissus adipeux", UMR UPS/CNRS/EFS 5273, U1031 INSERM, BP 84 225, 31432 Toulouse, France; Université de Toulouse, UPS; UMR 5241, BP 84 225, 31432 Toulouse, France

Christine Poitou Institut des maladies cardiométaboliques et de la nutrition (IHU ICAN), Paris, France; Nutriomique U872 équipe 7, Centre de recherche des

Cordeliers, INSERM, UPMC, 15, rue de l'école de médecine, 75006 Paris, France; Centre de Recherche en Nutrition Humaine Ile-de-France (CRNH IdF), Paris, France; Assistance Publique Hôpitaux de Paris, Service de Nutrition, Bâtiment Benjamin Delessert, Hôpital Pitié-Salpêtrière, 91 boulevard de L'hôpital, 75013 Paris, France

Daniel Ricquier Service de Biochimie Métabolique, Tour Lavoisier étage 4, Hôpital Necker-Enfants Malades, 149 rue de Sèvres, 75743 Paris Cedex 15, France; Faculté de Médecine de l'Université Paris Descartes, Institut Cochin pièce 3013, 24 Rue du Faubourg Saint-Jacques, 75014 Paris, France

Jean-Sébastien Saulnier-Blache Département Métabolisme et Obésité, Institut des maladies métaboliques et cardiovasculaires (I2MC), INSERM U1048/ Université Paul Sabatier Toulouse Equipe n°3, BP 84225, 31432 Toulouse Cedex 4, France

Jean-François Tanti INSERM U895, Centre Méditerranéen de Médecine Moléculaire, Université de Nice Sophia Antipolis, 151 route de St Antoine de Ginestière, BP 2 3194, 06204 Nice cedex 3, France

Joan Tordjman Institut des maladies cardiométaboliques et de la nutrition (IHU ICAN), Paris, France; Nutriomique U872 équipe 7, Centre de recherche des Cordeliers, INSERM, UPMC, 15, rue de l'école de médecine, 75006 Paris, France

Philippe Valet INSERM U1048 I2MC, 1 avenue Jean Poulhès BP 84 225, 31432 Toulouse Cedex 4, France

Camille Vatier Institut des maladies cardiométaboliques et de la nutrition (IHU ICAN), Paris, France; Faculté de Médecine Pierre et Marie Curie site Saint-Antoine, INSERM, UPMC, UMR_S938, 27 rue Chaligny, 75012 Paris, France

Hubert Vidal Unité INSERM U1060 - INRA 1235 - INSA-Lyon - Université-Lyon1, Laboratoire CarMeN et Centre Européen Nutrition Santé. Centre de Recherche en Nutrition Humaine Rhône-Alpes, Lyon, France; Faculté de Médecine Lyon Sud-BP 12, 165 Chemin du Grand Revoyet, 69921 Oullins Cedex, France

Nathalie Viguerie INSERM, UMR1048, Obesity Research Laboratory Team 4, I2MC, Institute of Metabolic and Cardiovascular Diseases, 1 avenue Jean Poulhès, BP 84225, 31432 Toulouse Cedex 4, France; University of Toulouse, UMR1048, Paul Sabatier University, Toulouse, France

Corinne Vigouroux Institut des maladies cardiométaboliques et de la nutrition (IHU ICAN), Paris, France; Faculté de Médecine Pierre et Marie Curie site Saint-Antoine, INSERM, UPMC, UMR_S938, 27 rue Chaligny, 75012 Paris, France; Service de Biochimie et Hormonologie, Hôpital Tenon, 4 rue de la Chine, 75970 Paris cedex 20, France

Phi Villageois Institute of Developmental Biology and Cancer, University of Nice Ophia-Antipolis, CNRS, UMR6543, 06107 Nice, France

Brigitte Wdziekonski Institute of Developmental Biology and Cancer, University of Nice Ophia-Antipolis, CNRS, UMR6543, 06107 Nice, France

Laure-Emmanuelle Zaragosi Institute of Developmental Biology and Cancer, University of Nice Ophia-Antipolis, CNRS, UMR6543, 06107 Nice, France

Olivier Ziegler Service de Diabétologie, Maladies Métaboliques, Nutrition, CHRU de Nancy—Hôpital Brabois Adultes, Rue du Morvan, Vandoeuvre-Lès-Nancy, 54500 Nancy, France

Part I
Adipose Tissue Development

Chapter 1
Development of Adipose Cells

Tala Mohsen-Kanson, Brigitte Wdziekonski, Phi Villageois, Anne-Laure Hafner, Nadège Lay, Pauline Martin, Laure-Emmanuelle Zaragosi, Nathalie Billon, Magali Plaisant, Pascal Peraldi, Annie Ladoux and Christian Dani

Introduction

Adipogenesis is generally described as a two-step process. The first step comprises the generation of adipocyte progenitors from pluripotent and multipotent mesenchymal stem cells (MSC). The second step involves the terminal differentiation of these progenitors into mature functional adipocytes. The differentiation of adipocyte progenitors into adipocytes has been extensively studied in vitro (Farmer 2006; Rosen et al. 2000) and will not be reviewed in this chapter. In contrast to terminal differentiation process, the molecular mechanisms controlling self-renewal of adipocyte progenitors remain largely unknown. In addition, the developmental origin of human fat cell progenitors has not yet been elucidated. We will present in this chapter data demonstrating that mouse adipocytes originate from both neuroectoderm and mesoderm, and will present factors regulating proliferation and differentiation of human adipose-derived stem cells (ASCs). Figure 1.1 indicates the different points addressed in this chapter.

The increase in adipose mass in normal development and in obesity is the result of an increase in size and number of adipocytes. As mature adipocytes do not divide in vivo, regeneration of adipocytes and the increase in adipocyte number depend on the self-renewal capacity of a pool of adipocyte progenitors which remains present during adult life (Hauner et al. 1989; Spalding et al. 2008). It is established that adipocyte progenitors are located in the stromal vascular fraction (SVF) of adipose tissue (AT). However, the SVF is a heterogeneous mixture of cells, and recently Zuk et al. (2002) reported for the first time that subpopulations

---

T. Mohsen-Kanson · B. Wdziekonski · P. Villageois · A.-L. Hafner · N. Lay · P. Martin
L.-E. Zaragosi · N. Billon · M. Plaisant · P. Peraldi · A. Ladoux · C. Dani (✉)
Institute of Developmental Biology and Cancer, University of Nice Sophia-Antipolis, CNRS, UMR6543, 06107 Nice, France
e-mail: Christian.Dani@unice.fr

J.-P. Bastard and B. Fève (eds.), *Physiology and Physiopathology of Adipose Tissue*,
DOI: 10.1007/978-2-8178-0343-2_1, © Springer-Verlag France 2013

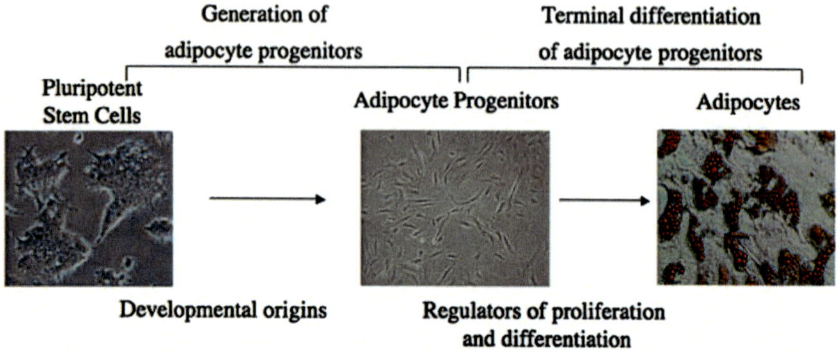

Fig. 1.1 Development of adipocytes. The first step consists in the generation of adipocytes progenitors from pluripotent stem cells. The second step is the terminal differentiation of adipocyte progenitors in mature adipocytes. In this chapter, we will address the developmental origins of adipocyte progenitors and factors regulating their self-renewal

of adipocyte progenitors in the SVF of human display mesenchymal stem cell features. Characterization of factors controlling self-renewal of human ASCs is at its infancy, partly due to the absence of appropriate cellular models. Identification of these factors is of fundamental importance, and could ultimately be translated into clinical interventions. The first challenge to identify regulators of ASC self-renewal has been to isolate human ASCs and to establish culture conditions to expand them. Recently, culture conditions have been set up allowing isolation and maintenance of human ASCs lines derived from the SVF of infant fats. These stem cells were termed human multipotent adipose-derived stem (hMADS) cells (Rodriguez et al. 2005). They exhibit the characteristics of MSC, i.e., the capacity to self-renew and to differentiate into several cell types at the clonal level. Cells can be expanded ex vivo for more than 160 population doublings (i.e., around 30 passages) while maintaining a normal diploid karyotype. Expanded hMADS cells are then able to differentiate under serum-free adipogenic condition into cells that exhibit characteristics of human fat cells. More recently, hMADs cells have been described as a faithful model to study human fat cell metabolism (Poitou et al. 2009; Bezaire et al. 2009). These data indicate that expanded hMADS cells enter the adipose lineage at a high rate and differentiate into cells that display a unique combination of properties similar, if not identical, to those of native human adipocytes. Thus, they provide a unique model to analyze human AT physiopathology. In this regard, they were further used to assess the molecular mechanisms underlying drug-induced at disorders such as lipodystrophy observed in HIV-infected patients receiving HAART therapy (Djedaini et al. 2009; Vernochet et al. 2005). After ex vivo expansion, hMADS cells maintain the ability to undergo differentiation into adipocytes, osteoblasts, and chondrocytes at the single cell level (Rodriguez et al. 2005; Zaragosi et al. 2006) without any loss of their therapeutic potential. When transplanted with a scaffold, hMADS cells are able to form ectopic bone in mouse (Elabd et al. 2007). Moreover, transplantation of

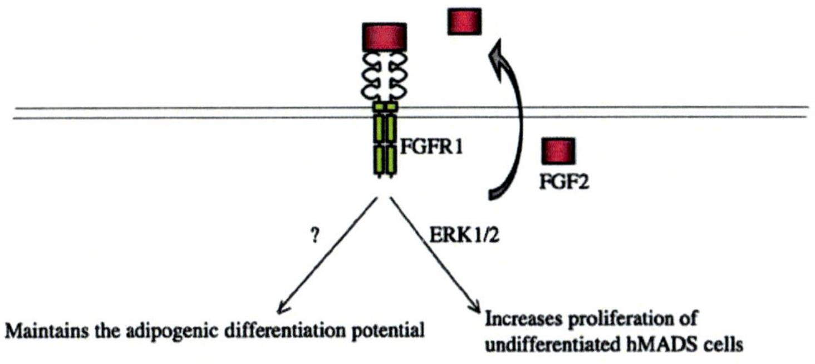

Fig. 1.2 FGF2 autocrine/paracrine loop regulating proliferation and differentiation of hMADS Cells. FGF2 secreted by undifferentiated hMADS cells promotes proliferation, via the ERK/2 pathway and maintains their adipogenic potential via an ERK1/2 independent pathway

hMADS cells into *mdx* mouse, an animal model for Duchenne muscular dystrophy, results in substantial expression of human dystrophin on a long-term basis and engraftment takes place in non-immunocompromised animals (Rodriguez et al. 2005). Altogether, hMADS cells appear to be a powerful cellular model to investigate human ASC self-renewal and differentiation.

FGF2 and Activin A are Secreted Factors Regulating Proliferation and Differentiation of hMADS Cells

During long-term ex vivo expansion, hMADS cells evolve from a spindle-shaped to a flat morphology. This morphological change is accompanied by a change in cell proliferation rate (at passage 15 the doubling time is around 2 days, and becomes 4 days at passage 20) and by a loss of differentiation potential. Interestingly, Zaragosi et al. reported that the increase in doubling time was concomitant to a decrease of fibroblast growth factor (FGF) 2 secretion by undifferentiated hMADS cells (Zaragosi et al. 2006). As cells also express FGF type 1 receptor, the high-affinity receptor for FGF2, it has been proposed that the FGF pathway plays an autocrine/paracrine role in ASC self-renewal. In support to this hypothesis, the addition of exogenous FGF2 was able to sustain proliferation and adipogenic potential over passages. Indeed, treatment of FGF2-expressing hMADS cells with PD173074, a specific FGF receptor inhibitor, strictly during the proliferation stage, dramatically decreases their differentiation potential, indicating that the FGF pathway is required for the maintenance of both proliferation and differentiation potential. It is interesting to note that epidermal growth factor (EGF), platelet-derived growth factor (PDGF), or FGF10 was not able to mimic FGF2 effects. Activation of extracellular signal-regulated kinase (ERK) 1/2 pathway is required to mediate FGF2 effects on hMADS cell proliferation. However, inhibition of mitogen-activated protein kinase (MEK)1 reduced the clonogenic potential of

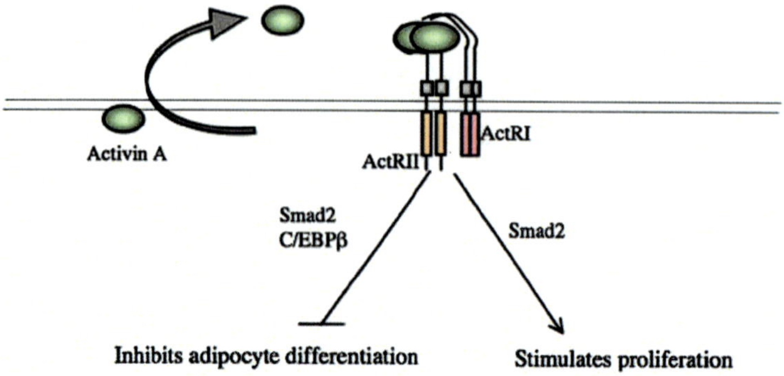

Fig. 1.3 Activin A is a critical regulator of human adipocyte progenitor proliferation and differentiation. Activin A secreted by undifferentiated hMADS cells promotes proliferation, via the Smad pathway, and inhibits adipogenic differentiation via Smad 2 pathway and C/EBPβ

hMADS cells but did not affect their differentiation potential, indicating that the ERK1/2 signaling pathway is partly involved in FGF2-mediated self-renewal (Fig. 1.2). FGF1 has also been reported to stimulate proliferation of human adipose progenitors and subsequently to increase their capacity to undergo differentiation (Widberg et al. 2009). The role of FGF pathway in the maintenance of ASC pool in AT remains to be investigated, but recently it has been proposed that FGFs could play a role in expansion of AT in obese patients (Mejhert et al. 2010).

Activin A, a member of the transforming growth factor (TGF)β family, is also secreted by undifferentiated hMADS cells and human ASCs isolated from various fat depots of donors of different ages. Its expression dramatically decreases as ASCs undergo differentiation into adipocytes. Activin A is not only a marker of undifferentiated cells but also plays a functional role in differentiation and proliferation. Sustained activation of activin A pathway promotes hMADS cell proliferation and impairs adipocyte differentiation, whereas its inhibition decreases proliferation and promotes differentiation. These effects are mediated via CCAAT/enhancer-binding protein (C/EBP)β and Smad2 pathways in an autocrine/paracrine manner (Zaragosi et al. 2010) (Fig. 1.3). Therefore, it has been proposed a model in which FGF2 and activin A regulate ASC proliferation and differentiation. The bone morphogenetic protein (BMP4), which is like activin A, a member of TGF β superfamily, regulates ASC proliferation in an autocrine and dose-dependent manner while maintaining their multipotent property (Vicente Lopez et al. 2010). Therefore, there is an emerging role of TGFβ pathway in self-renewal and differentiation of ASCs, as recently reviewed by Zamani and Brown (2011).

Proliferation and Differentiation of Human ASCs are Regulated by Macrophage-Secreted Factors

Investigating FGF2 and activin A gene expression in the obesity context revealed new regulators of ASC proliferation and differentiation. Obesity is associated with the presence of a higher number of hypertrophic adipocytes, with new macrophages being recruited into AT (Weisberg et al. 2003; Xu et al. 2003), and with an increased proportion of ASCs exhibiting proliferative potential (Maumus et al. 2008). This latter observation strongly suggests an important contribution of obese microenvironment is inducing ASC self-renewal. Interestingly, macrophages isolated from obese AT secrete factors that both inhibit differentiation of ASCs and stimulate expression of activin A and FGF2 genes in hMADS cells (Zaragosi et al. 2010). These observations fit well with a model proposing that macrophages play a key role in self renewal of ASCs in part through activin A and FGF2 pathways. Macrophage-secreted factors involved in the regulation of hMADS cell self-renewal are unknown so far. However, TNF-α could be one of them because it mimics the effects of macrophage-conditioned medium on ASC proliferation and differentiation. In addition, TNF-α stimulates expression of FGF2 and of activin A (Wdziekonski, unpublished data). Given the roles of tumor necrosis factor (TNF)α in non-adipose cells, it remains to be determined that chronic exposure of human adipocyte progenitors to TNF-α does not alter their adipogenic potential. Besides TNF-α, other macrophage factors, such as inflammatory cytokines or Wnt molecules (Christodoulides et al. 2009; Bilkovski et al. 2011) are likely to be involved in proliferation and differentiation of human ASCs. These observations suggest that macrophages constitute determinant niche components for ASC self-renewal. However, in addition to its stimulating effect on expression of pro-inflammatory cytokines in macrophages, activin A is positioned as a profibrotic factor in AT of obese patients (Keophiphath et al. 2009). Therefore, a question that should be addressed is related to the phenotype of adipocyte progenitors under conditions leading to increase production of activin A. Over-activity of activin A may turn adipocyte progenitors into myofibroblasts, which are a source of fibrosis observed in obese at (Divoux et al. 2010). This possibility remains to be analyzed in detail. (Fig. 1.4).

Hedgehog Pathway

Hedgehog (Hh) pathway affects self-renewal and differentiation of ASCs. Indeed, Hh signaling decreases during adipocyte differentiation of hMADS cells. Moreover, its activation inhibits adipocyte maturation, producing ill-differentiated adipocytes that are insulin resistant (Fontaine et al. 2008). Interestingly, there is a basal level of Hh signaling in undifferentiated cells that appears necessary for the maintenance of hMADS cell proliferation and clonogenic capacity, probably

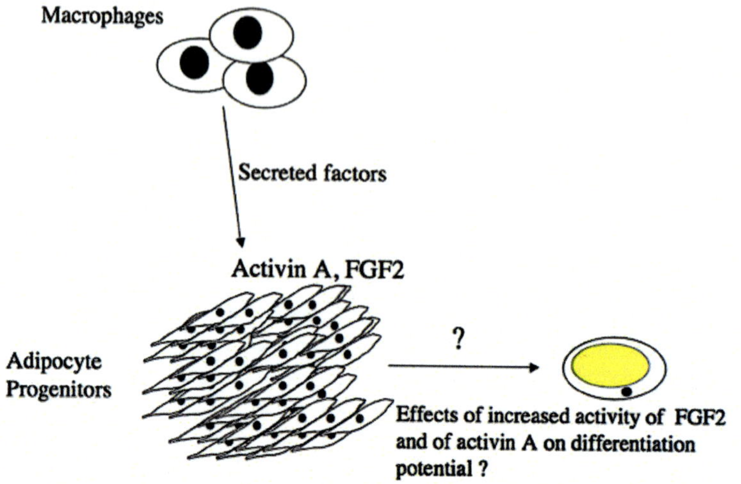

Fig. 1.4 Regulation of adipocyte progenitor proliferation and differentiation by macrophage-secreted factors. Macrophages secrete factors that promote proliferation of adipocyte progenitors and inhibit their differentiation, likely through increased expression of activin A and FGF2. The phenotype of adipocyte progenitors under an increased activity of activin A remains to be determined

through regulation of pRB phosphorylation and cyclin A expression (Plaisant et al. 2011). However, in contrast to the FGF pathway, inhibition of Hh signaling during proliferation did not alter the differentiation potential of hMADS cells.

MicroRNAs

MicroRNAs (miRNAs) also emerged as important players in adipogenesis (see for recent review (McGregor and Choi 2011)) and in ASC self-renewal and differentiation. Drosha and dicer are RNase III-family nucleases which are essential for miRNA biogenesis. Knockdown of both these enzymes using RNA interference, severely compromises hMADS cell proliferation and differentiation (Zaragosi and Dani, unpublished data). This observation is in agreement with the impairment of ASC survival consequent to disruption of the miRNA processing machinery reported recently by Sun Kim et al. (2011). Deep sequencing of small RNAs expressed in undifferentiated and differentiating hMADS cells has revealed an upregulation on the miR-30 family during adipogenic differentiation and a downregulation of this miRNA family during osteogenesis. Functional analysis indicates that inhibition of the miR-30 family blocks adipogenesis while promoting osteogenesis. Then, it has been additionally shown that Runx2 targeting is, at least in part, responsible for miR-30 positive effects on adipocyte differentiation (Zaragosi et al. 2011). In this context, it is interesting to remember that the transcription factor Runx2 is expressed in undifferentiated hMADS cells and that it represents the major regulator of osteogenesis. Altogether, these data support a model in which miR-30 regulates the fate of hMADS cells via modulation of runt-related transcription factor 2 (Runx2).

Developmental Origins of Adipocyte Progenitors

Although there have been attempts to characterize the distinct cellular intermediates between ASCs and mature adipocytes, such studies have been hampered by the lack of specific cell-surface markers to identify and prospectively isolate these cells in vivo. Some knowledge about mesenchymal cell fate decisions has been derived from studies on the immortalized mouse stromal cell line, C3H10T1/2, or mesenchymal precursor populations isolated from adult tissues. However, these cellular systems are not informative of the developmental origin of MSCs and adipocytes. Instead, the embryo might constitute a more suitable source of cells to address this issue. In particular, embryonic stem cells (ESCs) have provided an invaluable tool to model the earliest steps of mammalian development in vitro. Adipocyte progenitors are generally described to be derived from MSCs, which themselves are thought to arise from mesoderm. It is worth noting that during development of higher vertebrates, the mesoderm is not the only germ layer source of mesenchymal cells. In the head, for instance, the facial bones have been shown to derive from the neural crest (NC). The NC is a vertebrate cell population that arises from the neuroectoderm. After neural tube closure, NC cells undergo an epithelium-mesenchyme transition and migrate to diverse regions in the developing embryo, where they differentiate into various cell types. In the head and neck, the NC also yields mesenchymal precursors differentiating into connective tissue cells (reviewed in Dupin et al. (2006)). In vitro adipogenesis of mouse ESCs provided a powerful model to investigate the earliest steps of adipocyte development and revealed the surprising conclusions regarding the ontogeny of such cells in the NC.

Adipocyte Development in Mouse Embryonic Stem Cells

Mouse ESCs are proliferating, pluripotent stem cells that have been isolated from the epiblast of blastocyst-stage mouse embryos. They can be propagated indefinitely at the undifferentiated state in vitro. Furthermore, when transplanted into a mouse blastocyst, mESCs integrate into the embryo and contribute to all cell lineages, including germ cells (Smith et al. 1992). When aggregated to form embryoid bodies (EBs) in vitro, they undergo differentiation into ectodermal, mesodermal, and endodermal derivatives (Keller 1995). Directed differentiation of mESCs toward the adipocyte lineage has been accomplished (Dani et al. 1997), and showed that functional adipocytes could be obtained when mESCs were

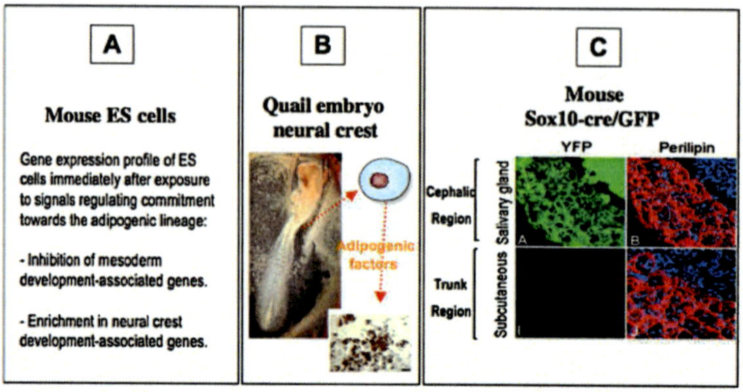

Fig. 1.5 Developmental origins of adipocytes. **a** Retinoic acid, which is required for commitment of mouse embryonic stem cells towards the adipogenic lineage, induces neural-crest associated gene expression and inhibits mesoderm derived cell types. **b** Neural crest isolated from quail embryos displays adipogenic potential. **c** Lineage tracing in mouse revealed two origins for adipocytes depending of their localisation

exposed to appropriate extracellular cues. In this system, the generation of adipocytes is dependent on an early and transient exposure of ESC-derived embryoid bodies to retinoic acid (RA) and a subsequent treatment with conventional adipogenic factors (e.g., insulin, triiodothyronine, and rosiglitazone). Both lipogenic and lipolytic activities, as well as high levels of expression of adipocyte-specific genes, could be detected in mESC-derived adipocytes. Remarkably, the sequence of expression of key transcription factors which are known to govern preadipocyte differentiation, such as members of the C/EBP and the peroxisome proliferator-activated receptor (PPAR) families was closely conserved during mESC adipogenesis. Thus, this model has provided a powerful system to address the different steps of adipocyte development (Wdziekonski et al. 2003, 2006; Billon et al. 2010; Schulz et al. 2009; Carnevalli et al. 2010; Tong et al. 2000; Takashima et al. 2007). In a first attempt to unravel the events underlying the formation of mesenchymal derivatives in RA-treated ES cells, Kawaguchi et al. examined the expression of various mesodermal and mesenchymal markers in early EBs. Surprisingly, they noticed that treatment with RA resulted in a sharp reduction in several mesodermal markers, as well as in the suppression of cardiomyocyte formation, suggesting that RA reduces overall mesoderm formation in ESCs (Kawaguchi et al. 2005). Since at high concentrations, RA was shown to promote neural differentiation of ESCs and since some mesenchymal tissues are known to be generated by the NC, which itself derives from neuroectoderm, these authors then analyzed the expression of various NC markers in ES cells. They showed that Sry-related high-mobility group (HMG) box (*sox*)9, *sox10*, Forkhead box D3 (*foxD3*), and *runx2*, which all play an important role in NC formation and/or mesenchymal condensation, were upregulated upon RA-treatment. Together, these data suggest that neuroectoderm/NC is the major source of adipocytes in RA-treated ESCs (Fig. 1.5a).

Study of Adipocyte Precursor Developmental Origins in Quail and Mouse Embryos

To better understand adipocyte lineage specification from the NC, our laboratory checked whether adipocytes could be obtained from NC cells isolated from a normal developing embryo. We used primary cultures of quail NC cells, since they have been instrumental in establishing the developmental potentialities of the NC. NC cells were isolated from both the cephalic and thoracic level and maintained in culture media permissive for adipocyte differentiation (Billon et al. 2007). This analysis revealed that typical mature adipocytes could readily be produced from cephalic NC cells, and, to a lesser extent, from truncal NC cells (Li et al. 2010). Therefore, quail NC cells from both the cephalic and the thoracic level exhibit an adipogenic potential in vitro (Fig. 1.5b). Finally, we have used a lineage tracing approach in mouse to address the origin of the adipocyte lineage in vivo and to provide direct evidence for the contribution of the neural crest. We have investigated whether subsets of adipocytes originate from the NC using *Sox10-cre*/yfp transgenic mice to map NC derivatives in vivo because to date, *Sox10* is considered as the best bona fide NC marker. Indeed, *Sox10* is strongly and specifically expressed in the NC from early embryonic development, and is not expressed in mesoderm. This study reveals adipocytes derived from NC in cephalic adipose depots, between the salivary gland and the ear area. In contrast, no NC-derived adipocytes can be detected in truncal adipose depots, including subcutaneous, perirenal, periepididymal, and interscapular tissues (Fig. 1.5c). These data therefore provide new information about the ontogeny of the adipocyte lineage and demonstrate that during normal development, a subset of adipocytes in the face originates from NC, and not from mesoderm (Billon et al. 2007). The role of RA in the early steps of adipocyte development remains to be demonstrated in vivo in mouse models. Interestingly, RA has recently been shown to be required for differentiation of cephalic NC cells into adipocytes in developing zebrafish embryos (Li et al. 2010), which is a reminiscence of the role of RA in mESC adipogenesis.

Cellular Origins and Tissue Localization of Adipocyte Progenitors

Recently, laboratories of Graff and of Friedman performed critical experiments to identify and localize adipocyte progenitors in mouse AT. Rodeheffer and colleagues used FACS analysis to isolate various cellular subpopulations from SVF and tested their adipogenic potential both in vitro and in vivo after transplantation in lipoatrophic A-Zip mice. By this approach, the authors identified mouse adipocyte progenitors in the SVF of AT as lin$^-$ /CD34$^+$/CD29$^+$/sca-1$^+$/CD24$^+$ cells (Rodeheffer et al. 2008). Whether adipocyte progenitors originated from NC or from mesoderm display the same immunophenotype remains to be determined. By a different approach based on the expression of PPARγ in SVF of AT, Tang and colleagues isolated undifferentiated cells able to undergo adipogenesis in vitro and in vivo in *nude* mice. These cells express markers of preadipocytes but not those of mature adipocytes, indicating that PPARγ can also be used to trace adipocyte progenitors. Interestingly, these cells are $CD45^{-}/Ter119^{-}/CD34^{+}/sca1^{+}$ indicating that they are similar, if not identical, to cells isolated by Friedman's laboratory. Thanks to the expression of a reporter gene under the control of PPARγ promoter, adipocyte progenitors have been localized in the mural cell compartment of AT vasculature in mice (Tang et al. 2008). The immunophenotype of human adipocyte progenitors has not yet been fully characterized, although they have been shown to reside in the $CD34^{+}/CD31^{-}$ subpopulation of stromal vascular cells of AT (Sengenes et al. 2005).

Do Adipocyte Progenitors of Different Developmental Origins Differ in Their Biological Properties?

Therefore, there are several lines of evidence showing two developmental origins of adipocytes: a subset of adipocytes in the cephalic region derives from the NC, whereas adipocytes in the trunk region originate likely from mesoderm. It is well established that adipocyte progenitors isolated from different depots display different features in terms of proliferation, differentiation, and gene expression profiles (Tchkonia et al. 2007). In addition, adipocytes derived from these adipocyte progenitors have different functional properties and have different contribution to metabolic diseases (Gesta et al. 2006; Montague et al. 1998). The cellular and molecular mechanisms underlying these fat depot-dependent differences are unknown today. However, several observations suggest that developmental mechanisms contribute to functional regional variations. Therefore, studies on the origins of adipocyte progenitors in mouse models open at least two questions: Are adipocytes derived from different developmental origins or cellular sources functionally different? What are the developmental origins of adipocyte progenitors in humans?

Human ES cells would be a powerful cellular model for investigating the earliest steps of human adipose cell development. Xiong and colleagues have reported that human ES cells have the potential to generate adipocytes (Xiong et al. 2005). However, human ES cells are difficult to work with in France because of ethical concerns and administrative regulations. A major discovery in the field of stem cells has been published by the end of 2007. Yamanaka's team demonstrated that ectopic expression of only a few defined transcription factors can reprogram human somatic cells to create pluripotent stem cells, resembling if not identical to human ES cells (Takahashi et al. 2007). Importantly, human-induced Pluripotent Stem (iPS) cells represent a valuable alternative to the use of embryonic cells, with no need to

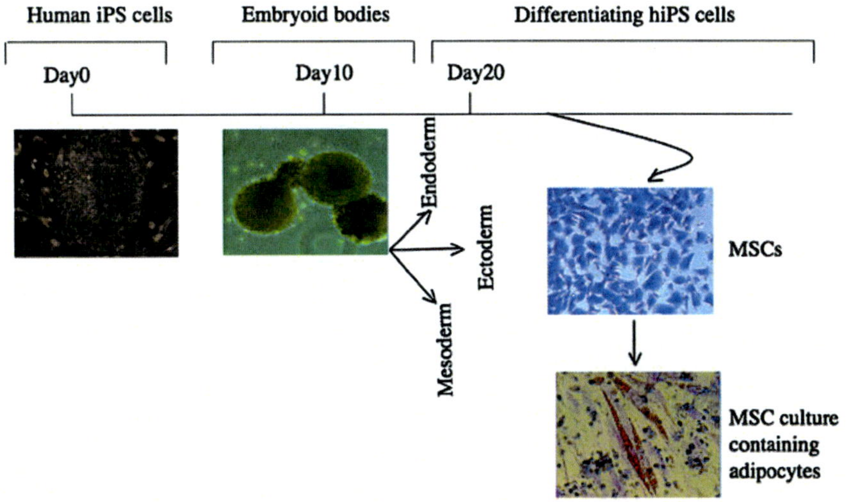

Fig. 1.6 Development of human iPS cells into adipocytes. Human iPS cells are induced to form embryoid bodies containing cells of the three germ layers. Human iPS-derived mesenchymal stem cells (MSCs) undergo adipocyte differentiation when maintained in appropriate conditions

destroy embryos, thus avoiding ethical concerns. In addition, patient-specific iPS cells can be generated, providing a powerful new tool to investigate adipocyte differentiation and further design and anticipate results from translational medicine including drug screening in inherited diseases. Recently, Taura and colleagues have reported that human iPS cells have an adipogenic potential comparable to human ES cells (Taura et al. 2009). More recently, we have generated iPS cells by reprogramming hMADS cells and established a culture system in order to derive adipocytes. As shown in Fig. 1.6, iPS cell-derived embryoid bodies contained cells of endoderm, ectoderm, and mesoderm origins and several days later cells filled with lipid droplets appear in the culture dishes. Mesenchymal stem cells, derived from differentiating hiPS cells, display an adipogenic phenotype when maintained in appropriate conditions (Mohsen-Kanson and Dani, unpublished data). Therefore, iPS cells appear as a promising cell model to investigate the earliest steps of human adipocyte development and to study adipocyte properties according to their developmental origins.

Conclusion

The information on AT development and the generation of adipocytes that arose from different embryonic and adult stem cell models revealed unsuspected findings. The better comprehension of the pathways important for adipocyte precursor cell expansion and differentiation opens perspectives on the possibilities to target them and to impact on the accumulation of AT and pathological signals that limit adipose expansion occurring in obese patients. Besides adipocytes derived from mesoderm, those arising from ectoderm also open the debate on their respective physiological roles. Do they exhibit equivalent potentials in AT remodelling? Do they display similar functions regarding lipid storage and metabolic regulations? These new models and the use of iPS cells will rush to knowledge in adipocyte development and will thus contribute to propose adapted therapeutic alternatives to both obese and lipodystrophic patients.

References

Bezaire V, Mairal A, Ribet C et al (2009) Contribution of adipose triglyceride lipase and hormone-sensitive lipase to lipolysis in hMADS adipocytes. J Biol Chem 284:18282–18291

Bilkovski R, Schulte DM, Oberhauser F et al (2011) Adipose tissue macrophages inhibit adipogenesis of mesenchymal precursor cells via wnt-5a in humans. Int J Obes (Lond) 35:1450–1454

Billon N, Iannarelli P, Monteiro MC et al (2007) The generation of adipocytes by the neural crest. Development 134:2283–2292

Billon N, Kolde R, Reimand J et al (2010) Comprehensive transcriptome analysis of mouse embryonic stem cell adipogenesis unravels new processes of adipocyte development. Genome Biol 11:R80

Carnevalli LS, Masuda K, Frigerio F et al (2010) S6K1 plays a critical role in early adipocyte differentiation. Dev Cell 18:763–774

Christodoulides C, Lagathu C, Sethi JK et al (2009) Adipogenesis and WNT signalling. Trends Endocrinol Metab 20(1):16–24

Dani C, Smith A, Dessolin S et al (1997) Differentiation of embryonic stem cells into adipocytes in vitro. J Cell Sci 110:1279–1285

Divoux A, Tordjman J, Lacasa D et al (2010) Fibrosis in human adipose tissue: composition, distribution, and link with lipid metabolism and fat mass loss. Diabetes 59:2817–2825

Djedaini M, Peraldi P, Drici MD et al (2009) Lopinavir co-induces insulin resistance and ER stress in human adipocytes. Biochem Biophys Res Commun 386:96–100

Dupin E, Creuzet S, Le Douarin NM (2006) The contribution of the neural crest to the vertebrate body. Adv Exp Med Biol 589:96–119

Elabd C, Chiellini C, Massoudi A et al (2007) Human adipose tissue-derived multipotent stem cells differentiate in vitro and in vivo into osteocyte-like cells. Biochem Biophys Res Commun 361:342–348

Farmer SR (2006) Transcriptional control of adipocyte formation. Cell Metab 4:63–273

Fontaine C, Cousin W, Plaisant M et al (2008) Hedgehog signaling alters adipocyte maturation of human mesenchymal stem cells. Stem Cells 26:1037–1046

Gesta S, Bluher M, Yamamoto Y et al (2006) Evidence for a role of developmental genes in the origin of obesity and body fat distribution. Proc Natl Acad Sci (USA) 103:6676–6681

Hauner H, Entenmann G, Wabitsch M et al (1989) Promoting effect of glucocorticoids on the differentiation of human adipocyte precursor cells cultured in a chemically defined medium. J Clin Invest 84:63–1670

Kawaguchi J, Mee PJ, Smith AG (2005) Osteogenic and chondrogenic differentiation of embryonic stem cells in response to specific growth factors. Bone 36:758–769

Keller GM (1995) In vitro differentiation of embryonic stem cells. Curr Opin Cell Biol 7:862–869

Keophiphath M, Achard V, Henegar C et al (2009) Macrophage-secreted factors promote a profibrotic phenotype in human preadipocytes. Mol Endocrinol 23:11–24

Kim BS, Jung JS, Jang JH et al (2011) Nuclear Argonaute 2 regulates adipose tissue-derived stem cell survival through direct control of miR10b and selenoprotein N1 expression. Aging Cell 10:277–291
Li N, Kelsh RN, Croucher P et al (2010) Regulation of neural crest cell fate by the retinoic acid and Pparg signalling pathways. Development 137:389–394
Maumus M, Sengenes C, Decaunes P et al (2008) Evidence of in situ proliferation of adult adipose tissue-derived progenitor cells: influence of fat mass microenvironment and growth. J Clin Endocrinol Metab 93:4098–4106
McGregor RA, Choi MS (2011) microRNAs in the regulation of adipogenesis and obesity. Curr Mol Med 11:304–316
Mejhert N, Galitzky J, Pettersson AT et al (2010) Mapping of the fibroblast growth factors in human white adipose tissue. J Clin Endocrinol Metab 95:2451–2457
Montague CT, Prins JB, Sanders L et al (1998) Depot-related gene expression in human subcutaneous and omental adipocytes. Diabetes 47:1384–1391
Plaisant M, Giorgetti-Peraldi S, Gabrielson M et al (2011) Inhibition of hedgehog signaling decreases proliferation and clonogenicity of human mesenchymal stem cells. PLoS One 6:e16798
Poitou C, Divoux A, Faty A et al (2009) Role of serum amyloid a in adipocyte-macrophage cross talk and adipocyte cholesterol efflux. J Clin Endocrinol Metab 94(5):1810–1817
Rodeheffer MS, Birsoy K, Friedman JM (2008) Identification of white adipocyte progenitor cells in vivo. Cell 135:240–249
Rodriguez AM, Pisani D, Dechesne CA et al (2005) Transplantation of a multipotent cell population from human adipose tissue induces dystrophin expression in the immunocompetent mdx mouse. J Exp Med 201:1397–1405
Rosen ED, Walkey CJ, Puigserver P et al (2000) Transcriptional regulation of adipogenesis. Genes Dev 14:1293–1307
Schulz H, Kolde R, Adler P et al (2009) The FunGenES database: a genomics resource for mouse embryonic stem cell differentiation. PLoS One 4:e6804
Sengenes C, Lolmede K, Zakaroff-Girard A et al (2005) Preadipocytes in the human subcutaneous adipose tissue display distinct features from the adult mesenchymal and hematopoietic stem cells. J Cell Physiol 205:114–122
Smith AG, Nichols J, Robertson M et al (1992) Differentiation inhibiting activity (DIA/LIF) and mouse development. Dev Biol 151:339–351
Spalding KL, Arner E, Westermark PO et al (2008) Dynamics of fat cell turnover in humans. Nature 453:783–787
Takahashi K, Tanabe K, Ohnuki M et al (2007) Induction of pluripotent stem cells from adult human fibroblasts by defined factors. Cell 131:861–872
Takashima Y, Era T, Nakao K et al (2007) Neuroepithelial cells supply an initial transient wave of MSC differentiation. Cell 129:1377–1388
Tang W, Zeve D, Suh JM et al (2008) White fat progenitor cells reside in the adipose vasculature. Science 322:583–586
Taura D, Noguchi M, Sone M et al (2009) Adipogenic differentiation of human induced pluripotent stem cells: comparison with that of human embryonic stem cells. FEBS Lett 583:1029–1033
Tchkonia T, Lenburg M, Thomou T et al (2007) Identification of depot-specific human fat cell progenitors through distinct expression profiles and developmental gene patterns. Am J Physiol Endocrinol Metab 292:E298–E307
Tong Q, Dalgin G, Xu H et al (2000) Function of GATA transcription factors in preadipocyte-adipocyte transition. Science 290:134–138
Vernochet C, Azoulay S, Duval D et al (2005) Human immunodeficiency virus protease inhibitors accumulate into cultured human adipocytes and alter expression of adipocytokines. J Biol Chem 280:2238–2243

Vicente Lopez MA, Vazquez Garcia MN, Entrena A et al (2010) Low doses of bone morphogenetic protein 4 increase the survival of human adipose-derived stem cells maintaining their stemness and multipotency. Stem Cells Dev 20:1011–1019

Wdziekonski B, Villageois P, Dani C (2003) Development of adipocytes from differentiated ES cells. Methods Enzymol 365:268–277

Wdziekonski B, Villageois P, Vernochet C et al (2006) Use of differentiating embryonic stem cells in pharmacological studies. Methods Mol Biol 329:341–351

Weisberg SP, McCann D, Desai M et al (2003) Obesity is associated with macrophage accumulation in adipose tissue. J Clin Invest 112:1796–1808

Widberg CH, Newell FS, Bachmann AW et al (2009) Fibroblast growth factor receptor 1 is a key regulator of early adipogenic events in human preadipocytes. Am J Physiol Endocrinol Metab 296:E121–E131

Xiong C, Xie CQ, Zhang L et al (2005) Derivation of adipocytes from human embryonic stem cells. Stem Cells Dev 14:671–675

Xu H, Barnes GT, Yang Q et al (2003) Chronic inflammation in fat plays a crucial role in the development of obesity-related insulin resistance. J Clin Invest 112:21–1830

Zamani N, Brown CW (2011) Emerging roles for the transforming growth factor-{beta} superfamily in regulating adiposity and energy expenditure. Endocr Rev 32:387–403

Zaragosi LE, Ailhaud G, Dani C (2006) Autocrine fibroblast growth factor 2 signaling is critical for self-renewal of human multipotent adipose-derived stem cells. Stem Cells 24:2412–2419

Zaragosi LE, Wdziekonski B, Villageois P et al (2010) Activin a plays a critical role in proliferation and differentiation of human adipose progenitors. Diabetes 59:2513–2521

Zaragosi LE, Wdziekonski B, Le Brigand K et al (2011) Small RNA sequencing reveals miR-642a-3p as a novel adipocyte-specific microRNA and miR-30 as a key regulator of human adipogenesis. Genome Biol 12:R64

Zuk PA, Zhu M, Ashjian P et al (2002) Human adipose tissue is a source of multipotent stem cells. Mol Biol Cell 13:4279–4295

Chapter 2
Emerging Roles of Cell Cycle Regulators in Adipocyte Metabolism

Carine Chavey, Sylviane Lagarrigue, Jean-Sébastien Annicotte and Lluis Fajas

Introduction

Cells can adapt their growth and metabolism to their needs and the extracellular signals they receive. Some stimuli, such as stress or nutrients, are not only proliferative but also metabolic signals, suggesting a close crosstalk between these two biological processes. Indeed, the response to a metabolic signal can require the activation of transcription factors and of signaling molecules that result in the inhibition of cell cycle progression and ultimately cell proliferation.

The cell cycle is a finely tuned process (Fig. 2.1) the progression of which is orchestrated by holoenzymes that are composed of regulatory subunits (i.e., cyclins) and catalytic subunits, i.e., cyclin-dependent kinases (Cdks). Cdks, which belong to the family of the serine/threonine kinases, are activated by phosphorylation and dephosphorylation events and interact with specific cyclins to form heterodimers (Malumbres and Barbacid 2005). The cyclin D-Cdk4 and cyclin D-Cdk6 complexes act during the G1 phase, whereas the cyclin E-Cdk2 complex regulates the G1/S transition. The main substrates of the cyclin-Cdk complexes are proteins of the retinoblastoma (known also as «pocket protein») family (i.e., pRb, p107 and p130) that regulate the G1/S transition of the cell cycle. Phosphorylation of retinoblastoma proteins by cyclin-Cdk complexes

C. Chavey · S. Lagarrigue · J.-S. Annicotte · L. Fajas (✉)
Métabolisme et Cancer, IGMM Institut de Génétique Moléculaire de Montpellier
CNRS-UMR 5535, 1919 Route de Mende, 34293 Montpellier-Cedex 5, France
e-mail: lluis.fajas@igmm.cnrs.fr; lluis.fajas@inserm.fr

C. Chavey
e-mail: carine.chavey@igmm.cnrs.fr

S. Lagarrigue
e-mail: sylviane.lagarrigue@igmm.cnrs.fr

J.-S. Annicotte
e-mail: jean-sebastien.annicotte@igmm.cnrs.fr

J.-P. Bastard and B. Fève (eds.), *Physiology and Physiopathology of Adipose Tissue*,
DOI: 10.1007/978-2-8178-0343-2_2,

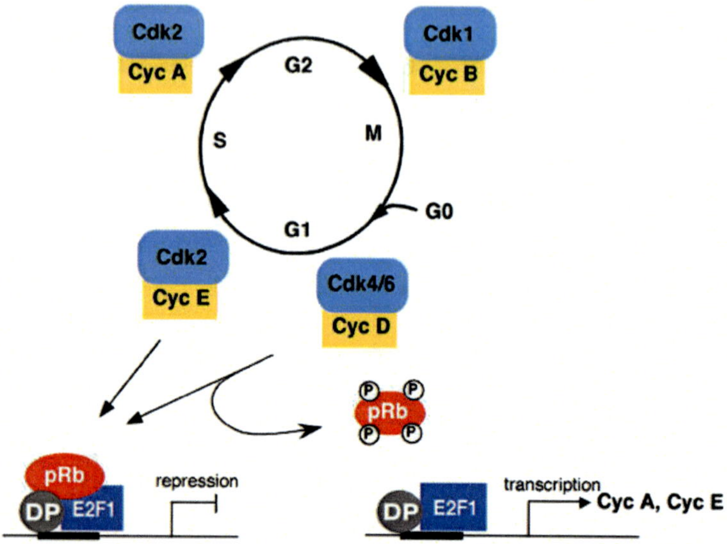

Fig. 2.1 The cell cycle machinery. During the G1/S transition, the Cdk4-cyclin D complex is activated by proliferative stimuli and then it phosphorylates pRb, p107, and p130. This phosphorylation allows the dissociation of the repressive complex constituted by the "pocket" proteins and E2F transcription factors. Once liberated, E2F factors can transactivate genes that promote cell cycle progression

abolishes their repressive effect on the E2F/DP transcription factors, thus allowing the transcription of genes required for cell cycle progression.

Indeed, E2F factors (E2F1-8) regulate the expression of genes involved in cell cycle progression, apoptosis, and DNA synthesis. The association of E2F proteins with proteins of the retinoblastoma family represses transcription following the recruitment of histone deacetylases (HDAC) (Brehm et al. 1998; Magnaghi-Jaulin et al. 1998) and of lysine/arginine methyl-transferases (Fabbrizio et al. 2002) and consequently the cell cycle is blocked in the G0/G1 phase.

Besides their established role in the control of cell proliferation and cell death, these cell cycle players are also key regulators of cell metabolism and of lipid and glucose homeostasis.

Cell Cycle Regulators in Adipose Tissue Physiology

Adipocyte Differentiation

Adipocytes are the main component of adipose tissue (AT). The elucidation of the molecular mechanisms underlying adipocyte differentiation is essential for understanding the physiology of AT.

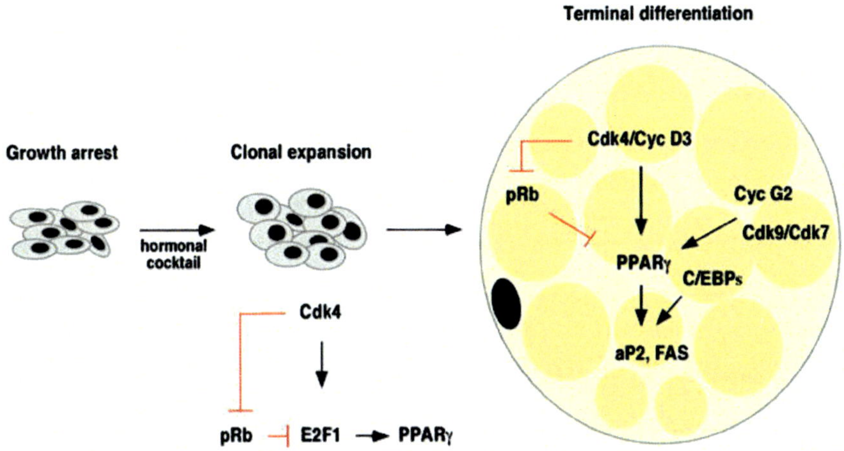

Fig. 2.2 Participation of cell cycle regulators in adipocyte differentiation. During the phase of clonal expansion, the Cdk4-pRb-E2F1 signaling pathway regulates the expression of genes involved in cell cycle entry. During the phase of terminal differentiation, the cell cycle regulators can influence the adipocyte biology by regulating directly the expression/activity of the key adipocyte transcription factors C/EBPα and PPARγ

Proliferation and differentiation are two extremely interconnected biological processes. During the development of the AT, adipocyte differentiation (adipogenesis) includes a proliferative step followed by a phase of differentiation, in which all the specific adipocyte markers are induced (Fajas 2003). Adipogenesis thus requires a very intimate crosstalk between cell cycle regulation and metabolic control.

The pre-adipocytic 3T3-L1 cell line is an experimental model used to identify molecular mechanisms involved in adipogenesis. In response to a hormonal cocktail, growth-arrested 3T3-L1 pre-adipocytes will re-enter the cell cycle (Fig. 2.2). This phase is called clonal expansion and cannot be dissociated from terminal differentiation. Indeed, blocking the phase of clonal expansion with DNA synthesis inhibitors fully arrests adipocyte differentiation (Richon et al. 1997). It has been shown that the Cdk-cyclin-E2F-Rb signaling cascade plays a major role during adipocyte differentiation. E2F1 is strongly upregulated during the first phases of adipogenesis, followed by the expression of its target genes, such as cyclin D1 and cyclin E (Richon et al. 1997). Moreover, E2F1 regulates adipocyte differentiation also by modulating the expression of peroxisome proliferator-activated receptor (PPAR)γ, the main transcription factor of adipogenesis (Fajas et al. 2002). However, other members of the E2F family seem to be involved in the negative regulation of adipogenesis. For instance, E2F4 inhibits the expression of PPARγ during terminal adipocyte differentiation (Fajas et al. 2002; Landsberg et al. 2003).

The retinoblastoma proteins (pRb, p107, and p130) regulate the activity of E2F transcription factors and play an important role during adipocyte differentiation as well. However, the findings are contradictory. pRb acts positively on terminal adipocyte differentiation by binding directly to the transcription factor CCAAT/

enhancer binding protein (C/EBP)β and facilitating its transactivation (Chen et al. 1996). Conversely, during cell cycle arrest in the G1 phase, pRb negatively regulates E2F1. Moreover, pRb can also act negatively during adipogenesis by forming a complex with PPARγ and HDAC3 on the promoter of target genes and thus blocking their induction (Fajas et al. 2002).

The cyclin-Cdk complexes, which regulate the activity of E2F factors by phosphorylating retinoblastoma proteins, also participate in the fine-tuning of adipogenesis. Cdk4 is an important regulator of adipocyte differentiation. We have shown that inactivation of Cdk4 blocks adipocyte differentiation, whereas the Cdk4 mutant R24C, which cannot be inhibited by p16^{INK4a}, increases the adipogenic potential of 3T3-L1 cells (Abella et al. 2005). Cdk4 directly regulates PPARγ by activating the transcription of its target genes through their direct phosphorylation.

Cdk7 and Cdk9 also are involved in adipocyte differentiation. Differently from other Cdks that act directly on cell cycle progression, these two kinases regulate transcription via phosphorylation of the RNA polymerase during the cell cycle. Cdk9 favors the transcription of PPARγ-target genes by direct interaction with and phosphorylation of PPARγ (Iankova et al. 2006). Conversely, Cdk7 phosphorylates PPARγ to inhibit its transcriptional activity (Helenius et al. 2009).

To carry out their activity, Cdks are associated with cyclins, which are the regulatory subunits of the Cdk/cyclin complexes. Cyclin D1 and D3 modulates adipogenesis in opposite ways. Both cyclins interact with PPARγ but, different from cyclin D3, cyclin D1 negatively regulates adipogenesis by recruiting HDACs on the promoter of PPARγ-target genes (Fu et al. 2005; Sarruf et al. 2005). On the other hand, cyclin G2, like cyclin D3, acts as a cofactor of PPARγ, thus favoring its transcriptional activity and consequently also adipocyte differentiation (Aguilar et al. 2010).

During the cell cycle, Cdk activity is also finely regulated by their physiological inhibitors, the CDKIs (CIP, KIP, and INK4). It is thus not surprising to see that CDKIs modulate adipocyte differentiation as well. The expression of p21/CIP and p27/KIP1 during adipocyte differentiation is very controversial. Nevertheless, disruption of p21 activity in a 3T3-L1 cell model (p21 knockdown by RNA interference) or in p21-/- MEFs (primary mouse embryonic fibroblasts) inhibits adipocyte differentiation, thus making p21 a pro-adipogenic factor (Inoue et al. 2008).

Overall, the studies carried out using the 3T3-L1 cell line show that many regulators of the cell cycle play a crucial role also in the process of adipocyte differentiation, either by modulating the cell cycle during the clonal expansion phase, or by regulating key adipogenic transcription factors.

Adipocyte Biology

Many works have described the key role of cell cycle players during adipogenesis, whereas very few studies have focused on the involvement of these regulators in adipocyte biology. We have reported that some cell cycle regulators are expressed in mature adipocytes, suggesting a role in the biology of adipocytes. Cdks could control the activity of enzymes required for the adipocyte functions, such as lipolysis or lipid synthesis. Indeed, we have shown that, in adipocytes, Cdk4 regulates glucose transport upon stimulation by insulin. Inhibition of Cdk4 by a pharmacological inhibitor blocks the glucose transporter and decreases the expression of genes involved in the signaling cascades of insulin and of the glucose transporter GLUT4 as well as of lipogenic genes, such as fatty acid synthase (FAS) and phosphoenol pyruvate carboxy kinase (PEPCK) (Abella et al. 2005). Similarly, Cdk5 plays a major role in glucose transport. Silencing of Cdk5 by siRNA or its pharmacological inhibition impairs the transport of glucose stimulated by insulin. Indeed, following its insulin-dependent activation, Cdk5 phosphorylates the protein E-Syt1, a partner of the GLUT4 transporter, thus favoring the transport of glucose (Lalioti et al. 2009; Muruais et al. 2009).

Moreover, in yeast, Kohlwein's group has shown that Cdk1 phosphorylates the lipase tgl4, which is an ortholog of adipose triglyceride lipase (ATGL), the key enzyme of lipolysis (Kurat et al. 2009). These results suggest that Cdks might also control lipolysis.

Although data on the involvement of the cell cycle regulators in the physiology of adipocytes are scarce, it seems that these factors could act directly on the biological functions of adipocytes, such as lipolysis and glucose transport.

Cell Cycle Regulators and AT Physiopathology

As cell cycle regulators are now considered novel major players in the control of lipid and glucose metabolism, they could also be involved in the development of associated pathologies, such as obesity, diabetes, and cancer.

Obesity and Type 2 Diabetes

Obesity is a major health problem. It is characterized by adipose mass accumulation resulting in an increase of the size and number of adipocytes. During weight gain, adipocytes become bigger (hypertrophy) and accumulate lipids. Since adipocytes cannot indefinitely accumulate lipids, they will then recruit adipocyte precursors to form new mature adipocytes. This process is called hyperplasia.

Type 2 diabetes is the most frequent complication of obesity. This pathology is characterized by fasting hyperglycemia due to the association of insulin resistance with destruction of pancreatic beta cells that produce insulin.

In the mouse, specific genetic ablation of cell cycle regulators in the AT has allowed demonstrating their importance in the physiopathology of AT, particularly during the development of hyperplasia (Table 2.1).

Table 2.1 Adipose tissue phenotype of cell cycle regulators knockout mice

Gene	Metabolic phenotype	References
Cyclin D3	Resistance to high fat diet, small adipocytes	(Sarruf et al. 2005)
Cdk4	Decrease body weight and adipose tissue mass	(Abella et al. 2005; Rane et al. 1999)
pRb	Resistance to high fat diet, switch from white to brown adipocyte	(Dali-Youcef et al. 2007)
P107	Decrease adipose tissue mass	(Scime et al. 2005)
P21	Increase adipose tissue mass by hyperplasia	(Naaz et al. 2004)
P27	Increase adipose tissue mass by hyperplasia	(Naaz et al. 2004)

Genetic ablation of cyclin D3 in the mouse protects these animals against obesity induced by a lipid-rich diet (Sarruf et al. 2005). Their adipocytes are smaller and express very low levels of adipocyte markers, such as aP2 and PPARγ, in comparison to wild-type controls. The presence of smaller adipocytes improves the general metabolism of these mice. Indeed, cyclin D3-/- mice are more glucose-tolerant and more sensitive to insulin in comparison to wild-type mice.

Since Cdks carry out the catalytic activity of the Cdk-cyclin complexes, it is not surprising that genetic ablation of cyclins or Cdks leads to similar phenotypes. In Cdk4-/- mice the adipose mass is reduced (Abella et al. 2005). Conversely, transgenic mice that express Cdk4R24C, the active mutant of Cdk4 (R24C), have increased body weight (Rane et al. 1999). These effects on the physiology of the AT are due, in part, to the direct effects of Cdk4 on PPARγ during adipocyte differentiation and also to the effects of Cdk4 on insulin secretion from the pancreas.

The metabolic phenotype of Cdk5-/- mice is unfortunately not available, but Cdk5 seems to play a primordial role in insulin resistance. Indeed, Nohara's group has shown that the effects of tumor necrosis factor (TNF)α on insulin resistance, particularly on glucose transport, are mediated through activation of Cdk5 (Nohara et al. 2011). Recently, a new anti-diabetic compound (SR1664) has been described. SR1664 binds to PPARγ, blocks its phosphorylation by Cdk5, and thus improves the biological parameters of diabetic mice (Choi et al. 2010; Choi et al. 2011). This compound has a powerful anti-diabetic activity without modifying the bone mass and without weight gain, the two main secondary effects of thiazolidinediones (TZD). Altogether these findings suggest that PPARγ phosphorylation by Cdk5 might be a major event in the development of type 2 diabetes.

The members (pRb, p107, p130) of the «pocket protein» family are key regulators of the physiology of AT. Deletion of pRb specifically in the adult AT leads to resistance to weight gain in animals fed on a lipid-rich diet, thanks to an increase of energy expenditure (Dali-Youcef et al. 2007). In these mice, both white and brown adipocytes have higher mitochondrial activity and the white AT shows morphological characteristics that are typical of the brown AT. In vitro, pRb -/- MEFs differentiate into adipose cells that have the phenotype of brown adipocytes and are characterized by the expression of PPAR gamma coactivator-1 (PGC-1), a major regulator of mitochondrial biogenesis, and of enzymes of the mitochondrial respiratory chain (Hansen et al. 2004). pRb thus regulates differentiation into the white or brown adipocyte lineage. A similar phenotype has been reported in p107-/- mice which shows a strong reduction of the adipose mass and an increased number of mitochondria (Scime et al. 2005).

P21 and p27 are important regulators of adipogenesis. Loss of expression of these Cdk inhibitors induces hyperplasia of the AT. The p21-/-, p27-/-, and p21-/- p27-/- mice develop obesity characterized by a larger number of adipocytes (Naaz et al. 2004). These observations are however controversial because two other studies have shown that p21 -/- mice are protected against obesity induced by lipid-rich diet (Inoue et al. 2008) and that p27-/- mice do not have an AT phenotype (Lin et al. 2003).

Altogether, these in vivo studies using mice in which cell cycle regulators have been genetically ablated demonstrate that these factors play a crucial role in the physiopathology of the AT.

Cancer

The activity of the Cdk-pRb-E2F signaling pathway is often altered in many human cancers, such as glioblastoma as well as lung, ovary, breast, and colon cancers (Chen et al. 2009). Few studies have investigated the involvement of this pathway in tumors of the AT.

Liposarcomas are tumors derived from the primitive cells that have differentiated into adipocytes. The term "liposarcoma" covers a huge variety of neoplastic processes, mainly benign lesions and also more aggressive, malignant lesions with a high rate of relapse and/or metastases. In comparison to other cancer types, these soft tissue sarcomas are relatively rare.

Only deregulation of the Cdk4 gene, among the many cell cycle regulators, has been described in these cancers (Helias-Rodzewicz et al. 2009; Chung et al. 2009). Differentiated and dedifferentiated liposarcomas are characterized by amplification of a region in chromosome 12 that carries the Cdk4 gene. This amplification can in some cases lead to overexpression of Cdk4 in tumor cells in comparison to mature adipocytes. Moreover, a study has demonstrated that loss of Cdk4 copies by chemical treatment in liposarcomas was correlated with an increase of adipocyte differentiation (Helias-Rodzewicz et al. 2009). These data, obtained in cancer cells, are reminiscent of those from studies on Cdk4 in the 3T3-L1 adipocyte cell line (Abella et al. 2005).

Conclusion

The role of the Cdk-cyclin-pRb-E2F signaling pathway in proliferation, cell cycle regulation, apoptosis, and cancer has been much studied. This review describes the involvement of this signaling pathway in the physiology and physiopathology of the AT as well. These cell cycle regulators act directly on key regulators of adipocytes, such as PPARγ and C/EBP.

However, their function is not limited to the AT. Indeed, activation of the Cdk-Cyclin-pRb-E2F signaling pathway has also been reported in other metabolic, non-proliferative tissues, such as pancreas and muscle. Moreover, an increasing number of studies have described the activity of these factors in glucose and energy homeostasis (Annicotte et al. 2009; Blanchet et al. 2011).

References

Abella A, Dubus P, Malumbres M et al (2005) Cdk4 promotes adipogenesis through PPARgamma activation. Cell Metab 2:239–249

Aguilar V, Annicotte JS, Escote X et al (2010) Cyclin G2 Regulates Adipogenesis through PPAR{gamma} Coactivation. Endocrinology 151:5247–5254

Annicotte JS, Blanchet E, Chavey C et al (2009) The CDK4-pRB-E2F1 pathway controls insulin secretion. Nat Cell Biol 11:1017–1023

Blanchet E, Annicotte JS, Lagarrigue S et al (2011) E2F transcription factor-1 regulates oxidative metabolism. Nat Cell Biol 13:1146–1152

Brehm A, Miska EA, Mccance DJ et al (1998) Retinoblastoma protein recruits histone deacetylase to repress transcription. Nature 391:597–601

Chen HZ, Tsai SyLeone G (2009) Emerging roles of E2Fs in cancer: an exit from cell cycle control. Nat Rev Cancer 9:785–797

Chen P, Riley DJ, Chen Y et al (1996) Retinoblastoma protein positively regulates terminal adipocyte differentiation through direct interaction with C/EBPs. Genes & Development 10:2794–2804

Choi JH, Banks AS, Estall JL et al (2010) Anti-diabetic drugs inhibit obesity-linked phosphorylation of PPARgamma by Cdk5. Nature 466:451–456

Choi JH, Banks AS, Kamenecka TM et al (2011) Antidiabetic actions of a non-agonist PPARgamma ligand blocking Cdk5-mediated phosphorylation. Nature 477:477–481

Chung L, Lau SK, Jiang Z et al (2009) Overlapping features between dedifferentiated liposarcoma and undifferentiated high-grade pleomorphic sarcoma. Am J Surg Pathol 33:1594–1600

Dali-Youcef N, Mataki C, Coste A et al (2007) Adipose tissue-specific inactivation of the retinoblastoma protein protects against diabesity because of increased energy expenditure. Proc Natl Acad Sci U S A 104:10703–10708

Fabbrizio E, El Messaoudi S, Polanowska J et al (2002) Negative regulation of transcription by the type II arginine methyltransferase PRMT5. EMBO Rep 3:641–645

Fajas L (2003) Adipogenesis: a cross-talk between cell proliferation and cell differentiation. Annals of Medicine 35:79–85

Fajas L, Landsberg RL, Huss-Garcia Y et al (2002a) E2Fs regulate adipocyte differentiation. Dev Cell 3:39–49

Fajas L, Egler V, Reiter R et al (2002b) The retinoblastoma-histone deacetylase 3 complex inhibits the peroxisome proliferator-activated receptor gamma and adipocyte differentiation. Developmental Cell 3:903–910

Fu M, Rao M, Bouras T et al (2005) Cyclin D1 inhibits peroxisome proliferator-activated receptor gamma-mediated adipogenesis through histone deacetylase recruitment. J Biol Chem 280:16934–16941

Hansen JB, Jorgensen C, Petersen RK et al (2004) Retinoblastoma protein functions as a molecular switch determining white versus brown adipocyte differentiation. Proc Natl Acad Sci U S A 101:4112–4117

Helenius K, Yang Y, Alasaari J et al (2009) Mat1 inhibits peroxisome proliferator-activated receptor gamma-mediated adipocyte differentiation. Mol Cell Biol 29:315–323

Helias-Rodzewicz Z, Pedeutour F, Coindre JM et al (2009) Selective elimination of amplified CDK4 sequences correlates with spontaneous adipocytic differentiation in liposarcoma. Genes Chromosomes Cancer 48:943–952

Iankova I, Rk Petersen, Annicotte JS et al (2006) PPAR{gamma} Recruits the P-TEFb Complex to Activate Transcription and Promote Adipogenesis. Mol Endocrinol 20:1494–1505

Inoue N, Yahagi N, Yamamoto T et al (2008) Cyclin-dependent kinase inhibitor, p21WAF1/CIP1, is involved in adipocyte differentiation and hypertrophy, linking to obesity, and insulin resistance. J Biol Chem 283:21220–21229

Kurat CF, Wolinski H, Petschnigg J et al (2009) Cdk1/Cdc28-dependent activation of the major triacylglycerol lipase Tgl4 in yeast links lipolysis to cell-cycle progression. Mol Cell 33:53–63

Lalioti V, Muruais G, Dinarina A et al (2009) The atypical kinase Cdk5 is activated by insulin, regulates the association between GLUT4 and E-Syt1, and modulates glucose transport in 3T3-L1 adipocytes. Proc Natl Acad Sci U S A 106:4249–4253

Landsberg RL, Sero JE, Danielian PS et al (2003) The role of E2F4 in adipogenesis is independent of its cell cycle regulatory activity. Proc Natl Acad Sci U S A 100:2456–2461

Lin J, Della-Fera MA, Li C et al (2003) P27 knockout mice: reduced myostatin in muscle and altered adipogenesis. Biochem Biophys Res Commun 300:938–942

Magnaghi-Jaulin L, Groisman R, Naguibneva I et al (1998) Retinoblastoma protein represses transcription by recruiting a histone deacetylase. Nature 391:601–605

Malumbres M, Barbacid M (2005) Mammalian cyclin-dependent kinases. Trends Biochem Sci 30:630–641

Muruais G, Lalioti V, Sandoval IV (2009) The Cdk5 inhibitor roscovitine strongly inhibits glucose uptake in 3T3-L1 adipocytes without altering GLUT4 translocation from internal pools to the cell surface. J Cell Physiol 220:238–244

Naaz A, Holsberger DR, Iwamoto GA et al (2004) Loss of cyclin-dependent kinase inhibitors produces adipocyte hyperplasia and obesity. Faseb J 18:1925–1927

Nohara A, Okada S, Ohshima K et al (2011) Cyclin-dependent kinase-5 is a key molecule in tumor necrosis factor-alpha-induced insulin resistance. J Biol Chem 286:33457–33465

Rane SG, Dubus P, Mettus RV et al (1999) Loss of Cdk4 expression causes insulin-deficient diabetes and Cdk4 activation results in beta-islet cell hyperplasia. Nat Genet 22:44–52

Richon VM, Lyle RE, McGehee RE Jr (1997) Regulation and expression of retinoblastoma proteins p107 and p130 during 3T3-L1 adipocyte differentiation. J Biol Chem 272:10117–10124

Sarruf DA, Iankova I, Abella A et al (2005) Cyclin D3 promotes adipogenesis through activation of peroxisome proliferator-activated receptor gamma. Mol Cell Biol 25:9985–9995

Scime A, Grenier G, Huh MS et al (2005) Rb and p107 regulate preadipocyte differentiation into white versus brown fat through repression of PGC-1alpha. Cell Metab 2:283–295

Chapter 3
Angiogenesis in Adipose Tissue

Anne Bouloumié and Jean Galitzky

Introduction

The primary role of adipose tissue (AT) is energy storage. Besides its critical role in energy homeostasis, AT is well recognized as an endocrine organ. Both the metabolic and secretory functions of AT focus attention upon the interactions existing between the blood compartment and the adipocytes and thereby the importance of the AT vasculature. In humans, the fat mass possesses a unique ability to grow and to develop throughout the life. The adipocyte changes its size dynamically according to the amount of lipid stored. Once hypertrophy is maximal, de novo adipocytes arise from the differentiation of adipocyte progenitor cells (i.e. adipogenesis) (Bjorntorp and Sjostrom 1972). Even if the number of adipocytes is considered to remain constant in adulthood, adipogenesis is essential for adipocyte turnover and maintenance of AT (Spalding et al. 2008). This AT plasticity requires constant vessel growth, regression, and remodeling. The present review describe the AT vasculature and its role in the control of AT metabolic and secretory activities in relation to AT blood flow. The potential mechanisms involved in the remodeling of the AT vascular system are exposed. Finally, the physiological and pathological consequences of the remodeling of AT vasculature are discussed.

A. Bouloumié · J. Galitzky (✉)
INSERM 1048, Team 1, I2MC, Batiment L4, CHU Rangueil,
1 Avenue Jean poulhès, BP 84225, 31432 Toulouse Cedex 4, France
e-mail: jean.galitzky@inserm.fr

A. Bouloumié
e-mail: anne.bouloumie@inserm.fr

J.-P. Bastard and B. Fève (eds.), *Physiology and Physiopathology of Adipose Tissue*,
DOI: 10.1007/978-2-8178-0343-2_3, © Springer-Verlag France 2013

The Adipose Tissue Vascular Network

Two main circulatory systems coexist in AT: the blood and the lymphatic vasculature. Both systems have distinct roles and are composed of distinct endothelial cell types.

The Lymphatic System

In the case of the AT lymphatic vasculature, few data are available, although the lymphatic system through its key role in lipid transport and inflammation is certainly a major player in the control of AT functions [for review, (Chakraborty et al. 2011)]. Lymph vessels appear to be mostly located in the interstitial space exterior to the fat lobules (Ryan 1995). It should be noted that the markers commonly used to identify human lymphatic endothelial cells include the plasma membrane surface markers CD206 (or mannose receptor) and Lymphatic Vessel Endothelial Receptor (LYVE-1) which are also expressed by human AT macrophages (Bourlier et al. 2008). Delineation of new specific markers is required to clearly identify AT lymphatic endothelial cells.

The role of the lymphatic system in AT development has been pinpointed by approaches using genetically modified mice. Indeed, in mice heterozygote for the master gene controlling lymphangiogenesis Prospero homeobox protein 1 (PROX-1 +/− mice), the abnormal lymphatic system is associated with an increase in subcutaneous AT (Harvey et al. 2005). A pro-adipogenic impact of lymph components on adipogenesis has been suspected (Harvey et al. 2005). However, the link between defects in the lymphatic system and AT development remains to be clearly established.

The role of the lymphatic system in AT secretory activity has been recently shown by the determination of the concentration and secretion of several adipokines [monocyte chemoattractant protein 1 (MCP-1), leptin, interleukin (IL)-6, IL-1beta, IL-8, tumor necrosis factor (TNF) alpha and adiponectin] in the afferent peripheral lymph and venous plasma from healthy men (Miller et al. 2011). Apart adiponectin, all studied adipokines exhibit higher concentrations in lymph than in plasma. When estimating the relative proportions of adipokines to be leaving the tissue via lymph and capillaries, leptin and MCP1 showed equally partitioning between both the lymphatic and blood system whereas IL-6 as well as IL-1beta were predominantly transported via the lymphatic route and IL-8 via the capillaries. Interestingly, the secretion of TNF alpha could not be demonstrated in human AT by arterio–venous differences measurements, but was found to be exclusively present in the lymph (Miller et al. 2011). A close anatomic relationship exists between AT and lymph nodes (Pond 2005). The lymph system might thus represent a “new” route of signaling between AT and immune cells from lymph node, that might be involved in the inflammation that takes place in AT with obesity.

The Blood System

The white AT is considered as a bradytrophic tissue, i.e., low metabolic rate and blood flow (by comparison human liver, heart or brain have blood flow that are 30–50-fold higher than subcutaneous AT from lean healthy subjects). In response to feeding, the adipose tissue blood flow (ATBF) increases in healthy individuals (Bulow et al. 1987). Concomitantly, the AT microvascular volume increases through the recruitment of capillaries as recently demonstrated using contrast-enhanced ultrasound imaging (Tobin et al. 2010). Such a process increases the surface of exposure of the lipoprotein lipase (LPL), located at the luminal side of the AT capillaries, to the circulating lipoproteins, and thus promotes lipid uptake and storage in AT. Recently, the glycosylphosphatidylinositol-anchored high density lipoprotein binding protein 1 (GPIHBP1), has been described as specifically expressed by the endothelial cells of tissues expressing LPL [for review: (Beigneux 2010)]. GPIHBP1 plays the role of a LPL shuttle to the luminal side of the endothelium. The dimerization of GPIHBP1 at the endothelial luminal side allows the binding of VLDL to GPIHBP1, in the vicinity of the LPL, facilitating triglyceride hydrolysis. Once hydrolyzed, non-esterified fatty acids (NEFA) are transferred, through the endothelial cell layer, to the adipocytes via a yet unidentified process. The transendothelial passage of hormones and adipokines in the AT is also unclear but plays certainly a major role in the hormonal control of the adipocyte metabolic activity but also in the AT secretory activity. The presence of fenestrated capillaries that may facilitate the transfer and secretion of adipokines has been shown in murine AT (Cao et al. 2001).

In humans, AT blood endothelial cells are characterized by the coexpression of both CD34 and CD31 cell surface markers and both have been used to isolate the endothelial cells from human AT. However, the use of CD31 alone to identify and to isolate the AT endothelial cells cannot be considered to be reliable since AT immune cells also express CD31. The same remark applies for the use of CD34 alone since CD34 is also expressed by the AT progenitor cells defined as $CD34^{+}/CD31^{-}$ (Miranville et al. 2004). Depletion of leukocytes from the AT stroma-vascular fraction (SVF), using for example CD45, is therefore necessary to obtain reliable AT endothelial cell preparations with the CD31 marker. The endothelial cells isolated from murine AT using such an approach and grafted into syngenic mice (Koh et al. 2011) or the implantation of AT microvessels in immunodeficient mice (Nunes et al. 2010) induce in the host a rapid formation of a vascular network through the reassembly of endothelial cells at the site of implantation. Three main dynamic events were involved in the formation of new blood vessels in the recipients: angiogenesis followed by the formation of immature vessel network and finally the maturation of the functional neovessels (Nunes et al. 2010). This vascular remodeling capacity was also observed for human AT vessels. Indeed, capillary branches emerge from human AT explants embedded in Matrigel (Gealekman et al. 2011), or from graft from human AT on the chicken chorioallantoic membrane. In this last model, the new vessels formed around and inside grafts were of both avian and human origin (Ledoux et al. 2008).

In the case of the architectural organization of the blood vascular network of the AT, few data are available. Concerning the coverage of the AT vessels in mice, pericytes were found to be mainly associated with the arterial vessels but remained less associated with the capillaries (Xue et al. 2008). In human AT, some progenitor cells are at perivascular location and predominantly near capillaries (Maumus et al. 2011). It has been suggested that they shared similar characteristics to pericytes (Traktuev et al. 2008). However, other study including ours showed no overlap in the expression of several cell markers between the both cell types (Maumus et al. 2011). Additional data are required to characterize the blood AT vascular network organization.

Angiogenic Factors Produced in Adipose Tissue

Angiogenesis is a complex process that requires a well-balanced production of factors impacting extracellular matrix remodeling, endothelial cell function (including permeability and vasodilation), endothelial cell proliferation and migration as well as the recruitment of mural cells. Several of these factors have been shown to be produced in AT. They include vascular endothelial growth factor (VEGF), monobutyrin, angiopoietin 1, hepatocyte growth factor (HGF), angiopoietin like factor 4 (angptl4), apelin, and fibroblast growth factors (FGFs) as well as matrix metalloproteases (MMPs). AT also produces angiogenic inhibitors required to control the end of the angiogenic process and to prevent further neovascularization. The list includes angiopoietin 2, thrombospondins, Secreted Protein Acidic and Rich in Cysteine (SPARC), osteopontin, resistin, interferon (IFN) gamma, tissue inhibitors of metalloproteases (TIMPs), and plasminogen activator inhibitor (PAI)-1 [for review, (Cao 2010; Christiaens and Lijnen 2010)].

Most of the pro-angiogenic factors have been described to be up-regulated during adipogenesis. Interestingly, several pro-angiogenic factors have been identified in the microvesicles produced by preadipocytes, suggesting that such microvesicles might represent a reservoir of angiogenic factors in AT (Aoki et al. 2010). A splice variant of the VEGF receptor, the soluble fms-like tyrosine kinase 1 (sFlt-1) has been identified in the conditioned media from human AT explants. This production is inversely correlated with body mass index (Herse et al. 2011). Since sFlt1 showed anti-angiogenic properties, increased adiposity appears to be associated in AT with decreased anti-angiogenic and increased pro-angiogenic capacities. In agreement with this concept, leptin, the secretion of which is increased in human obesity, possesses pro-angiogenic activity (Bouloumie et al. 1998). Both adipocytes and SVF cells are source of numerous factors involved in angiogenesis and vessel remodeling. Among the SVF cells, macrophages that accumulate with increased adiposity in the subcutaneous AT of healthy lean to overweight women exhibit pro-angiogenic activity. Indeed secretory products originating from human subcutaneous AT macrophages stimulated the migration and organization of human AT endothelial cells CD34⁺/CD31⁺ when embedded in Matrigel. Such angiogenic macrophages were characterized by the expression of LYVE-1 marker and their specific production of MMP-9 (Bourlier et al. 2008). Most cells located in the AT are also a reservoir of angiogenic factors and their degranulation is associated with strong stimulation of angiogenesis in the murine AT (Liu et al. 2009). The well-known condition linked to increased production of angiogenic factors is hypoxia. Culture of differentiated murine preadipocytes under low oxygen tension led to the up-regulation of the expression of VEGF, leptin and MMP-2 and -9 (Lolmede et al. 2003). Such an effect of low oxygen tension has also been reported in human mature subcutaneous adipocytes, at least for VEGF expression (Villaret et al. 2010). Sympathetic activation has been demonstrated to stimulate angiogenesis in murine AT via the up-regulation of VEGF, such as during cold adaptation (Xue et al. 2009) or stress involving the neuropeptide Y (NPY)/NPY2R pathway (Kuo et al. 2007).

Role of Angiogenesis in the Adipose Tissue Development

In species with fetal development of AT, the appearance of vascular structures precedes that of the fat cells (Crandall et al. 1997). In species with postnatal development of AT, as in rodents, the appearance of adipocytes in the epididymal depot is observed once vascularization is effective (Han et al. 2011). The post-natal angiogenesis in epididymal AT involves the VEGFA/VEGFR2 signaling pathway and the pro-angiogenic LYVE-1 + macrophages (Han et al. 2011). In adult mice, a dense vascular network is formed by angiogenesis in the end portion of epididymal AT and the newly formed vascular structures were associated with de novo formation of adipocytes (Cho et al. 2007). In human, a study based on the utilization of flow cytometry analyses of the SVF showed that the percentage of AT $CD34^+/CD31^+$ endothelial cells remained constant whatever the body mass index of adult healthy lean to overweight women (Miranville et al. 2004). Such a result demonstrates that there is an increase in endothelial cell number concomitant with an increase in the AT from lean to overweight women. Therefore, angiogenesis also occurs in healthy adult human AT.

The close relationship between angiogenesis and adipogenesis observed during AT development is further supported by grafting of murine preadipocyte cell line into immunodeficient mice which demonstrated that adipogenesis requires angiogenesis (Neels et al. 2004). Similarly, pharmacological-mediated stimulation of adipogenesis in mice with the peroxisome proliferator-activated (PPAR) gamma agonist rosiglitazone was associated with AT angiogenesis via increased VEGFA, VEGFB, and Angptl4 expression (Gealekman et al. 2008). The mechanisms underlying the link between angiogenesis and adipogenesis are still to be clearly defined. Some progenitor cells are in close interaction with endothelial cells within AT (Maumus et al. 2011). Secretions of AT endothelial cells affect not only directly the proliferation of AT progenitor cells (Hutley et al. 2001; Maumus et al. 2008) but also their adipogenic potential (Rajashekhar et al. 2008) as well as their migratory capacity (Sengenes et al. 2007a).Similarly, human AT progenitor cells themselves express angiogenic potentials as demonstrated by their promotion of the neovascularization of ischemic tissues after injection into athymic mice (Miranville et al. 2004; Planat-Benard et al. 2004). However, a more complete characterization of the exact nature of AT progenitor cells is necessary to reach any clear conclusions about the contribution of AT progenitor cells to vessel remodeling.

Angiogenesis and Obesity

Obese, insulin resistant, and type 2 diabetic subjects exhibit decreased fasting ATBF and alteration in the post-prandial increase of ATBF in subcutaneous AT (Karpe et al. 2002).

Reduced blood flow might lead to insufficient oxygen supply. Hypoxic area have been clearly described in the AT of obese mice in several models of murine obesity (genetically- or diet-induced obesity), [for review (Wood et al. 2009)] and associated with angiogenesis (Cho et al. 2007). However, in human, the data are still controversial. In overweight/obese human subjects, the partial oxygen tension (Po_2) in abdominal subcutaneous AT was found to be lower compared to lean subjects (Pasarica et al. 2009) though overlapping values were reported between both groups of subjects. Another approach using the continuous monitoring of Po_2 by microdialysis showed that the post-prandial increase in ATBF was associated with a concomitant increase in AT Po_2 in lean men, but these responses were blunted in obese insulin resistant men (Goossens et al. 2011). Surprisingly, despite a lower fasting ATBF in obese men compared with lean ones, fasting AT Po_2 was higher in obese men. These unexpected findings were explained by a lower consumption of oxygen in the obese AT compared with the lean AT (Goossens et al. 2011). This study suggests that adipocytes might adapt their consumption of oxygen according to its supply. To conclude, the presence of hypoxic areas consecutive to reduced blood flow within human obese subcutaneous AT remains to be clearly established.

Reduced blood flow might be due to severe structural and functional abnormalities of the microcirculation and/or a rarefaction of the vascular bed (i.e., insufficient angiogenesis). Type 2 diabetic overweight patients have a defective increase in the AT microvascular volume after a glucose load (Tobin et al. 2011). Moreover, blunted answer of the vasculature of obese to the autonomic cardiovascular control has been reported (Funada et al. 2011). An impaired endothelial transfer of insulin in AT from obese subjects with postprandial hyperglycemia has been suspected. Indeed, higher circulating insulin levels were needed to attain similar AT interstitial insulin levels than in lean subjects (Sandqvist et al. 2011). These observations suggest that the AT microcirculation exhibit functional abnormalities in obese patients. Controversial results have been reported concerning the potential rarefaction in capillary density in the subcutaneous AT from obese patients. Indeed, some studied reported decreased capillary density in subcutaneous AT in overweight/obese patients compared with lean ones as well as in morbid obese compared with overweight/obese (Gealekman et al. 2011; Pasarica et al. 2009), whereas others did not find any differences (Goossens et al. 2011). These results might be related to distinct cohort characteristics including the degree of obesity, the history of obesity, the age as well as the gender ratio and the presence or not of obesity-associated pathologies. The experimental approaches to estimate capillary density might also yield to inconsistent results. Indeed, capillaries are identified through the use of a single endothelial cell surface marker on histological sections of ATs and the number of positive signals is normalized per adipocyte or per surface. However, such approaches cannot produce reproducible quantification due to the large volume of adipocytes compared with capillaries. Indeed, depending on the studies, the maximum AT capillary densities ranged from 0.5 to 1.89 per adipocyte or from 35 to 300 capillaries/mm^2. Three-dimensional approaches of imaging using co-labeling of endothelial cells are needed to reliably assess the capillary density per volume of AT. Our approaches using flow cytometry analyses to quantify the number of endothelial cells $CD34^+/CD31^+$ per gram of AT did not show differences in endothelial cell number between subcutaneous AT from lean and obese subjects (Villaret et al. 2010). Large-scale transcriptomic analyses performed on AT from lean, overweight, obese, or obese with metabolic syndrome women showed that angiogenesis-related pathway was specifically enriched in AT from obese with metabolic syndrome compared with lean women (Klimcakova et al. 2010). Therefore, it appears that changes in the function of the microcirculation rather than insufficient angiogenesis might be responsible for the ATBF alterations observed in the subcutaneous AT of obese subjects. However, additional data are required to clearly support this hypothesis and to determine the mechanisms involved in the alterations of the AT endothelial cells with obesity.

Angiogenesis and Subcutaneous/Visceral Adipose Tissues

Increased visceral fat mass is closely linked with an increased risk in developing obesity-associated pathologies. The visceral AT is considered to have higher metabolic rate than the subcutaneous AT (see Chap. 23). However, conflicting results have been reported concerning potential differences in the fasting and postprandial blood flow in both AT locations. Concerning the secretory activity, secretome analyses of human subcutaneous and visceral AT and cells showed secretory capacity higher in visceral compared with subcutaneous AT with half of the products involved in angiogenesis (Hocking et al. 2011). Similarly, transcriptomic analyses showed specific enrichment in angiogenic-related pathway in the visceral compared with subcutaneous AT (Klimcakova et al. 2010). These observations suggest that the visceral AT constitutes a more marked proangiogenic environment compared with subcutaneous AT. However, when considering the angiogenic potential, the use of "in vitro" models leads to inconsistent results. Indeed, both depots stimulated angiogenesis in the same extent once grafted on the chick chorioallantoic membrane (Ledoux et al. 2008) but, in the model of AT explants embedded in Matrigel, higher capillary branch formation was observed for subcutaneous compared with visceral AT (Gealekman et al. 2011). Both approaches may not completely reflect angiogenic process occurring "in vivo". Using three-dimensional imaging approaches associated with flow cytometry analyses, we showed that visceral AT is more highly vascularized than subcutaneous AT from morbidly obese subjects (Villaret et al. 2010). In addition, visceral AT had a higher number of macrophages (Duffaut et al. 2009) and visceral adipocytes expressed more VEGF (Villaret et al. 2010), observations in agreement with the higher proangiogenic environment of visceral compared to subcutaneous AT. Visceral AT endothelial cells exhibited a pro-inflammatory and pro-angiogenic phenotype as well as markers for premature aging (or senescence) (Villaret et al. 2010). It is thus tempting to speculate that the strong pro-angiogenic pressure of visceral AT may lead to accelerated aging of endothelial cells leading to inflammation and endothelial cell dysfunction. In agreement with such an hypothesis, it was recently shown that endothelial cells cultured in the presence of obese AT secretomes and particularly the visceral secretome, showed increased proliferation, altered cell morphology and pro-inflammatory phenotype (Hanzu et al. 2011). Therefore, although angiogenesis is a physiological mechanism necessary to maintain the AT homeostasis during growth, it may become pathological under too strong pro-angiogenic microenvironment leading to endothelial cell senescence and to endothelial dysfunction.

Angiogenesis and Brown Adipose Tissue

Few data are available concerning the brown AT (BAT) vascular network, its regulation as well as its specific functions. In mice, VEGFB has been reported to stimulate NEFA transendothelial transport via its binding to VEGFR1 and neuropilin 1 with a concomitant increase in endothelial fatty acid transport protein (FATP) 3 and FATP4 expression (Hagberg et al. 2010). This specific effect observed in BAT (Hagberg et al. 2010), was not found in the white AT, a result showing that the endothelial cells from white and brown AT possess distinct characteristics. Brown adipocytes rely on oxygen to produce heat. Therefore, adequate perfusion is required for an optimal metabolic rate. Angiogenesis in BAT is enhanced after cold exposure (Xue et al. 2009), though it should be noted that angiogenesis is also reported in white AT under such a condition. Interestingly, the processes involved the BAT angiogenesis are independent of hypoxia but dependent on the VEGF/VEGFR2 pathway through the increase in the sympathetic nervous system-mediated stimulation of VEGF expression (Xue et al. 2009).

Anti-Angiogenic Therapy

Several approaches have been carried out in mice models of obesity (genetically- and diet-induced) using angiogenesis inhibitor treatment [for review, (Cao et al. 2001)]. Administration of endogenous angiogenesis inhibitors including angiostatin and endostatin reduce body weight in obese mice. Treatments with angiogenesis inhibitors such as TNP-470 and VEGFR2-specific inhibitors as well as MMP inhibitors prevent obesity. It must be noted however that TNP-470 has non-vascular effects including neurotoxicity and may affect food intake. Therefore, additional studies are necessary to understand how the antiangiogenic agents reduce body weight and to delineate their adverse effects.

Conclusion

Although many studies have clearly shown the importance of ATBF in the physiology of AT, the organization, and the cells composing the vascular bed of the AT have been long neglected. However, recently there has been increasing interest in the topic. The endothelial cells play a central role in AT lipid storage process as anchoring surface of the lipolytic platform and NEFA transfer to adipocytes. They also play a role in AT humoral responsiveness and secretory function as well as in AT/immune cell interactions [for review (Sengenes et al. 2007b)]. The modulation of AT endothelial cells remodeling and functions appears to be highly regulated by the microenvironment of the fat mass and as a consequence impact directly on the development and function of the AT. However, to clearly define and identify potential therapeutic strategies aimed to target AT vascular network function and/or extension, a better characterization of the endothelial cells of AT [depending on the AT location (subcutaneous and visceral) and function (white and brown)] and of the mechanisms involved in the control of vascular remodeling and function is necessary.

References

Aoki N, Yokoyama R, Asai N et al (2010) Adipocyte-derived microvesicles are associated with multiple angiogenic factors and induce angiogenesis in vivo and in vitro. Endocrinology 151:76–2567

Beigneux AP (2010) GPIHBP1 and the processing of triglyceride-rich lipoproteins. Clin Lipidol 5:575–582

Bjorntorp P, Sjostrom L (1972) Fat cell size and number in adipose tissue in relation to metabolism. Isr J Med Sci 8:320–324

Bouloumie A, Drexler HC, Lafontan M, Busse R (1998) Leptin, the product of Ob gene, promotes angiogenesis. Circ Res 83:66–1059

Bourlier V, Zakaroff-Girard A, Miranville A et al (2008) Remodeling phenotype of human subcutaneous adipose tissue macrophages. Circulation 117:806–815

Bulow J, Astrup A, Christensen NJ, Kastrup J (1987) Blood flow in skin, subcutaneous adipose tissue and skeletal muscle in the forearm of normal man during an oral glucose load. Acta Physiol Scand 130:657–661

Cao Y (2010) Adipose tissue angiogenesis as a therapeutic target for obesity and metabolic diseases. Nat Rev Drug Discov 9:15–107

Cao R, Brakenhielm E, Wahlestedt C et al (2001) Leptin induces vascular permeability and synergistically stimulates angiogenesis with FGF-2 and VEGF. Proc Natl Acad Sci U S A 98:6390–6395

Chakraborty S, Zawieja S, Wang W et al (2011) M Lymphatic system: a vital link between metabolic syndrome and inflammation. Ann N Y Acad Sci 1207(Suppl 1):E94–E102

Cho CH, Koh YJ, Han J et al (2007) Angiogenic role of LYVE-1-positive macrophages in adipose tissue. Circ Res 100:47–57

Christiaens V, Lijnen HR (2010) Angiogenesis and development of adipose tissue. Mol Cell Endocrinol 318:2–9

Crandall DL, Hausman GJ, Kral JG (1997) A review of the microcirculation of adipose tissue: anatomic, metabolic, and angiogenic perspectives. Microcirculation 211:4–32

Duffaut C, Zakaroff-Girard A, Bourlier V et al (2009) Interplay between human adipocytes and T lymphocytes in obesity: CCL20 as an adipochemokine and T lymphocytes as lipogenic modulators. Arterioscler Thromb Vasc Biol 29:1608–1614

Funada J, Dennis AL, Roberts R et al (2011) Regulation of subcutaneous adipose tissue blood flow is related to measures of vascular and autonomic function. Clin Sci (Lond) 119:313–322

Gealekman O, Burkart A, Chouinard M et al (2008) Enhanced angiogenesis in obesity and in response to PPARgamma activators through adipocyte VEGF and ANGPTL4 production. Am J Physiol Endocrinol Metab E1056:64–295

Gealekman O, Guseva N, Hartigan C et al (2011) Depot-specific differences and insufficient subcutaneous adipose tissue angiogenesis in human obesity. Circulation 123:94–186

Goossens GH, Bizzarri A, Venteclef N et al (2011) Increased adipose tissue oxygen tension in obese compared with lean men is accompanied by insulin resistance, impaired adipose tissue capillarization, and inflammation. Circulation 124:67–76

Hagberg CE, Falkevall A, Wang X et al (2010) Vascular endothelial growth factor B controls endothelial fatty acid uptake. Nature 464:917–921

Han J, Lee JE, Jin J et al (2011) The spatiotemporal development of adipose tissue. Development 5027:37–138

Hanzu FA, Palomo M, Kalko SG et al (2011) Translational evidence of endothelial damage in obese individuals: inflammatory and prothrombotic responses. J Thromb Haemost 9:1236–1245

Harvey NL, Srinivasan RS, Dillard ME et al (2005) Lymphatic vascular defects promoted by Prox1 haploinsufficiency cause adult-onset obesity. Nat Genet 37:1072–1081

Herse F, Fain JN, Janke J, Engeli S, Kuhn C, Frey N, Weich HA, Bergmann A, Kappert K, Karumanchi SA, Luft FC, Muller DN, Staff AC, Dechend R (2011) Adipose tissue-derived soluble fms-like tyrosine kinase 1 is an obesity-relevant endogenous paracrine adipokine. Hypertension 58:37–42

Hocking SL, Wu LE, Guilhaus M et al (2011) Intrinsic depot-specific differences in the secretome of adipose tissue, preadipocytes, and adipose tissue-derived microvascular endothelial cells. Diabetes 59:3008–3016

Hutley LJ, Herington AC, Shurety W et al (2001) Human adipose tissue endothelial cells promote preadipocyte proliferation. Am J Physiol Endocrinol Metab E1037:44–281

Karpe F, Fielding BA, Ilic V et al (2002) Impaired postprandial adipose tissue blood flow response is related to aspects of insulin sensitivity. Diabetes 2467:51–73

Klimcakova E, Roussel B, Marquez-Quinones A et al (2010) Worsening of obesity and metabolic status yields similar molecular adaptations in human subcutaneous and visceral adipose tissue: decreased metabolism and increased immune response. J Clin Endocrinol Metab 96:E73–E82

Koh YJ, Koh BI, Kim H et al (2011) Stromal vascular fraction from adipose tissue forms profound vascular network through the dynamic reassembly of blood endothelial cells. Arterioscler Thromb Vasc Biol 1141:31–50

Kuo LE, Kitlinska JB, Tilan JU et al (2007) Neuropeptide Y acts directly in the periphery on fat tissue and mediates stress-induced obesity and metabolic syndrome. Nat Med 803:11–13

Ledoux S, Queguiner I, Msika S et al (2008) Angiogenesis associated with visceral and subcutaneous adipose tissue in severe human obesity. Diabetes 3247:57

Liu J, Divoux A, Sun J et al (2009) Genetic deficiency and pharmacological stabilization of mast cells reduce diet-induced obesity and diabetes in mice. Nat Med 15:5–940

Lolmede K, de Saint Durand, Front V, Galitzky J et al (2003) Effects of hypoxia on the expression of proangiogenic factors in differentiated 3T3-F442A adipocytes. Int J Obes Relat Metab Disord 27:95–1187

Maumus M, Sengenes C, Decaunes P et al (2008) Evidence of in situ proliferation of adult adipose tissue-derived progenitor cells: influence of fat mass microenvironment and growth. J Clin Endocrinol Metab 4098:93–106

Maumus M, Peyrafitte JA, D'Angelo R et al (2011) Native human adipose stromal cells: localization, morphology and phenotype. Int J Obes (Lond) 35(9):1141–1153

Miller NE, Michel CC, Nanjee MN et al (2011) Secretion of adipokines by human adipose tissue in vivo: partitioning between capillary and lymphatic transport. Am J Physiol Endocrinol Metab 301:E659–E667

Miranville A, Heeschen C, Sengenes C et al (2004) Improvement of postnatal neovascularization by human adipose tissue-derived stem cells. Circulation 110:349–355

Neels JG, Thinnes T, Loskutoff DJ (2004) Angiogenesis in an in vivo model of adipose tissue development. FASEB J 983:5–18

Nunes SS, Greer KA, Stiening CM et al (2010) Implanted microvessels progress through distinct neovascularization phenotypes. Microvasc Res 79:10–20

Pasarica M, Sereda OR, Redman LM et al (2009) Reduced adipose tissue oxygenation in human obesity: evidence for rarefaction, macrophage chemotaxis, and inflammation without an angiogenic response. Diabetes 58:718–725

Planat-Benard V, Silvestre JS, Cousin B et al (2004) Plasticity of human adipose lineage cells toward endothelial cells: physiological and therapeutic perspectives. Circulation 656:63–109

Pond CM (2005) Adipose tissue and the immune system. Prostaglandins Leukot Essent Fat Acids 73:17–30

Rajashekhar G, Traktuev DO, Roell WC et al (2008) IFATS collection: adipose stromal cell differentiation is reduced by endothelial cell contact and paracrine communication: role of canonical Wnt signaling. Stem Cells 2674:26–81

Ryan TJ (1995) Lymphatics and adipose tissue. Clin Dermatol 13:493–498

Sandqvist M, Strindberg L, Schmelz M et al (2011) Impaired delivery of insulin to adipose tissue and skeletal muscle in obese women with postprandial hyperglycemia. J Clin Endocrinol Metab 96:E1320–E1324

Sengenes C, Miranville A, Maumus M et al (2007a) Chemotaxis and differentiation of human adipose tissue CD34 +/CD31− progenitor cells: role of stromal derived factor-1 released by adipose tissue capillary endothelial cells. Stem Cells 2269:25–76

Sengenes C, Miranville A, Lolmede K et al (2007b) The role of endothelial cells in inflamed adipose tissue. J Intern Med 262:415–421

Spalding KL, Arner E, Westermark PO et al (2008) Dynamics of fat cell turnover in humans. Nature 453:783–787

Tobin L, Simonsen L, Bulow J (2010) Real-time contrast-enhanced ultrasound determination of microvascular blood volume in abdominal subcutaneous adipose tissue in man. Evidence for adipose tissue capillary recruitment. Clin Physiol Funct Imaging 30:447–452

Tobin L, Simonsen L, Bulow J (2011) The dynamics of the microcirculation in the subcutaneous adipose tissue is impaired in the postprandial state in type 2 diabetes. Clin Physiol Funct Imaging 31:458–463

Traktuev DO, Merfeld-Clauss S, Li J et al (2008) A population of multipotent CD34-positive adipose stromal cells share pericyte and mesenchymal surface markers, reside in a periendothelial location, and stabilize endothelial networks. Circ Res 102:77–85
Villaret A, Galitzky J, Decaunes P et al (2010) Adipose tissue endothelial cells from obese human subjects: differences among depots in angiogenic, metabolic, and inflammatory gene expression and cellular senescence. Diabetes 2755:59–63
Wood IS, de Heredia FP, Wang B, Trayhurn P (2009) Cellular hypoxia and adipose tissue dysfunction in obesity. Proc Nutr Soc 68:370–377
Xue Y, Cao R, Nilsson D et al (2008) FOXC2 controls Ang-2 expression and modulates angiogenesis, vascular patterning, remodeling, and functions in adipose tissue. Proc Natl Acad Sci U S A 105:72–10167
Xue Y, Petrovic N, Cao R et al (2009) Hypoxia-independent angiogenesis in adipose tissues during cold acclimation. Cell Metab 9:99–109

Chapter 4
Prospects for Using Adipose Tissue in Regenerative Medicine

Louis Casteilla, Valérie Planat-Benard, Patrick Laharrague, Béatrice Cousin and Anne Lorsignol

Introduction

The traditionally widespread image of adipose tissues is the image associated with the roles attributed to them in obesity and in metabolic diseases. In parallel with this negative image, and in a more specialized field, there is a positive image tied to their old use in plastic and reconstructive surgery. Indeed, these tissues have long been used by doctors to reconstruct volumes following various deformations.

Much more recently, the identification and characterization of subpopulations of immature cells have drawn the attention of many investigators and is opening up broad prospects for regenerative medicine, especially since this tissue can be obtained in large quantities through liposuction.

Adipose Tissues as Filler and "Volumizing" Tissues

Long ago, the pragmatic approach to the problems of tissue reconstruction led surgeons to use adipose tissue as a filler tissue, with some success. Unfortunately, one of the often-debated recurring problems of this technique was: "how long the

L. Casteilla (✉) · V. Planat-Benard · P. Laharrague · B. Cousin · A. Lorsignol
Equipe 1 "Plasticité des tissus adipeux", UMR UPS/CNRS/EFS 5273,
U1031 INSERM, BP 84 225, 31432 Toulouse, France
e-mail: louis.casteilla@inserm.fr
http://www.stromalab.fr/

L. Casteilla · V. Planat-Benard · P. Laharrague · B. Cousin · A. Lorsignol
Université de Toulouse, UPS; UMR 5241, BP 84 225, 31432 Toulouse, France

P. Laharrague
Laboratoire d'Hématologie, CHU Toulouse, TSA 50032, 31 059 Toulouse, France

J.-P. Bastard and B. Fève (eds.), *Physiology and Physiopathology of Adipose Tissue*,
DOI: 10.1007/978-2-8178-0343-2_4,

reconstituted volume was maintained?" A decisive step was taken toward graft maintenance when Professor Colleman chose to sample and re-inject adipose tissues in the form of small samples (Mojallal and Foyatier 2004). This technique was so successful that it has become the current standard technique. Following anesthesia, usually local anesthesia, the adipose tissue is removed using cannulas, with the consequence of fragmenting the tissue sample in the form of small "noodles." These are then re-injected into the patient, after the excess liquid is removed through a brief centrifuging. The size of the cannulas, the centrifuging speeds, and sampling systems vary widely and are the subject of significant economic issues. It is worth noting that the development of these practices took place independent of the very small number of experimental or preclinical studies (Mojallal et al. 2009). It is only very recently that the use of tissue imaging tools has enabled the objective evaluation of the effects and protocols in humans.

The Various Cellular Fractions of Adipose Tissue

Following the work conducted by the group led by Zuk et al. (2002), the conception of what had been previously referred to as preadipocytes was profoundly changed. Described in the 1960s, a simple protocol was used from a sample of adipose tissue to purify a fraction of immature cells that were named preadipocytes (Gimble et al. 2007). In this protocol, the fat tissue is finely cut up and incubated with proteolytic enzymes in order to dissociate the extracellular matrix. Through centrifuging, it is then possible to separate the stroma-vascular fraction (SVF) from the fraction of mature adipocytes which, filled with lipids, float in the solution. This SVF is strongly heterogeneous, an element that we will come back to later. In a second step, the SVF is cultured in a plastic Petri dish in order to select an adherent cell population. The addition of a cocktail of differentiating agents induces the differentiation of these cells into adipocytes, later demonstrating the existence of adipocyte-precursor cells, or preadipocytes. These cells are found in all adipose tissues, regardless of the patient's age (Ailhaud et al. 1992).

How a Preadipocyte Becomes a Multipotent Cell with Regenerative Potentiality

Zuk et al. (2002) were the first to show that these preadipocytes were, in fact, multipotent because they were able to differentiate not only into adipocytes, but also into osteoblasts and chondrocytes. Preadipocytes were then re-christened adipose-derived stem cells (or ASCs), and then adipose-derived stromal cells

(Zuk et al. 2002; Gimble et al. 2007). In fact, since their capacity for self-renewal (one of the two criteria for determining the status of stem cell) has not been definitively established, in accordance with the International Society for Cellular Therapy (Phinney and Prockop 2007), it currently appears preferable to use the term "stromal" instead of "stem." Another difficulty resides in the fact that the term ASC is also used in the literature for the raw SVF. This fraction is strongly heterogeneous and contains several cellular subpopulations, including native ASCs, endothelial cells, and hematopoietic cells, which represent a large portion of the fraction (Prunet-Marcassus et al. 2006). From our point of view, the term ASC should be restricted to purified, multipotent stromal cells.

Nature and Properties of ASCs

The protocol of selection by adhesion, as well as their mutipotency, resulted in a parallel being drawn between ASCs and other stromal cells, mesenchymal stem cells (MSCs). These MSCs were first described as immature mesenchymal cells of the adult bone marrow, also capable of differentiating into osteoblasts, chondrocytes, and adipocytes (Pittenger et al. 1999). Identification at the clonal scale using the CFU-F technique (for Colony-Forming-Unit Fibroblast) shows that the frequency of MSCs is low, 0.01–0.0001 % of the nucleated cells in adult bone marrow. In 2005, the International Society for Cellular Therapy established a minimum definition of MSCs based on three criteria: adhesion to plastic, a phenotype matching surface antigens ($CD73^+$, $CD90^+$, $CD105^+$, $CD45^-$, $CD34^-$, CD14 or $CD11b^-$, $CD79^-$ or $CD19^-$, $HLA\text{-}DR^-$), and the capacity, described earlier, to give rise to adipocytes, osteoblasts, and chondrocytes, in vitro (Horwitz et al. 2005). In fact, it is also necessary to add to this definition the capacity of MSCs to maintain viability and the immaturity of other cells, such as hematopoietic stem cells, as well as their ability to modulate the immune response and the inflammatory reaction (Charbord 2010).

While very similar, ASCs and MSCs are not identical. Indeed, ASCs express the surface protein CD34, at least initially during culturing (Maumus et al. 2010), their frequency is much greater (100–500 times more) and it is possible to define, under identical cultural conditions, specific genomic and proteomic signatures (Noel et al. 2008). Furthermore, the differentiation protocols indicate schematically that MSCs are more easily oriented toward an osteoblastic and chondrogenic phenotype, whereas ASCs are more oriented toward an adipose phenotype (Noel et al. 2008). Another functional characteristic initially associated with MSCs, the ability to support hematopoiesis, also appears true for ASCs, although they appear to be less effective over the long term (Corre et al. 2006; De Toni et al. 2011). This unexpected property of adipose tissue could have very significant consequences, as will be discussed later.

The tissue location of ASCs remains a controversial subject. An initial series of experiments suggested that ASCs are pericytes, as was proposed for MSCs (Crisan et al. 2008; Traktuev et al. 2008). Our own work, in collaboration with Anne Bouloumié's group has arrived at different conclusions (Maumus et al. 2010): the immunohistological analysis in fact revealed a dual location of native ASCs: a portion in the perivascular position without being pericytary, but a majority dispersed in the stroma and not expressing, in vivo, pericytary markers. In fact, these markers appear during the culturing process. However, one may conclude that there are multiple populations of ASCs with varying degrees of immaturity. This idea would be in line with the work performed in mice by Friedman's and Graff's groups. The most immature population that contains true adipocyte precursors, and therefore ASCs, appears to be derived from cells present in the vascular wall, and is identified by surface markers (Lin^-, $CD29^+$, $CD34^+$, Sca-1^+, $CD24^+$) (Rodeheffer et al. 2008; Tang et al. 2008).

SVF and Regeneration

The possibility of obtaining this fraction extemporaneously at the "patient's bedside" presents a considerable advantage, through the flexibility of use and its quickness of implementation (see Table 4.1). Another advantage comes from the synergistic effects existing between the various cellular populations of this fraction, as we will illustrate later. These advantages are offset, however, by the heterogeneity of this fraction and the inability, in the brief time allotted, to establish quality controls to truly evaluate what is injected, since the cellular composition can vary considerably depending on the patient (effects of age, sex, metabolic status, etc.).

The first successful in vivo regeneration experiment was performed in mice, in the context of hematopoietic reconstitution (Cousin et al. 2003). Spectacularly, the injection of the SVF proved as effective as a bone marrow graft. This regeneration could be due to a population of hematopoietic stem cells present in the adipose tissue and reconstituting the destroyed hematopoietic compartment, and also to a stimulating effect of ASCs on endogenous hematopoiesis of irradiated mice.

In humans, the first study described the use of this cellular fraction to treat a massive bone defect in the cranium (Lendeckel et al. 2004). Three months after the application of the raw SVF mixed with a glue composed of fibrin, a boney neoformation and almost complete reconstitution of the cranial vault were observed. Unfortunately, no other publications followed this report.

Other trials are studying the effect of this SVF to treat acute myocardial infarction. Although it is necessary to wait for the full results to come out, an absence of improvement is probable, as was observed in similar protocols using mononuclear bone marrow cells (and not MSCs) (Menasche 2009).

Table 4.1 Advantages and disadvantages of the use of the raw stroma-vascular fraction or cells purified and multiplied in culture (ASCs)

	Raw stroma-vascular fraction	Cultivated ASCs
Preparation time	Extemporaneous to a few hours	Several days
Sample size	Strict dependance on the number of cells required	More limited because expansion is possible through culturing
Quantity of cells	Depends on the sample size	Depends on the sample size and culturing time
Heterogeneity	Strong	Limited
Phenotypic characterization	Impossible to perform within preparation times	Easy to perform
Biological effects	Synergistic effects possible from multiple cell populations	Relies on a single type of cells
Effects of culturing	None	Greater the longer the culturing time
Preconditioning	Limited	Possible
Quality controls	Limited	Strongly facilitated
Treatment of acute pathologies	Possible under autologous and allogenic conditions	Possible only under allogenic conditions
Treatment of chronic diseases	Possible under autologous and allogenic conditions	Possible under autologous and allogenic conditions
Regulations	Currently, no approval needed if cells are prepared in the operating theater	Approval mandatory All cultured products must be validated as a product associated with cell therapy (PTA standard)

ASCs and Regeneration

Zuk's studies on ASC mutipotency were published when the scientific community was fascinated with adult stem cells and their possible plasticity. At the same time, we showed, simultaneously with A. Bouloumié's group, that ASCs could regenerate a deficient vascular network in vivo (Planat-Benard et al. 2004a; Miranville et al. 2004). These results attracted even more attention since adipose tissue is abundant, easy to sample, and in a few cultures it is possible to obtain large quantities of ASCs. This quickness of obtainment, compared to MSCs derived from bone marrow, limits the risks of senescence and/or chromosomal abnormalities induced during culturing. As with MSCs derived from bone marrow, the analysis of works published shows that the benefits observed after the injection of ASCs can be explained in three non-exclusive ways (Charbord and Casteilla 2011):

- Direct participation of ASCs, through their differentiation potential, in neoformed tissues.
- Intense paracrine activities of ASCs that may have pleiotropic effects on the survival of endogenous cells, formation of neovessels, control of inflammation, etc.
- When differentiated cells from the injured tissue are multinucleated (e.g., skeletal muscle cells) regeneration may be due to fusion mechanisms.

Concerning the first explanation, apart from the traditional mesenchymal phenotype (adipocyte, osteoblast, chondrocytes), no studies have clearly demonstrated complete and functional differentiation toward other cell types. But usually the phenotype is established only by a few differentiation markers, and sometimes a partial functional analysis, which raises the question of the underlying cellular mechanisms.

For the second explanation centered on paracrine activity, we may note that adipose tissue is considered to be a true endocrine tissue, a property that will be developed in another chapter of this book. Given the complexity of effects, it is reasonable to suggest that these paracrine effects are due to a combination of many more or less redundant molecules. This multitude of factors may explain the contradictions between the various studies on the molecules involved (Gimble et al. 2007).

While many studies have been produced in the osteoarticular and cardiovascular fields, the most clinically successful studies are in the area of modulating the immune system and inflammation (therefore relying on paracrine properties). Most clinical trials with ASCs deal with the complications of fistulas that are or are not associated with Crohn's disease, which correspond to tissue degeneration after an uncontrolled inflammatory process. All these trials appear to indicate that ASCs are very effective for treating inflammation and improving the healing process (Constantin et al. 2009; Gonzalez et al. 2009; Garcia-Olmo et al. 2009). It should be noted that a test to use allogenic ASCs in this field is currently planned.

The results of this trial and the possibility of using ASCs sampled before hand from healthy donors could considerably open up the field of using ASCs in regenerative medicine.

ASCs, like MSCs from bone marrow, possess characteristic immunomodulating properties (Le Blanc and Ringden 2007; Puissant et al. 2005; Yanez et al. 2006) evaluated at the clinical level. The results of these trials with ASCs are awaited with confidence, considering the positive effects of MSCs (Le Blanc and Ringden 2007).

As for the capacities of ASCs to reconstitute a differentiated cell population, there is currently little published data. One remarkable result was obtained with the maxillary reconstruction in a patient through the implantation of ASCs committed to osteogenic differentiation using a pretreatment with the growth factor BMP2 combined with calcium phosphate (Mesimaki et al. 2009). The feasibility and efficacy of such treatments may not be proven, however, until after genuine phase I and II studies. Astonishingly, there are no clinical studies published to date regarding the use of ASCs to reconstitute adipose tissues, while it is a significant clinical issue both for mammary reconstruction as well as for lipodystrophy.

Since the initial studies described earlier regarding the angiogenic potential of ASCs for limb ischemia, the confirmed potential in myocardial infarction, and the healing process associated with radiation or not (Planat-Benard et al. 2004a; Ebrahimian et al. 2009; Mazo et al. 2008), only two trials, one using intramuscular administration, the other using intravenous administration, examine the effect of ASCs on critical hind limb ischemia.

These angiogenic properties, as well as the release of trophic factors or immunomodulation, can also be taken advantage of in order to facilitate the successful engraftment of whole tissues (including adipose tissue itself, as is starting to be done for mammary reconstruction) and also of exogenous cells, such as embryonic cells differentiated into cardiomyocytes (Bel et al. 2010).

Finally, we cannot talk about cell therapy without discussing the adverse effects and safety of these treatments. Two obstacles may arise: an undesirable differentiation process and a possible interaction between ASCs and resident cancer cells. Regarding the first point, only one study has reported cysts and microcalcifications after mammary reconstruction using lipoaspirates associated with the raw SVF (Yoshimura et al. 2008). To the best of our knowledge, no undesirable differentiation has ever been reported, such events are rare, or have not been sufficiently evaluated.

The immunosuppressive effects associated with the angiogenic properties raise the question of a possible interaction between ASCs and cancer cells. This question is crucial, given the positive correlation existing between obesity and cancer (Roberts et al. 2010). Contradictory studies have been published in this domain. Schematically, a positive effect of ASCs on tumor growth is observed when cells are co-injected with cancer cells or transplanted at the start of the tumoral process (Zimmerlin et al. 2011). On the other hand, a negative effect can be observed when ASCs are implanted in a pre-existing tumor (Cousin et al. 2009). Thus, we can suggest that, depending on the partner cell, a dynamic interaction may take place

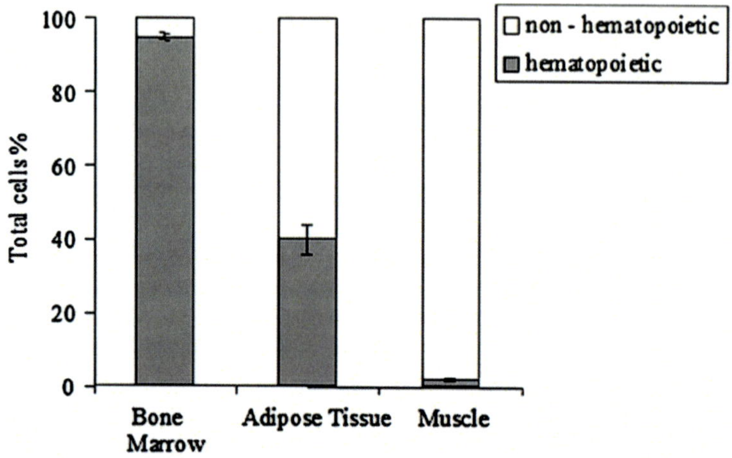

Fig. 4.1 Proportion of hematopoietic cells in stroma fractions of different tissues (bone marrow, anginal white adipose tissue, muscle)

between ASCs and other cell types to maintain the appropriate development of the tissue through a balance between proliferation and differentiation.

Other Immature Cells in Adipose Tissues

As has been stated several times in this chapter, the SVF is a heterogeneous fraction of which a significant number of the cells are poorly characterized. Analyses by flow cytometry reveal that 20–50 % of the cells of this fraction are hematopoietic cells. Except hematopoietic organs, no other tissue presents such a high proportion of hematopoietic cells (Fig. 4.1). An even more remarkable result is that approximately 1 % of these cells present an antigenic signature characteristic of hematopoietic stem cells. Furthermore, these cells are functional because they are able to participate, at least partially, in hematopoietic reconstruction following irradiation. However, this hematopoietic activity is special, above all, because it enables the reconstitution of non-hematopoietic organs (Poglio et al. 2010).

Another remarkable result is that we were able to show in mice a rare population capable of "spontaneously" differentiating into functional cardiomyocytes, the injection of which into an infarction zone of the heart will limit remodeling of the ventricle and provide a functional benefit (Planat-Benard et al. 2004b; Leobon et al. 2009). Unfortunately, the equivalent cells have not been identified in humans, at this point.

Be that as it may, the two types of results just cited well show that adipose tissue has not yet revealed all its secrets.

Conclusions or Contributions of Cellular Therapy to a Better Understanding of the Biology and Physiopathology of Adipose Tissues

The growing interest of plastic surgeons and the scientific and medical community in regeneration is attracting a growing number of teams previously focused on other domains of research and competency. While this relative lack of expertise in the domain sometimes poses problems for interpreting results, it leads to new questions and approaches, forcing the reconsideration of the entire biology of adipose tissues.

Thus has been revealed the astonishing complexity of the stromal fraction of adipose tissue, the presence of immature cells of multiple types, and with considerable potentialities, still to be definitively characterized, and vascular and hematopoietic compartments.

Furthermore, the abundance of several types of immature cells raises the question of adipose tissue as a physiological reservoir of cells that can be recruited in the event of an injury or tissue degeneration. This is particularly true for hematopoietic activity, since adipose tissue houses hematopoietic stem cells as well as stromal cells capable of maintaining this population, as is observed in bone marrow.

Adipose tissue, formerly decried for its involvement in metabolic diseases, has acquired an entirely different status that of a reservoir of useful regenerative cells. Even if other tissues possess similar properties, the odds are that the choice will be made to use adipose tissue and its cells, given the practical and ethical advantages of its use. There are already banks of frozen adipose tissues around the world, just like cord blood is frozen for later use.

Acknowledgments Our work receives financing from the following European projects: CASCADE and ADIPOA, the *Etablissement Français du Sang Pyrénées-Mediterranée*, the *Fondation pour la Recherche Médicale* (the "Normal and Pathological Cardiovascular Aging" Program) and the CTP project, Transpyreneen Stem Cells Group". No conflicts of interest.

References

Ailhaud G, Grimaldi P, Negrel R (1992) Cellular and molecular aspects of adipose tissue development. Annu Rev Nutr 12:207–233

Bel A, Planat-Bernard V, Saito A et al (2010) Composite cell sheets: a further step toward safe and effective myocardial regeneration by cardiac progenitors derived from embryonic stem cells. Circulation 122:S118–S123

Charbord P (2010) Bone marrow mesenchymal stem cells: historical overview and concepts. Hum Gene Ther 21:1045–1056

Charbord P, Casteilla L (2011) Human mesenchymal stem cell biology. Med Sci (Paris) 27: 261–267

Constantin G, Marconi S, Rossi B et al (2009) Adipose-derived mesenchymal stem cells ameliorate chronic experimental autoimmune encephalomyelitis. Stem Cells 27:2624–2635

Corre J, Barreau C, Cousin B et al (2006) Human subcutaneous adipose cells support hematopoietic differentiation in vitro. J Cell Physiol 206:282–288

Cousin B, Andre M, Arnaud E et al (2003) Reconstitution of lethally irradiated mice by cells isolated from adipose tissue. Biochem Biophys Res Commun 301:1016–1022

Cousin B, Ravet E, Poglio S et al (2009) Adult stromal cells derived from human adipose tissue provoke pancreatic cancer cell death both in vitro and in vivo. PLoS One 4:6278

Crisan M, Yap S, Casteilla L et al (2008) A perivascular origin for mesenchymal stem cells in multiple human organs. Cell Stem Cell 3:301–313

De Toni F, Poglio S, Youcef AB et al (2011) Human adipose-derived stromal cells efficiently support hematopoiesis in vitro and in vivo: a key step for therapeutic studies. Stem Cells Dev 20(12):2127–2138

Ebrahimian TG, Pouzoulet F, Squiban C et al (2009) Cell therapy based on adipose tissue-derived stromal cells promotes physiological and pathological wound healing. Arterioscler Thromb Vasc Biol 29:503–510

Garcia-Olmo D, Herreros D, Pascual I et al (2009) Expanded adipose-derived stem cells for the treatment of complex perianal fistula: a phase ii clinical trial. Dis Colon Rectum 52:79–86

Gimble JM, Katz AJ, Bunnell BA (2007) Adipose-derived stem cells for regenerative medicine. Circ Res 100:1249–1260

Gonzalez MA, Gonzalez-Rey E, Rico L et al (2009) Treatment of experimental arthritis by inducing immune tolerance with human adipose-derived mesenchymal stem cells. Arthr Rheum 60:1006–1019

Horwitz EM, Le Blanc K, Dominici M et al (2005) Clarification of the nomenclature for MSC: the international society for cellular therapy position statement. Cytotherapy 7:393–395

Le Blanc K, Ringden O (2007) Immunomodulation by mesenchymal stem cells and clinical experience. J Intern Med 262:509–525

Lendeckel S, Jodicke A, Christophis P et al (2004) Autologous stem cells (adipose) and fibrin glue used to treat widespread traumatic calvarial defects: case report. J Craniomaxillofac Surg 32:370–373

Leobon B, Roncalli J, Joffre C et al (2009) Adipose-derived cardiomyogenic cells: in vitro expansion and functional improvement in a mouse model of myocardial infarction. Cardiovasc Res 83:757–767

Maumus M, Peyrafitte J, D'Angelo R et al (2010) Native human adipose stromal cells: Localization, morphology and phenotype. Int J Obes (in press)

Mazo M, Planat-Benard V, Abizanda G et al (2008) Transplantation of adipose derived stromal cells is associated with functional improvement in a rat model of chronic myocardial infarction. Eur J Heart Fail 10:454–462

Menasche P (2009) Cell-based therapy for heart disease: a clinically oriented perspective. Mol Ther 17:758–766

Mesimaki K, Lindroos B, Tornwall J et al (2009) Novel maxillary reconstruction with ectopic bone formation by GMP adipose stem cells. Int J Oral Maxillofac Surg 38:201–209

Miranville A, Heeschen C, Sengenes C et al (2004) Improvement of postnatal neovascularization by human adipose tissue-derived stem cells. Circulation 110:349–355

Mojallal A, Foyatier JL (2004) Historical review of the use of adipose tissue transfer in plastic and reconstructive surgery. Ann Chir Plast Esthet 49:419–425

Mojallal A, Lequeux C, Shipkov C et al (2009) Improvement of skin quality after fat grafting: clinical observation and an animal study. Plast Reconstr Surg 124:765–774

Noel D, Caton D, Roche S et al (2008) Cell specific differences between human adipose-derived and mesenchymal-stromal cells despite similar differentiation potentials. Exp Cell Res 314:1575–1584

Phinney DG, Prockop DJ (2007) Concise review: mesenchymal stem/multipotent stromal cells: the state of transdifferentiation and modes of tissue repair—current views. Stem Cells 25:2896–2902

Pittenger MF, Mackay AM, Beck SC et al (1999) Multilineage potential of adult human mesenchymal stem cells. Science 284:143–147
Planat-Benard V, Silvestre JS, Cousin B et al (2004a) Plasticity of human adipose lineage cells toward endothelial cells: physiological and therapeutic perspectives. Circulation 109:656–663
Planat-Benard V, Menard C, Andre M et al (2004b) Spontaneous cardiomyocyte differentiation from adipose tissue stroma cells. Circ Res 94:223–229
Poglio S, De Toni-Costes F, Arnaud E et al (2010) Adipose tissue as a dedicated reservoir of functional mast cell progenitors. Stem Cells 28:2065–2072
Prunet-Marcassus B, Cousin B, Caton D et al (2006) From heterogeneity to plasticity in adipose tissues: site-specific differences. Exp Cell Res 312:727–736
Puissant B, Barreau C, Bourin P et al (2005) Immunomodulatory effect of human adipose tissue-derived adult stem cells: comparison with bone marrow mesenchymal stem cells. Br J Haematol 129:118–129
Roberts DL, Dive C, Renehan AG (2010) Biological mechanisms linking obesity and cancer risk: new perspectives. Annu Rev Med 61:301–316
Rodeheffer MS, Birsoy K, Friedman JM (2008) Identification of white adipocyte progenitor cells in vivo. Cell 135:240–249
Tang W, Zeve D, Suh JM et al (2008) White fat progenitor cells reside in the adipose vasculature. Science 322:583–586
Traktuev DO, Merfeld-Clauss S, Li J et al (2008) A population of multipotent CD34-positive adipose stromal cells share pericyte and mesenchymal surface markers, reside in a periendothelial location, and stabilize endothelial networks. Circ Res 102:77–85
Yanez R, Lamana ML, Garcia-Castro J et al (2006) Adipose tissue-derived mesenchymal stem cells have in vivo immunosuppressive properties applicable for the control of the graft-versus-host disease. Stem Cells 24:2582–2591
Yoshimura K, Sato K, Aoi N et al (2008) Cell-assisted lipotransfer for cosmetic breast augmentation: Supportive use of adipose-derived stem/stromal cells. Aesthet Plast Surg 32:48–55
Zimmerlin L, Donnenberg AD, Rubin JP et al (2011) Regenerative therapy and cancer: in vitro and in vivo studies of the interaction between adipose-derived stem cells and breast cancer cells from clinical isolates. Tissue Eng Part A 17:93–106
Zuk PA, Zhu M, Ashjian P et al (2002) Human adipose tissue is a source of multipotent stem cells. Mol Biol Cell 13:4279–4295

Chapter 5
Brown Adipose Tissue: Function and Development

Daniel Ricquier

Introduction

Brown adipose tissue (BAT) fulfils the criteria of an organ through its specific anatomical localisation, its functional organisation as well as its specific physiological role. BAT is mainly made of particular adipose cells, the brown adipocytes, which are able to rapidly oxidise fatty acids and produce heat in response to specific physiological situations. A total of 50 years after the discovery of the thermogenic function of BAT, and 35 years after the elucidation of the thermogenic mechanism, its activity as a powerful fat burner non observable in white adipocytes, makes this tissue still of interest since potential applications to treat metabolic diseases are expected. Recently, a close and unknown proximity between brown adipocytes and myocytes was discovered. In addition, recent data obtained through the use of modern medical imagery revaluates the importance of BAT in human adults.

BAT is Distributed in Specific Depots in Mammals

Due to fatty aspect and its brownish colour, BAT is easily observable. It is present in most mammals but its abundance varies according to the age of individuals. Whereas BAT is abundant during the whole life in rodents, it is particularly

D. Ricquier (✉)
Service de Biochimie Métabolique, Tour Lavoisier étage 4, Hôpital Necker-Enfants Malades, 149 rue de Sèvres, 75743 Paris Cedex 15, France
e-mail: daniel.ricquier@nck.aphp.fr; daniel.ricquier@parisdescartes.fr

D. Ricquier
Faculté de Médecine de l'Université Paris Descartes, Institut Cochin pièce 3013, 24 Rue du Faubourg Saint-Jacques, 75014 Paris, France

J.-P. Bastard and B. Fève (eds.), *Physiology and Physiopathology of Adipose Tissue*,
DOI: 10.1007/978-2-8178-0343-2_5, © Springer-Verlag France 2013

present in newborn rabbits as well as in a number of newborns of large-sized mammals (lamb, veal). The idea commonly accepted was that BAT was absent in adults of large-sized mammals, such as in the human species. Following reports describing brown adipocytes without quantitation in human adults around 1980, recent observations based on PET-Scan radiography of patients revaluate the importance of BAT in adults.

Anatomy of BAT Depots

BAT depots were well identified and characterised in rodents. They are localised in interscapular, peri-aortic, pericardiac, perirenal, axillary regions and between neck muscles. All BAT depots are innervated and vascularised. In interscapular BAT, each lobe is innervated by six sympathetic nerves including five intercostal nerves coming from cervical and thoracic ganglia. Actually, every brown adipocyte is directly innervated by fibres acting though noradrenaline release. This mediator plays an essential role in activation of brown adipocytes and in resulting thermogenesis. Several types of receptors including the ß3-adrenergic receptors participate in the activation of the thermogenic cells. It is important to recall that the innervation of brown adipocytes is under the control of the ventro-median part of the hypothalamus. BAT depots are highly vascularised and they present numerous arterio-venous anastomoses. Following exposure of a rat to the cold, the blood flow through BAT depots increases by a factor of ten within a few minutes, allowing the warmed blood to be sent to important regions such as the brain, heart and kidneys (Nicholls and Locke 1984; Himms-Hagen and Ricquier 1998; Cannon and Nedergaard 2004). In rodents and in humans, in addition to main BAT depots, a few islets of brown adipocytes, or eventually disseminated brown adipocytes are observable in white fat depots (Cousin et al. 1992; Cinti 2011).

Morphology of Brown Adipocytes

The most abundant and characteristic cells in BAT depots are brown adipocytes, the cells responsible for heat production. In addition to these cells, BAT depots contain interstitial cells, pericytes, endothelial cells and mastocytes. Brown adipocytes represent 50 % of the total cells present in BAT depots. Brown adipocytes contain many triglyceride droplets and perform lipogenesis and lipolysis, allowing these cells to be classified as adipocytes. However, the abundance of mitochondria in brown adipocytes is a striking feature, not observed in classical adipocytes. In addition to their high number, these mitochondria present a highly abundant inner membrane forming cristae. Therefore, the simple observation of brown adipocytes using electron microscopy indicates that such cells exhibit a very high oxidative respiration and substrate oxidation capacity, obviously linked to their elevated thermogenics capacity.

Physiological Data

Thermogenesis

Before presenting the thermogenic activity of BAT, it may be useful to briefly recall what is thermogenesis. Obligatory thermogenesis or resting thermogenesis at thermoneutrality and without exercise corresponds to the resting metabolic rate and to a series of metabolic reactions releasing heat in such a way that it permits body temperature to be close to 37 °C. In particular conditions such as exposure to a cold environment (for a non-hibernating mammal), the body temperature requires regulatory or adaptive thermogenesis, i.e. a supplement of thermogenesis. Only metabolic reactions, acceleration of metabolic pathways or futile cycles can generate thermogenesis. Indirect calorimetry was used to demonstrate that thermogenesis is linked to oxygen consumption and therefore to the velocity of cellular respiration operating in mitochondria. Skeletal muscles contain many mitochondria and are potential actors of adaptive thermogenesis, but their contribution, except for a role in shivering thermogenesis, was not really demonstrated.

A Thermogenic Organ at Birth, Upon Exposure to the Cold, or During Arousal from Hibernation

BAT is only present in mammals which are homeothermic. Homeotherms can dissipate heat in case of exposure to elevated temperatures and generate heat in situations presenting a risk of decrease of body temperature (cold environment, birth). Another interesting example is given by hibernators. These animals perfectly adapt to difficult situations (cold ambiance, reduced light, limited food supply) by a marked reduction of their energetic needs and metabolism leading to a decreased body temperature. However, hibernating animals require a thermogenic mechanism allowing them to arouse from hibernation.

A feature shared by different mammals is the necessity of the availability of a thermogenic system in three situations as different as chronic exposure to the cold, birth or arousal from hibernation. Several elegant and independent studies performed around 1962–1965 demonstrated that heat was produced by BAT depots in each of these three situations. These conclusions were confirmed by spectacular measurements of blood flow through BAT depots of rats either exposed to the cold or receiving catecholamines. The best demonstration of the thermogenic activity of BAT is its rapid activation upon exposure to the cold followed by its immediate inhibition as soon as the rat returns to an elevated ambient temperature (Nicholls and Locke 1984; Himms-Hagen and Ricquier 1998; Cannon and Nedergaard 2004).

An Organ for Diet-Induced Thermogenesis?

Obligatory thermogenesis is observed following food intake and implicating food ingestion, digestion, absorption and storage of nutrients. Although it was strongly debated, several studies led researchers to propose BAT as an actor of diet-induced thermogenesis (Rothwell and Stock 1979). This proposal was reinforced by the observation of body weight enlargement and fat gain in transgenic mice expressing a toxin targeted at BAT and destroying BAT (Lowell et al. 1993). However, the absence of fat gain in mice with non-thermogenic BAT contradicted the data (Enerbäck et al. 1997).

Quantitation of BAT Activity

The marked oxidative capacity of brown adipocytes was calculated knowing their oxygen consumption and rate of fatty acid oxidation, and also by directly quantitating heat produced either by brown fat pad (Seydoux and Girardier 1977), isolated brown adipocytes (Nedergaard et al. 1977) or respiring isolated BAT mitochondria (Ricquier et al. 1979). The energetic power of BAT is estimated at 300–500 watts per kilogram of tissue. Such numbers indicate that a small amount such as 50–100 g of rodent-type BAT would very significantly contribute to energetic expense, in such a way that the absence of BAT or its reduced activity would theoretically explain hypothermia and fattening over time.

Thermogenic Mechanism: Biochemical Aspects and Physiological Integration

Heat production by cells or organisms is associated to oxygen consumption and is correlated to the rate of oxidation of carbonated substrates. An oxidation reaction such as palmitate oxidation is written as follows:

$$C_{16}H_{32}0_2 + 23\ O_2 + 131\ ADP \rightarrow 16\ CO_2 + 16\ H_2O + 131\ ATP + \mathbf{Heat}$$

Such a formula indicates that although oxidation reactions facilitate energy conservation as ATP, a part of energy is dissipated as heat. Such an energy loss reduces available ATP but may be useful for a physiological maintenance of the necessary body temperature (Nicholls 2006).

Respiration Uncoupling

The largest part of ATP synthesised by cells comes from mitochondrial respiration "fed" with reduced NADH and FADH2 of which the reoxidation is coupled to oxygen reduction to water and ADP phosphorylation as ATP. When researchers observed the high number of mitochondria in brown adipocytes, the first conclusion was that these cells were producing a large number of ATP molecules subsequently hydrolysed by an ATP-ase generating heat. In fact, this hypothesis was never confirmed and was invalidated in 1967 when, independently, Smith and Lindberg observed a spontaneous respiration uncoupling, i.e. a rapid respiration not limited by the ability of mitochondria to respire according to their ability to phosphorylate ADP. In other words, the BAT mitochondria able to rapidly oxidise coenzymes do not make ATP and dissipate oxidation energy as heat. This uncoupling or loose coupling of respiration to ATP synthesis explains heat production by brown adipocytes and suggests that BAT mitochondria do contain a respiration uncoupler. The search for this component led to the discovery of the brown fat uncoupling protein referred to as UCP (Nicholls and Locke 1984). This UCP was renamed as UCP1 when UCP2 was identified (Fleury et al. 1997). What researchers were looking for was not only the understanding of a mechanism of a respiration uncoupling but also a mechanism physiologically regulated since induction of thermogenesis has to occur in response to heat requirement by organisms in specific conditions such as exposure to the cold. When there is no extra-heat requirement, brown adipocytes have to be quiescent and their mitochondria should not present respiration uncoupling, otherwise energy will be wasted and body temperature increased. Conversely, when heat production is required, brown adipocytes have to be stimulated and must express their ability to uncouple respiration.

UCP1 and Elucidation of the Thermogenic Mechanism

The spontaneous uncoupling of respiration of BAT mitochondria stimulated research on this mechanism and led to observations of an activation of uncoupling by fatty acids and an inhibition of uncoupling by certain nucleotides. The mechanism itself was elucidated by David Nicholls, a bioenergeticist inspired by the chimio-osmotic theory proposed by Peter Mitchell. This theory, a matter of dispute between bioenergeticists and mitochondriologists around 1970, proposed that an electro-chemical proton gradient and a circuit gradient through a membrane were controlling ADP phosphorylation. Mitchell proposed this mechanism for respiring mitochondria and also for chloroplasts exposed to light. His theory was first confirmed for chloroplast and then for mitochondria. Interestingly, Nicholls's work on BAT mitochondria substantially contributed to the validation of Mitchell's work (Nicholls and Locke 1984). According to Mitchell, complexes I, III and IV of respiratory chain function as proton pumps and generate a proton gradient ⊗ $\Delta\mu H^+$ of which energy, the proton motive force, consumed by endothermic ATP-synthase is used to phosphorylate ADP. Therefore, when respiratory chains are active, the proton gradient is high and slows down respiration and ATP synthesis. Conversely, in case or re-entry of protons from the intermembrane space to the mitochondrial matrix non operating through the proton channel of the ATP-synthase and not coupled to a mechanism requiring energy supply, the proton gradient collapses, respiratory chains are strongly activated and oxidation energy is dissipated as heat since ADP phosphorylation consuming energy is shunted (Fig. 5.1). The reason that respiratory chains are activated is that in such a situation, the fall of ⊗ $\Delta\mu H^+$ lowers the membrane potential which activates proton pumps inhibited when the membrane potential is elevated. It provokes thermogenesis since oxidation energy is no more consumed by ATP-synthase. Actually, such a situation is observed when respiring mitochondria are exposed to a chemical uncoupler such as 2,4-dinitrophenol or FCCP. Interested in Peter Mitchell's discovery, David Nicholls assayed the proton conductance of the inner membrane of brown adipocyte mitochondria and, interestingly, observed that it was unusually high suggesting the presence of a specific proton transporter acting as mimicking a chemical uncoupler. Following the finding of Rafael who observed that purine nucleotides (ATP, ADP, GTP, GDP) can inhibit the uncoupling mechanism in brown fat mitochondria, Nicholls verified that these nucleotides inhibit the high proton conductance of the inner membrane and understood that the nucleotides bind the proton pathway. The next step was the labelling of the binding site of nucleotides using azido-ATP and the identification of a 32-kD membranous protein (Heaton et al. 1978). This protein was previously described in 1976 by Ricquier and Kader who reported that a 32-kD membranous protein was present in brown adipocyte mitochondria but non existing in liver and that it was markedly and uniquely induced in brown adipocyte mitochondria upon exposure of rats to the cold; moreover, the level of this protein was decreased when cold-exposed animals returned to 24 °C (Ricquier and Kader 1976). Therefore, UCP1 is a physiologically important protein dedicated to metabolic thermogenesis and also is a membranous protein validating the Mitchell's chimio-osmotic theory (Fig. 5.2).

Ucp1-/- Mice

The full demonstration of the thermogenic activity of UCP1 came from transgenic mice made null for *Ucp1* using homologous recombination by Leslie Kozak and his colleagues, such mice being unable to maintain their body temperature at 37 °C in a cold environment (Enerbäck et al. 1997). These *Ucp1-/-* mice present a reduced metabolism in the cold but are not obese or fatty. However, the animals accumulate lipids when they are living at 30 °C, the thermic neutrality temperature of mice (Feldmann et al. 2009). While BAT is essentially an organ dedicated to the

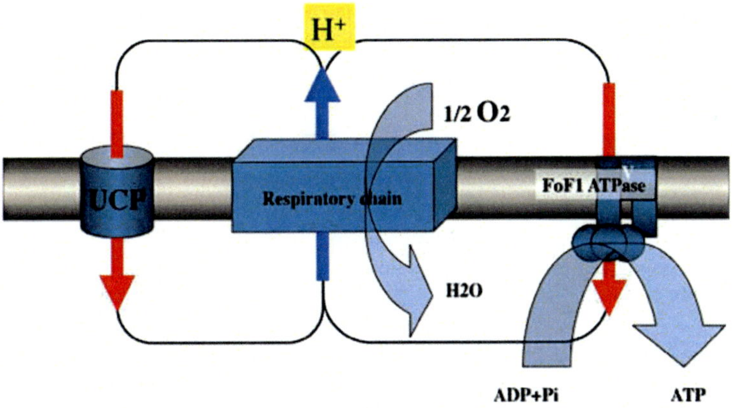

Fig. 5.1 The mitochondrial proton circuit in the inner membrane of mitochondria. In any type of mitochondria, complexes I, III and IV of respiratory chain build up an electrochemical proton gradient of which the energy is used by the ATP-synthase to phosphorylate ADP. This is achieved via proton circuit through the membrane linking exergonic respiration to endergonic ADP phosphorylation. This proton circuit explains the coupling between coenzyme reoxidation (respiration) and ATP synthesis. In the unique case of brown adipocyte mitochondria, the respiration rate is very high and not limited by ADP availability or the capacity of phosphorylation of ADP, since UCP1 a unique and additional component, shunts the proton re-entry through the ATP-synthase. When UCP1 is active, its lowers the membrane potential and the proton gradient which activate the respiratory chain of which oxidation energy, non consumed to phosphorylate ADP is dissipated as heat (figure kindly provided by Dr Frédéric Bouillaud, Institut Cochin)

fight against the fall of body temperature, a major and tightly controlled biological parameter, these data are in favour of a potential contribution of BAT to energy balance and body weight control.

UCP1, the Protein and the Gene

UCP1 has never been detected in significant amounts in cells other than brown adipocytes where it amounts to 1–4 % of total protein and 2–8 % of mitochondrial proteins. The primary structure of the protein was elucidated from its purification and sequencing (Aquila et al. 1985) and was also predicted from the sequence of cloned cDNA (Bouillaud et al. 1985; Bouillaud et al. 1986). Two major conclusions were made: (i) UCP1 shares a high level of identity with the mitochondrial ADP/ATP carrier (also referred to as the Adenine Nucleotide Translocator), (ii) UCP1 has a triplicated structure in such a way that every third (N-terminal, central, C-terminal), a hundred amino-acids long, can be partially aligned on the two other thirds. In fact, UCP1 sequence was the second identified sequence of the mitochondrial membranous metabolic carriers including the nucleotide exchanger,

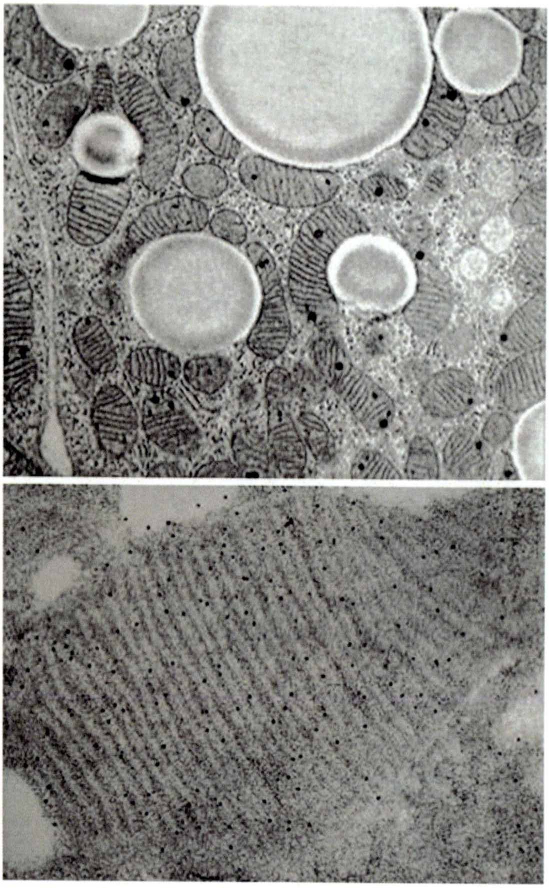

Fig. 5.2 Brown adipocyte morphology and mitochondrial localisation of UCP1. Brown adipocytes contain triglyceride droplets and many mitochondria with cristae (*upper panel*) in which UCP1 can be detected using antibodies prepared by the author (this figure was kindly given by Dr Saverio Cinti, University of Ancona)

the phosphate carrier, the citrate carrier, the oxoglutarate carrier and the acylcarnitine transporter. The genes encoding all these carriers derive from the same ancestor encoding a protein containing 100 amino acids. The human UCP1 gene was located on the long arm of chromosome 4 in q31 (Cassard et al. 1990). At the level of the UCP1 gene, in rodents as well as in man, a short region of 90 nucleotides localised a few kb upstream of the transcription start, controls both the transcriptional activation by catecholamines and cAMP, and the specificity of transcription in brown adipocytes (Cassard-Doulcier et al. 1993; Alvarez et al. 1995; Cassard-Doulcier et al. 1998; Rim and Kozak 2002).

Physiological, Adrenergic and Hormonal Regulation of Thermogenesis

Since BAT thermogenesis has to be activated only when necessary to protect body temperature in cold or to sharply rewarm awaking hibernators, its thermogenic activity is strongly controlled. The main system controlling activation of brown adipocytes is the sympathetic nervous system acting though sympathetic fibres which directly innervate individual adipocytes. Noradrenaline released by the sympathetic ends bind several subtypes of adrenergic receptors which activates a cascade implicating PKA and the activation of a hormone-sensitive lipase. This activation increases the level of free fatty acids in the cell which finally activate UCP1, overtaking the inhibition of the protein by nucleotides. Activated UCP1 translocates protons from the intermembrane space to the matrix bypassing the ATP-synthase unable to phosphorylate ADP. Simultaneously, the membrane potential is acutely decreased which activates the respiratory chain and oxidation energy, not used for ATP synthesis, is dissipated as heat. In addition to their control by the sympathetic nervous system, brown adipocytes are also dependent on thyroid hormones. Brown adipocytes contain a very active type II—deiodininase elevating T_3 level which participates in the positive control of the transcription of the UCP1 gene (Silva and Rabelo 1998).

Brown Adipocytes: Origin, Development, Plasticity

Precursors

Similar to white adipocytes, myocytes and chondrocytes, brown adipocytes are derived from mesodermic precursors. For a long time, it was assumed that, distinctly from pre-myocytes or pre-chondrocytes, other fibroblastic precursors were engaged in the adipocyte lineage, being able, according to the depot where they are, to develop either as white or brown adipocytes. The main idea was that the brown or white precursors were committed as adipocytes and unable to enter the myocyte or the chondrocyte pathway. Interestingly, this view, largely accepted, was questioned by Jan Nedergaard, Barbara Cannon and other Swedish researchers when they revealed a proximity, probably underestimated, between pre-brown adipocytes and pre-myocytes; their data suggested that the brown adipocytes were derived from precursors also able to differentiate as myocytes (Timmons et al. 2007; Walden et al. 2012). This proposal was confirmed by Seale et al. (2007). Other researchers, Saverio Cinti in particular, considering that a white fat depot is not entirely white and that a brown fat depot is not purely brown, described a marked plasticity of brown and white adipose tissues, proposing the existence of an interconvertibility between the two types of adipocytes (Cinti 2011). Moreover, Jean-Paul Giacobino and his colleagues were able to measure differences between brown adipocytes differentiated from brown fat depot fibroblasts and brown adipocytes differentiated from fibroblasts isolated from a white fat depot (Lehr et al. 2009). They proposed that there exist at least two types of brown adipocytes having distinct cell lineage.

PPARγ, PGC-1a, FOXC2, Rb, PRDM16

The major role of PPARγ in the differentiation processes of both white and brown adipocytes was demonstrated by Spiegelman (Puigserver et al. 1998). The same laboratory reported the key role played by PGC-1 in brown adipocyte maturation. PGC-1 is a transcriptional co-activator interacting with several transcription factors such as PPARγ, RXR, T3R. It is more abundant in brown adipocytes than in white adipocytes and when it is overexpressed in pre-adipocytes, it induces the brown adipocyte phenotype by inducing mitochondria and UCP1, as observed in murine cells (Puigserver et al. 1998) and in human cells (Tiraby et al. 2003). Other transcriptional regulators such as FOCX2 and the retinoblasma protein Rb participate in the brown adipocyte phenotype. However, these two proteins are unable to determine per se the differentiation as brown adipocyte. Looking for transcriptional regulators present in brown adipocytes and absent or weakly present in white adipocytes, Seale and Spiegelman identified PRDM16 as a major actor in the determination of brown adipocytes (Seale et al. 2007, 2008).

PRDM16 Controls the Determination of Brown Adipocytes

PRDM16 is 15-fold more abundant in brown adipocytes than in white adipocytes (Seale et al. 2007, 2008). It is highly expressed in established brown adipocyte cell lines and is induced during differentiation of brown adipocytes in culture. In 3T3-F442A preadipocytes and in immortalised pre-brown adipocytes, PRDM16 induces 200-fold UCP1 and also strongly induces PGC-1α. Conversely, PRDM16 suppression in brown adipocytes provokes the disappearance of their functional characteristics. However, PRDM16 activates the "brown programme" upstream of adipocyte differentiation and is unable to transform mature white adipocytes into brown adipocytes. Presently, PRDM16 is considered as a major actor in the differentiation of brown adipocytes and the establishment of their unique oxidative and thermogenic activity (Fig. 5.3).

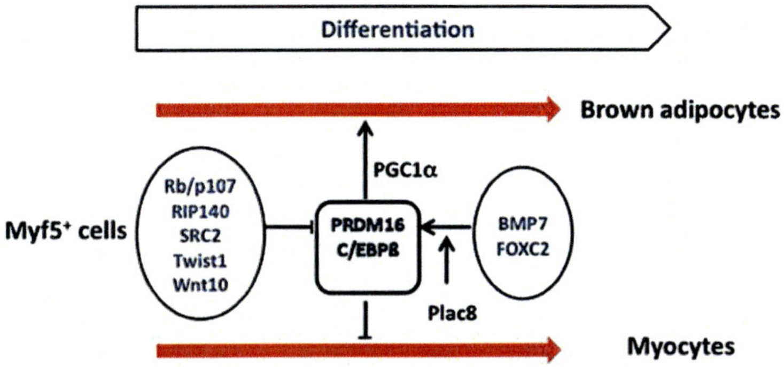

Fig. 5.3 Differentiation of brown adipocytes. Mesodermic cells expressing *Myf5* + , following expression or not of transcriptional factors or co-activators of transcription will, or not, express the zinc finger protein PRDM16. Expression of PRDM16 unequivocally orientates cells towards the brown adipocyte lineage and not towards the myocyte lineage (28, 32). PRDM16 synthesis is itself under the control of Plac8, which also controls expression of C/EBPß that also acts in favour of the differentiation of cells into brown adipocytes (Seale et al. 2008) (This figure was modified from a document kindly given by Dr E. Ravussin, University of Louisiana)

Pre-brown Adipocytes and Myoblasts

Simultaneous to the discovery of the role of PRDM16 in brown adipocyte differentiation, Stockholm researchers calculated an overlapping between gene signatures of pre-brown adipocytes and pre-myocytes, suggesting differentiation steps common to brown adipocytes and myocytes, both types of cells sharing abundant mitochondria and a marked oxidative capacity (Timmons et al. 2007; Walden et al. 2012). To further understand the mechanisms of determination of brown adipocytes, it was therefore logical to identify steps common to these cells and myocytes rather than events common to brown and white adipocytes. The discovery of the essential role of the zinc finger protein PRDM16 in brown adipocyte determination and inhibition of white adipocyte genes confirmed the Swedish observation since Seale et al. (2007, 2008) demonstrated that PRDM16 is controlling the switch between brown fat and muscle. These authors demonstrated that brown adipocytes, instead of the white adipocytes, derive from precursors expressing *myf5*, a gene implicated in muscular lineage. PRDM16 controls mechanisms guiding precursors towards myoblasts or brown adipocytes. The loss of PRDM16 in brown adipocyte precursors switches down the brown adipocyte phenotype and provokes the emergence of the myocyte characters when PRDM induction in myoblasts transform them into brown adipocytes (Seale et al. 2008). Therefore, PRDM16 dictates a brown adipocyte future to precursors expressing myoblast markers and not engaged in the white adipocyte lineage. The close proximity between differentiating brown adipocytes and myocytes was also noticed by Crisan et al. (2008) when they demonstrated that skeletal muscles are a reservoir of stem cells able to differentiate as typical brown adipocytes.

Brown Adipocytes of Brown or White Fat Depots

Comparing the in vitro differentiation of brown adipocytes coming from a brown fat depot to the differentiation of brown adipocytes originating from a white fat depot, Giacobino and his colleagues made interesting observations and concluded in favour of a distinct origin for the two types of brown adipocytes (Lehr et al. 2009). Importantly, the conclusion was that there is not a unique series of molecular events and steps leading to brown adipocyte appearance. In addition to the existence of specific brown and white adipocytes, morphological analysis incited Saverio Cinti to propose a possibility of interconversion for certain adipocytes able to differentiate in the two directions (Cinti 2011). Therefore, a certain plasticity may exist for a proportion of adipocytes. These observations and others led to the proposal of the existence of « brite cells » .

Human BAT

BAT in Human Newborn

For obvious reasons, most studies of BAT were achieved using cells and tissue taken from rodents. Whereas typical BAT was observed in human newborns, its presence in adults is more difficult to demonstrate and it was accepted that BAT is non important in human adults, in spite of ancient descriptions of UCP1-expressing brown adipocytes. This conclusion was in agreement with the fact that non-shivering thermogenesis is high in babies but is decreasing with age.

Adult Human BAT: Ancient and Recent Data

The generally accepted conclusion of the non presence of BAT in adult man was probably an overstatement since several studies made around 1980, more or less forgotten since, demonstrated the presence of typical brown adipocytes containing UCP1 in fat samples taken from adults undergoing surgery (Ricquier et al. 1982; Bouillaud et al. 1983; Lean et al. 1986). In a few studies, the analysis comprised functional test of isolated mitochondria revealing an uncoupled respiration sensitive to GDP (Ricquier et al. 1982; Bouillaud et al. 1983). A reason why these descriptions of brown adipocytes were neglected is probably related to the impossibility to quantitate brown fat depots in human adults and to estimate the contribution of the tissue to whole body oxygen consumption. Typical brown adipocytes were observed in perirenal biopsies obtained from patients bearing pheochromocytoma and undergoing surgery (Fig. 5.4). In addition, systematic analysis of perirenal fat biopsies obtained from various patients revealed the

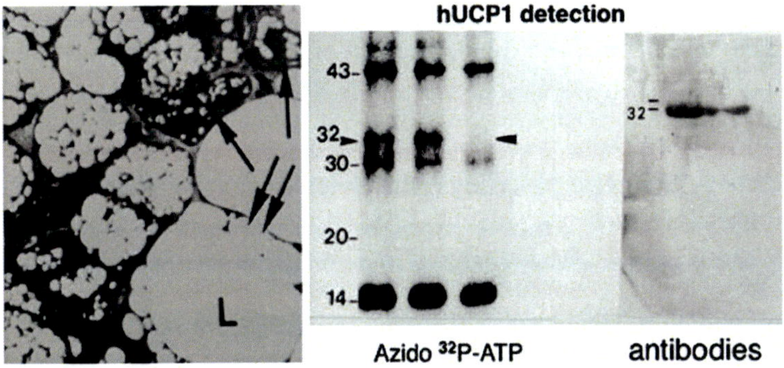

Fig. 5.4 Human brown adipocytes. *Left-hand side panel* brown and white adipocytes in the perirenal region of a patient operated for a pheochromocytoma. *Right-hand side panel* detection of UCP1 in the mitochondria isolated from these cells using either azido-ATP photo-labelling or detection with anti-UCP1 antibodies (Ricquier et al. 1982)

presence of UCP1 or its RNA, indicating that functional brown adipocytes were often present in adult man (Garruti and Ricquier 1992). The question of BAT in adult man resurged in 2009 from data obtained for years by nuclear medicine practitioners in many countries. Looking for tumour metastasis, radiologists assay glucose uptake by cancerous cells using PET-Scan and they became familiar with spots showing BAT depots in infants as well as in human adults [for review (Ravussin and Galgani 2011)]. Simultaneously, UCP1 immunodetection in cells located between neck muscles of adults confirmed the abundance of brown adipocytes (Zingaretti et al. 2009). These recent data changed the view of many researchers who presently claim that BAT is physiologically important in man and may contribute to body temperature regulation, fatty acid oxidation and to the control of body fat content, and even to resistance to obesity. These recent data re-open the question of the pharmacological induction of UCP1 and brown adipocytes in man to counteract fat enlargement. Recent attempts to induce brown adipocytes in man using ß3-adrenoceptor agonists failed, whereas studies in dogs were promising (Champigny et al. 1991). Taking into consideration this "rediscovery of brown fat in man" and the progresses in molecular mechanisms controlling brown fat differentiation such as PRDM16 and Plac8 just recently (Jimenez-Preitner et al. 2011), human BAT is re-entering a new period of interest.

Conclusion and Prospectives

After a large interest around 1970–1980, the research on BAT declined before coming again onto the stage. The four reasons for that are: (i) the identification of new factors controlling brown adipocyte differentiation, (ii) the discovery of the switch between pre-myocytes and pre-brown adipocytes and of stem cells of brown adipocytes in human muscles, (iii) the visualisation of active brown adipocytes in adult man and (iv) the concept of cells able to interconvert between brown and white adipocytes. The very recent identification by Bernard Thorens's team of Plac8 as a factor controlling brown adipocyte differentiation largely upstream of PRDM16 is an important progress toward the complete elucidation of signalisation pathways governing brown adipocyte differentiation (Jimenez-Preitner et al. 2011). Besides the fundamental mechanisms involved, this sum of very recent data leads to a re-evaluation of these particular cells uniquely able to burn fatty acids. It is reasonable to conceive applications based on brown adipocyte activation/induction helpful to the treatment of metabolic pathologies.

Acknowledgments The author thanks his remarkable colleagues and associates : Frédéric Bouillaud, Clotilde Alves-Guerra, Anne-Marie Cassard, Louis Casteilla, Saverio Cinti, Odette Champigny, Maria del Mar Gonzalez-Barroso, Brian Holloway, Marianne Larose, Corinne Lévi-Meyrueis, Bruno Miroux, Claire Pecqueur, Eduardo Rial, Denis Richard and Francesc Villarroya for their collaboration in studying UCP1. The author also thanks the Centre National de la Recherché Scientifique, the Institut pour la Santé et la Recherche Médicale, the European Community, the Human Frontier Science Program, the Servier Institute, the Délégation Générale à l'Armement and the Pierre et Marie Curie et René Descartes Universities for their support.

References

Alvarez R, de Andrés J, Yubero P et al (1995) A novel regulatory pathway of brown fat thermogenesis. Retinoic acid is a transcriptional activator of the mitochondrial uncoupling protein gene. J Biol Chem 270:5666–5673

Aquila H, Link TA, Klingenberg M (1985) The uncoupling protein from brown fat mitochondria is related to the mitochondrial ADP/ATP carrier. Analysis of sequence homologies and of folding of the protein in the membrane. EMBO J 4:2369–2376

Bouillaud F, Combes-George M, Ricquier D (1983) Mitochondria of adult human brown adipose tissue contain a 32 000-Mr uncoupling protein. Biosci Rep 3:775–780

Bouillaud F, Ricquier D, Thibault J, Weissenbach J (1985) Molecular approach to thermogenesis in brown adipose tissue: cDNA cloning of the mitochondrial uncoupling protein. Proc Natl Acad Sci USA 82:445–448

Bouillaud F, Weissenbach J, Ricquier D (1986) Complete cDNA-derived aminoacid sequence of rat brown fat uncoupling protein. J Biol Chem 261:1487–1490

Cannon B, Nedergaard J (2004) Brown adipose tissue: function and physiological significance. Physiol Rev 84:277–359

Cassard AM, Bouillaud F, Mattei MG et al (1990) Human uncoupling protein gene: structure, comparison with rat gene and assignment to the long arm of the chromosome 4. J Cell Biochem 43:255–264

Cassard-Doulcier AM, Gelly C, Fox N et al (1993) Tissue-specific and ß-adrenergic regulation of the mitochondrial uncoupling protein gene: Control by cis-acting elements in the 5′-flanking region. Mol Endocrinol 7:497–506

Cassard-Doulcier AM, Gelly C, Bouillaud F, Ricquier D (1998) The 211-bp enhancer of the rat UCP-1 gene controls specific and regulated expression in the brown adipose tissue. Biochem J 333:243–246

Champigny O, Ricquier D, Blondel O et al (1991) Beta 3-adrenergic receptor stimulation restores message and expression of brown-fat mitochondrial uncoupling protein in adult dogs. Proc Natl Acad Sci USA 88:10774–10777
Cinti S (2011) Between brown and white: novel aspects of adipocyte differentiation. Ann Med 43:104–115
Cousin B, Cinti S, Morroni M et al (1992) Occurrence of brown adipocytes in rat white adipose tissue: molecular and morphological characterization. J Cell Sci 103:931–942
Crisan M, Casteilla L, Lehr L et al (2008) A reservoir of brown adipocyte progenitors in human skeletal muscle. Stem Cells 26:2425–2433
Enerbäck S, Jacobsson A, Simpson EM et al (1997) Mice lacking mitochondrial uncoupling protein are cold-sensitive but not obese. Nature 387:90–94
Feldmann HM, Golozoubova V, Cannon B, Nedergaard J (2009) UCP1 ablation induces obesity and abolishes diet-induced thermogenesis in mice exempt from thermal stress by living at thermo neutrality. Cell Metab 9:203–209
Fleury C, Neverova M, Collins S et al (1997) Uncoupling protein-2: a novel gene linked to obesity and hyperinsulinemia. Nat Genet 15:269–272
Garruti G, Ricquier D (1992) Analysis of uncoupling protein and its mRNA in adipose tissue deposits of adult humans. Int J Obes Relat Metab Disord 16:383–390
Heaton GM, Wagenvoord RJ, Kemp A, Nicholls DG (1978) Brown-adipose-tissue mitochondria: photo affinity labelling of the regulatory site of energy dissipation. Eur J Biochem 82:515–521
Himms-Hagen J, Ricquier D (1998) Brown adipose tissue. In: Bray G, Bouchard C, James W (eds) Handbook of obesity. Marcel Dekker, New York, pp 415–441
Jimenez-Preitner M, Berney X, Uldry M et al (2011) Plac8 Is an inducer of C/EBPβ required for brown fat differentiation, thermoregulation, and control of body weight. Cell Metab 14:658–670
Lean ME, James WP, Jennings G, Trayhurn P (1986) Brown adipose tissue in patients with pheochromocytoma. Int J Obes 10:219–227
Lehr L, Canola K, Léger B, Giacobino JP (2009) Differentiation and characterization in primary culture of white adipose tissue brown adipocyte-like cells. Int J Obes 33:680–686
Lowell BB, S-Susulic V, Hamann A et al (1993) Development of obesity in transgenic mice after genetic ablation of brown adipose tissue. Nature 366:740–742
Nedergaard J, Cannon B, Lindberg O (1977) Microcalorimetry of isolated mammalian cells. Nature 267:518–520
Nicholls DG (2006) The physiological regulation of uncoupling proteins. Biochim Biophys Acta 1757:459–466
Nicholls DG, Locke RM (1984) Thermogenic mechanisms in Brown fat. Physiol Rev 64:1–64
Puigserver P, Wu Z, Park CW et al (1998) A cold-inducible coactivator of nuclear receptors linked to adaptive thermogenesis. Cell 92:829–839
Ravussin E, Galgani JE (2011) The implication of brown adipose tissue for humans. Annu Rev Nutr 31:33–47
Ricquier D, Kader JC (1976) Mitochondrial protein alteration in active brown fat. A sodium-dodecylsulfate-polyacrylamide gel electrophoretic study. Biochem Biophys Res Commun 73:577–583
Ricquier D, Gaillard JC, Turc JM (1979) Microcalorimetry of isolated mitochondria from brown adipose tissue Effect of guanine-diphosphate. FEBS Lett 99:203–206
Ricquier D, Néchad M, Mory G (1982) Ultrastructural and biochemical characterization of human brown adipose tissue in pheochromocytoma. J Clin Endocrinol Metab 54:803–807
Rim JS, Kozak LP (2002) Regulatory motifs for CREB-binding protein and Nfe2l2 transcription factors in the upstream enhancer of the mitochondrial uncoupling protein 1 gene. J Biol Chem 277:34589–34600
Rothwell NJ, Stock MJ (1979) A role for brown adipose tissue in diet-induced thermogenesis. Nature 281:31–35
Seale P, Kajimura S, Yang W et al (2007) Transcriptional control of brown fat determination by PRDM16. Cell Metab 6:38–54

Seale P, Bjork B, Yang W et al (2008) PRDM16 controls a brown fat/skeletal muscle switch. Nature 454:961–967

Seydoux J, Girardier L (1977) Control of brown fat thermogenesis by the sympathetic nervous system. Experientia 33:1128–1130

Silva JE, Rabelo R (1998) Regulation of the uncoupling protein gene expression. Eur J Endocrinol 136:51–64

Timmons JA, Wennmalm K, Larsson O et al (2007) Myogenic gene expression signature establishes that brown and white adipocytes originate from distinct cell lineages. Proc Natl Acad Sci USA 104:4401–4416

Tiraby C, Tavernier G, Lefort C et al (2003) Acquirement of brown fat cell features by human white adipocytes. J Biol Chem 278:33370–33376

Walden TB, Hansen IR, Timmons JA et al (2012) Recruited versus nonrecruited molecular signatures of brown, "brite" and white adipose tissues. Am J Physiol Endocrinol Metab 302:E19–E31

Zingaretti MC, Crosta F, Vitali A et al (2009) The presence of UCP1 demonstrates that metabolically active adipose tissue in the neck of adult humans truly represents brown adipose tissue. FASEB J 23:3113–3120

Chapter 6
Histology of Adipose Tissue

Joan Tordjman

Introduction

Adipose tissue (AT) is a connective tissue. It is classified as loose connective tissues and is composed of fat cells, or adipocytes, separated by a thin layer of extracellular matrix comprising a backbone of fiber, such as collagen fibers, and numerous vessels. There are two types of AT: White and brown. In this chapter, we will describe the physiological white AT and its histological changes during a disease directly related to the tissue, obesity, and its deleterious consequences that contribute to metabolic complications associated with this disease.

Histology of Normal Adipose Tissue

Depots of Adipose Tissue

In mammals, AT is composed of distinct non-contiguous deposits that have different structural and functional characteristics. In humans, white AT is one of the largest organs, which can reach 45 kg or more in obese subjects. In addition to the dermal layer of fat, the main deposits are distributed between the subcutaneous and intra-peritoneal or visceral area. The distribution of body fat is variable from one individual to another, according to sex, age, and in response to certain diseases.

J. Tordjman (✉)
Institut des maladies cardiométaboliques et de la nutrition (IHU ICAN), Paris, France
e-mail: joan.clement@upmc.fr

J. Tordjman
Nutriomique U872 équipe 7, Centre de recherche des Cordeliers, INSERM, UPMC,
15, rue de l'école de médecine, 75006 Paris, France

J.-P. Bastard and B. Fève (eds.), *Physiology and Physiopathology of Adipose Tissue*,
DOI: 10.1007/978-2-8178-0343-2_6,

The increase in visceral fat mass is associated with an increased risk of cardiovascular disease and type 2 diabetes. Conversely, increasing the mass of the subcutaneous AT is associated with a lower risk for these comorbidities. The expansion of the "safe" deposit would avoid the harmful effects of storage lipids in tissues not specialized. This hypothesis was tested in a mouse model in which glucose and lipid homeostasis has been improved by genetic manipulation causing overdevelopment of subcutaneous AT (Kim et al. 2007). This property is intrinsic to the tissue, as suggested by the improvement of glucose homeostasis induced by transplantation of subcutaneous AT in the visceral region in mice (Tran et al. 2008).

Adipocytes

The AT is a tissue organized into lobules separated by septa of loose connective tissue (Fig. 6.1a). These lobules are formed mainly by adipocytes. Adipocytes are specialized cells of AT in the synthesis, storage, and mobilization of triglycerides. Storage of triglycerides requires a particular structure, the lipid droplet, which, through its membrane consisting of a monolayer of phospholipids, allows the accumulation of hydrophobic molecules in the cytoplasm. Unlike the lipid droplets of hepatocytes or macrophages, which diameter does not exceed 2 μm, the diameter of the droplet of the adipocyte may vary in proportions from 10 to 150 μm. This droplet occupies almost the entire cell cytoplasm, the nucleus rejected at the periphery of the cell (Fig. 6.1b). Several proteins, including perilipin, are associated with this structure. Their role in the integrity of the droplet, the modification of its membrane, and the control of their size are largely studied. The secretion of several proteins depends on the adipocyte size. This implies the existence of a mechanism for detecting stock-specific adipocyte lipid which molecular players remain to be identified. Each adipocyte has a basement membrane through which it comes into contact with endothelial cells, each adipocyte is in contact with at least one capillary (Fig. 6.1c).

Other Adipose Tissue Cells

The non-adipocyte cell fraction is composed of stromal and endothelial cells, fibroblasts, and preadipocytes, and also cells of the innate immune system, primarily macrophages, and cells of the adaptive immune system, mainly T cells at the interface of the two systems, NK cells (Henegar et al. 2008), and mast cells (Liu et al. 2009) are also present in the murine and human AT. In humans, there are between 5 and 10 per 100 adipocytes macrophages, which represent about 15 % of stromal cells in normal AT (Aron-Wisnewsky et al. 2009; Duffaut et al. 2009). Macrophages are cells whose phenotype changes depending on the tissue microenvironment. Macrophages of "classically activated" M1-type are

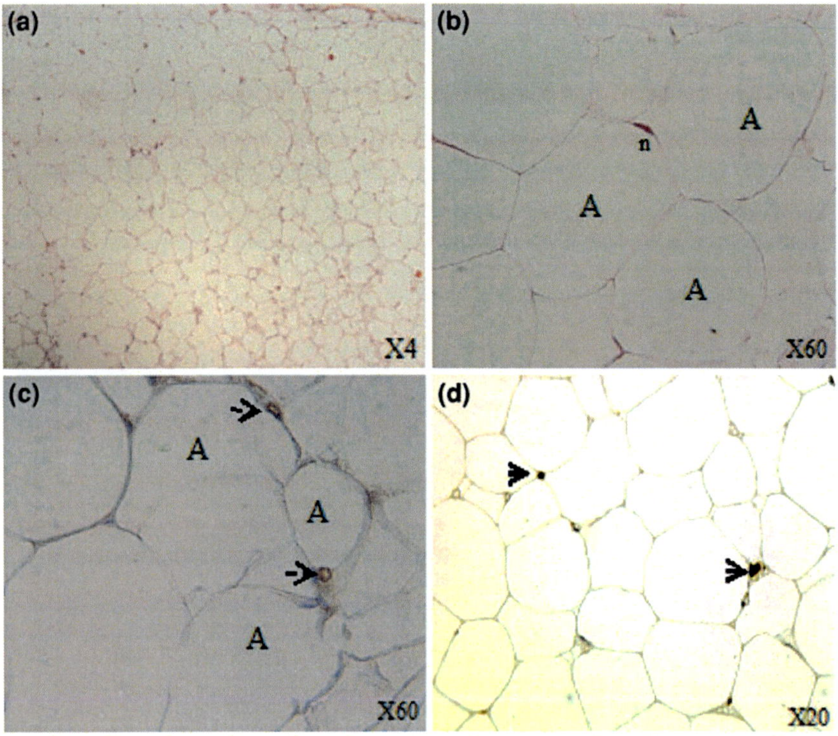

Fig. 6.1 Histology of normal adipose tissue. Biopsies of subcutaneous adipose tissue were obtained from control subjects. **a** and **b** staining with hematoxylin/eosin **c** biopsies were treated with an antibody specific of vessel Von Willerbrand Factor. The *dashed arrows* show the capillaries close to the adipocytes **d** biopsies were treated with an antibody specific for CD68 + macrophages. The arrows indicate the labeled macrophages in the adipose parenchyma. *A* adipocytes, *n* nucleus

stimulated by interferon-γ and TNF-α and produce proinflammatory cytokines such as IL-6 and IL-1. M2-type macrophages, "alternatively activated", are induced by IL-4 and IL-13 and, via the production of factors such as IL-10 and TGF-β, have an anti-inflammatory characteristic (see Chap. 20). These two types of macrophages are found in equivalent amount in the AT (Aron-Wisnewsky et al. 2009). They are located in the parenchyma fat between the adipocytes and near blood vessels (Fig. 6.1d). T cells, which are between 6 and 10 % of this fraction, are less common, whereas B cells are virtually undetectable (Aron-Wisnewsky et al. 2009; Duffaut et al. 2009; Feuerer et al. 2009).

Extracellular Matrix

During its development, the AT is the seat of a dynamic remodeling of the elements of the extracellular matrix (ECM). ECM is mainly composed of collagen

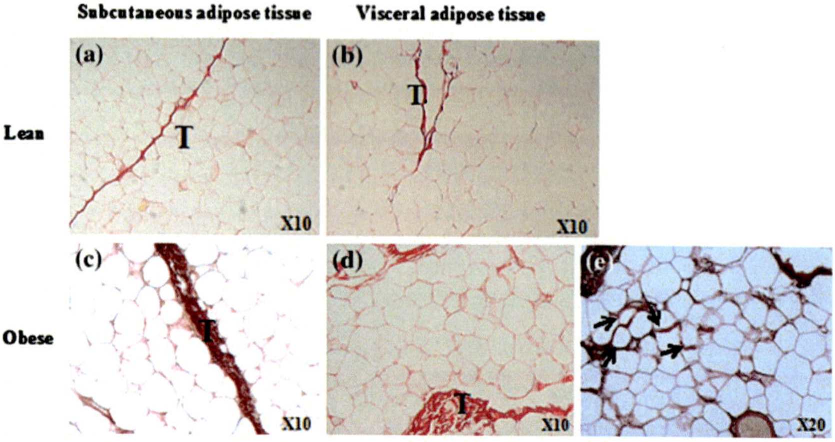

Fig. 6.2 Microscopic aspects of fibrosis deposition in adipose tissue of obese and control subjects. Surgical biopsies of subcutaneous and visceral adipose tissues of control (**a** and **b**) and obese (**c**, **d** and **e**) subjects were labeled with picrosirius red (marker for collagen) to reveal fibrotic areas. *T* bundles, the *arrows* indicate the pericellular fibrosis present only in obese subjects

fibers, including collagen I and III. Picrosirius red staining is used to visualize the total collagen on sections of AT. Collagen fibers are found in a thin bundles crossing the subcutaneous AT or surrounding the lobules. These structures are particularly visible in the visceral AT as they are of smaller sizes (Fig. 6.2a, b). In a non-pathologic tissue, there is a balance between synthesis and degradation of the ECM, degradation is necessary for the increased size of adipocytes.

Histopathology of Adipose Tissue During Obesity

Macrophages Accumulation in Adipose Tissue and Pathophysiological Consequences

One of the major cellular abnormalities characteristic of AT in obesity is the increased number of macrophages, which can reach 15–30 per 100 adipocytes. The macrophage content is not altered in muscle or liver in obese mice, indicating that AT is the main target of this accumulation. The macrophage infiltration of AT is only partially reversible with weight loss induced by gastric surgery (Aron-Wisnewsky et al. 2009; Cancello et al. 2005). Macrophages in AT are typically organized in a ring around an adipocyte, or crown-like structure, showing signs of cell death such as negativity to perilipin staining (Cancello et al. 2005; Cinti et al. 2005). This organization is specific to AT and more common in obese visceral AT (Fig. 6.3a, b and d). These observations suggest that macrophages play classical role of phagocytosis by surrounding the fat cells with their cytoplasmic processes

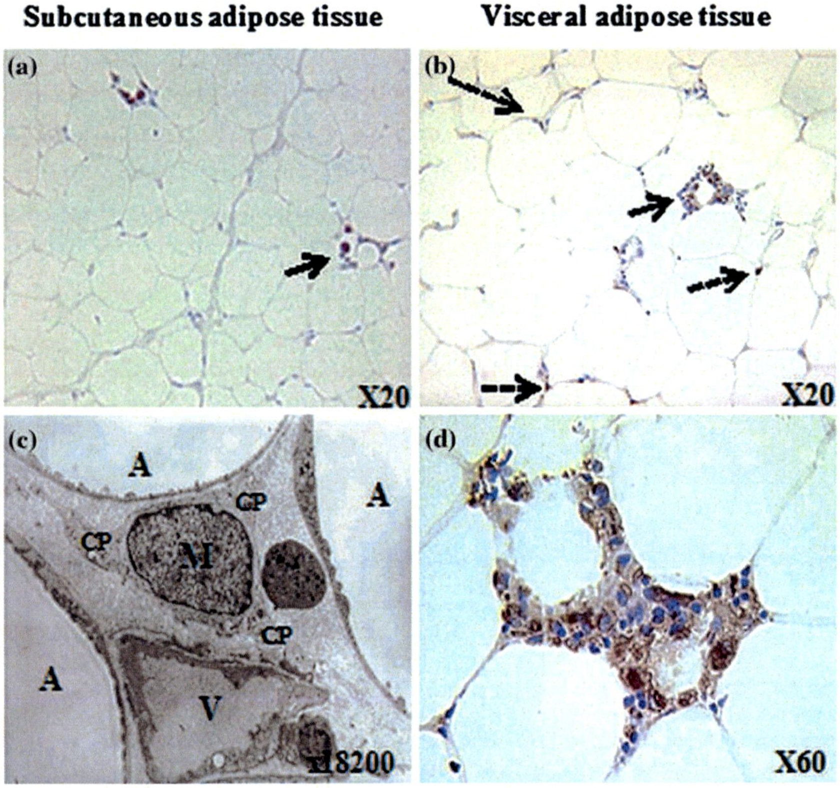

Fig. 6.3 Microscopic aspects of macrophages in the tissue of an obese subject. **a**, **b**, and **d** surgical biopsies of subcutaneous and visceral adipose tissue were treated with an antibody specific for CD68 + macrophages. The arrows indicate macrophages arranged in a crown-like structure, and the dashed arrows indicate macrophages located at the intersection of adipocytes in the parenchyma. **c** ultrathin sections of subcutaneous adipose tissue observed in transmission electron microscope (×18200). *A* Adipocyte; *M* macrophage; *V* vessel; *CP* cytoplasmic processes

(Fig. 6.3c), and elimination of metabolically deficient adipocytes in the AT of obese, especially those who have reached a critical size causing cell death. In mice, macrophages recruited into AT in response to the diet induction of obesity are characteristically M1 proinflammatory and will preferentially accumulate around the necrotic adipocytes (Lumeng et al. 2007; Nishimura et al. 2009). In massively obese patients, M1 macrophages are more numerous than in controls and more abundant in visceral AT (Aron-Wisnewsky et al. 2009). Other studies report an increase in mixed M1/M2 macrophage phenotype in the non-adipocyte fraction of obese AT or overweight (Bourlier et al. 2008; Zeyda et al. 2007). After weight loss, the number of M1 macrophages decreases for the benefit of anti-inflammatory M2 macrophages (Aron-Wisnewsky et al. 2009). No experimental data makes it possible to unambiguously determine if the macrophage phenotypes

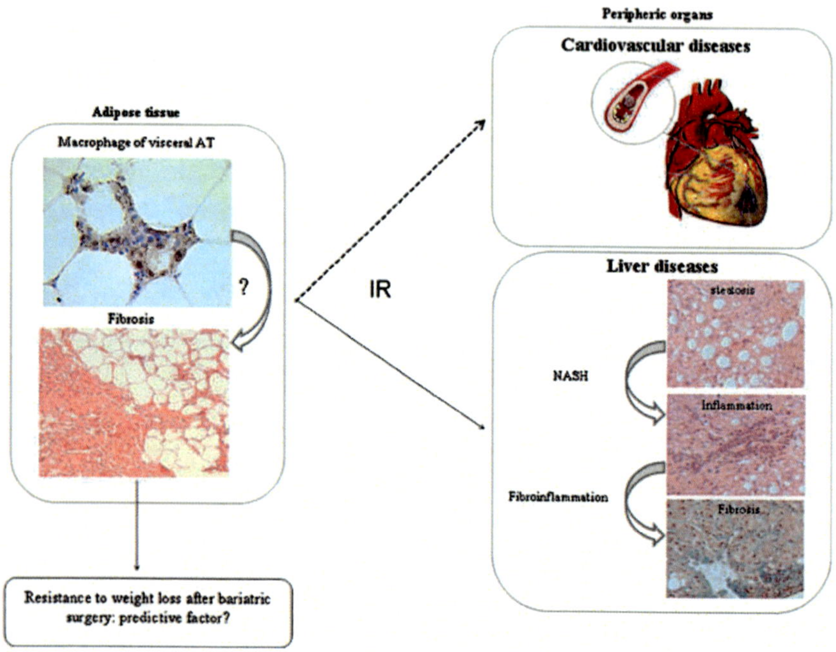

Fig. 6.4 Potential role of inflammation and fibrosis in adipose tissue in obesity-related comorbidities. Insulin resistance (IR) is a factor involved in the relationship between adipose tissue macrophages and liver diseases. The relative contribution of visceral fat *versus* subcutaneous is different depending on comorbidities. Adipocyte fibrosis is inversely correlated with the loss of fat mass after bariatric surgery and could therefore be considered as a factor predictive of diagnosis and weight loss. The dotted arrow indicates the lack of clinical and experimental evidence

in situ change or if new macrophages are recruited in response to changes in fat mass, the two phenomena being able to coexist. Following the discovery of the accumulation of macrophages in AT, the question of their role in metabolic complications and cardiovascular disease of obesity has rapidly risen. Pioneering work in mice showed that macrophage infiltration precedes hyperinsulinemia induced by a high fat diet, suggesting a cause and effect link. The modulation of the number of macrophages in AT by genetic manipulation in mouse models largely confirmed the causal relationship, reducing the number of macrophages consistently associated with improved glucose homeostasis and vice versa (Kanda et al. 2006). However, in morbid obesity, the number of macrophages is not correlated with parameters of insulin sensitivity (Cancello et al. 2005). Although widely believed, a link to increased cardiovascular risk has not been established in humans. However, a clinical study in a large population of morbidly obese subjects showed that the severity of liver histopathology (steatosis and fibro-inflammation) is related to the extent of macrophage accumulation in visceral AT (Tordjman et al. 2009) (Fig. 6.4). In addition to these systemic effects, the presence of

macrophages locally alters the biology of adipocytes. Adipocytes cultured in the presence of conditioned media of macrophages secrete pro-inflammatory factors and become resistant to insulin (Suganami et al. 2005). In addition, fatty acids released by adipocytes that rendered insensitive to the anti-lipolytic effect of insulin activate macrophages to a M1 phenotype via a paracrine loop between the two deleterious cell types. Another recognized effect of macrophages is the inhibition of differentiation of preadipocytes, which become proliferative (Bourlier et al. 2008; Lacasa et al. 2007). In addition, certain types of macrophages isolated from human AT exert a pro-angiogenic effect in vitro, which could contribute to the tissue
neo-vascularization during weight gain (Bourlier et al. 2008). Studies are still needed to clarify the full spectrum of consequences of macrophages infiltration on the metabolic functions and expansion of AT in obesity.

Modification of Extracellular Matrix, Fibrosis and Pathophysiological Consequences

In obesity, a deregulation of the extracellular matrix amount was observed, namely a decrease in its degradation in favor of its synthesis. Excessive rigidity induced experimentally by the deletion of a matrix protease in mice limits the growth of normal tissue which has reduced size of adipocytes. In obese mice, deregulation of various types of extracellular matrix proteins has been described, with an accumulation of collagen type VI (Khan et al. 2009). Unexpectedly, the deletion of the collagen gene improved the metabolic parameters despite the increase in body fat. Raising a mechanical stress, this manipulation led to the hypertrophy of adipocytes, thereby reducing the adverse metabolic impact of ectopic lipid deposition in the liver or muscles. Factors and cellular and molecular mechanisms involved in normal and pathological remodeling of the extracellular matrix of AT are poorly identified. Hypoxia and inflammatory factors stimulate the expression of proteases (cathepsins, metalloproteases) and other factors such as plasminogen activator inhibitor (PAI-1) in AT, which may contribute to alterations of the extracellular matrix in obesity. Associated with chronic inflammation, remodeling of the extracellular matrix can lead to abnormal accumulation of matrix elements that characterize fibrosis. Fibrotic deposits were detected in human AT, more abundant in obese subjects (Fig. 6.2c, d) than the control (Fig. 6.2a, b) (Henegar et al. 2008; Divoux et al. 2010). Fibrosis is organized in dense clusters and bundles traversing the parenchyma and is composed of type I and III collagens. In obese subjects, there are also pericellular fibrosis surrounding adipocytes and composed of type VI collagen (Fig. 6.2e). Several cell types are identifiable in the fibrotic bundles, including macrophages, mast cells, fibroblast cells, and preadipocytes, whereas lymphocytes are very rare (Divoux et al. 2010; Keophiphath et al. 2010). The culture of human preadipocytes in the presence of conditioned medium of macrophages gives them a pro-fibrotic phenotype (Keophiphath et al. 2009), suggesting their involvement in fibrosis, without excluding the role of other cell types (fibroblasts, mast cells). In a group of massively obese subjects who underwent gastric surgery, the amount of fibrosis measured in the subcutaneous AT before surgery is negatively associated with the loss of body fat mass at 3, 6, and 12 months post surgery. The presence of excess fibrosis may alter the remodeling of AT and limit the loss of fat mass (Divoux et al. 2010). Following these observations, studies are needed to elucidate the pathophysiological significance of AT fibrosis in human obesity and, in particular, its predictive role in weight loss (Fig. 6.4).

Conclusion

In obesity, white AT is the target of a major structural reorganization and cell whose components are probably not all identified. Cell types and signals involved are multiple and complex, creating a local microenvironment that promotes pro-inflammatory macrophage recruitment and formation of fibrosis. Although only partially reversible, these changes persist after weight loss. The identification of potentially protective mechanisms, such as the neutralization of certain types of immune cells, could open new therapeutic perspectives to contain the inflammation in AT.

References

Aron-Wisnewsky J, Tordjman J, Poitou C et al (2009) Human adipose tissue macrophages: m1 and m2 cell surface markers in subcutaneous and omental depots and after weight loss. J Clin Endocrinol Metab 94:4619–4623

Bourlier V, Zakaroff-Girard A, Miranville A et al (2008) Remodeling phenotype of human subcutaneous adipose tissue macrophages. Circulation 117:806–815

Cancello R, Henegar C, Viguerie N et al (2005) Reduction of macrophage infiltration and chemoattractant gene expression changes in white adipose tissue of morbidly obese subjects after surgery-induced weight loss. Diabetes 54:2277–2286

Cinti S, Mitchell G, Barbatelli G et al (2005) Adipocyte death defines macrophage localization and function in adipose tissue of obese mice and humans. J Lipid Res 46:2347–2355

Divoux A, Tordjman J, Lacasa D et al (2010) Fibrosis in human adipose tissue: composition, distribution, and link with lipid metabolism and fat mass loss. Diabetes 59:2817–2825

Duffaut C, Zakaroff-Girard A, Bourlier V et al (2009) Interplay between human adipocytes and T lymphocytes in obesity: CCL20 as an adipochemokine and T lymphocytes as lipogenic modulators. Arterioscler Thromb Vasc Biol 29:1608–1614

Feuerer M, Herrero L, Cipolletta D et al (2009) Lean, but not obese, fat is enriched for a unique population of regulatory T cells that affect metabolic parameters. Nat Med 15:930–939

Henegar C, Tordjman J, Achard V et al (2008) Adipose tissue transcriptomic signature highlights the pathological relevance of extracellular matrix in human obesity. Genome Biol 9:R14

Kanda H, Tateya S, Tamori Y et al (2006) MCP-1 contributes to macrophage infiltration into adipose tissue, insulin resistance, and hepatic steatosis in obesity. J Clin Invest 116: 1494–1505

Keophiphath M, Achard V, Henegar C et al (2009) Macrophage-secreted factors promote a profibrotic phenotype in human preadipocytes. Mol Endocrinol 23:11–24

Keophiphath M, Rouault C, Divoux A et al (2010) CCL5 promotes macrophage recruitment and survival in human adipose tissue. Arterioscler Thromb Vasc Biol 30:39–45

Khan T, Muise ES, Iyengar P et al (2009) Metabolic dysregulation and adipose tissue fibrosis: role of collagen VI. Mol Cell Biol 29:1575–1591

Kim JY, van de Wall E, Laplante M et al (2007) Obesity-associated improvements in metabolic profile through expansion of adipose tissue. J Clin Invest 117:2621–2637

Lacasa D, Taleb S, Keophiphath M et al (2007) Macrophage-secreted factors impair human adipogenesis: involvement of proinflammatory state in preadipocytes. Endocrinology 148:868–877

Liu J, Divoux A, Sun J et al (2009) Genetic deficiency and pharmacological stabilization of mast cells reduce diet-induced obesity and diabetes in mice. Nat Med 15:940–945

Lumeng CN, Deyoung SM, Bodzin JL et al (2007) Increased inflammatory properties of adipose tissue macrophages recruited during diet-induced obesity. Diabetes 56:16–23

Nishimura S, Manabe I, Nagasaki M et al (2009) CD8 + effector T cells contribute to macrophage recruitment and adipose tissue inflammation in obesity. Nat Med 15:914–920

Suganami T, Nishida J, Ogawa Y (2005) A paracrine loop between adipocytes and macrophages aggravates inflammatory changes: role of free fatty acids and tumor necrosis factor alpha. Arterioscler Thromb Vasc Biol 25:2062–2068

Tordjman J, Poitou C, Hugol D et al (2009) Association between omental adipose tissue macrophages and liver histopathology in morbid obesity: influence of glycemic status. J Hepatol 51:354–362

Tran TT, Yamamoto Y, Gesta S et al (2008) Beneficial effects of subcutaneous fat transplantation on metabolism. Cell Metab 7:410–420

Zeyda M, Farmer D, Todoric J et al (2007) Human adipose tissue macrophages are of an anti-inflammatory phenotype but capable of excessive pro-inflammatory mediator production. Int J Obes (Lond) 31:1420–1428

Part II
Adipose Tissue Metabolic Functions

Chapter 7
Glucose Transport in White Adipocyte

Mireille Cormont and Vincent Kaddai

Introduction

Glucose is transported through the plasma membrane of adipocyte in a facilitative manner. It is transported as function of the glucose gradient between the extracellular medium and the cytosol and without energy requirement. Glucose transport in adipose tissue is highly stimulated by insulin in vivo and in primary adipocyte or adipocyte cell lines. It is known since 1980 and before the cloning of molecules responsible for glucose transport activity that this is due to the redistribution of glucose transport activity from an intracellular compartment to the plasma membrane (Suzuki and Kono 1980; Cushman and Wardzala 1980). This control by insulin is perturbed in fasted condition, obesity, and type 2 diabetes. Adipocytes are hyperresponsive to insulin for glucose transport following refeeding after fasting or after a regimen with caloric restriction, and at the beginning of obesity development. Glucose transport in adipocyte is thus controlled both acutely and in long term.

M. Cormont (✉)
INSERM U895, Mediterranean Center of Molecular Medicine,
Team "Cellular and Molecular Physiopathology of obesity and diabetes",
Bâtiment Archimed. 151, route de St Antoine de Ginestière,
BP2 3194 06204 Nice Cedex 3, France
e-mail: cormont@unice.fr

V. Kaddai
Department of Pathology and Immunology, Geneva University,
Centre Médical Universitaire. 1, rue Michelet-Servet, 1211 Genève 4,
Switzerland
e-mail: Vincent.Kaddai@unige.ch

J.-P. Bastard and B. Fève (eds.), *Physiology and Physiopathology of Adipose Tissue*,
DOI: 10.1007/978-2-8178-0343-2_7, © Springer-Verlag France 2013

Physiopathological Control of Adipocyte Glucose Transport

Concerning its acute control and besides insulin, several molecules stimulate or inhibit glucose transport in cultured or isolated adipocyte. The counterregulatory hormones such as glucagon, adrenaline, cortisol, and growth hormone are released during hypoglycemia and under other stress conditions. These hormones have insulin antagonist effects in insulin sensitive tissues, including the adipose tissue. Glucagon, growth hormone, and cortisol inhibit insulin-induced glucose transport in cultured adipocytes. Stimulation of β adrenergic receptors induces by itself glucose transport but counteracts insulin-induced glucose transport. Most of these results have been obtained in white adipose tissue in vivo or in isolated adipocytes and the described effects on glucose transport were related to circulating catecholamines but not to local sympathetic innervation of the white adipose tissue. The effect of these insulin counterregulatory hormones in adipose cells certainly contributes to their full action on glucose homeostasis.

Orexin A, known to modulate food intake, energy expenditure, and body weight was recently found to acutely stimulate glucose uptake in adipocytes. It also induces lipid accumulation and adiponectin secretion, and thus could act as an insulin sensitizer (Skrypski et al. 2011). The adipokine adiponectin that possesses insulin-sensitizing properties on liver and muscle also stimulates glucose uptake in isolated adipocyte at least in its globular form (Wu et al. 2003).

Adenosine through the activation of its specific A1 receptors enhances the effect of insulin on glucose transport in rodent and human adipocytes (Heseltine et al. 1995). Adenosine is produced by adipocytes and would play an autocrine role on adipocyte metabolism. In accordance with this hypothesis, transgenic mice that specifically overexpress the A1 adenosine receptor in adipocyte are protected from obesity-related insulin resistance (Dong et al. 2001). Also, a selective partial A1 adenosine receptor agonist improves insulin sensitivity in obese animals determined in hyperinsulinemic euglycemic clamp studies (Dhalla et al. 2007). Furthermore, insulin sensitivity is significantly impaired in mice with a general knockout of the A1 adenosine receptor, due to reduced glucose uptake in muscle and adipocyte. However, the knockout mice are significantly heavier than wild type mice because of an increased fat mass (Faulhaber-Walter et al. 2011). Thus, the impairment of insulin sensitivity could be the consequence of the increase in body weight rather than a direct effect of the loss of adenosine signaling. The adenosine A2B receptor also increases insulin resistance in diabetes not by directly targeting adipocyte glucose transport but rather by mediating inflammation (Figler et al. 2011).

Numerous natural or chemical compounds are able to induce glucose transport. Some of them act on the insulin signaling pathway, and thus will not be efficient in insulin resistant situations. Others act on glucose transporters expression, whereas only few act as acute inducers of glucose transport by recruiting glucose transporters at the plasma membrane without a requirement for molecules of insulin signaling. The nitric oxide-donating derivative of acetylsalicylic acid stimulates glucose transport and glucose transporters translocation in adipocytes. Its effect is additive to low concentration of insulin and involved S-nitrosylation (Kaddai et al. 2008).

Numerous stresses within the adipose tissue from obese subjects, inflammation, hypoxia, reticulum stress, contribute to glucose transport defects of the adipose tissue. These long-term changes essentially involve alterations of glucose transporter genes expression in the adipocyte. They also inhibit insulin signaling pathway and inflammation could also perturb the mechanisms involved in the processes leading to the intracellular sequestration of the glucose transporters (Regazzetti et al. 2009; Hoehn et al. 2008; Kaddai et al. 2009).

The Adipocyte Glucose Transporters

Membrane transporters belonging to the Major Facilitator Superfamily are required for adipocyte glucose transport. They are part of the Glut protein family that comprises 14 members (for review (Thorens and Mueckler 2009)). Gluts are proteins of around 500 amino acids and are predicted to possess 12 transmembrane-spanning α helices and a single N-linked oligosaccharide. The most studied glucose transporters expressed by adipocytes are Glut4 encoded by the gene slc2a4 (solute carrier family 2, facilitated glucose transporter member 4). But adipocytes also express to lower extend Glut1 (encoded by the gene slc2a1) and other Glut, like Glut12 (slc2a12) that presents functional similarity with Glut4 (Purcell et al. 2011).

Glut4

Glut4 is expressed in white and brown adipocytes, skeletal and smooth muscles, and the heart (Thorens and Mueckler 2009). It is also expressed in less classical insulin-sensitive cells like podocytes (Lewko et al. 2005) and some discrete neurons in several brain regions (Kobayashi et al. 1996). Glut4 expression is turned on at late stage of adipocyte differentiation. In white adipocyte, its expression is under hormonal and nutritional control and it is widely affected in pathological conditions like obesity and type 2 diabetes (Graham and Kahn 2007). Changes in Glut4 expression are largely parallel to perturbations in adipocyte glucose transport and principally occur at the transcriptional level. The capacity of insulin to acutely control glucose transport in adipocyte, like in muscle cells, is mainly due to the redistribution of the glucose transporter Glut4 from intracellular compartment to the plasma membrane. This is the results of interplay between a complex intracellular trafficking of the glucose transporters and insulin signaling (Rowland et al. 2011). The current knowledge of these processes will be discussed in a following section.

Glut1

Glut1 is the ubiquitously expressed glucose transporter first cloned from an hepatoma DNA library (Thorens and Mueckler 2009). It is expressed at low level in primary adipocytes. The adipocyte cell lines frequently used in laboratories to study adipocyte glucose transport expressed larger amount of Glut1. The down regulation of Glut4 expression has a bigger impact on adipocyte glucose transport in isolated mice adipocyte than in 3T3-L1 adipocyte (Liao et al. 2006; Abel et al. 2001). Thus, by using adipocyte cell lines, the major role of Glut4 for adipocyte glucose transport is under evaluated.

Glut12

Glut12 was cloned in 2012 from breast cancer cells for its homology to Glut4. It is expressed in adipose tissue, skeletal muscle, heart, small intestine, and prostate. Thus it has a restricted tissue expression, but interestingly with overlapping into insulin-sensitive tissues. Transgenic mice expressing Glut12 under ubiquitous promoter have improved whole-body insulin sensitivity. It is the result of an increased glucose clearance rate in insulin-sensitive tissues under insulin stimulated, but not basal conditions (Purcell et al. 2011). The exact contribution of Glut12 in adipocyte glucose transport is currently unknown. No study is available concerning its expression control, but like Glut4 its subcellular distribution is controlled by insulin in fat and skeletal muscle (Purcell et al. 2011).

Other Glut

Another Glut, Glut5, is found in adipose tissue but it is not a glucose transporter. Glut5 is present at the surface of adipocyte and transports fructose (Hajduch et al. 1998). HMIT, Glut3, and Glut8 have been detected at the mRNA levels in adipose tissues. Their expression in adipocyte at the protein level is not demonstrated. Furthermore, HMIT is an H +/myo-inositol symporter (Thorens and Mueckler 2009).

Role of Adipocyte Glucose Transport

An important role of adipocyte glucose transport has been discovered thanks to the development of mice that do not express Glut4 in the adipose tissue. Although it is accepted that the principal consumer of glucose in fed conditions are muscles, mice without Glut4 in the adipose tissue surprisingly develop adipocyte and systemic insulin resistance and are glucose intolerant (Abel et al. 2001). On the contrary, overexpression of Glut4 in the adipose tissue prevents insulin resistance and glucose intolerance of mice that do not express Glut4 in the muscle (Carvalho et al. 2005). Taken together, these studies prove that adipocyte glucose transport is required for glucose homeostasis at least in mice. They also lead to the hypothesis that adipocyte glucose transport is linked to its endocrine function in order to mediate insulin resistance in other peripheral tissues. Serum retinol binding protein 4 secretion is increased in adipocytes without Glut4 and contributes to induce gluconeogenesis in liver and to inhibit insulin signaling in liver (Yang et al. 2005).

The Mechanisms of the Metabolic and Hormonal Control of Adipocyte Glucose Transport in Normal and Pathological Situations

This chapter will essentially concern glucose transport due to the glucose transporter Glut4 that is, as state above, the most important in adipocytes and also the most studied.

Control of Glut4 Gene Expression

Transcriptional Activation of Glut4 Gene Expression

The regions of the Glut4 promoter necessary for a specific expression in adipocyte have been characterized by functional analysis of the human Glut4 promoter in transgenic mice. A promoter of 900 base pairs upstream the transcription initiation site is required to obtain the correct pattern of Glut4 expression, comprising the adipocytes. Within this sequence a binding site for a transcription factor of the MEF2 family is required (Fig. 7.1a). But, the expression level of adipocyte MEF2 (the isoform MEF2A) is unchanged during adipocyte differentiation, thus suggesting that additional events are required for MEF2A-dependent transcriptional activation. The Krüppel-like factor KLF15 transactivates the murine Glut4 promoter, binds to it close to the MEF2A binding site, and this effect requires an intact MEF2 binding site. Furthermore, KLF15 interacts with MEF2A and its expression is turned on during adipocyte differentiation. Taken together, this suggests that MF2A and KLF15 cooperate to allow Glut4 gene transcription (Im et al. 2007).

The high expression of Glut4 mRNA is specifically lost in adipose tissue, but not in muscle, when 165 base pairs are deleted at the 5′ end of the 900 base pairs promoter. Within this region, some domains are conserved across species, but the functional sequences within them are not fully characterized. There is an element

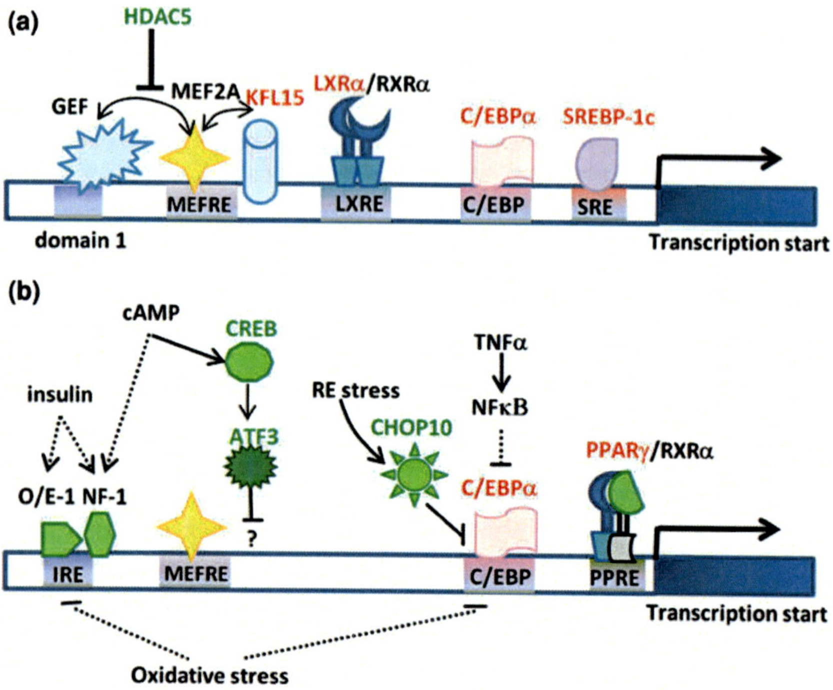

Fig. 7.1 Control of Glut4 gene expression. **a** Transcriptional activation; **b** Transcriptional inhibition. Proteins written in red have an increased expression during adipocyte differentiation. In green, the inhibitory pathways and the transcriptional repressors. The dashed lines represent hypothetic pathways

responsible for insulin-mediated down-regulation of Glut4, but this could not explain the importance of this region for adipocyte specific expression. The Glut4 enhancer factor (GEF) binds to a domain called domain I within this region of the Glut4 promoter. Binding of MEF2A to its binding site increases GEF binding to domain I and GEF forms a complex with MEF2A. More importantly, expression of GEF increases the MEF2A-dependent transcriptional activation of the Glut4 promoter. Thus MEF2A and GEF cooperate to increase Glut4 mRNA expression in adipocyte, but it remains unclear whether GEF is the factor responsible for the role of domain I to trigger adipocyte specific Glut4 expression. Indeed, GEF is also expressed in muscle in which it also cooperates with MEF2A to increase Glut4 mRNA expression (Sparling et al. 2008). HDAC5, a known co-repressor of MEF2A, is expressed in adipocytes and its nuclear amount is decreased during adipocyte differentiation. Because it is able to inhibit MEF2A- and GEF-dependent Glut4 gene transcription, this could contribute to the induction of Glut4 expression during adipocyte differentiation. However, the mechanism involved in the decrease of its nuclear amount is unknown (Sparling et al. 2008).

Additional elements within the Glut4 promoter contribute to Glut4 expression in adipocyte. A LXR responsive element is present in the mouse and human promoter, which binds LXRα/RXRα. Activation of LXR by specific agonists increases adipocyte Glut4 expression in wild type mice but not in LXRα and β double knockout and Glut4 expression is decreased in LXRα -/- mice (Im et al. 2007). CCAAT/enhancer binding protein (C/EBP-α) and sterol response element binding protein-1-c (SREBP-1c) bind the Glut4 promoter and activates Glut4 transcription in in vitro experiments. Their role in the control of Glut4 expression in vivo have not been investigated and still remains unclear. Indeed, SREBP-1c is activated by insulin but insulin inhibits Glut4 gene transcription (cf. "Transcriptional Inhibition of Glut4 Gene Expression").

Transcriptional Inhibition of Glut4 Gene Expression

Generally, Glut4 gene expression is down regulated in states of insulin deficiency (type 1 diabetes and chronic fasting) and increased in animals chronically treated with insulin (Im et al. 2007). In cultured cells, and by contrast with the in vivo data, insulin decreases Glut4 mRNA expression (Im et al. 2007). Thus, the insulin effect on Glut4 gene expression in vivo perhaps results of an indirect effect of the hormone. The inhibitory effect of insulin on Glut4 gene transcription involved an insulin responsive element (IRE) in the 5' region of the Glut4 promoter. Insulin signaling would activate transcriptional repressors including O/E gene family members and NF-1 resulting in the inhibition of Glut4 gene expression (Fig. 7.1b).

The expression of Glut4 is decreased in adipocyte from obese and diabetic subjects. Several stresses that developed within the adipose tissue could result in an inhibition of the expression of Glut4 mRNA in cultured adipocytes. Proinflammtory cytokines produced by macrophages inhibit adipocyte Glut4 mRNA expression when the two cell types are co-cultured (Lumeng et al. 2006). Several cytokines probably inhibit the expression of Glut4 mRNA, TNFα being the most studied. NF-κB is an obligatory mediator of TNFα effect on Glut4 expression (Ruan et al. 2002). But NF-κB binding sites are not found within the Glut4 promoter, thus favoring an indirect role requiring other molecular events. One of them could be related to the capacity of TNFα to inhibit adipocyte differentiation in an early step. In accordance, the effect of TNFα on Glut4 expression is counteracted by PPARγ agonist, PPARγ being a master transcription factor for adipocyte differentiation. Although in vivo treatment with PPARγ agonist increases Glut4 expression in adipose tissue because of its ability to induce adipocyte differentiation, the PPARγ response elements identified in the Glut4 promoter are involved in the down regulation of Glut4 expression (Koumanov et al. 2005). In accordance, we observed that treatment of mature adipocytes or human adipose tissue explants with the PPARγ agonist rosiglitazone inhibits Glut4 mRNA and protein expression (personal communication). It is not excluded that TNFα signaling interferes with the activity and/or expression of other transcription factors involved in adipocyte Glut4 expression such as C/EBPα (Fig. 7.1a).

Oxidative stress inhibits Glut4 mRNA and protein expression. It impairs nuclear proteins binding to the insulin responsive element and to the C/EBPα binding element of the Glut4 promoter. In a reporter gene assay, the inhibitory effect of oxidative stress is partly dependent on the Glut4 promoter region containing the C/EBP biding site (Pessler-Cohen et al. 2006). Activation of the endoplasmic reticulum (ER) stress in cultured adipocytes decreases Glut4 expression at the level of gene transcription. ER stress increases the expression of an inhibitor of the activity and expression of C/EBPα, CHOP10.

Glut4 gene expression is also inhibited when cAMP level increases in adipocytes. The binding of NF-1 to IRE appears to be involved in this effect. In the insulin resistant state catecholamine effects on cAMP signaling in adipocytes is potentiated by chronic elevation of insulin. This leads to an increase in active phosphorylated CREB (cAMP-responsive transcription factor) that induces the expression of the transcriptional repressor ATF3. Forskoline that increases cAMP levels inhibits Glut4 gene transcription an effect that is blocked when ATF3 expression or CREB activity are inhibited. Furthermore, inhibition of CREB activity increases Glut4 gene expression, whereas ATF3 overexpression inhibited it (Qi et al. 2009). A Chip assay shows recovery of Glut4 promoter from immunoprecipitates of ATF3 that is further increases by forskoline treatment (Qi et al. 2009). The region of ATF3 binding and the partner(s) of ATF3 involved in the repression of Glut4 gene transcription are presently unknown.

Post Transcriptional Control of Glut4 Gene Expression

In some situations, the expression level of Glut4 is also controlled by acting on the stability of the mRNA and of the protein. TNFα inhibits Glut4 gene transcription and also inhibits the stability of Glut4 mRNA. The mechanism involved is not entirely defined, but would involved protein binding in the 3'untranslated region and inhibition of the translation (Long and Pekala 1996). The half-life of the Glut4 transporter is decreased by threefold by a chronic treatment with insulin. Hyperinsulinemia could, thus, induce the degradation of the protein Glut4. This could results from a more efficient targeting of the glucose transporter Glut4 toward the lysosomes where they could be degraded.

Ubc9, a SUMO conjugating enzyme, binds Glut4 and controls its degradation. However, this is independent of its enzymatic activity and will be related to its potential to disrupt Glut4 trafficking toward lysosomal degradation (Bogan and Kandror 2010). It is, however, unknown whether Ubc9 expression/function is altered in insulin resistance to explain changes in the half-life of Glut4.

Control of the Intracellular Localization of Glut4

Insulin-induced glucose transport is mainly due to the redistribution of Glut4 from intracellular compartments to the plasma membrane. In nonstimulated adipocyte, Glut4 is sequestered into the cell in small insulin-responsive vesicles. Intracellular Glut4-containing vesicles are, however, not static but continuously recycle between several intracellular organelles in a way that they are mainly excluded from the plasma membrane. Insulin stimulates the translocation of these vesicles to the cell surface, inserting Glut4 within the plasma membrane to enhance glucose transport. Multiple insulin signaling pathways have been implicated in the control of Glut4 localization. Signaling pathways intersect with Glut4 trafficking routes at distinct steps. In other words, insulin signaling is going to act on molecules that control Glut4 intracellular trafficking steps. The most studied pathways concern signaling through Akt2 to AS160/TBC1D4, a Rab-GTPase activating protein that controls Glut4 localization in adipocyte, a pathway involving atypical protein kinase C isoforms, and one that involves the Rho-family GTPase, TC10α. These pathways are reviewed elsewhere (Rubin and Bogan 2009).

The Intracellular Glut4 Containing Compartments

To understand the intracellular Glut4 trafficking routes, it has been necessary to characterize the intracellular compartments that contain Glut4. By using various technical approaches (i.e. subcellular fractionation, endosomes ablation, electronic, and immunofluorescence microscopy), it is now established that Glut4 is present in early endosomes (EE), in endosomal recycling compartment (ERC), and in a specific subdomain of the *trans* Golgi network (TGN). However, the main localization of Glut4 in basal condition is in highly sensitive tubulovesicular compartments from which Glut1 and the endosomal markers, such as transferrin receptors and cellubrevin, are excluded. This compartment is often called IRVs for insulin responsive vesicles. IRVs are enriched in Glut4, IRAP (insulin-responsive aminopeptidase), sortilin, LRP1 (low density lipoprotein receptor-related protein), and VAMP2 (vesicle-associated membrane protein) [for review (Bogan and Kandror 2010; Kaddai et al. 2008b)]. Intracellular compartments that contain the small GTPase Rab4b are more overlapped with Glut4 than with transferrin receptor. Although not specific to the IRV, Rab4b is thus a small GTPase of this specialized compartment needed for Glut4 intracellular retention (Kaddai et al. 2009b).

The Glut4 Trafficking Routes in Basal and Insulin Conditions

In basal condition, Glut4 is in equilibrium between IRV and other organelles located in the perinuclear region, probably the TGN and/or the ERC (Rowland et al. 2011; Bogan and Kandror 2010). Intracellular Glut4 has nearly no access to

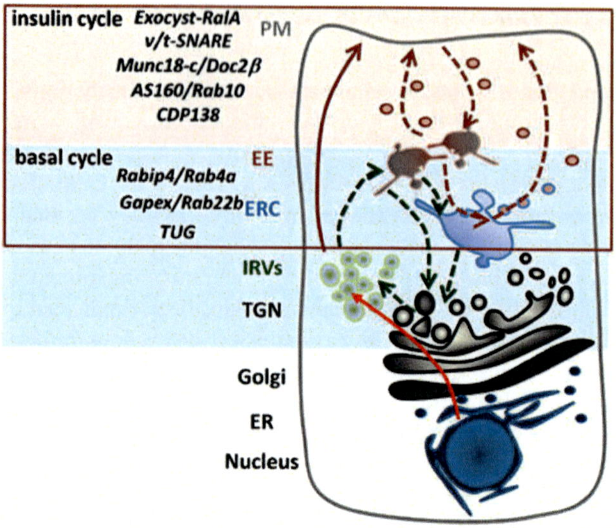

Fig. 7.2 Glut4 cycles in absence and in presence of insulin. Green arrows are for trafficking steps in basal condition. Brown arrows are for trafficking steps occurring in presence of insulin. The steps that are not definitively demonstrated are shown with dashed lines. *PM* plasma membrane; *EE* early endosomes; *ERC* endosomal recycling compartment; *IRVs* insulin responsive vesicles; *TGN* trans Golgi network; *ER* endoplasmic reticulum

the plasma membrane in basal condition (Fig. 7.2). Newly synthetized Glut4, together with the other IRV-specific proteins, are sorted at the level of the TGN directly into the IRVs. Interestingly, Glut4, IRAP, sortilin, and LRP1 can interact with each other through their luminal domain bringing these proteins together, and thus facilitating their sorting into the IRVs. Indeed, an experimental decrease in the expression of one of these proteins perturbed the sorting of the others into IRVs. The signaling sequences on Glut4 as well as proteins involved in this process are reviewed in (Bogan and Kandror 2010). Briefly, it involves the formation of clathrin coats on donor membrane through clathrin adaptors ACAP1 that binds Glut4, and GGA that binds sortilin. The recruitment of ACAP1 and GGA on donor membranes could possibly involve the small GTPase Arf6. The other way for Glut4 to enter in the IRVs is after insulin withdrawal, when Glut4 is internalized from the plasma membrane and directed back to the IRVs. From the plasma membrane Glut4 is internalized in peripheral early endosomes, then in a compartment enriched in Rabip4 an effector of Rab4 to eventually reach a perinuclear compartment labeled with markers of a subdomain of the TGN (Syntaxin6/16). It is not clearly established whether Glut4 sorting toward TGN requires trafficking through ERC or occurs directly from early endosomes (Fig. 7.2). Glut4 targeting toward perinuclear compartments is perturbed when a form of Rabip4 unable to bind Rab4 is overexpressed (Mari et al. 2006). In human adipocyte, this trafficking step requires the clathrin heavy chain CHC22, a form of clathrin that is not expressed in mice (Bogan and Kandror 2010). CHC22 functions in an endosomal sorting step for several cargoes not only in insulin sensitive tissues.

In absence of insulin, IRVs are retained very efficiently within adipocytes. The mechanisms involved in this retention process are not entirely known but certainly require a complex intracellular trafficking between TGN and IRVs. It is also not excluded that some class of endosomes could play a role in this intracellular retention process. Indeed in absence of the endosomal small GTPase Rab4b, Glut4 amount is decreased in endosomes while it increases in IRVs as well as at the plasma membrane (Kaddai et al. 2009b). A trafficking route between endosomes and nonendosomal containing compartment, thus, exists in absence of insulin. Each time a trafficking step of this intracellular retention process will be experimentally affected, Glut4 will be targeted toward the plasma membrane. Several different situations have, however, been observed. Glut4 vesicles could fuse with the plasma membrane but Glut4 will transport or not glucose. Either, vesicles could accumulate close to the plasma membrane primed for a following stimulation by insulin. Depending on the intracellular trafficking step affected, the nature of the Glut4 containing vesicles directed toward the plasma membrane would, thus, not have the same requirement for fusion with the plasma membrane. We will not review here all the trafficking proteins identified that play a role in Glut4 intracellular retention but only those that appear more interesting because they are targeted by insulin, Rabip4, Gapex-5, and TUG.

Before going more in depth on the role of these proteins regarding to insulin action, we need to know what happens on the Glut4 intracellular trafficking route in presence of insulin (Fig. 7.2). It is known for a long time that insulin recruits to the plasma membrane about half of the intracellular Glut4, but it is now established that it recruits nearly all the IRVs (Bogan and Kandror 2010). This indicates that insulin drives the mobilization of the IRVs and then their fusion with the plasma membrane. In the continuous presence of insulin, Glut4 is still internalized from the plasma membrane and one of the open questions is whether Glut4 then recycles by the classical endosomal routes (involving the early and the recycling endosomes) or whether it also traffics through the IRVs like in basal condition. A recent publication reported that the size of the IRVs that fuse with the plasma membrane just after insulin stimulation compared to those that fuse next are not the same. The later one is similar to the size of endosomes that contains TfR (transferrin receptor) and that fuse with the plasma membrane, suggesting that there is a switch in vesicular traffic between two distinct circuits (Xu et al. 2011). The fusion process was here analyzed by total internal reflection fluorescence microscopy (TIRFM) by using a Vamp2-pHfluorin fusion protein as an IRVs markers. But it remains unclear whether Glut4 is always colocalized with Vamp2.

Rabip4 is an effector of the small GTPase Rab4a, the two playing a role in Glut4 intracellular trafficking (Kaddai et al. 2008b). Rabip4 is an endosomal protein and is not detected in the IRVs. Interestingly, insulin induces its translocation to the plasma membrane. The overexpression of a mutated form of Rabip4 unable to bind Rab4 triggers the translocation of Glut4 to the plasma membrane in absence of insulin. Uncoupling of Rab4 and Rabip4 could, thus, contribute to

Glut4 translocation. Insulin-induced delocalization of Rabip4 could, thus, be a way to uncouple it from endosomal Rab4. Further investigations will be necessary to understand the mechanisms involved in insulin-induced Rabip4 delocalization and its role in Glut4 trafficking.

Gapex-5 is a GTP-exchange factor for the Rab5 family members. In insulin-treated adipocyte Gapex-5 is recruited to the plasma membrane far away from its intracellular target Rab22b. In absence of insulin, Gapex-5 will maintain Rab22b activated to favor intracellular trafficking between early endosomes and TGN, contributing to the intracellular retention machinery. In presence of insulin, Rab22b will be maintained in its inactive form due to the delocalization of Gapex-5. Accordingly, either Gapex-5 or active Rab22b overexpression inhibits insulin-induced Glut4 translocation to the plasma membrane. However, Rab22b down regulation by siRNA does not increase basal plasma membrane Glut4. Although the absence of a Rab protein is not necessary mimicked by the Rab inactive form, further investigations are necessary to define the requirement of Gapex-5 on Rab22b for intracellular Glut4 retention (Kaddai et al. 2008b).

TUG (Tether containing a UBX domain, for Glut4) was identified in a functional screen for regulators of Glut4 trafficking. TUG is a binding partner for nonendosomal Glut4 (Bogan and Kandror 2010), but is ubiquitously expressed and thus is not specific of the Glut4 trafficking pathway. TUG interacts with the N-Terminus of Glut4, which is required for Glut4 targeting to the IRVs. On TUG, the regions involved in Glut4 binding are its N-terminal and central regions. The C-terminal part of TUG is essential to retain Glut4 intracellularly possibly because it interacts with TGN anchoring protein(s). Overexpression of this region disrupts intracellular retention of Glut4 by competing with the entire TUG/Glut4 complex. Down regulation of TUG by siRNA redistributes Glut4 to the plasma membrane and increases glucose uptake in absence of insulin, whereas TUG overexpression causes Glut4 to accumulate in nonendosomal vesicles. Insulin stimulates the dissociation of Glut4 from TUG in adipocytes. The signaling pathway and the mechanisms involved in insulin-induced dissociation of the complex are unknown, but certainly represent key events to further recruit IRVs at the plasma membrane.

Thus, the cycle of Glut4 intracellular trafficking provides an efficient way to retain Glut4 inside the adipocytes and insulin possibly acts at different steps to brake this cycle in order to efficiently target Glut4 at the plasma membrane.

Glut4 Containing Vesicles Fusion with the Plasma Membrane Principal Site of Insulin Action

Fusion of Glut4 containing vesicles with the plasma membrane could be broken down in multiple steps: (1) transport of IRVs along microtubules and cortical actin; (2) tethering of the vesicles with the plasma membrane; (3) fusion of the vesicles with the plasma membrane; (4) dispersal of Glut4 from the site of fusion (Rowland et al. 2011; Bryant and Gould 2011) (Fig. 7.3). All these steps could be followed since the development of TIRF microscopy and the establishment of an

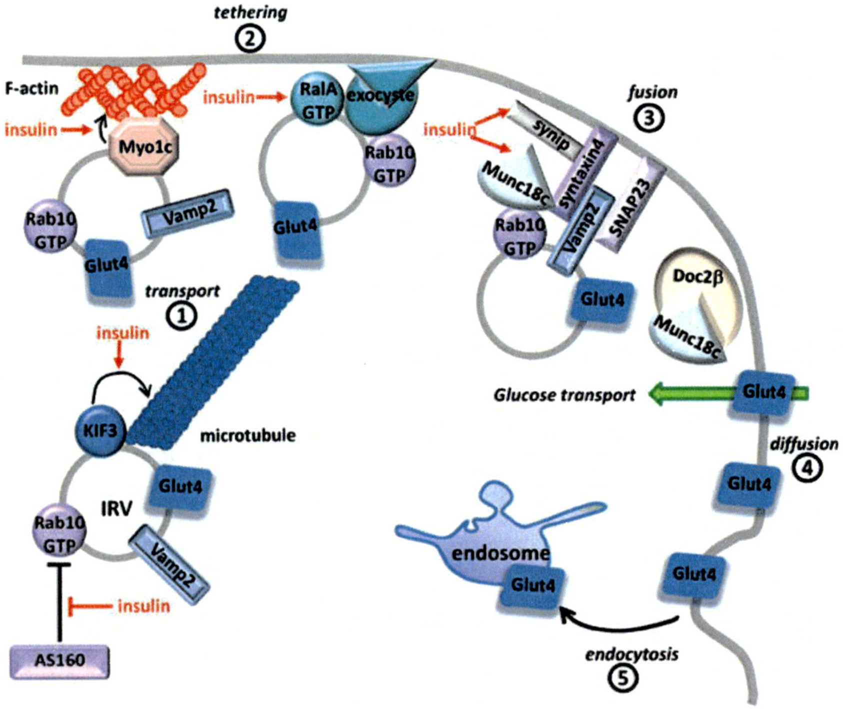

Fig. 7.3 Schematic representation of molecular events of insulin-induced Glut4 translocation. IRVs are transported on microtubules, captured by cortical actin, tethered at the plasma membrane then fused with it. The Glut4 diffuses at the plasma membrane prior to its endocytosis. The proteins targeted by insulin are indicated

elegant cell-free system that recapitulates the final steps of Glut4 translocation (Koumanov et al. 2005). It has been possible to analyze what are the steps affected by insulin. Prior fusion, the docking step is more efficient in presence of insulin, but the post-docking step that prepares Glut4-vesicles for fusion is the major insulin-stimulated step. The dispersion of Glut4 after fusion is also widely affected by insulin. Two types of Glut4-containing vesicles could fuse with the plasma membranes: the IRVs rapidly recruited after insulin treatment and the vesicles that recycle between the plasma membrane and endosomes in the continuous presence of insulin. The molecular mechanisms are not necessary the same but this question has not been at this day carefully addressed. But in accordance, the insulin-induced increase in the rate of Glut4 vesicle fusion with the plasma membrane was found to be transient with a peak during 2–3 min after insulin treatment and then it declines to levels comparable to basal situations.

Glut4 vesicles are associated with both the microtubule and actin cytoskeleton. The kinesin KIF3 has been involved in insulin-induced Glut4 translocation and insulin increases KIF3 association with microtubules. Actin also plays a role in insulin-induced Glut4 translocation in adipocytes. Insulin rapidly remodels cortical actin cytoskeleton. Cortical actin cytoskeleton is not required for the recruitment of Glut4 vesicles beneath the plasma membrane but for their fusion with the plasma membrane. The molecular motor Myo1c that drives vesicles movement along F-actin is required for insulin-induced Glut4 translocation and is phosphorylated in response to insulin. All together, these observations indicate that the cytoskeleton is targeted by insulin to promote Glut4 translocation to the plasma membrane, but the exact role of these cytoskeleton changes for Glut4 vesicles targeting, tethering, or fusion is not known. It could promote Glut4 vesicles release from its internal location by guiding vesicles on cytoskeleton elements, by positioning Glut4 vesicles near the plasma membrane, by aiding them to dock and fuse with the plasma membrane, or also perhaps by positioning signaling intermediates.

Glut4 vesicles tethering involved the organization of the exocyst complex at the interface of the IRVs and the plasma membrane. The formation of the exocyst complex requires the activation of the small GTPase RalA present at the surface of IRVs. Insulin could trigger the activation of RalA via the phosphorylation of a RalA GAP activity.

The IRVs fusion event is mediated by the action of the tSNAREs Syntaxin4 and SNAP23 and the vSNARE VAMP2 in concert with the inhibitory regulatory molecules Munc18c. From in vitro studies using liposomes, it is thought that interaction between vSNAREs and their cognate tSNARE is sufficient for vesicle fusion. Munc18c is a binding partner of syntaxin4 that blocks the interaction with VAMP2 and thus the fusion process. Munc18c is phosphorylated on tyrosine residue 521 by the insulin receptor and this phosphorylation abrogates its binding to syntaxin 4 (Aran et al. 2011). Tyrosine phosphorylation of Munc18c is required for insulin induced-Glut4 translocation (Jewell et al. 2011). Munc18c is also phosphorylated on residue 219 in response to insulin. Munc18c binds to Doc2β when phosphorylated on this site and Doc2β is essential for Glut4 vesicles translocation (Rowland et al. 2011). Doc2β is a SNARE-related protein without a transmembrane region that is recruited to the plasma membrane in response to insulin where it interacts with syntaxin 4 in order to increase the fusion step. Thus, its interaction with phosphorylated Munc18c could be part of the mechanism that induces the binding switch from an inhibitor to an activator of v/tSNARE-mediated fusion. Another partner of Syntaxin4, Synip, is phosphorylated by insulin-activated PKB. Its role in Glut4 vesicles fusion is still controversial (Rowland et al. 2011).

Molecular coordination of tethering and fusion process is known to be controlled at least in yeast by tomosyn-like molecule, a common partner of the exocyst complex and the v/tSNARE complex. A Tomosyn was shown to form a complex with the tSNAREs Syntaxin 4/SNAP23 in adipocyte and to play a role in Glut4 translocation. But it is unknown whether it also binds an exocyst subunits and whether this is controlled by insulin.

A previously uncharacterized protein named CDP138 was recently discovered to play a role in insulin-stimulated insertion of Glut4 into the plasma membrane.

CDP138 is a C2 domain-containing protein, phosporylated in response to insulin by PKB, and thus it could link insulin signaling to Glut4 vesicle fusion to the plasma membrane. The C2 domain is capable to bind Ca^{++} and a mutant deleted for this domain blocks insulin-induced Glut4 vesicle insertion into the plasma membrane. This confirms previous results suggesting that Ca^{++} plays a role in insulin-induced glucose transport (Xie et al. 2011).

Numerous studies from yeast to mammals indicate that the tethering/fusion of exocytic vesicles are controlled by Rab GTPases. Exocytosis in yeast requires the Rab GTPase sec4p that is needed to organize the tethering exocyst complex and to link it to the SNARE fusion machinery through its interaction with a tomosin-like molecule (Novick et al. 2006). The Munc-like inhibitor of the plasma membrane v/tSNARE complex also binds active sec4p and this is required for polarized secretion (Weber-Boyvat et al. 2011). Thus, in this model one Rab protein is involved in exocytosis and binds, once activated, a subunit of the exocyst, a common partner of the exocyst and SNARE complexes, and the Munc-like inhibitor of SNARE-induced fusion.

Is there such a Rab protein in IRVs fusion with the plasma membrane in adipocytes? A candidate is Rab10. Rab10 is supposed to be maintained inactive in basal condition by the action of the Rab GTPase activating protein (GAP) AS160 (also known as TBC1D4). Insulin triggers the phosphorylation of AS160 on several residues most importantly Ser-588 and Thr-642 and it is accepted that this will lead to inactivation of its GAP activity, although the mechanism is unknown (Rowland et al. 2011; Bryant and Gould 2011). Indeed, overexpression of a nonphosphorylated form of AS160 inhibits insulin-induced Glut4 translocation but only when its GAP domain is functional. Inactivation of the GAP activity would yield to activation of one or several Rab. Rab10 is an in vitro target of the GAP domain of AS160. Furthermore, down regulation of AS160 increases Glut4 at the plasma membrane, an effect lost when Rab10 is also depleted. Thus, in presence of AS160, Rab10 is inactivated and Glut4 can traffic through its intracellular retention cycle. In absence of AS160, active Rab10 is increased and could drive Glut4 targeting to the plasma membrane thus releasing Glut4 vesicles from their intracellular location. In accordance, overexpression of active Rab10 partially mimics insulin action and the GEF for Rab10 Dennd4C is also required (Sano et al. 2011). All these observations clearly point to AS160 inactivation/Rab10 activation as being an important point of intersection of signaling and trafficking. The use of the form of AS160 mutated on its phosphorylation site in total internal reflection fluorescence microscopy indicates that it blocks Glut4 tethering and indirectly the downstream fusion, but does not affected Glut4 arrival in the evanescent field. Down regulation of AS160 in adipocytes causes incorporation of Glut4 to the plasma membrane but to lower extent than insulin like overexpression of active Rab10. It also causes Glut4 to accumulate in compartments that are primed for fusion. Such experiments with constitutive active Rab10 are at this time not available. Further, works are required to determine whether active Rab10 controls the same molecular mechanisms as sec4p in yeast by identifying its effectors in adipocytes. It remains also to be determined whether inactivation of the GAP activity by insulin affects the first burst of IRVs fusion or the second round of fusion. This will be technically difficult until the establishment of tools allowing for rapid and transient activation/inactivation of the studied proteins. Whatever it is clear that activation of Rab10 by insulin is not the only event controlled by insulin in order to induce Glut4 vesicles tethering and fusion. This is evidence by the numerous insulin-induced phosporylated targets among the proteins of the tethering/fusion machinery as described above (Fig. 7.3).

Effects on Glut4 Trafficking of Insulin-Mimetic Agents and Insulin Counterregulatory Hormones

The effect of insulin-mimetic agents on Glut4 trafficking has not been extensively explored. Often, these agents act on insulin signaling. Thus we could assume that they will trigger the same mechanisms as insulin. Sometimes they act through activation of the AMP kinase pathway and this pathway has not been extensively studied regarding Glut4 trafficking in adipocytes. Concerning the nitric oxide-donating derivative of acetylsalicylic acid, we assume that it recruits to the plasma membrane the endosomal transporters because it recruits as efficiently Glut4 and Glut1, but the step of trafficking it affects remains unknown (Kaddai et al. 2008a).

The inhibitory effect β adrenergic on insulin-induced glucose transport involved a blockade of Glut4 function at the plasma membrane. The amount of Glut4 at the plasma membrane is not changed by β adrenergic, but its accessibility to extracellular specific photolabeling reagents is decreased. β-adrenergic would, thus, render Glut4 occluded for participation on glucose transport and this involves changes in local pH.

Glut4 Trafficking in Insulin Resistance

We will not describe here in details the mechanisms that induce the inhibition of insulin signaling and that participate in defects of insulin-induced Glut4 translocation in adipocytes. An important deregulated signaling molecule in mice models of insulin resistance and in human is IRS1 (Gual et al. 2003). However, a partial loss of IRS1 is not sufficient to mimic defect in insulin-induced glucose transport. In accordance, it has recently been established by studying various in vitro and in vivo insulin resistant models that the most deleterious defects in insulin action occur independently of IRS1 (Hoehn et al. 2008). The decrease in Glut4 expression described previously in this review contributes to insulin-induced glucose transport occurring in adipocytes from insulin-resistant subjects. In addition, perturbations of Glut4 trafficking were suggested to play a role. Indeed, it is clearly demonstrated in cultured adipocyte and in genetically modified mice that alterations in intracellular Glut4 trafficking could result in altered insulin-induced Glut4 recruitment from inside the cell to the plasma membrane.

Glut4 Localization in Insulin-Resistant Adipocytes

The group of Garvey was the first to demonstrate that the subcellular distribution of Glut4 is not similar in adipose tissues from control and insulin-resistant patients. It used subcellular fractionation of adipocytes to demonstrate that Glut4 is located in compartments with different sedimentation properties in insulin-resistant adipocytes from patients with type 2 diabetes or gestational diabetes compared to control adipocytes (Garvey et al. 1993; Maianu et al. 2001). The dynamic of Glut4 compartments was analyzed in 3T3-L1 rendered insulin resistant by using endothelin 1. The results confirm the existence of different trafficking parameters in basal adipocytes. In this experimental insulin resistance, a significant increase in the event rate of Glut4 appearance in the TIRF zone is observed regardless insulin stimulation, whereas many few Glut4 remains in the TIRF zone (Fujita et al. 2010).

Defect of Proteins Involved in Glucose Trafficking in Adipocyte Insulin Resistance Development

The identification of the molecular mechanisms involved in the deregulation of Glut4 trafficking in insulin resistant adipocyte is just beginning and only few data are available. Either mutations in or deregulated expression of proteins involved in Glut4 trafficking have been described.

Rab4a and b that play a role in Glut4 trafficking have their mRNA expression decreased in the adipose tissue from insulin-resistant obese db/db mice and morbidly obese patients compared to their controls. Rab4a (mRNA and protein) is decreased in the adipose tissue from murine models of obesity with insulin resistance but not in morbidly obese patients (Kaddai et al. 2009b). Interestingly, Rab4a amount is increased in the phase of obesity installation in obese Zucker rats, when adipocytes are hyperesponsive to insulin (Kaddai et al. 2008b). The mechanisms involved in the deregulation of Rab4 s expression are unknown but could be related to adipogenic differentiation stage, because their expression is increased during adipocyte differentiation.

We described that sortilin, a major protein of the IRVs involved in Glut4 sorting to the IRVs in adipocyte is down regulated in the adipose tissue of db/db mice and in morbidly obese diabetic patients. Sortilin expression in adipose tissue is strongly correlated with TNFα expression. Interestingly, TNFα inhibits the expression of sortilin at the mRNA and protein level in culture adipocytes from mouse and human origin. Furthermore, injection of TNFα to mice also induces the down regulation of adipose tissue sortilin mRNA expression. Chronic inflammation within the adipose tissue of obese subjects could, thus, contribute to insulin resistance by modulating proteins that control Glut4 trafficking in addition to inhibit insulin signaling (Kaddai et al. 2009a). AS160 mRNA and protein expression is strongly affected by TNFα in cultured adipocytes and it is decreased in the adipose tissue from db/db obese diabetic mice and from mice injected with TNFα [(Hoehn et al. 2008) and nonpublished observation].

AS160 mutations have also been described in relation to insulin resistance. A human truncated variant of AS160 leads to insulin resistance and post-prandial hyperinsulinemia in a family with acanthosis nigricans. The effect of this truncation is to perturb insulin-induced Glut4 recruitment at the plasma membrane in insulin target cells. Furthermore, families with AS160 variants have been identified and these mutations may contribute to varying degree of insulin resistance (Dash et al. 2010).

It appears evident that current studies will, in the next future, provide other mechanisms to explain where and how Glut4 trafficking is altered in insulin resistant adipocytes.

Conclusion

Glucose transport by adipocyte plays an important role in glucose homeostasis. A global view of the mechanisms allowing for the integration of insulin signaling and Glut4 trafficking is still not available because of the complexity of both the insulin signaling pathway and of Glut4 trafficking. We discover even today direct substrates of the insulin receptor (like Munc18c) and also numerous substrates of PKB, the main pathway for insulin-induced glucose transport. A current objective for better understanding insulin action on glucose transport is to succeed in the identification of each trafficking step controlled by each of the insulin-activated intermediaries identified. It will be important to set up more powerful technique of imaging (high resolution and multiple colors TIRF microscopy) to characterize the protein complexes involved in the tethering and fusion of the IRVs. This step is particularly important because it is the most activated by insulin. Another objective will be to characterize the molecular mechanisms that lead to defects in Glut4 expression and trafficking in insulin resistant adipocyte. This will certainly help us to find novel putative targets for therapeutic intervention in patients with insulin resistance and type 2 diabetes.

References

Abel ED, Peroni OD, Kim JK et al (2001) Adipose-selective targeting of the GLUT4 gene impairs insulin action in muscle and liver. Nature 409:729–733

Aran V, Bryant NJ, GG W (2011) Tyrosine phosphorylation of Munc18c on residue 521 abrogates binding to syntaxin 4. BMC Biochem 12:19

Bogan JS, Kandror KV (2010) Biogenesis and regulation of insulin-responsive vesicles containing Glut4. Curr Opin Cell Biol 22:506–512

Bryant NJ, Gould GW (2011) SNARE proteins underpin insulin-regulated GLUT4 traffic. Traffic 12:657–664

Carvalho E, Kotani K, Peroni OD et al (2005) Adipose-specific overexpression of GLUT4 reverses insulin resistance and diabetes in mice lacking GLUT4 selectively in muscle. Am J Physiol Endocrinol Metab 289:E551–E561
Cushman SW, Wardzala LJ (1980) Potential mechanism of insulin action on glucose transport in the isolated rat adipose cell. Apparent translocation of intracellular transport systems to the plasma membrane. J Biol Chem 255:4758–4762
Dash S, Langenberg C, Fawcett KA et al (2010) Analysis of TBC1D4 in patients with severe insulin resistance. Diabetologia 53:1239–1342
Dhalla AK, Wong MY, Voshol PJ et al (2007) A1 adenosine receptor partial agonist lowers plasma FFA and improves insulin resistance induced by high-fat diet in rodents. Am J Physiol Endocrinol Metab 292:E1358–E1363
Dong Q, Ginsberg HN, Erlanger BF (2001) Overexpression of the A1 adenosine receptor in adipose tissue protects mice from obesity-related insulin resistance. Diabetes Obes Metab 3:360–366
Faulhaber-Walter R, Jou W, Mizel D et al (2011) Impaired glucose tolerance in the absence of adenosine A1 receptor signaling. Diabetes 60:2578–2587
Figler RA, Wang G, Srinivasan S et al (2011) Links between insulin resistance, adenosine A2B receptors, and inflammatory markers in mice and humans. Diabetes 60:669–679
Fujita H, Hatakeyama H, Watanabe TM et al (2010) Identification of three disting functional sites of insulin-mediated GLUT4 trafficking in adipocytes using quantitative single molecule imaging. Mol Biol Cell 21:2721–2731
Garvey WT, Maianu L, Zhu J-H et al (1993) Multiple defects in the adipocyte glucose transport system cause cellular insulin resistance in gestational diabetes. Heterogeneity in the number and a novel abnormality in subcellular localization of GLUT4 glucose transporters. Diabetes 42:1773–1785
Graham TE, Kahn BB (2007) Tissue-specific alterations of glucose transport and molecular mechanisms of intertissue communication in obesity and type 2 diabetes. Horm Metab Res 39:717–721
Gual P, Le Marchand-Brustel Y, Tanti JF (2003) Positive and negative regulation of glucose uptake by hyperosmotic stress. Diabetes Metab 29:566–575
Hajduch E, Darakhshan F, Hundal HS (1998) Fructose uptake in rat adipocytes: GLUT5 expression and the effects of streptozotocin-induced diabetes. 41:821–828
Heseltine L, Webster JM, Taylor R (1995) Adenosine effects upon insulin action on lipolysis and glucose transport in human adipocytes. Mol Cell Biochem 144:147–151
Hoehn KL, Hohnen-Behrens C, Cederberg A et al (2008) IRS1-independent defects define major nodes of insulin resistance. Cell Metab 7:233–421
Im S–S, Kwon S-K, Kim T-H et al (2007) Regulation of glucose transporter type 4 isoform gene expression in muscles and adipocytes. IUBMB Life 59:134–145
Jewell JL, Oh E, Ramalingam L et al (2011) Munc18c phosphorylation by insulin receptor links cell signaling directly to SNARE exocytosis. J Cell Biol 193:185–199
Kaddai V, Gonzalez T, Bolla M et al (2008a) The nitric oxide-donating derivative of acetylsalicylic acid, NCX 4016, stimulates glucose transport and glucose transporters translocation in 3T3-L1 adipocytes. Am J Physiol Endocrinol Metab 192:E162–E169
Kaddai V, Le Marchand-Brustel Y, Cormont M (2008b) Rab proteins in endocytosis and Glut4 trafficking. Acta Physiol 192:75–88
Kaddai V, Jager J, Gonzalez T et al (2009a) Involvement of TNF-alpha in abnormal adipocyte and muscle sortilin expression in obese mice and humans. Diabetologia 2009:932–940
Kaddai V, Gonzalez T, Keslair F et al. (2009) Rab4b is a small GTPase involved in the control of the glucose transporter GLUT4 localization in adipocyte. PLoS One 4(4):e5257
Kobayashi M, Nikami H, Morimatsu M et al (1996) Expression and localisation of insulin-regulatable glucose transporter (GLUT4) in rat brain. Neurosci Lett 213:103–106
Koumanov F, Jin B, Yang J et al (2005) Insulin signaling meets vesicle traffic of GLUT4 at a plasma-membrane-activated fusion step. Cell Metab 2:179–189

Lewko B, Bryl E, Witkowski JM et al (2005) Characterization of glucose uptake by cultured rat podocytes. Kidney Blood Press Res 28:1–7

Liao W, Nguyen MT, Imamura T et al (2006) Lentiviral short hairpin ribonucleic acid-mediated knockdown of GLUT4 in 3T3-L1 adipocytes. Endocrinology 147:2245–2252

Long SD, Pekala PH (1996) Regulation of GLUT4 mRNA stability by tumor necrosis factor-alpha: alterations in both protein binding to the 3′ untranslated region and initiation of translation. Biochem Biophys Res Commun 220:949–953

Lumeng CN, Deyoung SM, Saltiel AR (2006) Macrophages block insulin action in adipocytes by altering expression of signaling and glucose transport proteins. Am J Pysiol Endocrinol Metab 292:E166–E174

Maianu L, Keller SR, Garvey WT (2001) Adipocytes exhibit abnormal subcellular distribution and translocation of vesicles containing glucose transporter 4 and insulin-regulated aminopeptidase in type 2 diabetes mellitus: implication regarding defects in vesicle trafficking. J Clin Endocrinol Metab 86:5450–5456

Mari M, Monzo P, Kaddai V et al (2006) The Rab4 effector Rabip4 plays a role in intracellular trafficking of Glut 4 in 3T3-L1 adipocytes. J Cell Sci 119:1297–1306

Novick P, Medkova M, Dong G et al (2006) Interactions between Rabs, tethers, SNAREs and their regulators in exocytosis. Biochem Soc Trans 34:683–686

Pessler-Cohen D, Pekala PH, Kosvan J et al (2006) GLUT4 repression in response to oxidative stress is associated with reciprocal alterations in C/EBP alpha and delta isoforms in 3T3-L1 adipocytes. Arch Physiol Biochem 112:3–12

Purcell SC, Aerni-Flessner LB, Willcockson AR et al (2011) Improved insulin sensitivity by GLUT12 overexpression in mice. Diabetes 60:1478–1482

Qi L, Saberi M, Zmuda E et al (2009) Adipocyte CREB promotes insulin resistance in obesity. Cell Metab 9:277–286

Regazzetti C, Peraldi P, Grémeaux T et al (2009) Hypoxia decreases insulin signaling pathways in adipocytes. Diabetes 58:95–103

Rowland AF, Fazakerley DJ, James DE (2011) Mapping insulin/GLUT4 circuitry. Traffic 12:672–681

Ruan H, Hacohen N, Golub TR et al (2002) Tumor necrosis factor-alpha suppresses adipocyte-specific genes and activates expression of preadipocyte genes in 3T3-L1 adipocytes: nuclear factor-kappaB activation by TNF-alpha is obligatory. Diabetes 51:1319–1336

Rubin BR, Bogan JS (2009) Intracellular retention and insulin stimulated mobilization of GLUT4 glucose transporters. Vitam Horm 80:155–192

Sano H, Peck GR, Kettenbach AN et al (2011) Insulin-stimulated GLUT4 protein translocation in adipocytes requires the Rab10 guanine nucleotide exchange factor Dennd4C. J Biol Chem 286:16541–16545

Skrypski M, Le T, Kaczmarek P T et al (2011) Orexin A stimulates glucose uptake, lipid accumulation and adiponectin secretion from 3T3-L1 adipocytes and isolated primary rat adipocytes. Diabetologia 54:1841–1852

Sparling DP, Griesel BA, Weems J et al (2008) GLUT4 enhancer factor (GEF) interacts with MEF2A and HDAC5 to regulate GLUT4 promoter in adipocytes. J Biol Chem 283:7429–7434

Suzuki K, Kono T (1980) Evidence that insulin causes translocation of glucose transport activity to the plasma membrane from an intracellular storage site. Proc Natl Acad Sci USA 77:2542–2545

Thorens B, Mueckler M (2009) Glucose transporters in the 21st century. Am J Physiol Endocrinol Metab 298:E141–E145

Weber-Boyvat M, Aro N, Chernov KG et al (2011) Sec1p and Mso1p C-terminal tails cooperate with SNAREs and Sec4 in ploarized exocytosis. Mol Biol Cell 22:230–244

Wu X, Motoshima H, Mahadev K et al. (2003) Involvement of AMP-activated protein kinase in glucose uptake stimulated by the globular domain of adiponectin in primary rat adipocytes. Diabetes 1355–1363

Xie X, Gong Z, Mansuy-Aubert V et al (2011) C2 domain-containing phosphoprotein CDP138 regulates Glut4 insertion into the plasma membrane. Cell Metab 14:378–389

Xu Y, Rubin BR, Orme CM et al (2011) Dual-mode of insulin action controls GLUT4 vesicle exocytosis. J Cell Biol 193:643–653

Yang Q, Graham TE, Mody N et al (2005) Serum retinol binding protein 4 contributes to insulin resistance in obesity and type 2 diabetes. Nature 436:337–338

Chapter 8
Mechanism of Storage and Synthesis of Fatty Acids and Triglycerides in White Adipocytes

Fabienne Foufelle and Pascal Ferré

Introduction

In adipose tissue, fatty acids are stored as triglycerides formed from a backbone of glycerol on which three fatty acids are esterified. In a lean young adult human, the weight of triglycerides stored represents about 10–20 kilograms i.e. 90,000–180,000 kcal. This energy can be released as fatty acids in case of energy shortage for the needs of oxidative organs, such as skeletal muscles (red fibers), heart, kidney cortex, and liver. The brain does not use fatty acids, but their metabolites, ketone bodies, produced by the liver. Fat storage allows to survive up to 60–70 days of starvation in humans as recorded in some extreme situations. In case of obesity, the survival period can be extended to 90–100 days (3 months). This capacity of fasting for long periods was probably a condition for the survival of our species and unfortunately is still important in countries facing famine periods.

The origin of the fatty acids stored as triglycerides in adipocytes is for an important part the diet. Triglycerides contained in the diet are delivered as chylomicrons into the circulation after their intestinal hydrolysis and re-synthesis in the enterocytes. Fatty acids in adipocytes can also originate from the de novo synthesis of fatty acids from glucose in the liver (lipogenesis) delivered as "Very Low density Lipoproteins" (VLDL) in the circulation. Finally, fatty acids can originate from lipogenesis in the adipocyte itself. Whatever their origin, exogenous or endogenous, free fatty acids must be activated into acyl-CoAs in adipocytes and esterified to finally reach the lipid droplet.

F. Foufelle · P. Ferré (✉)
Team 8 UMRS-872, Centre de Recherches des Cordeliers,
15 rue de l'Ecole de Médecine, 75270 Paris cedex 06, France
e-mail: pascal.ferre@crc.jussieu.fr

F. Foufelle
e-mail: fabienne.foufelle@crc.jussieu.fr

J.-P. Bastard and B. Fève (eds.), *Physiology and Physiopathology of Adipose Tissue*,
DOI: 10.1007/978-2-8178-0343-2_8, © Springer-Verlag France 2013

In this review, we will address the process of fatty acids delivery to the adipocytes from lipoproteins (chylomicrons and VLDL) and the specific role of lipoprotein lipase, the uptake of fatty acids by the adipocyte and their activation, their de novo synthesis and finally their esterification into triglycerides.

Fatty Acids Coming From Lipoproteins

Triglyceride-Rich Lipoproteins

Triglycerides are highly hydrophobic and thus not soluble in the plasma in which they would coalesce and form large oily droplets if delivered as such. Thus, in the post-prandial state triglycerides originating from the diet are delivered by enterocytes as particles called chylomicrons (CM) in the lymphatic ducts. They will thereafter enter into the blood. CM are very large spherical particles (75–450 nm) (Hussain 2000) with a core of triglycerides and cholesterol esters surrounded by a monolayer membrane of phospholipids, free cholesterol, and apolipoproteins with the main one being apoB48. In the blood, chylomicrons exchange proteins with High-Density Lipoproteins (HDL) and acquire apolipoprotein C-II (ApoC2), and apolipoprotein E (ApoE). Triglycerides produced by the liver are delivered into the blood as Very Low Density Lipoproteins (VLDL). VLDL have a structure similar to that of chylomicrons except that the main apolipoprotein is ApoB100 and that their size is smaller (30–80 nm). They are also enriched in ApoC2 and ApoE by protein exchange from HDL.

Lipoprotein Lipase

Chylomicrons and VLDL (abbreviated as TRL for Triglyceride-Rich Lipoproteins) are too large to cross the endothelium of blood vessels surrounding adipocytes. Since triglycerides cannot enter into adipocytes as such and must be hydrolyzed into fatty acids, an enzyme, lipoprotein lipase (LPL) is synthesized by the adipocytes and exported to the luminal side of the capillary endothelium (Fig. 8.1). LPL then hydrolyzes triglycerides contained in the TRL into fatty acids and 2-monoacylglycerol.

LPL is a member of the triglyceride lipase family with pancreatic lipase and hepatic lipase. It is synthesized by parenchymal cells of several tissues including adipose tissue, skeletal muscle, heart, and lactating mammary gland. The LPL mRNA is translated in the rough endoplasmic reticulum. LPL undergoes a series of post-translational modifications including dimerization and N-glycosylation. The oligosaccharide chains are further rearranged and modified in the various Golgi compartments (Braun and Severson 1992).

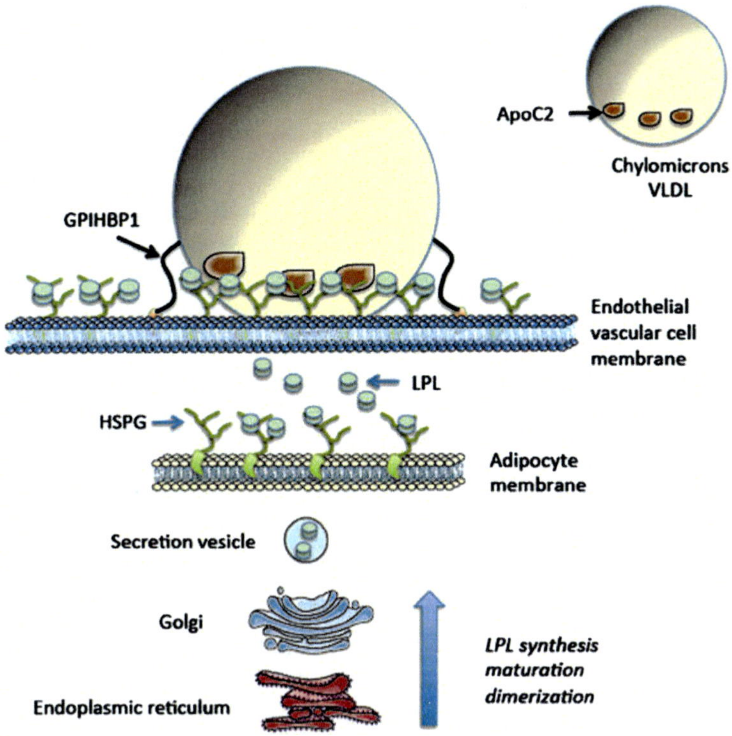

Fig. 8.1 Lipoprotein lipase and the hydrolysis of lipoprotein triglycerides. Lipoprotein lipase (LPL) is synthesized in the rough endoplasmic reticulum and further modified in the Golgi apparatus where it dimerizes. It is secreted and temporarily stored on heparin sulfate proteoglycan (HSPG) on the adipocyte cell membrane. It is then exported to the endothelial vascular cell membrane where it binds to HSPG. The protein glycosylphosphatidylinositol-anchored high density lipoprotein binding protein 1 (GPIHBP1) could serve as a platform allowing the interaction between LPL and TRL. The apoprotein ApoC2 on mature lipoproteins is an obligatory activator of LPL

The mature LPL is then addressed to the parenchymal cell surface where it binds to heparan sulfate proteoglycans (HSPG). The glycosaminoglycan (heparan sulfate) part of HSPG is a highly negatively charged molecule which can interact with positively charged proteins such as LPL. Finally LPL is translocated by a still unclear mechanism to the luminal surface of capillary endothelium where it binds to HSPG (Fig. 8.1). A novel protein has been involved in the localization of LPL at the luminal side of endothelial cells. This protein called glycosylphosphatidylinositol-anchored high density lipoprotein binding protein 1 (GPIHBP1) could serve as a scaffolding platform allowing the interaction between LPL and TRL (Dallinga-Thie et al. 2007). TRL can then interact with LPL and HSPG and/or GPIHBP1 (Fig. 8.1).

Regulation of LPL in Adipose Tissue

At a transcriptional level, LPL gene promoter is activated in adipose tissue by transcription factors, Sterol regulatory element binding protein (SREBP) 1 and 2 (see the paragraph on lipogenesis for further details on SREBP-1), and Peroxisome Proliferator-Activated Receptor (PPAR) γ (Schoonjans et al. 1996; Schoonjans et al. 2000). At a post-translational level, a membrane-bound protein of the endoplasmic reticulum called Lipase Maturation Factor 1 (LMF1), is necessary for the maturation of catalytically active LPL (Peterfy et al. 2007). Once bound to the endothelial wall, LPL activity can still be regulated through association with proteins. ApoC2, which is present on the surface of mature CM and VLDL is an essential cofactor/activator of LPL (Connelly et al. 1987). In addition, two secreted proteins of the angiopoietin-like (Angptl) family, Angptl3, and Angptl4 have been shown to inhibit LPL activity. Angptl4 binds to the active LPL dimer and favors the dissociation into less active monomers (Sukonina et al. 2006). The inhibitory mechanism for Angptl3 is less clear.

Interestingly, the expression of Angptl4 is under nutritional control. Under fasting conditions, transcription of Angptl4 in adipose tissue is increased (it was previously named FIAF for fasting-induced adipose factor).

LPL activity is increased in adipose tissue in the fed state and decreased during fasting whereas the converse is observed in muscles and heart. In adipose tissue, insulin increases LPL mRNA (probably through a stabilization mechanism), LPL activity, and secretion. In contrast, catecholamines decrease LPL transcription and activity.

In summary, during prandial states LPL is actively hydrolyzing triglycerides from TRL in adipose tissues and insulin has an important role in these activations. In contrast in post-absorptive or fasting periods, LPL activity is decreased due to the decreased insulin/increased catecholamine and Angptl4 concentrations. At the same time LPL activity in oxidative tissues (skeletal muscles, heart) is increased. LPL has thus a major role in the tissue partitioning of fat fuels according to the nutritional state. The reader can find further details on LPL in the following reviews (Fielding and Frayn 1998; Wang and Eckel 2009; Preiss-Landl et al. 2002).

Fatty Acid Transport in Adipose Tissue

Once liberated from lipoproteins, fatty acids and 2-monoacylglycerols must enter into adipocytes. Although fatty acids are lipophilic and could theoretically passively diffuse through the plasma membrane, it has been shown that this process would be too slow due to the fact that free fatty acid concentrations are extremely low in the presence of albumin. It is now well established that a protein-mediated saturable uptake of fatty acids exists in adipose tissue and a number of proteins have been involved in this process, including fatty acid transport proteins (FATP), the human scavenger receptor CD36 (or rodent fatty acid translocase), and plasma membrane fatty acid binding protein (FABPpm).

FATPs are a family of six transport proteins (Kazantzis and Stahl 2011; Gimeno 2007). They are integral membrane proteins with an extracellular/luminal N-terminal and a cytosolic C-terminal domain and their expression varies according to the tissue with FATP1 being the most expressed in adipose tissue. Interestingly, FATP1 translocates from an intracellular location to the plasma membrane in response to insulin concomitantly with an increased inwards flux of fatty acids (Wu et al. 2006). FATP1 deletion in mice confirms its potential role as a major fatty acid transporter in adipose tissue. FATP3 and 4 could be the members expressed in the endothelial capillary cells, allowing the transfer of fatty acids across the capillary to the peri-adipocyte space. The mechanism involved in the transfer by FATPs is not entirely clear since in contrast with transporters of hydrophilic molecules such as the glucose transporter family, their structure is not compatible with a transmembrane channel. Since FATP show 20–40 % sequence identity with long-chain acyl-CoA synthetase, it has been also suggested that they could facilitate fatty acid uptake by transforming incoming fatty acids into their CoA derivative. Indeed a long-chain fatty acyl-CoA synthetase activity was found associated with partially purified FATP1.

The scavenger receptor CD36 is present on many mammalian cell types, such as macrophages, myocytes, hepatocytes, adipocytes, enterocytes, and endothelial cells (Hajri and Abumrad 2002; Silverstein and Febbraio 2009). It is an integral membrane protein with an heavily glycosylated extracellular domain, two trans-membrane domains and two short intracellular domains. Scavenger receptors were initially involved in the recognition and elimination of foreign organisms. CD36 recognizes specific lipid and lipoprotein components of pathogens or of endogenous structures (e.g. oxidized lipoproteins). It has pleiotropic effects depending upon the cells in which it is expressed, such as inhibition of angiogenesis in endothelial cells, promotion of endocytosis in macrophages, and activation of platelets. CD36 is a marker of preadipocyte differentiation into adipocytes. In mature adipocytes as well as in muscle cells, CD36 facilitates fatty acid transport by a mechanism which remains elusive. In CD36 null mice, fatty acid uptake is reduced by 60 % in adipocytes (Coburn et al. 2000).

FABPpm is expressed in numerous tissues, adipose tissue, endothelial cells, liver, heart, skeletal muscles, and intestine (Hajri and Abumrad 2002). In contrast with FATPs and CD36, FABPpm is not a transmembrane protein but is associated to the plasma membrane. Paradoxically, it is undistinguishable from mitochondrial aspartate aminotransferase and this has raised questions concerning the identity and real role of FABPpm. Some experiments nevertheless tend to confirm the reality of FABPpm involvement in fatty acid transport. An antibody against FABPpm partially inhibits fatty acid uptake in adipocytes. Expression of FABPpm in Xenopus laevis oocytes and in fibroblasts induces an increase in fatty acid uptake. Like CD36, FABPpm expression is increased during adipocyte differentiation.

In summary, a number of fatty acid transporters have been described in adipocytes, which can account for the high transport activity although their respective importance and the mechanisms involved are not yet clear.

Intracellular Fatty Acid Transport

Intracellular fatty acids are transported in the cytoplasm to various compartments by specific proteins called cytoplasmic fatty acid-binding proteins (FABPs). FABPs are a family of small proteins (14–15 Kd) that bind lipophilic ligands such as saturated and unsaturated long-chain fatty acids or eicosanoids with high affinity inside an hydrophobic pocket formed by two orthogonal five-stranded β-sheets (Furuhashi and Hotamisligil 2008; Storch and McDermott 2009). The preferential expression of specific forms of FABPs has been described in various tissues, such as intestine, liver, adipose tissue, brain heart, skin, etc. In adipocytes, the major FABP expressed is FABP4 (or A-FABP), which is also expressed in macrophages and dendritic cells. It is also known as adipocyte P2 (aP2) for its similarity with peripheral myelin protein 2. A-FABP expression is considerably increased during adipocyte differentiation (Hunt et al. 1986). In the mature adipocyte its expression is enhanced by fatty acids (Amri et al. 1991). It is one of the most abundant cytoplasmic protein. Mice deleted for FABP4 do not show obvious phenotypes but this can be due to the compensatory expression of another FABP (FABP5) which is normally a minor form in adipocytes. It is interesting to note that although the primary structure of FABP4 does not show any nuclear localization signal (NLS), a signal can be found when considering the 3D structure which would be rearranged in the presence of specific ligands. This would allow the transfer of these ligands in the nucleus allowing for instance the activation of nuclear receptors such as PPARs (Furuhashi and Hotamisligil 2008). In addition, a physical association has been found between FABP4 and hormone sensitive lipase possibly allowing FABP4 to rapidly remove and channel fatty acids liberated from triglycerides by the action of the lipase (Furuhashi and Hotamisligil 2008).

De Novo Synthesis of Fatty Acids in the Adipocyte: The Lipogenic Pathway

In rodents, lipogenesis (de novo synthesis of fatty acids) is one of the major metabolic pathway leading to triglyceride accumulation in adipose tissue. Lipogenesis metabolizes an excess of glucose into fatty acids which are then esterified in triglycerides and stored in the lipid droplet of the adipocytes (Fig. 8.2). In rodents, lipogenesis is particularly active in the liver and in white adipose tissue and its activity is highly dependent on the nutritional status of the animal. The

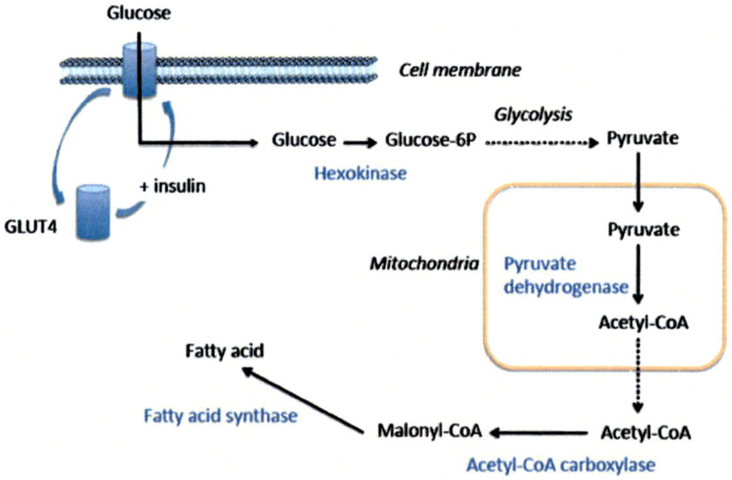

Fig. 8.2 Scheme of adipocyte lipogenesis. In the presence of insulin, the glucose transporter GLUT4 migrates at the cell membrane. Glucose transport is stimulated and glucose enters into the adipocyte. After phosphorylation into glucose-6-phosphate, it enters glycolysis to produce pyruvate. Pyruvate is transported inside the mitochondria where it is converted to acetyl-CoA by pyruvate dehydrogenase. Acetyl-CoA leaves the mitochondria by a shuttle system. It is then transformed into the cytoplasm in malonyl-CoA by acetyl-CoA carboxylase. Malonyl-CoA is the substrate for fatty acid synthase that produces fatty acids

lipogenic activity is low in adipose tissue of fasted rodents or in rodents fed a high fat diet (Ballard 1972). Conversely, lipogenesis is very active in the adipose tissue of animals fed a diet enriched in carbohydrates. Regulation of the lipogenic activity involves both short-term (changes of enzymes activity) and long-term mechanisms (modification of the transcription rate).

Short-Term Control of Lipogenic Activity

As underlined before, glucose is the main lipogenic substrate in rodents and humans. The entry of glucose into the adipocyte is thus the first regulated step in the short-term control of lipogenesis. In adipose tissue and skeletal muscle, glucose enters into the cell by the GLUT4 glucose transporter which has the particularity to re-localize from an intracellular compartment to the plasma membrane in response to insulin (see chap. 7). In adipocytes, a major role of insulin will be therefore to increase glucose transport by inducing the translocation of GLUT4 to the plasma membrane.

The activity of the lipogenic enzymes can also be modified by phosphorylation/dephosphorylation events allowing acute control of the lipogenic flux. Hormones of which concentrations are modified by the nutritional status (insulin, catecholamines) play a key role in this process.

The pyruvate dehydrogenase (PDH) complex and acetyl CoA carboxylase (ACC) are two important lipogenic enzymes. PDH catalyzes the oxidative decarboxylation of pyruvate to acetyl-CoA. PDH complex is mainly controlled by phosphorylation/dephosphorylation mechanisms. Phosphorylation of the E1 subunit of the PDH complex is catalyzed by PDH kinases (PDKs), leading to its inactivation (Peters 2003). There are four different isoforms of PDK (PDK1-4). PDK 4 is the major isoform expressed in white adipose tissue. PDK4 expression is activated in adipose tissue by fasting, ß adrenergic hormones, and exercise (Wan et al. 2010). Conversely, insulin activates PDH by dephosphorylating multiple sites in the E1 subunit. Insulin acts by stimulating a phosphatase activity rather than by inhibiting PDK4 (Denton and Brownsey 1983). ACC, which catalyzes the carboxylation of acetyl-CoA to malonyl-CoA is phosphorylated on many serine residues by at least seven different protein kinases (Kim 1997). Several hormones of which concentrations change with nutritional status (insulin and beta adrenergic hormones) can lead to ACC phosphorylation in isolated adipocytes. However, the consequences of these phosphorylations on the in vivo activity of ACC and on the lipogenic activity are not always clear. More recently, the AMP-activated protein kinase (AMPK) has emerged as a major protein controlling ACC activity in different tissues. ACC has been the first AMPK substrate identified (Winder et al. 1997). AMPK is considered as the sensor of intracellular levels of energy. In stress conditions, such as fasting or exercise, AMPK phosphorylates ACC leading to its inactivation (Kahn et al. 2005). This mechanism allows to inhibit the lipogenic pathway, which consumes energy, and to activate ß oxidation that produces energy (the product of the reaction catalyzed by ACC, malonyl-CoA is an inhibitor of the fatty acid oxidation).

Transcriptional Control of Lipogenesis

An important regulation of the lipogenic pathway occurs at the level of transcription. This transcriptional mechanism is very important since some enzymes of the pathway are controlled exclusively at a transcriptional level, such as fatty acid synthase (FAS) for example. It has been clearly shown that a high carbohydrate diet increases the transcription of genes encoding enzymes of lipogenic and esterification pathways, such as ACC, FAS, DGAT, or GPAT (Foufelle and Ferre 2002). Studies in vitro in different models, white adipose tissue explants, adipocytes in primary culture, or 3T3-L1 cells, have shown that insulin and glucose are the main drivers of the expression of enzymes of lipogenesis (Foufelle et al. 1992; Moustaid et al. 1994). Glucose and insulin act synergistically to stimulate the expression of lipogenic enzymes and their simultaneous presence is essential for the maximal induction of the expression of these enzymes. The effects of insulin are transmitted by the transcription factor SREBP-1c and the effects of glucose by the transcription factor ChREBP (Carbohydrate Responsive Element Binding Protein) (Foufelle and Ferre 2002; Postic et al. 2007). Although most of the work on the molecular control of lipogenesis has been performed in hepatocytes, these transcription factors also appear to play an important role in the control of lipogenesis in the adipose tissue. SREBP-1c is a transcription factor of the b-HLH/LZ (basic Helix-Loop helix/leucine zipper) family, which is produced by an alternative splicing of the SREBP-1 gene. In rodents, adipose tissue is the second tissue after the liver in terms of SREBP-1c expression (Shimomura et al. 1997). SREBP-1c was first identified as a key factor in adipocyte differentiation (SREBP-1c is also called ADD1 for Adipocyte Differentiation and Determination) (Tontonoz et al. 1993). The expression of a dominant negative form of SREBP-1c in 3T3-L1 preadipocytes significantly reduces adipogenesis (Kim and Spiegelman 1996). Conversely, the presence of a constitutively active form of SREBP-1c leads to NIH-3T3 preadipocytes differentiation as long as they are cultured under conditions permissive for differentiation (presence of a PPARγ ligand) (Kim and Spiegelman 1996). Finally, it was shown that ADD1/SREBP-1c has an important role in adipocyte differentiation by producing an endogenous ligand for PPARγ (Kim et al. 1998a). As in the liver, SREBP-1c expression is induced in adipose tissue by refeeding mice with a carbohydrate-enriched diet or by insulin treatment in 3T3-L1 and in primary cultured adipocytes (Kim et al. 1998b). The expression of a dominant negative form of SREBP-1c precludes the stimulatory effect of insulin on the expression of lipogenic genes demonstrating that in vitro SREBP-1c is absolutely necessary for the control of adipocyte lipogenesis (Le Lay et al. 2002). The role of SREBP-1c in the in vivo control of adipocyte lipogenesis is more controversial. Several groups have reported that the expression of SREBP-1c does not vary according to the expression of lipogenic enzymes in adipose tissue of rats fasted or fed ad libitum (Letexier et al. 2003; Bertile and Raclot 2004). Experiments of genetic manipulations of SREBP-1c in adipose tissue have also produced some surprising results. Indeed, while the invalidation of the SREBP-1 gene (this gene encodes two isoforms, SREBP-1c and SREBP-1a) strongly decreases hepatic lipogenesis, lipogenic enzyme expression is not affected in adipose tissue, which has a normal weight (Shimano et al. 1997; Sekiya et al. 2007). Overexpression of SREBP-1c in adipocytes using an aP2 promoter leads to a phenotype completely unexpected since this mouse develops a severe lipodystrophy (Shimomura et al. 1998). This phenotype is due partly to a problem of adipocyte differentiation with an increase in preadipocyte markers (PREF1) and a decrease in mature adipocyte markers such as C/EBPα and PPARγ. Interpretation of the role of SREBP-1c is further complicated since the overexpression of SREBP-1a, (which activates the same spectrum of genes that SREBP-1c) leads to the expected phenotype i.e. adipocyte hypertrophy due to activation of adipocyte lipogenesis (Horton et al. 2003). In conclusion, while the role of SREBP-1c in the molecular control of hepatic lipogenesis and in adipocyte differentiation is well established, its role in the control of adipocyte lipogenesis appears controversial.

The transcription factor ChREBP acts in synergy with SREBP-1c to control lipogenic enzyme expression (Postic et al. 2007). ChREBP is a transcription factor of the b-HLH/LZ family of which transcriptional activity is induced by high glucose concentrations (Yamashita et al. 2001). In the liver, it has been shown that glucose increases ChREBP mRNA expression, its nuclear translocation and binding to promoters of its target genes. Only a few studies have addressed the role of ChREBP in the control of lipogenesis in the adipose tissue. ChREBP is expressed in white adipose tissue in rodents and its expression is stimulated by refeeding mice with carbohydrates. In 3T3-L1 adipocytes, ChREBP expression is increased by glucose and insulin and inhibited by fatty acids (He et al. 2004). Global invalidation of ChREBP leads to a decrease in hepatic lipogenesis (Iizuka et al. 2004). These mice also present smaller adipose tissue fat pads but it is unclear whether this is due directly to a reduced adipose tissue lipogenesis or indirectly to a reduced hepatic lipogenesis. Interestingly, crossing ChREBP-/- mice with ob/ob mice causes a reduction in the size of fat depots in knockout mice compared with ob/ob mice (Iizuka et al. 2006).

Lipogenesis in Human Adipose Tissue

The contribution of de novo synthesis of fatty acids in the accumulation of triglycerides in human adipocytes is very controversial. It was previously established that lipogenesis is nearly absent in human adipose tissue (Shrago et al. 1969; Patel et al. 1975; Galton 1968). More sophisticated measures using radioactive tracers have shown that there is indeed a lipogenic capacity in human white adipose tissue but much lower than in the liver (Diraison et al. 2003). Since lipogenesis is induced by carbohydrate-rich diets, several groups have studied how the lipogenic flux changes in the liver and adipose tissue of individuals eating such a diet for a few days. The results are contradictory. Indeed, some studies show that de novo lipogenesis is more active in liver than in adipose tissue and that its activity is not stimulated by a high carbohydrate diet in the latter (Diraison et al. 2003). Other studies show instead that over-eating carbohydrates in healthy individuals leads to an activation of the whole body lipogenesis and that this increase can only be explained by an increased lipogenesis in adipose tissue (Acheson et al. 1988; Aarsland et al. 1997). The results on the expression of lipogenic enzymes are also controversial as Minehira et al. (2004) reported an increased expression of SREBP-1c and FAS in adipose tissue of subjects fed a high-carbohydrate diet (Minehira et al. 2003) while two other studies have shown that the expression of these two genes is not affected (Letexier et al. 2003; Diraison et al. 2002). However, it is important to underline that human adipocytes express all enzymes and transcription factors required for de novo synthesis of fatty acids. Indeed, human preadipocytes are able to differentiate into mature adipocytes in the absence of any exogenous source of fatty acids in the culture medium (Hauner et al. 2001). In this cell culture model, the expression of SREBP-1c and of lipogenic enzymes is induced in the presence of insulin and glucose.

It is also important to point out that the de novo synthesis of fatty acids is much lower in adipose tissue of obese individuals compared to healthy subjects suggesting that lipogenesis contributes weakly to triglycerides accumulation in obesity (Diraison et al. 2002; Minehira et al. 2004). The group of K. Frayn has shown that within the same fat depot of a healthy subject, the expression of lipogenic enzymes as well as end-products of lipogenesis (palmitic acid, stearic acid) are inversely correlated with adipocyte size (Roberts et al. 2009). These authors suggest that small adipocytes use de novo lipogenesis to begin the process of lipid accumulation, with pathways for uptake of extracellular fatty acids becoming more important as the cells develop. In support of this hypothesis, it has indeed been shown that the lipogenic capacity of human fetal preadipocytes was very high (Dunlop and Court 1978).

In conclusion, the role of lipogenesis in the storage of triglycerides in human adipocytes is still debated although as underlined above, the human adipocyte possesses all the enzymes necessary for de novo synthesis of fatty acids.

Fatty Acid Activation into Acyl-CoA

In order to be metabolized either for triglyceride synthesis, phospholipid remodeling or oxidation, fatty acids must be transformed into acyl-CoA. In addition to the FATP family (see above), another family of fatty acid activating enzymes, acyl-CoA synthetase (ACS) has been characterized (Ellis et al. 2010). Five different isoforms of ACS were identified which are expressed differentially according to the tissue considered. They can activate fatty acids with chain lengths from 12 to 20 carbons and their structure predicts that they are membrane proteins. In adipocytes, ACS1 is the major ACS isoform. It has been localized in various subcellular compartments, plasma membranes, lipid droplets, microsomes, mitochondrial membranes, and even GLUT4 containing vesicles (Soupene and Kuypers 2008). ACS1 expression is dramatically increased during adipocyte differentiation and is a target of PPARγ (Tontonoz et al. 1995). Invalidation of ACS1 in the adipocyte cell line 3T3-L1 did not affect long chain fatty acid uptake, triglyceride concentration, and lipid droplet size but enhances the lipolytic rate suggesting that an important role of ACS1 could be to re-esterify fatty acids arising from lipolysis (Lobo et al. 2009). The authors of this study thus suggested that FATP could be the major fatty acid synthetase for fatty acids coming from the plasma. This remains to be confirmed in adipocytes from adipose tissue.

Triglyceride Synthesis

Once activated into acyl-CoA, fatty acids can be stored in the form of triglycerides. Two synthetic pathways are used in cells to synthesize triglycerides (Fig. 8.3). The first one uses monoacylglycerol as a starting point. It begins with the acylation of monoacylglycerol with an acyl-CoA by a monoacylglycerol transferase. This pathway is predominant in enterocytes. It must be underlined that LPL produces monoacylglycerol molecules which can be transported inside adipocytes and are usually converted there into glycerol and fatty acids by a monoacylglycerol lipase (Fredrikson et al. 1986). Thus, the main biosynthetic route in adipocytes involves as a starting point glycerol-3-phosphate and was first described by E. Kennedy and colleagues in the 1950s. This pathway starts with the esterification of a first acyl-CoA on a molecule of glycerol-3-phosphate to form lysophosphatidic acid, then of a second acyl-CoA to form phosphatidic acid. Phosphatidic acid is dephosphorylated to form a diacylglycerol and the third acyl-CoA is then added to yield a triglyceride (Fig. 8.3). We will review successively the enzymes involved as well as the pathways allowing the production of glycerol-3-phosphate. This topic cannot be fully detailed due to the huge amount of data generated on this pathway and we have tried to be as synthetic as possible. It must be pointed out that in addition to their role as intermediates in the synthesis of triglycerides, lysophosphatidic acid, phosphatidic acid, and diacylglycerol have cellular functions as structural molecules, second messenger or precursor of glycerophospholipids, such as phosphatidylinositol, phosphatidylcholine, phosphatidylserine, cardiolipin, phosphatidylethanolamine which are all important components of cellular membranes. This aspect will not be developed here.

Glycerol-3-phosphate Acyltransferase

The first step in triglyceride synthesis is catalyzed by Glycerol-3-phosphate acyltransferases (GPAT) and is the acylation of sn-glycerol-3-phosphate (sn for stereospecific numbering due to the presence of the chiral carbon 2) to form lysophosphatidic acid or 1-acyl-sn-glycerol-3-phosphate. Four isoforms of GPAT have been cloned. GPAT1 and 2 are localized at the outer mitochondrial membrane whereas GPAT3 and 4 are localized in the endoplasmic reticulum (Takeuchi and Reue 2009; Wendel et al. 2009). The four isoforms have putative transmembrane domain and the transmembrane structure has been confirmed for the GPAT1 isoform. Their respective importance in adipose tissue triglyceride synthesis is not clear. The highest expression of GPTA1 is observed in liver and adipose tissue. However, in the latter it does not represent more than 10 % of total GPAT activity. Although localized on the outer mitochondrial membrane it could nevertheless contribute to triglyceride synthesis as shown in the liver. GPTA1 has a preference for saturated fatty acids. GPAT1 mRNA increases 10-fold during

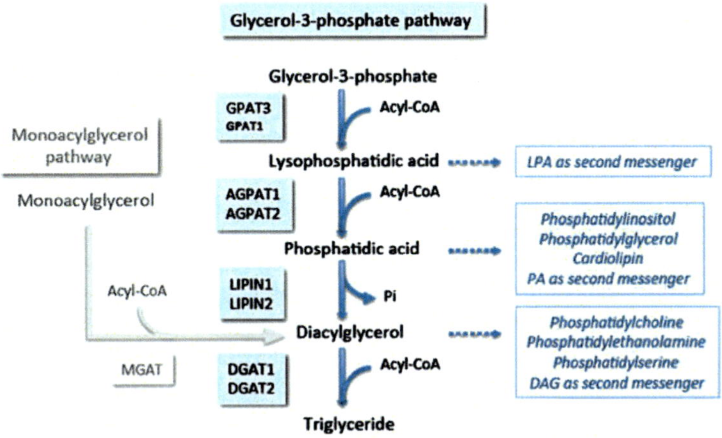

Fig. 8.3 Triglyceride synthesis pathway in adipocytes. In adipocytes, the glycerol-3-phosphate pathway is predominant over the monoacylglycerol pathway whereas it is the converse in enterocytes. The most relevant isoforms of the different enzymes are indicated. It is underlined that intermediates of the pathway are precursors of other lipid species or can serve as second cellular messenger. AGPAT, 1-acylglycerol-3-phosphate O-acyltransferase; DAG, diacylglycerol; DGAT, diacylglycerol acyl transferase; GPAT, Glycerol-3-phosphate acyltransferase; MGAT, monoacylglycerol transferase; LIPIN, lipin or phosphatidic acid phosphatase; LPA, lysophosphatidic acid; PA, phosphatidic acid

adipose tissue differentiation and its expression is upregulated by insulin probably through the activation of the transcription factor SREBP-1c (Takeuchi and Reue 2009). The mitochondrial isoform GPAT2 is poorly expressed in adipose tissue.

GPAT3 is localized to the endoplasmic reticulum and is highly expressed in rodent adipose tissue but to a lower level in human adipose tissue. It has a broad substrate specificity for long-chain fatty acyl-CoAs since it accommodates saturated and unsaturated fatty acids. Its expression increases 60-fold during the differentiation of 3T3-L1 cells into adipocytes. It is also increased in the adipose tissue of ob/ob mice by a PPAR γ ligand (thiazolidinedione). Interestingly, a decreased expression of GPAT3 in 3T3-L1 adipocyte using siRNA induces a large decrease in the incorporation of fatty acids into lysophosphatidic acid (Takeuchi and Reue 2009). GPAT4 is also localized to the endoplasmic reticulum but its quantitative role in triglyceride synthesis is questionable in adipose tissue.

In summary, GPAT3 and to a lower extent GPAT1 could represent the main enzymes involved in the synthesis of lysophosphatidic acid.

1-acylglycerol-3-phosphate O-acyltransferase

1-acylglycerol 3-phosphate O-acyltransferase (AGPAT) adds an acyl-CoA to the sn-2 position of 1-acylglycerol-3-phosphate and converts lysophosphatidic acid into phosphatidic acid. Two main isoforms have been described and characterized

(AGPAT1 and 2) although a number of other isoforms exist but probably with lower specificity for lysophosphatidic acid (Takeuchi and Reue 2009). AGPAT1 and 2 are transmembrane proteins localized in the endoplasmic reticulum. AGPAT1 is expressed in most tissues including adipose tissue and its overexpression in 3T3-L1 adipocytes increases oleate incorporation into phosphatidic acid.

AGPAT2 is clearly involved in triglyceride synthesis in human adipose tissue. Mutations in its gene are responsible for one type of the Berardinelli-Seip congenital generalized lipodystrophy characterized by an absence of adipose tissue present from birth or occurring in early infancy, an hypertriglyceridemia, an hepatic steatosis and a severe insulin resistance (Agarwal et al. 2002; Magre et al. 2003). Interestingly, mechanic adipose tissues (e.g. retro-orbital, periarticular) are not affected suggesting that other AGPAT isoforms can be involved in these specialized adipose tissues (Simha and Garg 2003). A model of AGPAT2 knockout mice has been developed which confirms the findings in humans. One obvious question is the reason why an hepatic steatosis (accumulation of triglycerides) is nevertheless present despite the fact that residual hepatic AGPAT activity is only 10 % in AGPAT2 knockout mice. In fact in AGPAT2 knockout mice there is a huge induction of monoacylglycerol acyl transferase 1 activity suggesting that a monoacylglycerol pathway similar to that present in intestine could compensate for the AGPAT2 deficiency (Cortes et al. 2009).

Phosphatidate Phosphohydrolase (Phosphatidic Acid Phosphatase)

The next step is then the Mg^{2+}-dependent dephosphorylation of phosphatidic acid (PA) to form diacylglycerols. The phosphatidic acid phosphatase activity is achieved by a family of proteins called lipins with three isoforms (Takeuchi and Reue 2009). Lipins are cytoplasmic proteins and they need to translocate to the endoplasmic reticulum since their substrate, phosphatidic acid is an insoluble lipid found in membranes. In adipose tissue, lipin 1 is the most important isoform. An alternative splicing leads to three proteins with only two possessing a phosphatase activity. Lipin 1 alpha can be found in the nucleus whereas lipin 1 beta is cytoplasmic. The role of lipin 1 in triglyceride synthesis was demonstrated in rodents since its gene was initially identified as responsible for the BALB/cByJ-fld mouse phenotype (fatty liver dystrophy) leading to a severe lipodystrophy and fatty liver (Peterfy et al. 2001) (in the liver the phosphatidic acid phosphatase activity is only partially reduced due probably to the presence of lipin 2). Conversely, specific overexpression of lipin 1 in the adipose tissue in mice leads to a marked increase in triglyceride storage (Phan and Reue 2005). In humans, mutations affecting both lipin 1 and 2 have been described. They are not concomitant with lipodystrophy but with rhabdomyolysis (muscle disease) for the former (Zeharia et al. 2008) and an inflammatory disorder, the Majeed syndrome, for the latter (Al-Mosawi et al. 2007). In fact, it has been suggested that an accumulation of phosphatidic acid (the substrate of lipins) could be responsible for the phenotypes due to its numerous roles in cell signaling pathways.

Lipin 1 mRNA is strongly induced during 3T3–L1 adipocyte differentiation (Peterfy et al. 2001). It is also regulated positively by glucocorticoids in adipocytes and negatively by proinflammatory cytokines. Translocation of lipin 1 from the cytosol to its site of action is stimulated in hepatocytes by fatty acids (specially unsaturated fatty acids) by an unknown mechanism (Gomez-Munoz et al. 1992). Paradoxically insulin has the reverse effect possibly through an mTOR-mediated phosphorylation (see also below) but this needs to be confirmed. The exact mechanism for the translocation of lipin 1 in adipocytes thus remains to be elucidated. The role of Lipin 3 which is expressed at low levels in visceral tissues is probably minor in adipose tissue.

As mentioned above, lipin 1 is also found in the nucleus and a role for lipins as co-activators/repressors of transcription factors has been described. Lipin 1 increases the transcriptional activity of PPARγ 2 via a direct physical interaction (Finck et al. 2006). A decreased expression of lipin using siRNA impedes the differentiation of 3T3–L1 preadipocytes (Koh et al. 2008). A recent paper (Peterson et al. 2011) suggests that when phosphorylated in an mTOR-dependent way, lipin 1 is excluded from the nucleus. This favors the expression of lipogenic enzymes since lipin 1 represses SREBP1c transcriptional activity.

Diacylglycerol Acyltransferase

In the final reaction of the pathway, a diacylglycerol and a fatty acyl-CoA molecule are condensed at the sn-3 position to form a triglyceride. This reaction catalyzed by a diacylglycerol acyl transferase (DGAT) is common to both monoacylglycerol and glycerol-3-phosphate pathways. A DGAT activity is found essentially in the endoplasmic reticulum. The most common view is that the triglycerides synthesized are released in the membrane lipid bilayer where they will participate in the formation of lipid droplets (see chap. 9).

Two non-homologous genes have been described, DGAT1 and DGAT2 (Yu and Ginsberg 2004; Yen et al. 2008). DGAT1 protein possesses several putative transmembrane domain and works as a tetramer. DGAT 2 has one or two transmembrane domains with an active site on the cytoplasmic side of the endoplasmic reticulum (ER) membrane. DGAT1 and 2 proteins are highly expressed in adipose tissue both in humans and rodents as well as in other tissues with a high rate of triglyceride synthesis, such as liver, intestine, and lactating mammary gland. DGAT1 and 2, although both at the ER membrane have different locations in the cell. It has been suggested (but not fully demonstrated) that DGAT2 acts on products of the lipogenic pathway, whereas DGAT1 would rather esterify exogenous fatty acids (Yen et al. 2008). In addition, it has been suggested from competition assays that DGAT 2 has a substrate preference for monounsaturated substrate when compared to saturated ones, whereas it is not the case for DGAT1 (Yen et al. 2008).

DGAT1 and 2 mRNA strongly increase during 3T3-L1 preadipocyte differentiation and DGAT1 expression is increased in adipocytes by a PPARγ agonist. In addition, insulin regulates positively DGAT1 expression in 3T3-L1 adipocytes and glucose increases the expression of both DGAT1 and 2 in the same model (Yen et al. 2008). Finally, acylation stimulating protein (ASP) a protein produced by adipocytes, released upon activation of the complement pathway by interaction of complement factor C3 with factor B and adipsin, increases the DGAT activity in adipose tissue (Yasruel et al. 1991; Saleh et al. 2011).

To our knowledge, there is no adipose tissue-specific deletion of DGATs. Increasing the expression of DGAT1 specifically in adipose tissue in mice using an aP2 promoter leads to higher fat pad weights and larger adipocytes.

In summary, enzymes involved in triglyceride synthesis have for most of them several isoforms with slightly different biochemical properties, expression, regulation, and localization (tissue and subcellular). This is probably the consequence of the fact that they act on a large number of different substrates (the different fatty acids found in triglycerides) and that the Kennedy pathway also feeds other important biochemical routes such as phospholipid synthesis and lipid second messenger production (diacylglycerol and lysophosphatidic acid for instance) (Fig. 8.3).

Availability of Glycerol-3-phosphate

Since there is a negligible glycerol kinase activity in adipose tissue, it was accepted that in adipocytes glycerol-3-phosphate comes from glucose metabolized in glycolysis through the reduction of the glycolytic intermediate dihydroxyacetone phosphate. However, an alternative pathway has emerged from the initial finding that some enzymes of the liver gluconeogenic pathway, namely pyruvate carboxylase and phosphoenolpyruvate carboxykinase (PEPCK) were also present in adipose tissue (Ballard et al. 1967) and that pyruvate could be incorporated into glycerol-3-phosphate. This pathway, which utilizes the first steps of gluconeogenesis was named glyceroneogenesis. In addition to liver cells, PEPCK was found expressed in all cells with a high fatty acid esterification activity, such as adipocytes, enterocytes, and lactating mammary gland. Glyceroneogenesis was initially seen as allowing to provide glycerol-3-phosphate in fasting conditions since part of the fatty acids released by the lipolytic process are re-esterified at a time (fasting) when glucose availability and transport into the adipocytes are low (Nye et al. 2008). It was later demonstrated that even when mice are fed with a high carbohydrate diet, about 30 % of glycerol-3-phosphate arises from glyceroneogenesis, a percentage which rises to 70 % when mice are fed with a high fat diet (Bederman et al. 2009). The PEPCK promoter contains a response element for PPAR gamma and is activated by thiazolidinediones in adipose tissue (Tontonoz et al. 1995; Hallakou et al. 1997). Finally, overexpression of PEPCK in mouse adipose tissue induces an increase in fatty acid esterification, adipocyte size, and fat mass (Franckhauser et al. 2002).

Conclusion

Lipid storage in adipocytes is essential for energy homeostasis and has probably played a major role in the survival of our species. As described in this chapter, it involves complex mechanisms, which for some of them are also implicated in other metabolic pathways.

It is tempting to pharmacologically target one or several steps of triglyceride synthesis to modulate fat accretion in the context of the increasing prevalence of obesity. However, since the adipocyte is the only cell really specialized in fat storage this may have dramatic consequences by forwarding fatty acids toward other tissues and triggering lipotoxicity (as seen for instance in lipoatrophic disorders linked to DGAT mutation). In addition, this may strongly affect the concentration of metabolic lipid intermediates themselves involved in side pathways including cell signaling. This is perfectly illustrated, as described above, by mutations in lipins which induce rhabdomyolysis or inflammatory syndrome.

References

Aarsland A, Chinkes D, Wolfe RR (1997) Hepatic and whole-body fat synthesis in humans during carbohydrate overfeeding. Am J Clin Nutr 65:1774–1782

Acheson KJ, Schutz Y, Bessard T et al (1988) Glycogen storage capacity and de novo lipogenesis during massive carbohydrate overfeeding in man. Am J Clin Nutr 48:240–247

Agarwal AK, Arioglu E, De Almeida S et al (2002) AGPAT2 is mutated in congenital generalized lipodystrophy linked to chromosome 9q34. Nat Genet 31:21–23

Al-Mosawi ZS, Al-Saad KK, Ijadi-Maghsoodi R et al (2007) A splice site mutation confirms the role of LPIN2 in Majeed syndrome. Arthritis Rheum 56:960–964

Amri EZ, Bertrand B, Ailhaud G, Grimaldi P (1991) Regulation of adipose cell differentiation I. Fatty acids are inducers of the aP2 gene expression. J Lipid Res 32:1449–1456

Ballard FJ (1972) Effects of fasting and refeeding on the concentrations of glycolytic intermediates and the regulation of lipogenesis in rat adipose tissue in vivo. Biochim Biophys Acta 273:110–118

Ballard FJ, Hanson RW, Leveille GA (1967) Phosphoenolpyruvate carboxykinase and the synthesis of glyceride-glycerol from pyruvate in adipose tissue. J Biol Chem 242:2746–2750

Bederman IR, Foy S, Chandramouli V et al (2009) Triglyceride synthesis in epididymal adipose tissue: contribution of glucose and non-glucose carbon sources. J Biol Chem 284:6101–6108

Bertile F, Raclot T (2004) mRNA levels of SREBP-1c do not coincide with the changes in adipose lipogenic gene expression. Biochem Biophys Res Commun 325:827–834

Braun JE, Severson DL (1992) Regulation of the synthesis, processing and translocation of lipoprotein lipase. Biochem J 287(Pt 2):337–347

Coburn CT, Knapp FF Jr, Febbraio M et al (2000) Defective uptake and utilization of long chain fatty acids in muscle and adipose tissues of CD36 knockout mice. J Biol Chem 275:32523–32529

Connelly PW, Maguire GF, Hofmann T, Little JA (1987) Structure of apolipoprotein C-IIToronto, a nonfunctional human apolipoprotein. Proc Natl Acad Sci U S A 84:270–273

Cortes VA, Curtis DE, Sukumaran S et al (2009) Molecular mechanisms of hepatic steatosis and insulin resistance in the AGPAT2-deficient mouse model of congenital generalized lipodystrophy. Cell Metab 9:165–176

Dallinga-Thie GM, Zonneveld-de Boer AJ, van Vark-van der Zee LC et al (2007) Appraisal of hepatic lipase and lipoprotein lipase activities in mice. J Lipid Res 48:2788–2791

Denton RM, Brownsey RW (1983) The role of phosphorylation in the regulation of fatty acid synthesis by insulin and other hormones. Philos Trans R Soc Lond B Biol Sci 302:33–45

Diraison F, Dusserre E, Vidal H et al (2002) Increased hepatic lipogenesis but decreased expression of lipogenic gene in adipose tissue in human obesity. Am J Physiol Endocrinol Metab 282:E46–E51

Diraison F, Yankah V, Letexier D et al (2003) Differences in the regulation of adipose tissue and liver lipogenesis by carbohydrates in humans. J Lipid Res 44:846–853

Dunlop M, Court JM (1978) Lipogenesis in developing human adipose tissue. Early Hum Dev 2:123–130

Ellis JM, Frahm JL, Li LO, Coleman RA (2010) Acyl-coenzyme A synthetases in metabolic control. Curr Opin Lipidol 21:212–217

Fielding BA, Frayn KN (1998) Lipoprotein lipase and the disposition of dietary fatty acids. Br J Nutr 80:495–502

Finck BN, Gropler MC, Chen Z et al (2006) Lipin 1 is an inducible amplifier of the hepatic PGC-1alpha/PPARalpha regulatory pathway. Cell Metab 4:199–210

Foufelle F, Ferre P (2002) New perspectives in the regulation of hepatic glycolytic and lipogenic genes by insulin and glucose: a role for the transcription factor sterol regulatory element binding protein-1c. Biochem J 366:377–391

Foufelle F, Gouhot B, Pegorier JP et al (1992) Glucose stimulation of lipogenic enzyme gene expression in cultured white adipose tissue. A role for glucose 6-phosphate. J Biol Chem 267:20543–20546

Franckhauser S, Munoz S, Pujol A et al (2002) Increased fatty acid re-esterification by PEPCK overexpression in adipose tissue leads to obesity without insulin resistance. Diabetes 51:624–630

Fredrikson G, Tornqvist H, Belfrage P (1986) Hormone-sensitive lipase and monoacylglycerol lipase are both required for complete degradation of adipocyte triacylglycerol. Biochim Biophys Acta 876:288–293

Furuhashi M, Hotamisligil GS (2008) Fatty acid-binding proteins: role in metabolic diseases and potential as drug targets. Nat Rev Drug Discov 7:489–503

Galton DJ (1968) Lipogenesis in human adipose tissue. J Lipid Res 9:19–26

Gimeno RE (2007) Fatty acid transport proteins. Curr Opin Lipidol 18:271–276

Gomez-Munoz A, Hamza EH, Brindley DN (1992) Effects of sphingosine, albumin and unsaturated fatty acids on the activation and translocation of phosphatidate phosphohydrolases in rat hepatocytes. Biochim Biophys Acta 1127:49–56

Hajri T, Abumrad NA (2002) Fatty acid transport across membranes: relevance to nutrition and metabolic pathology. Annu Rev Nutr 22:383–415

Hallakou S, Doare L, Foufelle F et al (1997) Pioglitazone induces in vivo adipocyte differentiation in the obese Zucker fa/fa rat. Diabetes 46:1393–1399

Hauner H, Skurk T, Wabitsch M (2001) Cultures of human adipose precursor cells. Methods Mol Biol 155:239–247

He Z, Jiang T, Wang Z et al (2004) Modulation of carbohydrate response element-binding protein gene expression in 3T3-L1 adipocytes and rat adipose tissue. Am J Physiol Endocrinol Metab 287:E424–E430

Horton JD, Shimomura I, Ikemoto S et al (2003) Overexpression of sterol regulatory element-binding protein-1a in mouse adipose tissue produces adipocyte hypertrophy, increased fatty acid secretion, and fatty liver. J Biol Chem 278:36652–36660
Hunt CR, Ro JH, Dobson DE et al (1986) Adipocyte P2 gene: developmental expression and homology of 5′-flanking sequences among fat cell-specific genes. Proc Natl Acad Sci U S A 83:3786–3790
Hussain MM (2000) A proposed model for the assembly of chylomicrons. Atherosclerosis 148:1–15
Iizuka K, Bruick RK, Liang G et al (2004) Deficiency of carbohydrate response element-binding protein (ChREBP) reduces lipogenesis as well as glycolysis. Proc Natl Acad Sci U S A 101:7281–7286
Iizuka K, Miller B, Uyeda K (2006) Deficiency of carbohydrate-activated transcription factor ChREBP prevents obesity and improves plasma glucose control in leptin-deficient (ob/ob) mice. Am J Physiol Endocrinol Metab 291:E358–E364
Kahn BB, Alquier T, Carling D, Hardie DG (2005) AMP-activated protein kinase: ancient energy gauge provides clues to modern understanding of metabolism. Cell Metab 1:15–25
Kazantzis M, Stahl A (2011) Fatty acid transport proteins, implications in physiology and disease. Biochim Biophys Acta Sep 25. [Epub ahead of print]
Kim KH (1997) Regulation of mammalian acetyl-coenzyme a carboxylase. Annu Rev Nutr 17:77–99
Kim JB, Spiegelman BM (1996) ADD1/SREBP1 promotes adipocyte differentiation and gene expression linked to fatty acid metabolism. Genes Dev 10:1096–1107
Kim JB, Wright HM, Wright M, Spiegelman BM (1998a) ADD1/SREBP1 activates PPARgamma through the production of endogenous ligand. Proc Natl Acad Sci U S A 95:4333–4337
Kim JB, Sarraf P, Wright M et al (1998b) Nutritional and insulin regulation of fatty acid synthetase and leptin gene expression through ADD1/SREBP1. J Clin Invest 101:1–9
Koh YK, Lee MY, Kim JW et al (2008) Lipin1 is a key factor for the maturation and maintenance of adipocytes in the regulatory network with CCAAT/enhancer-binding protein alpha and peroxisome proliferator-activated receptor gamma 2. J Biol Chem 283:34896–34906
Le Lay S, Lefrere I, Trautwein C et al (2002) Insulin and sterol-regulatory element-binding protein-1c (SREBP-1C) regulation of gene expression in 3T3-L1 adipocytes. Identification of CCAAT/enhancer-binding protein beta as an SREBP-1C target. J Biol Chem 277:35625–35634
Letexier D, Pinteur C, Large V et al (2003) Comparison of the expression and activity of the lipogenic pathway in human and rat adipose tissue. J Lipid Res 44:2127–2134
Lobo S, Wiczer BM, Bernlohr DA (2009) Functional analysis of long-chain acyl-CoA synthetase 1 in 3T3-L1 adipocytes. J Biol Chem 284:18347–18356
Magre J, Delepine M, Van Maldergem L et al (2003) Prevalence of mutations in AGPAT2 among human lipodystrophies. Diabetes 52:1573–1578
Minehira K, Bettschart V, Vidal H et al (2003) Effect of carbohydrate overfeeding on whole body and adipose tissue metabolism in humans. Obes Res 11:1096–1103
Minehira K, Vega N, Vidal H et al (2004) Effect of carbohydrate overfeeding on whole body macronutrient metabolism and expression of lipogenic enzymes in adipose tissue of lean and overweight humans. Int J Obes Relat Metab Disord 28:1291–1298
Moustaid N, Beyer RS, Sul HS (1994) Identification of an insulin response element in the fatty acid synthase promoter. J Biol Chem 269:5629–5634
Nye C, Kim J, Kalhan SC, Hanson RW (2008) Reassessing triglyceride synthesis in adipose tissue. Trends Endocrinol Metab 19:356–361
Patel MS, Owen OE, Goldman LI, Hanson RW (1975) Fatty acid synthesis by human adipose tissue. Metabolism 24:161–173
Peterfy M, Phan J, Xu P, Reue K (2001) Lipodystrophy in the fld mouse results from mutation of a new gene encoding a nuclear protein, lipin. Nat Genet 27:121–124

Peterfy M, Ben-Zeev O, Mao HZ et al (2007) Mutations in LMF1 cause combined lipase deficiency and severe hypertriglyceridemia. Nat Genet 39:1483–1487
Peters SJ (2003) Regulation of PDH activity and isoform expression: diet and exercise. Biochem Soc Trans 31:1274–1280
Peterson TR, Sengupta SS, Harris TE et al (2011) mTOR complex 1 regulates lipin 1 localization to control the SREBP pathway. Cell 146:408–420
Phan J, Reue K (2005) Lipin, a lipodystrophy and obesity gene. Cell Metab 1:73–83
Postic C, Dentin R, Denechaud PD, Girard J (2007) ChREBP, a transcriptional regulator of glucose and lipid metabolism. Annu Rev Nutr 27:179–192
Preiss-Landl K, Zimmermann R, Hammerle G, Zechner R (2002) Lipoprotein lipase: the regulation of tissue specific expression and its role in lipid and energy metabolism. Curr Opin Lipidol 13:471–481
Roberts R, Hodson L, Dennis AL et al (2009) Markers of de novo lipogenesis in adipose tissue: associations with small adipocytes and insulin sensitivity in humans. Diabetologia 52:882–890
Saleh J, Al-Wardy N, Farhan H et al (2011) Acylation stimulating protein: a female lipogenic factor? Obes Rev 12:440–448
Schoonjans K, Peinado-Onsurbe J, Lefebvre AM et al (1996) PPARalpha and PPARgamma activators direct a distinct tissue-specific transcriptional response via a PPRE in the lipoprotein lipase gene. EMBO J 15:5336–5348
Schoonjans K, Gelman L, Haby C et al (2000) Induction of LPL gene expression by sterols is mediated by a sterol regulatory element and is independent of the presence of multiple E boxes. J Mol Biol 304:323–334
Sekiya M, Yahagi N, Matsuzaka T et al (2007) SREBP-1-independent regulation of lipogenic gene expression in adipocytes. J Lipid Res 48:1581–1591
Shimano H, Shimomura I, Hammer RE et al (1997) Elevated levels of SREBP-2 and cholesterol synthesis in livers of mice homozygous for a targeted disruption of the SREBP-1 gene. J Clin Invest 100:2115–2124
Shimomura I, Shimano H, Horton JD et al (1997) Differential expression of exons 1a and 1c in mRNAs for sterol regulatory element binding protein-1 in human and mouse organs and cultured cells. J Clin Invest 99:838–845
Shimomura I, Hammer RE, Richardson JA et al (1998) Insulin resistance and diabetes mellitus in transgenic mice expressing nuclear SREBP-1c in adipose tissue: model for congenital generalized lipodystrophy. Genes Dev 12:3182–3194
Shrago E, Spennetta T, Gordon E (1969) Fatty acid synthesis in human adipose tissue. J Biol Chem 244:2761–2766
Silverstein RL, Febbraio M (2009) CD36, a scavenger receptor involved in immunity, metabolism, angiogenesis, and behavior. Sci Signal 2: re3
Simha V, Garg A (2003) Phenotypic heterogeneity in body fat distribution in patients with congenital generalized lipodystrophy caused by mutations in the AGPAT2 or seipin genes. J Clin Endocrinol Metab 88:5433–5437
Soupene E, Kuypers FA (2008) Mammalian long-chain acyl-CoA synthetases. Exp Biol Med (Maywood) 233:507–521
Storch J, McDermott L (2009) Structural and functional analysis of fatty acid-binding proteins. J Lipid Res 50(Suppl):S126–S131
Sukonina V, Lookene A, Olivecrona T, Olivecrona G (2006) Angiopoietin-like protein 4 converts lipoprotein lipase to inactive monomers and modulates lipase activity in adipose tissue. Proc Natl Acad Sci U S A 103:17450–17455
Takeuchi K, Reue K (2009) Biochemistry, physiology, and genetics of GPAT, AGPAT, and lipin enzymes in triglyceride synthesis. Am J Physiol Endocrinol Metab 296:E1195–E1209
Tontonoz P, Kim JB, Graves RA, Spiegelman BM (1993) ADD1: a novel helix-loop-helix transcription factor associated with adipocyte determination and differentiation. Mol Cell Biol 13:4753–4759

Tontonoz P, Hu E, Spiegelman BM (1995a) Regulation of adipocyte gene expression and differentiation by peroxisome proliferator activated receptor gamma. Curr Opin Genet Dev 5:571–576

Tontonoz P, Hu E, Devine J et al (1995b) PPAR gamma 2 regulates adipose expression of the phosphoenolpyruvate carboxykinase gene. Mol Cell Biol 15:351–357

Wan Z, Thrush AB, Legare M et al (2010) Epinephrine-mediated regulation of PDK4 mRNA in rat adipose tissue. Am J Physiol Cell Physiol 299:C1162–C1170

Wang H, Eckel RH (2009) Lipoprotein lipase: from gene to obesity. Am J Physiol Endocrinol Metab 297:E271–E288

Wendel AA, Lewin TM, Coleman RA (2009) Glycerol-3-phosphate acyltransferases: rate limiting enzymes of triacylglycerol biosynthesis. Biochim Biophys Acta 1791:501–506

Winder WW, Wilson HA, Hardie DG et al (1997) Phosphorylation of rat muscle acetyl-CoA carboxylase by AMP-activated protein kinase and protein kinase A. J Appl Physiol 82:219–225

Wu Q, Ortegon AM, Tsang B et al (2006) FATP1 is an insulin-sensitive fatty acid transporter involved in diet-induced obesity. Mol Cell Biol 26:3455–3467

Yamashita H, Takenoshita M, Sakurai M et al (2001) A glucose-responsive transcription factor that regulates carbohydrate metabolism in the liver. Proc Natl Acad Sci U S A 98:9116–9121

Yasruel Z, Cianflone K, Sniderman AD et al (1991) Effect of acylation stimulating protein on the triacylglycerol synthetic pathway of human adipose tissue. Lipids 26:495–499

Yen CL, Stone SJ, Koliwad S et al (2008) Thematic review series: glycerolipids. DGAT enzymes and triacylglycerol biosynthesis. J Lipid Res 49:2283–2301

Yu YH, Ginsberg HN (2004) The role of acyl-CoA:diacylglycerol acyltransferase (DGAT) in energy metabolism. Ann Med 36:252–261

Zeharia A, Shaag A, Houtkooper RH et al (2008) Mutations in LPIN1 cause recurrent acute myoglobinuria in childhood. Am J Hum Genet 83:489–494

Chapter 9
Adipocyte Lipid Droplet Physiology

Isabelle Dugail and Soizic Le Lay

Introduction

The management of energy is a crucial issue for cells, and the intermittent storage is a fruitful strategy to survive short periods of nutritional scarcity. In this regard, all eukaryotic cells, from yeasts to mammalians, have the ability to accumulate fat and built up lipid droplets when faced with lipid load. Those lipid droplets are filled with neutral lipids, which are the best molecules for energy storage, as totally hydrophobic structures with a fatty acid backbone containing a large number of carbons to be oxidized in mitochondria for ATP production. By analogy with glycogen granules, which also constitute a form of energy storage from glucose, lipid droplets were long considered as inert intracytoplasmic deposits. However, in the last years, our knowledge on the biology of intracellular lipid droplets improved dramatically, and it became apparent that lipid droplets could no longer be considered as simple reservoir of lipids, buffering and sequestering neutral lipids from aqueous compartments but dynamic intracellular organelles. Several recent reviews provide extensive updated information on this subject (Farese and Walther 2009; Murphy et al. 2009; Thiele and Spandl 2008; Goodman 2008). During evolution, the widespread ability to form lipid droplets evolved toward the emergence of a specialized cell type, the adipocyte, whose function is entirely devoted to the management, and packaging of lipid stores. In adipocytes, a striking feature is the

I. Dugail (✉)
Centre de Recherche des Cordeliers, INSERM, U872—Équipe 8 Pathologies métaboliques et nutritionnelles, Diabète, Obésité, 15rue de l'école de médecine, 75006 Paris, France
e-mail: isabelle.dugail@crc.jussieu.fr

S. Le Lay
INSERM U1063, «Stress Oxydatif et pathologies métaboliques»—IBS-IRIS, Rue des Capucins, 49100 Angers, France
e-mail: soazig.lelay@inserm.fr

J.-P. Bastard and B. Fève (eds.), *Physiology and Physiopathology of Adipose Tissue*,
DOI: 10.1007/978-2-8178-0343-2_9, © Springer-Verlag France 2013

over importance of the lipid droplet that becomes the most prominent intracellular organelle, occupying a central cell position, and filling almost all the cytoplasm. This makes the adipocyte lipid droplet a prototype for the study of lipid storage organelles, and this chapter will review how the study of adipocytes contributed to increased knowledge on the general process of lipid storage. In addition, we will also focus on distinctive properties of adipocyte lipid droplets, and especially their unique ability to efficiently sequester fatty acids, that emerged as an important parameter to limit the spilling out of lipids outside of adipose tissue. In this regard, adipose tissue lipid storage is now at the center of the stage, with the burst of epidemic obesity and related metabolic diseases. In this perspective, consequences of increased filling of lipid droplets in adipocyte biology need to be fully estimated. In particular, the question of whether metabolic complications of obesity, leading to low grade obesity-associated inflammatory response, type 2 diabetes and cardio-vascular diseases are related to a failure of adipocyte to expand their lipid droplets, or a breakdown of the lipid droplet barrier leading to the spillover of fatty acids with lipotoxic effects in peripheral tissues has to be investigated.

Birth and Death of a Lipid Droplet

Lipid Droplet Organelle: A Protein-Decorated Phospholipid Monolayer Surrounding a Neutral Lipid Core

The development of the electron microscopy as a tool to examine intracellular structural organization with extraordinary resolution has revealed extreme compartmentalization of intracellular space, and defined cell organelles as membrane-limited structures. In this regard, intracytoplasmic lipid deposits did not meet the criteria required for organelle definition, as they do not appear to be surrounded by clear membrane structures resembling electron dense double leaflets. At the best, close apposition with other organelles such as endoplasmic reticulum (ER) or mitochondria has been frequently observed, which could not suffice to decipher on their organelle nature. However, from a physical point of view, it can be hardly conceivable that neutral (nonpolar) lipids could float freely in the cytoplasm, without a protein shell or a polar lipid interface.

In 2002, observation of an hemi-phospholipid membrane at the lipid droplet surface by the use of cry-oelectron microscopy at liquid-helium temperature on isolated lipid droplet preparations (without any treatment with stain or fixative) clearly establishes the unique organelle status of lipid droplets. Under these conditions, the lipid droplet surface could be seen as an electron dense thin line (2,5 nm of thickness) closely resembling one of the two parallel leaflets seen at the surface of a vesicle (Tauchi-Sato et al. 2002). Analysis of fatty acids composition of lipid droplets from hepatoma cells (HepG2) pointed out an original lipid composition differing from those of the ER and plasma membranes with a unique

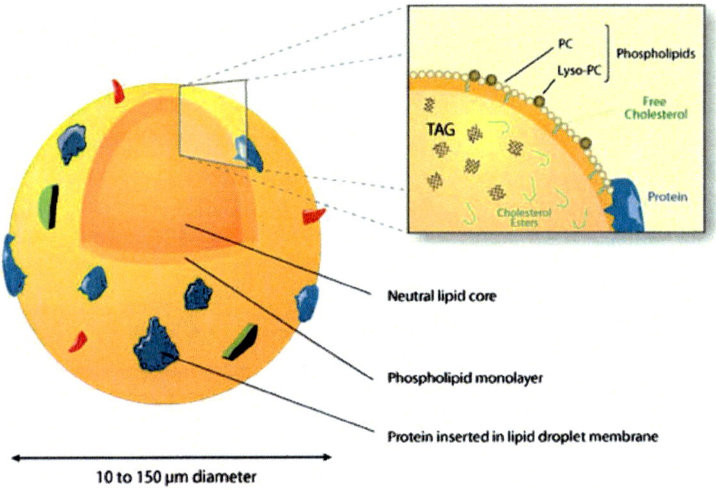

Fig. 9.1 Lipid droplet structure. Lipid droplet surface is a phospholipid monolayer where diverse proteins are inserted. This envelope has a specific lipid composition with phosphatidylcholine (PC) being the major phospholipid and presents a specific enrichment in free cholesterol. This lipid droplet membrane surrounds the neutral lipid droplet core where triacylglycerol and cholesterol esters are stored

phospholipids arrangement and specific enrichment with free cholesterol. Mass spectrometry analysis also revealed that phosphatidylcholine (PC) is the major phospholipid of the lipid droplet surface, characteristics which are conserved over species ranking from yeast to mammalian cells (Grillitsch et al. 2011; Blouin et al. 2010). This membrane-like structure delimits the lipid droplet core where triacylglycerols are the predominant lipid ester in adipocytes (Fig. 9.1).

Identification of specific proteins selectively decorating the phospholipid monolayer illustrates the unique structure of this membrane organelle (Fig. 9.1). Most abundant lipid droplet-associated proteins constitute the PAT (Perilipin-ADRP-TIP47) family, renamed perilipin family (Kimmel et al. 2010). Perilipin1, as the founding member, was identified in 1991 (Greenberg et al. 1991) in adipocyte fat cakes, and served to define a larger family by sequence homology, conserved from drosophila to humans. Interestingly, the conserved PAT domain is not required for lipid droplet targeting, but all perilipin proteins identified so far by sequence homology have been shown in morphological studies to interact with lipid droplets. Perilipin proteins are generally believed to operate as shells on lipid droplets, with packaging properties that protect the neutral lipid core from hydrolysis by cytoplasmic lipases (Bickel et al. 2009).

Finally, proteomic studies on adipocytes lipid droplets, (Brasaemle et al. 2004) but also isolated from other cell types (for review (Zehmer et al. 2009)], have defined a larger than expected panel of proteins in the lipid droplet proteome. Indeed, proteins that associate to the lipid droplet surface are not restricted to the perilipin family, or even only related to lipid metabolism. Indeed, a large variety of signaling molecules (caveolins, PKC, and Ras), membrane trafficking proteins (Arf1, Rab, and Rho proteins), cytoskeleton proteins (actin, filamin A, and myosin heavy chain), enzymes (alcohol dehydrogenase, DGAT), chaperones, and proteins associated with cellular organelles have been repeatedly described in independent proteomic analyses using different cell types as lipid droplet sources. It is likely that the list of lipid droplet-associated protein will further increase since new mass spectrometry technologies (such iTraq) are emerging; they will allow direct identification of proteins skipping the electrophoretic separation step. Although the precise function of many of these new lipid droplet proteins has not yet been clearly defined, this observation underlines the dynamic aspect of lipid droplets and the importance of communications with other intracellular compartments through specific interactions (Murphy et al. 2009).

Lipid Droplet Importance in Adipocyte Organization and Relation to Other Organelles

Due to their large unilocular lipid droplet, adipocytes present a unique morphology where the cytoplasm is under represented. Such an anatomy makes the lipid droplet stuck in the cell, with limited movements. Whereas nonadipocyte lipid droplets (within 1 μm of diameter) have been shown to move along microtubules (Welte 2009), adipocytes lipid droplets are enwrapped in a vimentin meshwork (Franke et al. 1987) which might guarantee the structural integrity of this big organelle.

Large lipid droplets in adipocytes are in close vicinity with the plasma membrane and electron microscopy images often show a less than 200 nm cytoplasmic space to the plasma membrane. Thus, privileged connections might exist favoring lipid storage. Especially, abundant caveolae 50–100 nm invaginations, specific subclasses of lipid rafts, could be viewed as portal entry for lipids (Parton et al. 2002). Several receptors involved in fatty acid uptake, like CD36 or FATP4, are enriched in caveolar membranes [for review (Pilch and Liu 2011)]. Moreover, by analogy with neutral lipid synthesis in bacteria, it has been suggested that triglycerides might be synthesized within these structures that were shown to contain DGAT (Ost et al. 2005), subsequently contributing to the unilocular neutral lipid pool by an unknown mechanism. In addition to the obvious link between lipid droplets and ER (see above), adipocyte lipid droplet connection with other organelles is also suggested by morphological studies. Mitochondria are often closely apposed to lipid storage compartment, but no specific protein complex indicative of the presence of contact points, like those linking ER to mitochondria referred as MAM (mitochondria-associated membrane) has been described so far.

Lipid Droplet Biogenesis

Although no direct observation of nascent lipid droplet has ever been made, many evidence suggest that lipid droplets would emerge from the ER membrane. Electron micrographs clearly show a close opposition between ER membrane and the lipid droplet, most of the time adipocyte lipid droplet being enwrapped in ER membranes (Blanchette-Mackie et al. 1995). The image of the egg (lipid droplet) laying on its egg cup (ER enwrapping membrane) has even been proposed following freeze-fracture views of lipid droplet/ER membrane association (Robenek et al. 2006). Such close apposition is likely to favor a transfer of lipids to growing lipid droplets, considering the presence of lipid ester synthesizing enzymes in the ER.

Assuming that droplets emerge from the ER, the obvious question is: how can a membrane bilayer give rise to a phospholipid monolayer? The prevailing model hypothesizes that lipid esters will accumulate between the two leaflets of the ER and when reaching a critical size, the nascent lipid droplet would be pinched off from the outer leaflet of the ER membrane to become an independent organelle [for review see (Murphy and Vance 1999; Martin and Parton 2006)] (Fig. 9.2). However, this model has been questioned by the fact that transmembrane proteins (like class I MHC molecule) or ER proteins with their amino-terminal domain luminally exposed (Calnexin or BiP) have been identified on lipid droplets surface. This lead Ploegh H. to propose an alternative model speculating that lipid droplets can form from both leaflets of the ER membrane by excision of a "bicelle" (Ploegh 2007). Therefore, the excised lipid droplet would, then, be surrounded by cytoplasmic and luminal leaflets of the ER membranes leaving a transient hole in the ER phospholipid monolayer. Such an excision mechanism is compatible with the escape of viruses (like polyoma or simian virus) on large assembled particles. This model also allows the creation of a « wrinkled » surface which would explain the presence of integral proteins on the lipid droplet surface (Fig. 9.2). However, this questioned about the consequences of transient pores formed in the ER, which could alter the local redox environment and the targeting of newly formed lipid droplets toward cytoplasm rather than the ER lumen.

So far, no direct evidence has allowed favoring one of these models. Moreover, observations of membranous structures and/or hydrophilic proteins (like caveolins) within the lipid esters core by freeze-fracture electron microscopy (Robenek et al. 2004) still questioned the lipid droplet biogenesis process.

Death of a Lipid Droplet by Autophagy: Lipophagy

To establish the dynamics within the stable population of adipocytes in adults, Peter Arner's team developed a method that is based on the incorporation of ^{14}C from nuclear bomb tests into genomic DNA and adipocyte lipids. By this means,

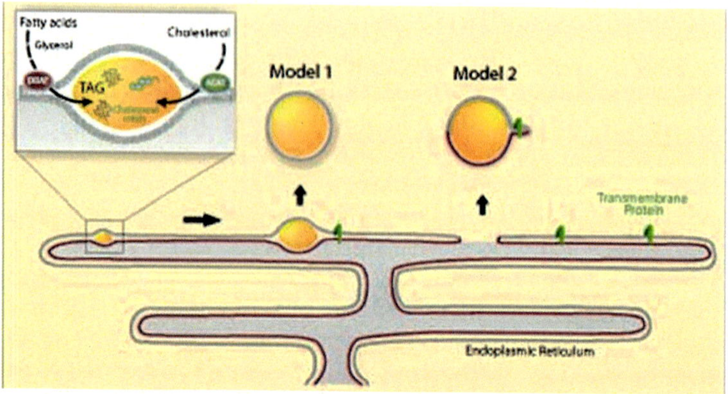

Fig. 9.2 Models of lipid-droplet biogenesis from the endoplasmic Reticulum. Newly synthesized neutral lipids accumulate between the two leaflets of the ER. Nascent lipid droplet may emerge by budding from the outer leaflet of the ER (*model 1*) or by excision of the entire lipid lens (*model 2*) leaving a transient hole in the ER membrane

it has been calculated that a human adipocyte can live for about 10 years (Spalding et al. 2008), and that during this lifespan, its triglycerides are renewed six times (Arner et al. 2011). Moreover, obesity neither changes the adipocyte half live, nor cell death rate in adipose tissues. Thus, increasing lipid droplet size does not promote apoptosis in adipocytes.

What are the main pathways leading to lipid droplet fat mobilization? It has been recognized for a long time that cytoplasmic lipases, that can be activated by fasting conditions, are the main players in this process called lipolysis. Basically, our knowledge on fat mobilization is now quite extensive, due to the discovery of perilipins acting as lipases anchoring proteins on lipid droplets, and of new lipases such as ATGL (Zimmermann et al. 2004). The detailed description of the lipolysis process is extensively discussed in Chap. 10.

Although the importance of the use of cytoplasmic lipases for lipid droplet degradation is not questionable, another lipid droplet degradation process has been recently described and referred to lipophagy (Singh et al. 2009a). It is well-known that autophagy is a key mechanism to recycle damaged organelles within cells by sequestering cargos in autophagosomes for delivery to lysosomes. Similarly, lipophagy uses the autophagic machinery to degrade lipid droplet organelles in which lipids are hydrolyzed by lysosomal lipases. On the basis of lipid accumulation phenotype in mice with liver-specific invalidation of key autophagic genes (atg5, atg7), this process was first demonstrated to regulate lipid metabolism in fasting hepatocytes, which comprise abundant lipid droplets and have low cytoplasmic lipase activities (Singh et al. 2009a). Similar tissue-specific invalidation approaches failed to demonstrate active lipophagy in adipose tissue, since it compromised white adipocyte differentiation and led to brown-like fat cells (Singh et al. 2009b; Zhang et al. 2009; Baerga et al. 2009). Noteworthy, simultaneous knockout of the two main adipocyte lipases HSL and ATGL can abrogate more than 90 % of lipolytic activity in fat, making it unlikely that lipophagy plays a major role in lipid mobilization from adipocytes (Schweiger et al. 2006). However, it cannot be excluded that lipophagy might be a substitute to impaired lipolysis. Indeed, caveolin null mice are unable to mobilize fatty acids upon adrenergic stimulation exhibit active autophagy in adipocytes (Le Lay et al. 2010). Additionally, lipophagy has also been shown to be active in cholesterol ester-loaded macrophages, and promote cholesterol efflux (Ouimet et al. 2011). Therefore, increasing evidence is suggesting that lipid mobilization by autophagy might play an important physiological role in different cell types, still requiring further investigations.

Lipid Droplets as Intracellular Organelles with Changing Sizes

Lipid Droplet Filling With Neutral Lipids

Fatty acids to be stored in lipid droplets generally derive from extracellular sources in most cell types, except in the liver and adipose tissue of rodents, which exhibit a noticeable ability for de novo lipogenesis from glucose precursors. Thus, the extraction of fatty acids from the blood stream, where they are incorporated in lipoproteins, is a major determinant for cellular utilization and potential ending in the lipid droplet storage compartment. Free fatty acids uptake after triglyceride rich lipoprotein hydrolysis by endothelium-associated lipoprotein lipase in the vessel walls is a critical step, that can also participate in the channeling of fatty acids to particular cell types such as adipocytes or muscles. Alternatively, some fatty acids can also be provided by endocytosis of lipoprotein particles such as LDL (Low Density Lipoprotein) in cells that expose endocytic lipoprotein receptors on their surface. Processes controlling fatty acid availability are tightly regulated by nutritional conditions and hormones, whereas those regulating fatty acids entry by specific transport are much less clearly elucidated. Subsequently, intracellular fatty acids need to be activated (by addition of a coenzymeA group) for subsequent cell utilization. Potential storage is related to their stepwise esterification onto a glycerol backbone to form triacylglycerol. Noteworthy, whereas fatty acids that are to be used for oxydation are diverted from this metabolic route at early steps after activation, the metabolic fate of fatty acids which will be stored as triacylglycerols or which will be used for membrane phospholipid synthesis follows first a common pathway which diverges only at its very last step from diacylglycerols to triacylglycerols or phospholipid synthesis. Thus, energy storage and glycerol-based phospholipid synthesis are tightly linked.

Coupling Lipid Storage and Phospholipid Synthesis for Lipid Droplet Expansion

Lipid droplets provide an interesting example of rapid and reversible changes in organelle size. During storage and lipid droplet expansion, massive amounts of neutral lipids are deposited in the lipid droplet core and phospholipids have to be added on their surface to maintain organelle structure. Although it has been suggested that lipid droplet growth might occur by fusion, this event is not frequently observed in living cells. Moreover, the fusion of two lipid droplets to give rise to a larger one would not increase the absolute amount of neutral lipid since the core volume of the resulting one would be equal to the sum of the unfused droplets (Ohsaki et al. 2009), but would rather result in an excess of surface phospholipids and extra monolayer by volume/surface ratios.

The current view on lipid droplet growth rather suggests that it occurs through neutral lipid accretion, which implies a coupling with phospholipid surface addition. In agreement, DGAT2, the triacylglycerol forming enzyme has been shown to localize not only in ER but also to lipid droplets (Kuerschner et al. 2008). Moreover, the presence of phospholipid remodeling enzymes has been demonstrated on lipid droplets (Moessinger et al. 2011), and a recently published study (Krahmer et al. 2011) provides a fascinating explanation of the way how expanding lipid droplets accommodate more neutral lipids and consequently increase their phospholipid surface. According to this work, increasing the core volume of the lipid droplet disrupts surface phospholipid (mainly PC) organization thus leading to surface exposure of core lipids. This serves as a signal for the recruitment of a protein, CTP: phosphocholine cytidylyltransferase (CCT) onto lipid droplets, an event that causes the activation of the enzyme which catalyzes the rate limiting step in PC biosynthesis. As a result, more PC is produced when lipid droplets expand, which can serve to maintain organelle homeostasis for lipid packaging. Although this model was established in nonadipose cells, it provides a view in which lipid droplet storage function tightly connects with lipid metabolism and intracellular membrane maintenance. Such a lipid droplet-centered conception of the overall cell organization might be of particular importance in adipocyte biology with respect to alterations that follow excessive lipid deposition in obesity and metabolic diseases.

Lipid Droplet Fusion and Fragmentation

Neutral lipids particles have a natural propensity to coalesce in a hydrophilic environment. However, this process is rarely observed in living cells, even when sophisticated imaging techniques are used (Murphy et al. 2010). This suggests that some mechanisms exist to prevent lipid droplet coalescence in cells. Surfactant properties of surface phospholipids and particularly PC might operate to counteract this process. This was shown in vitro on reconstituted lipid droplets with neutral lipids mixed to different phospholipid ratios study (Krahmer et al. 2011). In vivo, a lipid droplet morphology-based screen identified CCT1, a key enzyme in PC biosynthesis as a key regulator of lipid droplet fusion, since invalidation of this enzyme induced giant lipid droplets (Guo et al. 2008). One chapter (Boström et al. 2007) described lipid droplet fusion involving SNARE proteins, indicating a process similar to vesicular fusion but this still remains controversial.

A noticeable exception to this scheme is the unilocular adipocyte lipid droplet found in adipose tissue. Even in vitro models of differentiated adipocytes (like 3T3-L1) exhibit multiple lipid droplets never leading to a unique vacuole. Such morphological differences have been so far attributed to the importance of a 3-D environment, since the importance of stiffness exerted by the extracellular matrix (ECM) in adipose tissue is of crucial importance for tissue-specific gene expression and morphology (Chun et al. 2006). Another possibility, considering the data from Krahmer et al., could be insufficient PC to triglyceride ratios that would favor unilocular lipid droplet (Krahmer et al. 2011).

Unlike fusion, lipid droplet fragmentation is well recognized to occur under conditions favoring lipid mobilization (Marcinkiewicz et al. 2006). It is believed that such a process would favor the access of lipases to their substrates by increasing lipid droplet surface. It has been particularly described during lipolysis in adipocytes suggesting a connection with cAMP signaling pathway. However, the precise mechanism underlying this fragmentation process remains to be elucidated.

Specificity of Adipocyte Lipid Droplets and Relevance to Obesity-Associated Disorders

The Lipotoxic Hypothesis in the Pathogenesis of Metabolic Diseases

Obesity is defined as the result of excessive adipose tissue lipid storage, and represents a well-known risk factor for the development of type 2 diabetes mellitus and cardiovascular diseases. Importantly, considerable differences exist between obese individuals regarding concomitant lipid accumulation in nonadipose tissues. In other words, some obese subjects maintain excessive lipid deposition within their adipose tissue, whereas others also develop ectopic fat deposition within nonadipose tissues, generally liver, muscle, or pancreas. In the recent years, ectopic lipid accumulation was proposed as a possible link between obesity and its metabolic complications (van Herpen and Schrauwen-Hinderling 2008). In this respect, many deleterious metabolic effects have been documented since lipids species such as free (nonesterified) fatty acids, ceramides, or diacylglycerols are produced as intermediates in lipid synthesis. These molecules directly impede the transduction of the insulin signal, and are therefore potential causing factors for peripheral insulin resistance. Moreover, lipid accumulation can alter insulin production in pancreatic islets associated with the development of overt diabetes, or produce cardiac dysfunction. Similarly, lipid overloading of macrophages from cholesterol-rich lipoproteins drives the accumulation of spumous cells in the artery wall, a key pathogenic event in atherosclerosis. Thus, an emerging paradigm in metabolic disease points toward selective outcomes depending on the site where the excess of fat is accumulated, with a protective role of fat deposition in the specialized highly expandable adipose tissue, but detrimental effects of lipid accretion in other sites.

Adipocyte-Specific Lipid Droplet-Associated Proteins with Direct Functional Links to Metabolic Syndrome

Central to lipotoxic hypothesis is the notion that it is not equivalent to store fatty acids within adipocyte or ectopic lipid droplets. A possible explanation for this might be that intrinsic differences exist in lipid droplet composition between these two situations. Accordingly, at least three classes of lipid droplet-associated proteins are known to be present as specific isoforms in adipocytes and are not found in other cells that ectopically accumulate lipids. These abundant adipocyte proteins (perilipin1, FSP27 and caveolin-1) are described in further detail below with particular attention to their role in lipid droplet biology. Noteworthy, their essential role is underlined by the fact that their invalidation by gene targeting in mice (Razani et al. 2002; Nishino et al. 2008; Tansey et al. 2001) severely impair adipocyte lipid storage and produce lean animals that cannot expand their adipose tissue properly. Moreover, patients with nonfunctional perilipin1 (Gandotra et al. 2011), caveolin-1 (Kim et al. 2008) or FSP27 (Rubio-Cabezas et al. 2009) mutations develop severe lipodystrophic syndromes, and quasi absence of adipose tissue, despite unaltered preadipocyte differentiation potential.

Perilipin1 in the Shielding of Lipid Droplets and the Control of Fatty Acid Release

Perilipin1 is the best-characterized member in the PAT family, renamed perilipin family (Kimmel et al. 2010). In relation to its adipose-specific expression and its close association to the lipid droplet surface, it is now recognized as a central regulator of lipid metabolism (Brasaemle 2007). Indeed, the absence of perilipin expression in adipocytes severely impairs lipid mobilization by increasing fatty acid release under basal unstimulated conditions (Tansey et al. 2001). Conversely, ectopic expression of perilipin in cultured fibroblasts (in which it is not endogenously expressed) increases cell triacylglycerol content by slowing the rate of triacylglycerol breakdown (Brasaemle et al. 2000). Thus, it appears that perilipin can protect lipid droplets from accelerated degradation by cytoplasmic lipases. Interestingly, in nonadipose cells that do not express perilipin, endogenous lipid droplets are coated by ADRP (renamed perilipin2), another PAT family member with ubiquitous expression (Brasaemle et al. 1997). So, in perilipin null adipocytes, ADRP is likely to become the main lipid droplet coating protein (Tansey et al. 2001). Changes observed between ADRP and perilipin coated lipid droplets clearly indicate that ADRP cannot entirely substitute the perilipin function to limit basal fatty acid release from lipid droplets. This difference might be important when considering lipotoxic effects of lipid accumulation in nonadipose tissues, since wrapping lipid droplets with ADRP instead of perilipin would be less efficient at preventing free fatty acid spill over. In this regard, the expression profile of the different members of the PAT family is intertwined and rather complex and little is known of the factors that govern PAT protein-specific expression in different cell types. As a salient feature, perilipin expression is controlled by PPARgamma, a master gene of adipocyte differentiation (Dalen et al. 2004). In agreement, perilipin protein has been detected in steatotic human hepatocytes, where PPARgamma expression is turned on (Straub et al. 2008). Whether perilipin expression in some steatotic conditions protects liver cells from lipotoxic effects or is linked to hepatic insulin resistance is not clear. An interesting possibility might be that a gradual ability of PAT protein to protect cells from lipotoxicity exists from the most efficient perilipin to less active ADRP and TIP 47 (perilipin3). In agreement, a recent study shows that down regulation of ADRP and TIP47 by siRNA in AML12 liver cells that do not express perilipin, induced insulin signaling defects (Bell et al. 2008).

The role of perilipin is not limited to a protective scaffold against lipases. Upon stimulation of lipolysis by catecholamines, perilipin is heavily phosphorylated by protein kinase A (Greenberg et al. 1991), a feature that is not shared by other PAT family members. Perilipin phosphorylation, through conformational changes, facilitates maximal lipolytic activity of hormone sensitive lipase and adipose triglyceride lipase (Brasaemle 2007). This suggests a model in which perilipin serves as a dynamic scaffold to coordinate the access of enzymes to the lipid droplet in a manner that is responsive to the metabolic status of the adipocyte. In agreement, perilipin null mice exhibit exacerbated fatty acid release in the basal state underlining the importance of the shielding function of perilipin, and blunted stimulated lipolysis indicating its requirement for maximal lipolytic activity. As a result of these defects, perilipin null mice display a metabolic lean phenotype despite normal food intake, due to aberrant adipose tissue lipolysis (Tansey et al. 2001). Most interestingly, perilipin null mice are prone to develop glucose intolerance and peripheral insulin resistance (Tansey et al. 2004), which fits well with the idea that sequestration of lipids into adipose tissue is a critical process in the control of glucose homeostasis. In humans, little information is available on the regulation of adipose tissue perilipin. Apart from the recent report of perilipin1 mutation in lipodystrophic patients (Gandotra et al. 2011), a study reported the association of a minor allele in the perilipin gene with higher risk of metabolic syndrome in a population of obese children and adolescents (Deram et al. 2008).

Fsp27 and the Unilocular Adipocyte Morphology

Fsp27 was originally identified in 1992 as a 27 kD fat specific rodent protein in a screen for adipocyte differentiation-dependent transcripts (Danesch et al. 1992). It remained completely unexplored until 2003–2004 when its expression was reported in the steatotic mouse liver due to PPAR gamma overexpression (Matsusue et al. 2008). Concomitantly, its human homolog CIDE3, was cloned as a member of the ≪ cell-death-inducing-DNA-fragmentation-factor (DFF45)-like-effector ≫ family that could induce cell apoptosis in 293 cells (Liang et al. 2003). In these early reports, indirect evidence suggests that fsp27/CIDE3 could be localized on intracytoplasmic lipid droplets, since a GFP-CIDE3 was distributed on ≪ cytosolic corpuscules ≫ and fsp27 was identified as a member of the 3T3-L1 adipocyte lipid droplet proteome (Brasaemle et al. 2004). Direct morphological evidences in confocal microscopy are now available that fsp27, expressed as a GFP fusion can decorate the surface of lipid droplets with nice rings in 3T3-L1 adipocytes (Nishino et al. 2008; Keller et al. 2008) or colocalize with Nile red positive structures in 293 cells (Puri et al. 2007). Furthermore, short regions of 20–30 amino acids, with partial homologies to mouse perilipin (20–40 %) were found in the fsp27 sequence (Puri et al. 2008), identifying fsp27 as a bona fide lipid droplet-associated protein. Most interestingly, several recent studies provide new insights in the function of fsp27 and suggest an important role in metabolic diseases. First, ectopic fsp27 overexpression can stimulate triacylglycerol accumulation in several cell types, likely resulting from down-regulated fatty acid oxidation (Nishino et al. 2008; Keller et al. 2008; Puri et al. 2007) Second, the silencing of fsp27 expression by siRNA in cultured adipocytes revealed an unexpected role as a determinant of intracellular lipid droplet morphology. Indeed, cultured fat cells with fsp27 knockdown remarkably accumulated smaller and more lipid droplets than controls (Nishino et al. 2008). Lipid packaging in fragmented small lipid droplets was also visible in the adipose tissue of Fsp27 KO mice, in which the white adipocytes filled with multiple small lipid droplets instead of the unique large unilocular lipid droplet usually found in white fat cells (Nishino et al. 2008). Thus, specific fsp27 expression in white adipose tissue contributes to the formation of supersized unilocular lipid droplet, a unique feature for lipid accumulation in this tissue. Importantly, fragmentation of lipid droplets in adipocytes lacking fsp27 was shown to be associated with significant metabolic changes (Nishino et al. 2008). Such a fragmentation process is thought to increase the lipid interface with the cytoplasm, and thus might facilitate lipolysis by favoring access of lipases to their substrates. In agreement, fsp27-deficient mice mobilize more fatty acids than controls upon activation of lipolysis, and are consequently leaner. Whether adipocyte lipid droplet fragmentation represents a primary function for fsp27 is presently unknown.

Caveolin-1: Adipocyte Specific Structural Scaffolds on Expanding Lipid Droplet Surface

Caveolin proteins are primarily known as the scaffolding units that coat plasmalemnal caveolae (Rothberg et al. 1992). Caveolin-1 is particularly abundant in adipocytes, which contain numerous caveolae in their surface (Parton et al. 2002). Adipocyte caveolin-1 can localize to lipid droplets upon lipid loading of adipocytes (Le Lay et al. 2006) or other cell types (Fujimoto et al. 2001; Ostermeyer et al. 2001; Pol et al. 2001). In fat cells, caveolin association to lipid droplets is a regulated process related to caveolar endocytosis, that can be induced by exogenous addition of cholesterol or fatty acids (Le Lay et al. 2006) or lipid droplet maturation during adipocyte differentiation (Blouin et al. 2008). However, no experimental evidence exists that caveolin-1 can transport lipids to lipid droplets, and the current view is that they might participate in the maintenance of adipocyte lipid droplets as structural scaffolds. Because cholesterol staining of the lipid droplet shell by filipin is lost in caveolin-1 null adipocytes, caveolin-1 appears to be involved in surface free cholesterol regulation onto lipid droplets (Le Lay et al. 2009). In addition, detailed analysis of lipid droplet composition in caveolin deficient lipid droplets indicated normal triacylglycerol composition but revealed alterations in surface phospholipid composition (Blouin et al. 2010). Moreover, the presence of a fluorescent caveolin construct on the surface of individual lipid droplets formed in fibroblasts in response to a fatty acid load has been shown to favor their expansion. Although it is not known precisely what caveolin exactly does at the lipid droplet surface, they are likely to participate in the remodeling of the adipocyte lipid droplet surface, and to facilitate the efficient packing of fatty acids that characterize adipocyte lipid storage.

Concluding Remarks: Other Emerging Views on Lipid Droplet Biology

To conclude, we would like to stress that this chapter mainly focused on adipocyte lipid droplet, which led to neglect some novel aspects of lipid droplet biology that may appear of less importance in a metabolic context. In particular, the newly discovered role of lipid droplets as a refuge for proteins to escape degradation was not developed here but is reviewed elsewhere (Walther and Farese 2009). Similarly, increasing evidence is provided that lipid droplets may be associated with virus assembly (Miyanari et al. 2007; Boulant et al. 2008). Finally, due to their lipophilic properties, pollutants of the environment accumulate in lipid droplets within the adipose tissue (Kim et al. 2011) and Bourez et al. (2012) questioning about protective or deleterious role of lipid droplet storage regarding toxicity of these compounds. These novel data altogether participate in the drawing of a broad landscape in which lipid droplets no longer appear as inert storage deposits.

Acknowledgments The authors thank Cédric M. Blouin for its major contribution for the iconography.

References

Arner P, Bernard S, Salehpour M et al (2011) Dynamics of human adipose lipid turnover in health and metabolic disease. Nature 478:110–113

Baerga R, Zhang Y, Chen PH et al (2009) Targeted deletion of autophagy-related 5 (atg5) impairs adipogenesis in a cellular model and in mice. Autophagy 5:1118–1130

Bell M, Wang H, Chen H et al (2008) Consequences of lipid droplet coat protein downregulation in liver cells: abnormal lipid droplet metabolism and induction of insulin resistance. Diabetes 57:2037–2045

Bickel PE, Tansey JT, Welte MA (2009) PAT proteins, an ancient family of lipid droplet proteins that regulate cellular lipid stores. Biochim Biophys Acta 1791:419–440

Blanchette-Mackie EJ, Dwyer NK, Barber T et al (1995) Perilipin is located on the surface layer of intracellular lipid droplets in adipocytes. J Lipid Res 36:1211–1226

Blouin CM, Le Lay S, Lasnier F et al (2008) Regulated association of caveolins to lipid droplets during differentiation of 3T3-L1 adipocytes. Biochem Biophys Res Commun 376:331–335

Blouin CM, Le Lay S, Eberl A et al (2010) Lipid droplet analysis in caveolin-deficient adipocytes: alterations in surface phospholipid composition and maturation defects. J Lipid Res 51:945–956

Boström P, Andersson L, Rutberg M et al (2007) SNARE proteins mediate fusion between cytosolic lipid droplets and are implicated in insulin sensitivity. Nat Cell Biol 9:1286–1293

Boulant S, Douglas MW, Moody L et al (2008) Hepatitis C virus core protein induces lipid droplet redistribution in a microtubule- and dynein-dependent manner. Traffic 9:1268–1282

Bourez S, Le Lay S, Van Den Daelen C et al (2012) Accumulation of polychlorinated biphenyls in adipocytes: selective targetting to lipid droplets and role of caveolin-1. PLoS One. 7:e31834

Brasaemle DL (2007) Thematic review series: adipocyte biology. The perilipin family of structural lipid droplet proteins: stabilization of lipid droplets and control of lipolysis. J Lipid Res 48:2547–2559

Brasaemle DL, Barber T, Wolins NE et al (1997) Adipose differentiation-related protein is an ubiquitously expressed lipid storage droplet-associated protein. J Lipid Res 38:2249–2263

Brasaemle DL, Rubin B, Harten IA et al (2000) Perilipin A increases triacylglycerol storage by decreasing the rate of triacylglycerol hydrolysis. J Biol Chem 275:38486–38493

Brasaemle DL, Dolios G, Shapiro L, Wang R (2004) Proteomic analysis of proteins associated with lipid droplets of basal and lipolytically stimulated 3T3-L1 adipocytes. J Biol Chem 279:46835–46842

Chun TH, Hotary KB, Sabeh F et al (2006) A pericellular collagenase directs the 3-dimensional development of white adipose tissue. Cell 125:577–591

Dalen KT, Schoonjans K, Ulven SM et al (2004) Adipose tissue expression of the lipid droplet-associating proteins S3–12 and perilipin is controlled by peroxisome proliferator-activated receptor-gamma. Diabetes 53:1243–1252

Danesch U, Hoeck W, Ringold GM (1992) Cloning and transcriptional regulation of a novel adipocyte-specific gene, FSP27. CAAT-enhancer-binding protein (C/EBP) and C/EBP-like proteins interact with sequences required for differentiation-dependent expression. J Biol Chem 267:7185–7193

Deram S, Nicolau CY, Perez-Martinez P et al (2008) Effects of perilipin (PLIN) gene variation on metabolic syndrome risk and weight loss in obese children and adolescents. J Clin Endocrinol Metab 93:4933–4940

Farese RV Jr, Walther TC (2009) Lipid droplets finally get a little R-E-S-P-E-C-T. Cell 139: 855–860

Franke WW, Hergt M, Grund C (1987) Rearrangement of the vimentin cytoskeleton during adipose conversion: formation of an intermediate filament cage around lipid globules. Cell 49:131–141

Fujimoto T, Kogo H, Ishiguro K et al (2001) Caveolin-2 is targeted to lipid droplets, a new "membrane domain" in the cell. J Cell Biol 152:1079–1085
Gandotra S, Le Dour C, Bottomley W et al (2011) Perilipin deficiency and autosomal dominant partial lipodystrophy. N Engl J Med 364:740–748
Goodman JM (2008) The gregarious lipid droplet. J Biol Chem 283:28005–28009
Greenberg AS, Egan JJ, Wek SA et al (1991) Perilipin, a major hormonally regulated adipocyte-specific phosphoprotein associated with the periphery of lipid storage droplets. J Biol Chem 266:11341–11346
Grillitsch K, Connerth M, Kofeler H et al (2011) Lipid particles/droplets of the yeast Saccharomyces cerevisiae revisited: Lipidome meets Proteome. Biochim Biophys Acta 1811:1165–1176
Guo Y, Walther TC, Rao M et al (2008) Functional genomic screen reveals genes involved in lipid-droplet formation and utilization. Nature 453:657–661
Keller P, Petrie JT, De Rose P et al (2008) Fat-specific protein 27 regulates storage of triacylglycerol. J Biol Chem 283:14355–14365
Kim CA, Delepine M, Boutet E et al (2008) Association of a homozygous nonsense caveolin-1 mutation with Berardinelli-Seip congenital lipodystrophy. J Clin Endocrinol Metab 93: 1129–1134
Kim MJ, Marchand P, Henegar C et al (2011) Fate and complex pathogenic effects of dioxins and polychlorinated biphenyls in obese subjects before and after drastic weight loss. Environ Health Perspect 119:377–383
Kimmel AR, Brasaemle DL, McAndrews-Hill M et al (2010) Adoption of Perilipin as a unifying nomenclature for the mammalian PAT-family of intracellular lipid storage droplet proteins. J Lipid Res 51:468–471
Krahmer N, Guo Y, Wilfling F et al (2011) Phosphatidylcholine synthesis for lipid droplet expansion is mediated by localized activation of CTP: phosphocholine cytidylyltransferase. Cell Metab 14:504–515
Kuerschner L, Moessinger C, Thiele C (2008) Imaging of lipid biosynthesis: how a neutral lipid enters lipid droplets. Traffic 9:338–352
Le Lay S, Hajduch E, Lindsay MR et al (2006) Cholesterol-induced caveolin targeting to lipid droplets in adipocytes: a role for caveolar endocytosis. Traffic 7:549–561
Le Lay S, Blouin CM, Hajduch E, Dugail I (2009) Filling up adipocytes with lipids. Lessons from caveolin-1 deficiency. Biochim Biophys Acta 1791:514–518
Le Lay S, Briand N, Blouin CM et al (2010) The lipoatrophic caveolin-1 deficient mouse model reveals autophagy in mature adipocytes. Autophagy 6:754–763
Liang L, Zhao M, Xu Z et al (2003) Molecular cloning and characterization of CIDE-3, a novel member of the cell-death-inducing DNA-fragmentation-factor (DFF45)-like effector family. Biochem J 370:195–203
Marcinkiewicz A, Gauthier D, Garcia A, Brasaemle DL (2006) The phosphorylation of serine 492 of perilipin a directs lipid droplet fragmentation and dispersion. J Biol Chem 281: 11901–11909
Martin S, Parton RG (2006) Lipid droplets: a unified view of a dynamic organelle. Nat Rev Mol Cell Biol 7:373–378
Matsusue K, Kusakabe T, Noguchi T et al (2008) Hepatic steatosis in leptin-deficient mice is promoted by the PPARgamma target gene Fsp27. Cell Metab 7:02–311
Miyanari Y, Atsuzawa K, Usuda N et al (2007) The lipid droplet is an important organelle for hepatitis C virus production. Nat Cell Biol 9:1089–1097
Moessinger C, Kuerschner L, Spandl J et al (2011) Human lysophosphatidylcholine acyltransferases 1 and 2 are located in lipid droplets where they catalyze the formation of phosphatidylcholine. J Biol Chem 286:21330–21339
Murphy DJ, Vance J (1999) Mechanisms of lipid-body formation. Trends Biochem Sci 24: 109–115
Murphy S, Martin S, Parton RG (2009) Lipid droplet-organelle interactions; sharing the fats. Biochim Biophys Acta 1791:441–447

Murphy S, Martin S, Parton RG (2010) Quantitative analysis of lipid droplet fusion: inefficient steady state fusion but rapid stimulation by chemical fusogens. PLoS One 5:e15030
Nishino N, Tamori Y, Tateya S et al (2008) FSP27 contributes to efficient energy storage in murine white adipocytes by promoting the formation of unilocular lipid droplets. J Clin Invest 118:2808–2821
Ohsaki Y, Cheng J, Suzuki M et al (2009) Biogenesis of cytoplasmic lipid droplets: from the lipid ester globule in the membrane to the visible structure. Biochim Biophys Acta 1791:399–407
Ost A, Ortegren U, Gustavsson J et al (2005) Triacylglycerol is synthesized in a specific subclass of caveolae in primary adipocytes. J Biol Chem 280:5–8
Ostermeyer AG, Paci JM, Zeng Y et al (2001) Accumulation of caveolin in the endoplasmic reticulum redirects the protein to lipid storage droplets. J Cell Biol 152:1071–1078
Ouimet M, Franklin V, Mak E et al (2011) Autophagy regulates cholesterol efflux from macrophage foam cells via lysosomal acid lipase. Cell Metab 13:655–667
Parton RG, Molero JC, Floetenmeyer M et al (2002) Characterization of a distinct plasma membrane macrodomain in differentiated adipocytes. J Biol Chem 277:46769–46778
Pilch PF, Liu L (2011) Fat caves: caveolae, lipid trafficking and lipid metabolism in adipocytes. Trends Endocrinol Metab 22:318–324
Ploegh HL (2007) A lipid-based model for the creation of an escape hatch from the endoplasmic reticulum. Nature 448:435–438
Pol A, Luetterforst R, Lindsay M et al (2001) A caveolin dominant negative mutant associates with lipid bodies and induces intracellular cholesterol imbalance. J Cell Biol 152:1057–1070
Puri V, Konda S, Ranjit S et al (2007) Fat-specific protein 27, a novel lipid droplet protein that enhances triglyceride storage. J Biol Chem 282:34213–34218
Puri V, Ranjit S, Konda S et al (2008) Cidea is associated with lipid droplets and insulin sensitivity in humans. Proc Natl Acad Sci U S A 105:7833–7838
Razani B, Combs TP, Wang XB et al (2002) Caveolin-1-deficient mice are lean, resistant to diet-induced obesity, and show hypertriglyceridemia with adipocyte abnormalities. J Biol Chem 277:8635–8647
Robenek MJ, Severs NJ, Schlattmann K et al (2004) Lipids partition caveolin-1 from ER membranes into lipid droplets: updating the model of lipid droplet biogenesis. FASEB J 18:866–868
Robenek H, Hofnagel O, Buers I et al (2006) Adipophilin-enriched domains in the ER membrane are sites of lipid droplet biogenesis. J Cell Sci 119:4215–4224
Rothberg KG, Heuser JE, Donzell WC et al (1992) Caveolin, a protein component of caveolae membrane coats. Cell 68:673–682
Rubio-Cabezas O, Puri V, Murano I et al (2009) Partial lipodystrophy and insulin resistant diabetes in a patient with a homozygous nonsense mutation in CIDEC. EMBO Mol Med 1:280–287
Schweiger M, Schreiber R, Haemmerle G et al (2006) Adipose triglyceride lipase and hormone-sensitive lipase are the major enzymes in adipose tissue triacylglycerol catabolism. J Biol Chem 281:40236–40241
Singh R, Kaushik S, Wang Y et al (2009a) Autophagy regulates lipid metabolism. Nature 458:1131–1135
Singh R, Xiang Y, Wang Y et al (2009b) Autophagy regulates adipose mass and differentiation in mice. J Clin Invest 119:3329–3339
Spalding KL, Arner E, Westermark PO et al (2008) Dynamics of fat cell turnover in humans. Nature 453:783–787
Straub BK, Stoeffel P, Heid H et al (2008) Differential pattern of lipid droplet-associated proteins and de novo perilipin expression in hepatocyte steatogenesis. Hepatology 47:1936–1946
Tansey JT, Sztalryd C, Gruia-Gray J et al (2001) Perilipin ablation results in a lean mouse with aberrant adipocyte lipolysis, enhanced leptin production, and resistance to diet-induced obesity. Proc Natl Acad Sci U S A 98:6494–6499
Tansey JT, Sztalryd C, Hlavin EM et al (2004) The central role of perilipin a in lipid metabolism and adipocyte lipolysis. IUBMB Life 56:379–385

Tauchi-Sato K, Ozeki S, Houjou T et al (2002) The surface of lipid droplets is a phospholipid monolayer with a unique Fatty Acid composition. J Biol Chem 277:44507–44512
Thiele C, Spandl J (2008) Cell biology of lipid droplets. Curr Opin Cell Biol 20:378–385
van Herpen NA, Schrauwen-Hinderling VB (2008) Lipid accumulation in non-adipose tissue and lipotoxicity. Physiol Behav 94:231–241
Walther TC, Farese RVJr (2009) The life of lipid droplets. Biochim Biophys Acta 1791:459–466
Welte MA (2009) Fat on the move: intracellular motion of lipid droplets. Biochem Soc Trans 37:991–996
Zehmer JK, Huang Y, Peng G et al (2009) A role for lipid droplets in inter-membrane lipid traffic. Proteomics 9:914–921
Zhang Y, Goldman S, Baerga R et al (2009) Adipose-specific deletion of autophagy-related gene 7 (atg7) in mice reveals a role in adipogenesis. Proc Natl Acad Sci U S A 106:19860–19865
Zimmermann R, Strauss JG, Haemmerle G et al (2004) Fat mobilization in adipose tissue is promoted by adipose triglyceride lipase. Science 306:1383–1386

Chapter 10
Adipose Tissue Lipolysis

Dominique Langin and Etienne Mouisel

Introduction

White adipose tissue (WAT) is the major body energy repository in mammals. WAT exerts a buffering activity for energy imbalance at the cellular and whole-organism levels, storing energy in the form of triacylglycerol (TAG) in period of excess energy intake and releasing it in the form of non-esterified fatty acids (NEFA) for other organs during fasting. As such, the pathways controlling fat accumulation (cellular FA uptake, de novo lipogenesis, and esterification into TAG) and mobilization (lipolysis) in adipocytes are tightly and co-ordinately regulated. While TAG synthesis occurs in various tissues, such as the liver for very low density lipoprotein production, lipolysis for the release of FA as energy provider for other tissues is unique to adipocytes. The understanding of the cellular and molecular factors regulating these metabolic processes is in constant evolution, and recent discoveries have dramatically altered the view of TAG lipolysis and highlighted the importance of additional molecular actors regulating this process. Elucidating their mode of action may lead to novel therapeutic targets for the treatment and prevention of obesity and related metabolic disorders.

D. Langin (✉) · E. Mouisel
I2MC, INSERM UPS U1048, Obesity Research Laboratory, Team 4,
CHU Rangueil, Institute of Metabolic and Cardiovascular Diseases,
1 avenue Jean Poulhès, BP 84225, 31432 Toulouse Cedex 4, France
e-mail: dominique.langin@inserm.fr
URL: www.i2mc.inserm.fr/d_langin

D. Langin · E. Mouisel
UMR1048, University of Toulouse, Paul Sabatier University, Toulouse, France

D. Langin
Clinical Biochemistry Department, Toulouse University Hospitals, Toulouse, France

J.-P. Bastard and B. Fève (eds.), *Physiology and Physiopathology of Adipose Tissue*, 141
DOI: 10.1007/978-2-8178-0343-2_10, © Springer-Verlag France 2013

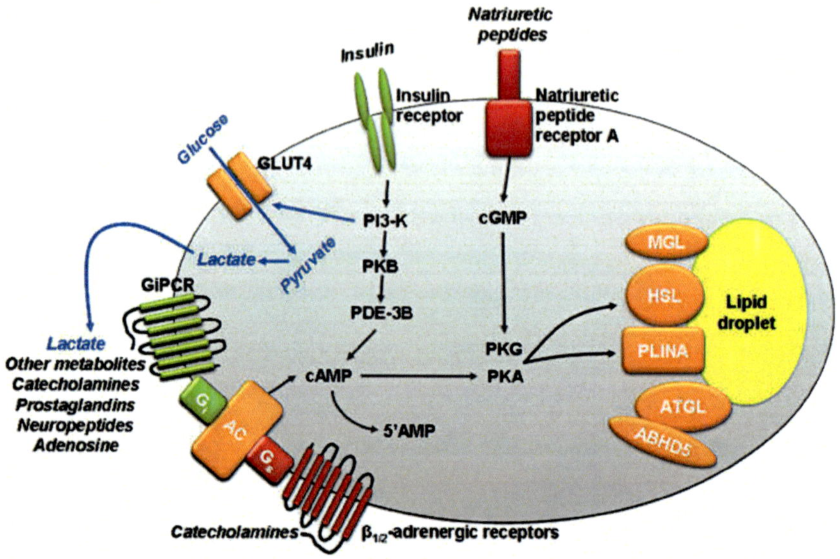

Fig. 10.1 Control of human adipocyte lipolysis. Binding of catecholamines to Gs protein-coupled $\beta_{1/2}$-adrenoceptors stimulate cAMP production by adenylyl cyclase (AC) and activate protein kinase A (PKA). Conversely, stimulation of Gi protein-coupled receptors reduces cyclic AMP (cAMP) and PKA-activation. Insulin favors cAMP degradation through activation of phosphatidylinositol-3 phosphate kinase (PI3-K), and protein kinase B (PKB), and stimulation of phosphodiesterase 3B (PDE-3B) activity. Natriuretic peptides promote cGMP accumulation and protein kinase G (PKG) activation. PKA and PKG phosphorylate hormone-sensitive lipase (HSL) and perilipin A (PLINA). Adipose triglyceride lipase (ATGL) and its cofactor ABHD5 and monoglyceride lipase (MGL) are also participating in the hydrolysis of triglycerides. A new pathway shown by blue arrows involving the glucose transporter GLUT4, glycolysis-mediated lactate production, and the Gi protein-coupled lactate receptor, GPR81, has been proposed in insulin-induced antilipolytic effect

Hormonal Control of Lipolysis

Catecholamines (the neurotransmitter, noradrenaline, and the hormone, adrenaline), natriuretic peptides, and insulin are considered to represent the major regulators of lipolysis in humans (Fig. 10.1). However, the physiological significance of a number of other lipolytic and antilipolytic agents, especially paracrine and autocrine factors, remains to be elucidated. Sympathetic and sensory nerve fibers were shown to innervate adipose tissue (AT) and to modulate lipolysis (Bartness et al. 2010). In addition, parasympathetic innervation was also postulated but this remains a highly debated question. Lipolytic and antilipolytic molecules activate receptors present at the surface of the fat cell. However, their action can be indirect. Notably, it has recently been shown that insulin known to directly activate fat cell insulin receptors exerts part of its antilipolytic action through hypothalamic control of the sympathetic nervous system (Scherer et al. 2011).

G Protein-Coupled Receptors and Adenylyl Cyclase Regulation

Activation or inhibition of adenylyl cyclase activity via receptors from the seven transmembrane domain G protein-coupled receptor family controls the formation of cyclic AMP (cAMP) from ATP. An increase in intracellular cyclic AMP enhances lipolysis while inhibition of lipolysis is associated with the lowering of cyclic AMP levels.

Receptors Coupled to Gs Protein

There are three beta-adrenergic receptor subtypes (beta-AR) (beta$_1$-AR, beta$_2$-AR, and beta$_3$-AR), each of which is coupled to the G-alpha subunit of the Gs protein. Their stimulation increases intracellular cAMP levels in fat cells of various species although considerable species-specific differences exist (Lafontan and Berlan 1993). Strikingly, catecholamines at high concentrations stimulate lipolysis in white adipocytes of mice without beta-ARs (Tavernier et al. 2005). However, the nature of this putative receptor is still unknown. While beta$_1$-AR and beta$_2$-AR are expressed in many body tissues, the beta$_3$-AR is predominantly found in white and brown adipocytes in rodents. In human white fat cells, both beta$_1$- and beta$_2$-ARs are known to stimulate lipolysis in vitro and in vivo (Langin 2006). The physiological role of the beta$_3$-AR in human WAT remains questionable, but, importantly, this receptor does not contribute to catecholamine-induced lipolysis in human subcutaneous adipocytes (Langin 2006). The recent (re)discovery of BAT in human adults will undoubtedly stimulate new investigation on beta$_3$-AR in human brown adipocytes (Lönnqvist et al. 1993). Molecules other than cathecholamines exert potent lipolytic effects through Gs protein-coupled receptors in rodent fat cells. However, they are ineffective or weak activators of lipolysis in human fat cells.

Receptors Coupled to Gi Proteins

Surprisingly, the number of molecules and receptors involved in inhibition of lipolysis through Gi-protein-coupled receptors is very large. Ligands include neuropeptides, paracrine factors, and autacoid agents (adenosine, prostaglandins and their metabolites, and other small molecules such as short-chain FA, beta-hydroxybutyrate, and lactate) originating from the adipocytes themselves, and also from preadipocytes, endothelial cells, macrophages, and sympathetic nerve terminals. Catecholamines have a special status since they are able to stimulate both beta-ARs and a major antilipolytic pathway involving alpha$_2$-ARs. In fact, adrenaline and noradrenaline stimulate and/or inhibit adenylyl cyclase and lipolysis depending on their relative affinity for the $beta_{1/2}$- and $alpha_2$-ARs and the relative number of $beta_{1/2}$- and $alpha_2$-ARs expressed in the fat cell (Lafontan and Berlan 1995).

The in vivo relevance of other Gi protein-coupled receptors is more elusive. It is however recognized that agonists leading to activation of Gi protein-coupled receptors of the adipocytes will limit NEFA release and represent putative anti-hyperlipidemic drugs. The best example is provided by nicotinic acid (niacin), an old lipid-lowering drug, which acts through GPR109a (Ahmed et al. 2009). The natural ligand for this receptor may be the ketone body, beta-hydroxybutyrate, which concentration increases during fasting. GPR81 which shows high expression in AT mediates the antilipolytic action of lactate. Evidence has recently been provided that lactate is involved in the inhibitory effect of insulin on lipolysis (Ahmed et al. 2010). Insulin stimulates glucose uptake by fat cells. Glucose is then metabolized by the glycolytic pathway into lactate which is released in significant amounts by adipocytes. In an autocrine or paracrine loop, the metabolite limits TAG hydrolysis by activating GPR81 (Fig. 10.1). The relevance of this pathway in human physiology has not been demonstrated yet (Langin 2010). Finally, GPR43 in rodent adipocytes leads to inhibition of lipolysis and suppression of plasma FA (Ge et al. 2008). GPR43 belongs to a subfamily of related GPCRs, including GPR40 and GPR41 that have been identified as receptors with specificity for FA. Short-chain FA activate GPR41 and GPR43 while medium-chain FA activate GPR40 (Hirasawa et al. 2008).

Neuropeptide Y (NPY) and peptide YY (PYY) though the NPY-Y1 receptor subtype inhibits adenylyl cyclase activity, cAMP production and lipolysis in human fat cells (Lafontan and Langin 2009). Secretion of NPY by neurons is dramatically up-regulated by the presence of adipocytes in co-culture and appears to be mediated by an adipocyte-derived soluble factor (Turtzo et al. 2001). Dipeptidyl peptidase IV through hydrolysis of NPY (1–36) into NPY (3–36) diminishes the activation of NPY-Y1. Therefore, dipeptidyl peptidase IV inhibitors augment the antilipolytic effect of NPY in human AT (Kos et al. 2009). Further studies are required to show the impact of dipeptidyl peptidase IV inhibitors on NPY effects in WAT in type 2 diabetic patients treated with these drugs.

Despite rapid turnover, adenosine and prostaglandins are found in substantial amounts in the vicinity of adipocytes. Binding of adenosine to a small fraction of the large population of adipocyte A1-adenosine receptors is sufficient to cause a marked antilipolytic effect (Lafontan and Langin 2009). The most relevant prostaglandin for the control of lipolysis is prostaglandin E_2 (PGE_2) produced by an adipocyte phospholipase A_2. PGE_2 acts through the EP3 receptor. Genetic ablation of the enzyme results in blunted WAT PGE_2 levels and increased lipolysis (Jaworski et al. 2009).

Natriuretic Peptide Pathway

For a long time, it has been considered that cAMP constituted the only second messenger involved in the control of AT lipolysis. However, the discovery of a new hormonal lipolytic pathway in human fat cells challenged that view (Lafontan et al. 2005). Atrial natriuretic peptide (ANP) and brain natriuretic peptide (BNP) stimulate human fat cell lipolysis through type A natriuretic peptide receptors (Fig. 10.1). These receptors possess guanylyl cyclase activity. Cyclic GMP-dependent activation of protein kinase G leads to phosphorylation of perilipin and hormone-sensitive lipase (HSL) as does activation of protein kinase A by cAMP.

Tyrosine-Kinase Receptors and Signaling Pathways

A critical point in the understanding of AT lipolysis is the potent antilipolytic effect of insulin (Fig. 10.1). Failure to suppress NEFA in response to the ingestion of a meal and the subsequent rise in insulinemia leads to abnormal elevations of plasma NEFA. The resistance of adipocytes to the antilipolytic effect of insulin is an important element of AT biology which could be at the origin of metabolic risks. In human adipocytes, insulin inhibits fat cell glycerol and FA release. Insulin controls cyclic AMP levels and lipolysis through the activation of cyclic nucleotide phosphodiesterase 3B (PDE3B). The importance of PDE3B in the regulation of lipolysis and of insulin-induced antilipolysis was confirmed in PDE3B-null mice (Choi et al. 2006). Insulin activates PDE3B and initiates PDE3B-dependent degradation of cAMP to 5'AMP, leading to a decrease in cAMP, an inactivation of protein kinase A and a subsequent reduced phosphorylation of HSL and perilipins, thus inhibiting lipolysis. The activation cascade leading to PDE3B activation is rather complex and some points remain unclear. When insulin binds to its receptor, the receptor is activated by phosphorylation on tyrosine residues, which causes tyrosine phosphorylation on intracellular substrates such as insulin receptor substrates and, binding and activation of phosphatidyl inositol kinase-3. This step is followed by protein kinase B (Akt) phosphorylation and activation, and PDE3B activation.

Other Lipolytic Pathways

Growth hormone-dependent stimulation of lipolysis probably represents a physiological adaptation to stress (e.g., during fasting and exercise) (Moller and Jorgensen 2009). In human adipocytes, the effect is delayed when compared to that of catecholamines and the exact mechanism of action is not fully established (Lafontan and Langin 2009). The transducing pathways are suspected to involve those used by catecholamines (cAMP- and protein kinase A-dependent pathways).

Interleukin 6 which is produced by AT could also stimulate lipolysis. However, its exact physiological role in vivo in humans is not clear. Conversely, multiple impacts of tumor necrosis factor alpha on fat cells have been identified (Langin and Arner 2006). Tumor necrosis factor alpha pathways interact with lipase expression, perilipin expression and phosphorylation, insulin effects and Gi-dependent inhibitory signaling pathways. In vivo data in humans suggest that, in conditions associated with low grade inflammation, tumor necrosis factor alpha could play a role in lipolysis (Plomgaard et al. 2008). Finally, zinc-alpha2-glycoprotein is an abundant circulating protein that enhances the lipolytic effect of catecholamines on fat cells. Besides inflammatory cytokines, zinc-alpha2-glycoprotein may be involved in the enhanced WAT lipolysis observed in cancer cachexia (Bing 2011).

Lipases and Triacylglycerol Hydrolysis

During lipolysis, intracellular TAG is sequentially hydrolyzed into diacylglycerol (DAG), monoacylglycerol (MAG), and glycerol, releasing one molecule of FA at each step. Three major lipases are involved: adipose triglyceride lipase (ATGL), hormone-sensitive lipase (HSL), and monoacylglycerol lipase (MGL). Since the discovery of ATGL in 2004, and because both ATGL and HSL possess TAG hydrolase activity, the relative contribution of these two lipases to adipose TAG hydrolysis in vivo has been assessed in numerous mouse and human models. Other lipases may play a minor role (Lafontan and Langin 2009; Girousse and Langin 2011). NEFA and glycerol efflux from the fat cells is followed by transport of these metabolites in the bloodstream to other tissues (mainly liver for glycerol, and skeletal muscle, liver and heart for NEFA). Some of the NEFA that are formed during lipolysis do not leave the fat cell and can be re-esterified into intracellular TAG. The glycerol formed during lipolysis is not re-utilized to a major extent by white fat cells because they contain minimal amounts of the enzyme glycerol kinase necessary for its metabolism. However, expression of the enzyme can be stimulated thereby allowing a futile cycle of lipolysis and reesterification as observed in brown adipose tissue (BAT) (Ribet et al. 2010; Mazzucotelli et al. 2007).

The mature white adipocyte comprises a large lipid droplet occupying the major part of the cell. Lipid droplets are considered as dynamic organelles that are critical for the management of cellular lipid stores and lipolytic processes (Brasaemle 2007). Lipolysis requires soluble cytosol lipases (i.e., ATGL and HSL) that can access the highly hydrophobic TAG substrates coated by proteins surrounding the lipid droplet. Moreover, glycerol and the hydrophobic NEFA released by lipolysis must be removed from the fat cell. During lipolysis, adipocyte lipid droplets undergo an important structural reorganization involving lipid-droplet associated proteins (e.g., perilipin), lipases (e.g., ATGL and HSL), and cofactors (e.g., CGI-58/ABHD5, a coactivator of ATGL) (Fig. 10.1).

Lipases

Hormone-Sensitive Lipase

Adipocyte HSL is composed of an N-terminal domain and a C-terminal catalytic domain that is identical in all known HSL isoforms (Lafontan and Langin 2009). This catalytic domain contains the active site, including residues of the catalytic triad (Ser, Asp, His), as well as a regulatory module with all the known phosphorylation sites of HSL. The N-terminal domain interacts with FABP4 as mentioned below. In vitro, HSL catalyzes the hydrolysis of TAG into DAG and DAG into MAG. The relative acylglycerol hydrolase activity of HSL in vitro is tenfold greater against DAG than TAG and MAG. HSL shows a preference for activity against FA in the sn-1 and sn-3 positions. The enzyme is also responsible for the hydrolysis of cholesterol and retinyl esters.

Unlike other known mammalian TAG lipases, HSL is regulated by reversible phosphorylation of Serine residues, being induced by protein kinase A and protein kinase G-mediated phosphorylations and inhibited by AMP-activated protein kinase-induced phosphorylation (Bezaire and Langin 2009; Kolditz and Langin 2010; Krintel et al. 2008). An important step in lipolysis activation is the translocation of HSL from a cytosolic compartment to the surface of the lipid droplet. Moreover, protein kinase A-induced phosphorylation promotes an increase in the hydrophobic surface area of HSL (Krintel et al. 2009).

HSL disruption results in blunted stimulated lipolysis. However, when fed a high fat diet, HSL-null mice failed to become obese (Girousse and Langin 2011). Growth curves indicated that HSL-null mice gain as much weight as the wild-type mice during the early stages of the diet but suddenly stop putting on more weight suggesting a limitation in WAT expandability. HSL-null mice were also resistant to genetic-induced obesity when they were bred on the ob/ob background. Reduced weight gain in HSL-null mice is not a result of reduced food intake. Fat absorption was also reported to be unchanged in HSL-null mice whereas energy expenditure was significantly increased (Strom et al. 2008). WAT shows metabolic brown adipocyte-like features. Reduced fat deposition could also result from impaired adipogenesis and/or adipocyte maturation due to a defect in production of PPARgamma ligands (Shen et al. 2011). DAG accumulation, retinoic acid metabolites and local inflammation can also interfere with adipocyte differentiation (Zimmermann et al. 2009; Strom et al. 2009). HSL-null mice is indeed an unusual model of pronounced WAT inflammation not associated with obesity.

Adipose Triglyceride Lipase

ATGL belongs to a family of proteins containing a patatin-like domain; it is a lipid hydrolase with an unusual folded topology that differs from classical lipases. ATGL enzymology, gene, and protein structure have recently been reviewed

(Zechner et al. 2009). The patatin domain harbors the active site of the enzyme. Structural domains of human ATGL have been described. The C-terminal protein region is essential for localization to the lipid droplet. However, the C-terminal region suppresses enzyme activity and interferes with ABHD5 interaction and enzyme activation (Schweiger et al. 2008). Two phosphorylation sites have been localized in the C-terminal region. The functional role of such sites is unknown. ATGL exhibits tenfold higher substrate specificity for TAG than DAG. Extensive studies in of ATGL- and HSL-deficient mice provided strong support for designating ATGL as the major TAG lipase in WAT, and assigned the primary function of HSL as a DAG lipase in vivo (Zechner et al. 2009; Zimmermann et al. 2004). The pivotal role of ATGL in both basal and stimulated lipolysis has been demonstrated in mouse and human fat cells (Ahmadian et al. 2009; Bezaire et al. 2009). In human fat cells, translocation of ATGL from the cytosol to smaller lipid droplets increases its colocalization with HSL under stimulated conditions (Bezaire et al. 2009).

ATGL-null mice show blunted fat cell lipolysis. ATGL-deficient mice rapidly become obese (Haemmerle et al. 2006). The major phenotypic consequence of ATGL disruption is massive TAG accumulation in adipose and non-adipose organs. Supra-physiological TAG accumulation in cardiac muscle leads to lethal impairment of cardiac function but can be completely prevented in ATGL-null mice with specific rescue in the heart (Haemmerle et al. 2006; Schoiswohl et al. 2010). Global energy metabolism is characterized by an inability to mobilize enough FA as fuel as exemplified by defective thermoregulation, reduced energy expenditure during fasting, and impaired exercise-stimulated lipolysis (Huijsman et al. 2009). Adipose-specific ablation of ATGL in mice converts BAT to a WAT-like tissue. The mice exhibit severely impaired thermogenesis revealing the requirement of ATGL-catalyzed lipolysis for maintaining a brown fat phenotype (Ahmadian et al. 2011).

Monoglyceride Lipase

MGL belongs to the serine hydrolase superfamily with a catalytic triad composed of the active site Serine, and Aspartic acid and a Histidine. The enzyme is required in the final hydrolysis of the 2-monoacylglycerols produced by HSL. It hydrolyses the 1(3) and 2-esters bonds of MAG at equal rates and is without in vitro catalytic activity against DAG, TAG, or cholesteryl esters. Due to its abundance in WAT, it was thought not to be limiting. However, ex vivo stimulated lipolysis is decreased in MGL-null mice (Taschler et al. 2011). In this mouse model, MAG hydrolase activity is not abolished due to partial compensation by HSL.

Lipid Droplet-Associated Proteins and Lipid Binding Proteins

Fatty Acid-Binding Protein-4

FABP4 is a cytosolic lipid-binding protein highly expressed in adipocytes that is involved in FA intracellular trafficking. It acts as a molecular chaperone, facilitating FA uptake and lipolysis (Furuhashi and Hotamisligil 2008). FABP4-null mice have decreased lipolytic capacity (Scheja et al. 1999; Coe et al. 1999). FABP4 and HSL form a physical complex when HSL is phosphorylated and FABP4 is bound to a FA. This complex translocates to lipid droplets upon protein kinase A activation (Smith et al. 2007). Interestingly, FABP4 inhibitors have been proposed in the treatment for type 2 diabetes and atherosclerosis (Furuhashi et al. 2008). In line with data from knock out mice, this class of compounds inhibits lipolysis in vitro (Lan et al. 2011).

ABHD5/CGI-58

ABHD5 (alpha/beta-hydrolase domain-containing protein 5) also known as CGI-58 (Comparative Gene Identification 58) has been shown to specifically activate ATGL (Lass et al. 2006). ABHD5 binds to lipid droplets by interacting with perilipin in non-stimulated adipocytes. With increases in intracellular cyclic AMP levels, ABHD5 dissociates from perilipin and interacts with ATGL and activates TAG hydrolysis (Miyoshi et al. 2007; Granneman et al. 2007; Yamaguchi et al. 2007). ABHD5 null mice die soon after birth due to impaired development of the skin permeability barrier (Radner et al. 2010). TAG hydrolase activity and lipolysis measured in mouse embryonic fibroblasts were markedly reduced in the absence of ABHD5 leading to increased TAG accumulation. Ectopic TAG accumulation in several tissues is reminiscent of neutral lipid storage disease or Chanarin Dorfman syndrome, a rare genetic disease caused by ABHD5 mutations.

G0/G1 Switch Gene 2

G0/G1 switch gene 2 (G0S2) encodes a protein first described in human mononuclear cells as possibly involved in cell cycle regulation. G0S2 is expressed at high levels in WAT and BAT and to a lesser extent in liver, skeletal muscle, and heart. In adipocytes, G0S2 colocalizes with ATGL on lipid droplet and plays an important role in lipolysis inhibiting ATGL activity (Yang et al. 2010). G0S2 binds directly to ATGL and dose dependently inhibits its TAG hydrolase activity regardless of the presence of ABHD5. G0S2 could 1) affect substrate accessibility to ATGL, 2) alter ATGL conformation, or 3) divert ATGL primary TAG hydrolase function to an acyl transferase activity. The role of G0S2 at the lipid droplet and its competition with ABHD5 require further study.

Perilipins

The perilipin family is proteins covering the lipid droplets in adipocytes and other cell types that regulate the coordination of lipid storage and utilization (Brasaemle et al. 2009; Ducharme and Bickel 2008). Perilipin A, the most important perilipin isoform in adipocytes, possesses highly hydrophobic domains which allow targeting and anchoring to lipid droplets. It is also the most highly phosphorylated protein following stimulation by lipolytic hormones. Perilipin ablation causes a dramatic attenuation of hormone-stimulated lipolysis while increased rates of lipolysis are observed under unstimulated conditions (Girousse and Langin 2011). The reduction of basal lipolysis may be due to the sequestration of ABHD5 by perilipin. Phosphorylation of perilipin on Serine residues rapidly releases ABHD5 from perilipin, allowing interaction with ATGL and stimulation of lipolysis (Granneman et al. 2009). However, the precise mechanisms of these interactions remain to be described. It has recently been shown that another member of the perilipin family, OXPAT (oxidative tissues-enriched PAT protein), which is expressed in BAT, interacts both with ATGL and ABHD5 whereas perilipin only binds to ABHD5 (Granneman et al. 2011). The balance between perilipin and OXPAT could control the channeling of FA in the fat cell toward esterification, oxidation, or lipolysis.

Caveolin-1

Caveolin-1 is abundant in adipocytes where it is an important component of the fat cell plasma membrane and is also localized to the lipid droplet (Pilch et al. 2007). Caveolin-1 null mice exhibit attenuated lipolytic activity. The phosphorylation of perilipin was dramatically reduced in these mice, suggesting a role for caveolin-1 in this process (Cohen et al. 2004). Highlighting their roles in lipid droplets, inactivating mutations of perilipin and caveolin-1 have been reported in patients with lipodystrophy (Vigouroux et al. 2011).

Cide-Domain-Containing Proteins

Members of the cell death-inducing DNA fragmentation factor-alpha-like effector (CIDE) gene family have been shown to regulate lipid metabolism. CIDEA is a protein that regulates lipolysis in human adipocytes through cross-talk involving tumor necrosis factor alpha which negatively regulates transcription of this gene (Nordstrom et al. 2005). Depletion of CIDEA by RNA interference in human adipocytes leads to increased lipolysis. CIDEA, known to be a mitochondrial protein in brown adipocytes, co-localizes with perilipin around lipid droplets in white fat cells (Puri et al. 2008; Zhou et al. 2003). Another CIDE protein, CIDEC (FSP27, fat-specific protein 27) is expressed in adipocytes. Small interfering RNA-mediated knockdown of CIDEC resulted in an increased basal release of NEFA, and decreased responsiveness to adrenergic stimulation of lipolysis (Magnusson et al. 2008). Accordingly, CIDEC null mice show increased basal lipolysis (Nishino et al. 2008; Toh et al. 2008).

Dysregulation of Lipolysis in Obesity

The dynamics of lipid turnover is an essential process determining the development of obesity and its complications. Recently, the age of lipids was determined in human subcutaneous WAT (Arner et al. 2011). The incorporation of atmospheric ^{14}C into WAT lipid was used to estimate lipid age which reflects the irreversible removal of lipids from fat stores consisting in lipolysis followed by fatty acid oxidation and/or ectopic deposition. From lipid age and total fat mass, the authors calculated the net lipid storage, which represents the amount of lipid stored in WAT each year and reflects fat incorporation from exogenous sources (e.g., derived from food) and endogenous synthesis (e.g., de novo synthesis of fatty acids from glucose) subtracted from the irreversible removal of lipids. Obesity is characterized by both increased lipid storage and decreased lipid removal (Fig. 10.2). Increased capacity to store fat with low net mobilization leads to expansion of fat mass and may also be viewed as a way to do a safe deposit of lipids into a harmless compartment (Langin 2011). Indeed, according to the WAT expandability hypothesis, as long as an individual has the capacity to store fat in WAT, there is no ectopic deposition of lipids and resulting metabolic complications (Virtue and Vidal-Puig 2008). However, when the storage capacity is overcome (McQuaid et al. 2011), ectopic deposition of lipids in liver and skeletal muscle may favor the development of insulin resistance through lipotoxic mechanisms (Samuel et al. 2010). This is exemplified in familial combined hyperlipidemia, a hereditary lipid disorder predisposing to premature coronary heart disease. In that condition, both triglyceride storage and lipid removal rates were low (Arner et al. 2011). This defect induces a routing of fatty acids to the liver where fatty acid overflow contributes to the mixed dyslipidemia characteristic of this condition. Similarly, lipodystrophic patients who have a defect of triglyceride storage in adipose tissue resulting in lipid accumulation elsewhere in the body develop severe insulin resistance (Langin 2011).

Against conventional wisdom, fasting plasma NEFA concentration is largely unrelated to body fat mass (Karpe et al. 2011). In the fasting state, plasma NEFA arise almost entirely from hydrolysis of TAG within the adipocyte. In the obese state, the lack of increase in plasma NEFA is partly explained by a decrease in subcutaneous WAT NEFA production; the majority of NEFA originates from this depot. Impairment in the catecholamine-induced lipolysis in subcutaneous WAT is a common feature of obese subjects (Lafontan and Langin 2009; Kolditz and

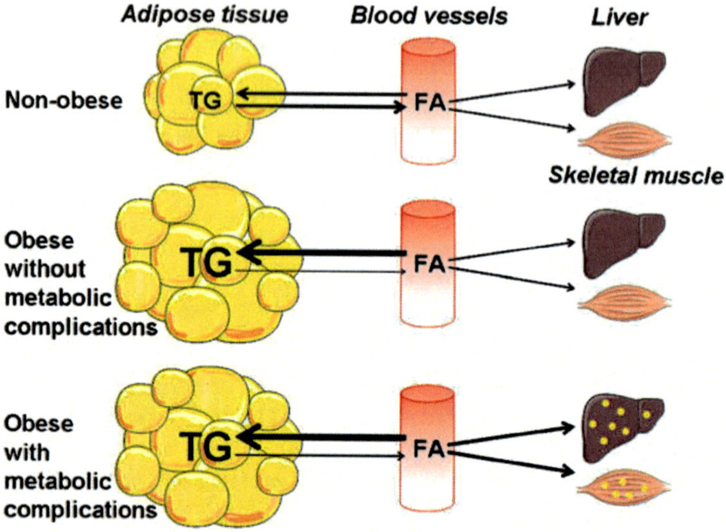

Fig. 10.2 Lipid turnover in adipose tissue of non-obese, obese with and without metabolic complications individuals. Increased lipid storage and decreased lipid removal favors the expansion of fat mass in obesity. If the capacity of storage is exceeded, there is ectopic lipid deposition in other organs which favors metabolic complications leading to diabetes and cardiovascular diseases

Langin 2010). This lipolytic resistance has been related to alteration in the lipolytic cascade at multiple levels. Defects in expression of HSL, ATGL, beta-AR, perilipin, or the regulatory subunit of protein kinase A have been described in obese subjects. In addition to changes in the stimulatory system, lipolytic resistance to catecholamines may involve increased antilipolytic responsiveness of the alpha$_2$-ARs. The release of NEFA from WAT is determined not only by lipolysis but also by WAT blood flow which facilitates the removal of NEFA and glycerol. WAT blood flow is decreased in obesity and is one of the components contributing to decreased NEFA delivery in obese individuals (McQuaid et al. 2011).

Conclusions

The last decade has been marked by the discovery of a number of mechanisms able to clarify the control of lipid mobilization. Lipases and lipolytic and antilipolytic receptor dysfunction may play a noticeable role in the development of obesity complications.

The impact of altered sympathetic nervous system on its target cells could lead to the development of obesity (Bartness and Song 2007). The reduced efficiency of beta-AR-dependent lipolysis and/or enhanced alpha$_2$–AR-mediated antilipolysis could both impair lipolysis and lead to catecholamine-resistance and promote the development and/or stabilization of obesity. In addition, loss of efficiency of the ANP-dependent lipolytic pathway could also be another element that could explain altered lipid mobilization (Lafontan et al. 2008). The reduced lipolysis commonly reported in the subcutaneous WAT of obese subjects could also be viewed as an adaptive mechanism limiting an excess rate of lipolysis. This could lead not only to a protection against lipotoxicity in non-adipose organs but also to the mitigation of WAT inflammation.

Knowledge of the lipolytic pathways has fostered the identification of new targets for lipolysis modulators (Langin 2006). The recent (re)discovery of BAT in human adults may reactivate research on beta$_3$-AR agonists as candidates to activate thermogenesis (Langin 2010). However, the modest clinical results obtained with this class of drugs must be kept in mind. As HSL and ATGL are involved in TAG hydrolysis, selective inhibitors targeting this step are likely promising drugs (Wang and Fotsch 2006), although the distribution of these enzymes in other tissues than AT could be a limitation. Drug discovery efforts are now focusing on therapeutics directed toward targeting antilipolytic pathways (Lukasova et al. 2011). In addition to the nicotinic acid receptor, other Gi protein-coupled receptors with antilipolytic activity might also be targeted, such as alpha$_2$-ARs, the A1-adenosine and EP3-receptors, and metabolite-activated Gi protein-coupled receptors (Lafontan and Langin 2009). ANP has a potent lipolytic effect in abdominal subcutaneous WAT of healthy and obese subjects. Therefore, ANP antagonists could be potential antilipolytic compounds for the treatment of dyslipidemia and NEFA-related insulin resistance. The question remains largely open in the absence of suitable pharmacological compounds. Unfortunately, the field is also facing major limitations since the efficacy of ANP antagonists cannot be assessed in rodent adipocytes, which are devoid of natriuretic peptide receptor A. To conclude, with the accumulating knowledge of the lipolytic pathways, continued efforts in this area will hopefully lead to new chemical entities for the treatment of obesity-related problems such as insulin resistance, dyslipidemia, and cardiovascular risk.

Acknowledgments Lipolysis studies from our laboratories are supported by Inserm, Université Paul Sabatier, Hôpitaux de Toulouse, FRM, SFD, DHOS, GlaxoSmithKlineBeecham, Région Midi-Pyrénées and the Commission of the European Communities Collaborative Project ADAPT (www.adapt-eu.net), Contract No. HEALTH-F2-2008-2011 00.

References

Ahmadian M, Duncan RE, Varady KA et al (2009) Adipose Overexpression of Desnutrin Promotes Fatty Acid Use and Attenuates Diet-Induced Obesity. Diabetes 58:855–866

Ahmadian M, Abbott MJ, Tang T et al (2011) Desnutrin/ATGL is regulated by AMPK and is required for a brown adipose phenotype. Cell Metab 13:739–748

Ahmed K, Tunaru S, Offermanns S (2009) GPR109A, GPR109B and GPR81, a family of hydroxy-carboxylic acid receptors. Trends Pharmacol Sci 30:557–562

Ahmed K, Tunaru S, Tang C et al (2010) An autocrine lactate loop mediates insulin-dependent inhibition of lipolysis through GPR81. Cell Metab 11:311–319

Arner P, Bernard S, Salehpour M et al (2011) Dynamics of human adipose lipid turnover in health and metabolic disease. Nature 478:110–113

Bartness TJ, Song CK (2007) Thematic review series: adipocyte biology. Sympathetic and sensory innervation of white adipose tissue. J Lipid Res 48:1655–1672

Bartness TJ, Shrestha YB, Vaughan CH et al (2010) Sensory and sympathetic nervous system control of white adipose tissue lipolysis. Mol Cell Endocrinol 318:34–43

Bezaire V, Langin D (2009) Regulation of adipose tissue lipolysis revisited. Proc Nutr Soc 68:350–360

Bezaire V, Mairal A, Ribet C et al (2009) Contribution of Adipose Triglyceride Lipase and Hormone-sensitive Lipase to Lipolysis in hMADS Adipocytes. J Biol Chem 284: 18282–18291

Bing C (2011) Lipid mobilization in cachexia: mechanisms and mediators. Curr Opin Support Palliat Care 5:356–360

Brasaemle DL (2007) Thematic review series: adipocyte biology. The perilipin family of structural lipid droplet proteins: stabilization of lipid droplets and control of lipolysis. J Lipid Res 48:2547–2559

Brasaemle DL, Subramanian V, Garcia A et al (2009) Perilipin A and the control of triacylglycerol metabolism. Mol Cell Biochem 326:15–21

Choi YH, Park S, Hockman S et al (2006) Alterations in regulation of energy homeostasis in cyclic nucleotide phosphodiesterase 3B-null mice. J Clin Invest 116:3240–3251

Coe NR, Simpson MA, Bernlohr DA (1999) Targeted disruption of the adipocyte lipid-binding protein (aP2 protein) gene impairs fat cell lipolysis and increases cellular fatty acid levels. J Lipid Res 40:967–972

Cohen AW, Razani B, Schubert W et al (2004) Role of caveolin-1 in the modulation of lipolysis and lipid droplet formation. Diabetes 53:1261–1270

Ducharme NA, Bickel PE (2008) Lipid droplets in lipogenesis and lipolysis. Endocrinology 149:942–949

Furuhashi M, Hotamisligil GS (2008) Fatty acid-binding proteins: role in metabolic diseases and potential as drug targets. Nat Rev Drug Discov 7:489–503

Furuhashi M, Fucho R, Gorgun CZ et al (2008) Adipocyte/macrophage fatty acid-binding proteins contribute to metabolic deterioration through actions in both macrophages and adipocytes in mice. J Clin Invest 118:2640–2650

Ge H, Li X, Weiszmann J et al (2008) Activation of G protein-coupled receptor 43 in adipocytes leads to inhibition of lipolysis and suppression of plasma free fatty acids. Endocrinology 149:4519–4526

Girousse A, Langin D (2011) Adipocyte lipases and lipid droplet-associated proteins: insight from transgenic mouse models. Int J Obes (Lond) Jun 14. doi: 10.1038/ijo.2011.113. [Epub ahead of print]

Granneman JG, Moore HP, Granneman RL et al (2007) Analysis of lipolytic protein trafficking and interactions in adipocytes. J Biol Chem 282:5726–5735

Granneman J, Moore H, Krishnamoorthy R, Rathod M (2009) Perilipin controls lipolysis by regulating the interactions of AB-hydrolase containing 5 (Abhd5) and adipose triglyceride lipase (Atgl). J Biol Chem 284:34538–34544

Granneman JG, Moore HP, Mottillo EP et al (2011) Interactions of perilipin-5 (Plin5) with adipose triglyceride lipase. J Biol Chem 286:5126–5135

Haemmerle G, Lass A, Zimmermann R et al (2006) Defective lipolysis and altered energy metabolism in mice lacking adipose triglyceride lipase. Science 312:734–737

Hirasawa A, Hara T, Katsuma S et al (2008) Free fatty acid receptors and drug discovery. Biol Pharm Bull 31:1847–1851

Huijsman E, van de Par C, Economou C et al (2009) Adipose triacylglycerol lipase deletion alters whole body energy metabolism and impairs exercise performance in mice. Am J Physiol Endocrinol Metab 297:E505–E513

Jaworski K, Ahmadian M, Duncan RE et al (2009) AdPLA ablation increases lipolysis and prevents obesity induced by high-fat feeding or leptin deficiency. Nat Med 15:159–168

Karpe F, Dickmann JR, Frayn KN (2011) Fatty acids, obesity, and insulin resistance: time for a revaluation. Diabetes 60:2441–2449

Kolditz CI, Langin D (2010) Adipose tissue lipolysis. Curr Opin Clin Nutr Metab Care 13: 377–381

Kos K, Baker AR, Jernas M et al (2009) DPP-IV inhibition enhances the antilipolytic action of NPY in human adipose tissue. Diabetes Obes Metab 11:285–292

Krintel C, Osmark P, Larsen MR et al (2008) Ser649 and Ser650 are the major determinants of protein kinase A-mediated activation of human hormone-sensitive lipase against lipid substrates. PLoS One 3:e3756

Krintel C, Morgelin M, Logan DT, Holm C (2009) Phosphorylation of hormone-sensitive lipase by protein kinase A in vitro promotes an increase in its hydrophobic surface area. FEBS J 276:4752–4762

Lafontan M, Berlan M (1993) Fat cell adrenergic receptors and the control of white and brown fat cell function. J Lipid Res 34:1057–1091

Lafontan M, Berlan M (1995) Fat cell α2-adrenoceptors: the regulation of fat cell function and lipolysis. Endocrine Rev 16:716–738

Lafontan M, Langin D (2009) Lipolysis and lipid mobilization in human adipose tissue. Prog Lipid Res 48:275–297

Lafontan M, Moro C, Sengenes C et al (2005) An unsuspected metabolic role for atrial natriuretic peptides: the control of lipolysis, lipid mobilization, and systemic nonesterified fatty acids levels in humans. Arterioscler Thromb Vasc Biol 25:2032–2042

Lafontan M, Moro C, Berlan M et al (2008) Control of lipolysis by natriuretic peptides and cyclic GMP. Trends Endocrinol Metab 19:130–137

Lan H, Cheng CC, Kowalski TJ et al (2011) Small-molecule inhibitors of FABP4/5 ameliorate dyslipidemia but not insulin resistance in mice with diet-induced obesity. J Lipid Res 52: 646–656

Langin D (2006) Adipose tissue lipolysis as a metabolic pathway to define pharmacological strategies against obesity and the metabolic syndrome. Pharmacol Res 53:482–491

Langin D (2010a) Adipose tissue lipolysis revisited (again!): lactate involvement in insulin antilipolytic action. Cell Metab 11:242–243

Langin D (2010b) Recruitment of brown fat and conversion of white into brown adipocytes: Strategies to fight the metabolic complications of obesity? Biochim Biophys Acta 1801: 372–376

Langin D (2011) In and out: adipose tissue lipid turnover in obesity and dyslipidemia. Cell Metab 14:569–570

Langin D, Arner P (2006) Importance of TNF alpha and neutral lipases in human adipose tissue lipolysis. Trends Endocrinol Metab 17:314–320

Lass A, Zimmermann R, Haemmerle G et al (2006) Adipose triglyceride lipase-mediated lipolysis of cellular fat stores is activated by CGI-58 and defective in Chanarin-Dorfman Syndrome. Cell Metab 3:309–319

Lönnqvist F, Krief S, Strosberg AD et al (1993) Evidence for a functional β3-adrenoceptor in man. Br J Pharmacol 110:929–936

Lukasova M, Hanson J, Tunaru S, Offermanns S (2011) Nicotinic acid (niacin): new lipid-independent mechanisms of action and therapeutic potentials. Trends Pharmacol Sci 32: 700–707

Magnusson B, Gummesson A, Glad CA et al (2008) Cell death-inducing DFF45-like effector C is reduced by caloric restriction and regulates adipocyte lipid metabolism. Metabolism 57: 1307–1313

Mazzucotelli A, Viguerie N, Tiraby C et al (2007) The transcriptional coactivator PGC-1alpha and the nuclear receptor PPARalpha control the expression of glycerol kinase and metabolism genes independently of PPARgamma activation in human white adipocytes. Diabetes 56:2467–2475

McQuaid SE, Hodson L, Neville MJ et al (2011) Downregulation of adipose tissue fatty acid trafficking in obesity: a driver for ectopic fat deposition? Diabetes 60:47–55

Miyoshi H, Perfield JW, Souza SC et al (2007) Control of adipose triglyceride lipase action by serine 517 of perilipin A globally regulates protein kinase A-stimulated lipolysis in adipocytes. J Biol Chem 282:996–1002
Moller N, Jorgensen JO (2009) Effects of growth hormone on glucose, lipid, and protein metabolism in human subjects. Endocr Rev 30:152–177
Nishino N, Tamori Y, Tateya S et al (2008) FSP27 contributes to efficient energy storage in murine white adipocytes by promoting the formation of unilocular lipid droplets. J Clin Invest 118:2808–2821
Nordstrom EA, Ryden M, Backlund EC et al (2005) A human-specific role of cell death-inducing DFFA (DNA fragmentation factor-alpha)-like effector A (CIDEA) in adipocyte lipolysis and obesity. Diabetes 54:1726–1734
Pilch PF, Souto RP, Liu L et al (2007) Cellular spelunking: exploring adipocyte caveolae. J Lipid Res 48:2103–2111
Plomgaard P, Fischer CP, Ibfelt T et al (2008) Tumour necrosis factor-alpha modulates human in vivo lipolysis. J Clin Endocrinol Metab 93:543–549
Puri V, Ranjit S, Konda S et al (2008) Cidea is associated with lipid droplets and insulin sensitivity in humans. Proc Natl Acad Sci U S A 105:7833–7838
Radner FP, Streith IE, Schoiswohl G et al (2010) Growth retardation, impaired triacylglycerol catabolism, hepatic steatosis, and lethal skin barrier defect in mice lacking comparative gene identification-58 (CGI-58). J Biol Chem 285:7300–7311
Ribet C, Montastier E, Valle C et al (2010) PPARalpha control of lipid and glucose metabolism in human white adipocytes. Endocrinology 151:123–133
Samuel VT, Petersen KF, Shulman GI (2010) Lipid-induced insulin resistance: unravelling the mechanism. Lancet 375:2267–2277
Scheja L, Makowski L, Uysal KT et al (1999) Altered insulin secretion associated with reduced lipolytic efficiency in aP2-/- mice. Diabetes 48:1987–1994
Scherer T, O'Hare J, Diggs-Andrews K et al (2011) Brain insulin controls adipose tissue lipolysis and lipogenesis. Cell Metab 13:183–194
Schoiswohl G, Schweiger M, Schreiber R et al (2010) Adipose triglyceride lipase plays a key role in the supply of the working muscle with fatty acids. J Lipid Res 51:490–499
Schweiger M, Schoiswohl G, Lass A et al (2008) The C-terminal region of human adipose triglyceride lipase affects enzyme activity and lipid droplet binding. J Biol Chem 283: 17211–17220
Shen WJ, Yu Z, Patel S et al (2011) Hormone-sensitive lipase modulates adipose metabolism through PPARgamma. Biochim Biophys Acta 1811:9–16
Smith AJ, Thompson BR, Sanders MA, Bernlohr DA (2007) Interaction of the adipocyte fatty acid-binding protein with the hormone-sensitive lipase: regulation by fatty acids and phosphorylation. J Biol Chem 282:32424–32432
Strom K, Hansson O, Lucas S et al (2008) Attainment of brown adipocyte features in white adipocytes of hormone-sensitive lipase null mice. PLoS One 3:e1793
Strom K, Gundersen TE, Hansson O et al (2009) Hormone-sensitive lipase (HSL) is also a retinyl ester hydrolase: evidence from mice lacking HSL. FASEB J 23:2307–2316
Taschler U, Radner FPW, Heier C et al (2011) Monoglyceride lipase-deficiency in mice impairs lipolysis and attenuates diet-induced insulin resistance. J Biol Chem 286:17467–17477
Tavernier G, Jimenez M, Giacobino JP et al (2005) Norepinephrine induces lipolysis in beta1/ beta2/beta3-adrenoceptor knockout mice. Mol Pharmacol 68:793–799
Toh SY, Gong J, Du G et al (2008) Up-regulation of mitochondrial activity and acquirement of brown adipose tissue-like property in the white adipose tissue of fsp27 deficient mice. PLoS One 3:e2890
Turtzo LC, Marx R, Lane MD (2001) Cross-talk between sympathetic neurons and adipocytes in coculture. Proc Natl Acad Sci U S A 98:12385–12390
Vigouroux C, Caron-Debarle M, Le Dour C et al (2011) Molecular mechanisms of human lipodystrophies: from adipocyte lipid droplet to oxidative stress and lipotoxicity. Int J Biochem Cell Biol 43:862–876

Virtue S, Vidal-Puig A (2008) It's not how fat you are, it's what you do with it that counts. PLoS Biol 6:e237

Wang M, Fotsch C (2006) Small-molecule compounds that modulate lipolysis in adipose tissue: targeting strategies and molecular classes. Chem Biol 13:1019–1027

Yamaguchi T, Omatsu N, Morimoto E et al (2007) CGI-58 facilitates lipolysis on lipid droplets but is not involved in the vesiculation of lipid droplets caused by hormonal stimulation. J Lipid Res 48:1078–1089

Yang X, Lu X, Lombes M et al (2010) The G(0)/G(1) switch gene 2 regulates adipose lipolysis through association with adipose triglyceride lipase. Cell Metab 11:194–205

Zechner R, Kienesberger PC, Haemmerle G et al (2009) Adipose triglyceride lipase and the lipolytic catabolism of cellular fat stores. J Lipid Res 50:3–21

Zhou Z, Yon Toh S, Chen Z et al (2003) Cidea-deficient mice have lean phenotype and are resistant to obesity. Nat Genet 35:49–56

Zimmermann R, Strauss JG, Haemmerle G et al (2004) Fat mobilization in adipose tissue is promoted by adipose triglyceride lipase. Science 306:1383–1386

Zimmermann R, Lass A, Haemmerle G, Zechner R (2009) Fate of fat: The role of adipose triglyceride lipase in lipolysis. Biochimica Et Biophysica Acta-Molecular and Cell Biology of Lipids 1791:494–500

Chapter 11
The Adipose Tissue: Storage, Source, and Target of Pollutants

Robert Barouki and Karine Clément

Introduction

While the adipose tissue (AT) was initially considered as a relatively passive storage tissue with mostly physical and energetic functions, recent studies have shown that this tissue carries other regulatory, endocrine, and metabolic functions. This tissue is also clearly implicated in a number of diseases, including obesity, metabolic, and systemic diseases. Thus, more attention was given to the AT in the recent literature. In the present review, we wish to describe and discuss another as yet poorly described

R. Barouki (✉)
INSERM UMR-S 747, Paris, France
e-mail: robert.barouki@parisdescartes.fr

R. Barouki
Centre Universitaire des Saints-Pères, Université Paris Descartes,
45 Rue des Saints-Pères, 75270, Paris Cedex 06, France

R. Barouki
Assistance Publique-Hôpitaux de Paris, Hôpital Necker-Enfants Malades, 75015, Paris, France

K. Clément
Institut des maladies cardiométaboliques et de la nutrition (IHU ICAN), Paris, France
e-mail: karine.clement@psl.aphp.fr

K. Clément
Nutriomique U872 équipe 7, Centre de recherche des Cordeliers, INSERM, UPMC,
15, rue de l'école de médecine, 75006 Paris, France

K. Clément
Centre de Recherche en Nutrition Humaine Ile-de-France (CRNH IdF), Paris, France

K. Clément
Assistance Publique Hôpitaux de Paris,Service de Nutrition, Bâtiment Benjamin Delessert,
Hôpital Pitié-Salpêtrière, 91 boulevard de L'hôpital, 75013, Paris, France

J.-P. Bastard and B. Fève (eds.), *Physiology and Physiopathology of Adipose Tissue*,
DOI: 10.1007/978-2-8178-0343-2_11, © Springer-Verlag France 2013

function of the AT, i.e., its toxicological function. This comprises several distinct roles: a protective function toward certain toxicants, a possibly deleterious endogenous source of toxicants, and a target of toxicant effects. Evidence for this newly described function is certainly not as solid as that of the other functions. Nevertheless, this concept has elicited recent work and discussion and has gained support lately.

Common Mechanisms of Metabolic Detoxification

Detoxification primarily consists in the metabolism and transport of xenobiotics which are small exogenous molecules such as medical drugs, pollutants, contaminants, and certain food constituents. Detoxification mechanisms are best understood in the context of general cellular defense mechanisms. Indeed, an important fraction of the genetic program and of cellular protein content is devoted to adaptation to stressful conditions within the framework of cellular stress signaling. For example, oxidative, hypoxic, osmotic, heat, conformational, and mechanical stress help cells adapt to a variety of chemical, biological, and physical stressors (Goldberg 2003; Morel and Barouki 1999; Morel et al. 2000; Semenza 2001). Exposure to xenobiotics leads to a significant shift of normal cellular conditions leading to an adaptive mechanism called "xenobiotic stress". While all stress pathways are adaptive in nature, they can also lead to some toxicity in the long run. This concept also applies to xenobiotic stress, since, in some cases, detoxification pathways can generate toxic metabolites. In addition, the xenobiotic stress concept accounts for the inducibility of the detoxification pathways, a common property of adaptive mechanisms. One should bear in mind that stress is by essence an adaptive response mechanism rather than an insult. Xenobiotics are chemically heterogeneous. Detoxification mechanisms are best understood for hydrophobic xenobiotics. In fact, this is the case of many xenobiotics and this property explains at least part of their toxicity. Indeed, hydrophobic chemicals can interact with certain proteins and incorporate into membranes thereby altering their fluidity, their structure, and the function of resident proteins. Furthermore, hydrophobic xenobiotics tend to localize in adipose tissue and to persist in the organism in the absence of an adequate metabolizing system. Thus, one of the main functions of the detoxification system is to detect hydrophobic xenobiotics and to increase their hydrophilicity in order to eliminate them through bile and urine. Different protein players contribute to this function: intracellular receptors detect these xenobiotics and induce metabolic enzymes and transporters, cytochromes P450 add a reactive chemical function to these hydrophobic molecules, thereby allowing phase two enzymes such as transferases to add highly hydrophilic groups, and finally transporters allow these molecules to cross membranes (Barouki 2010). Remarkably, certain aromatic halogenated xenobiotics are excellent inducers of the metabolic system but are not themselves metabolized. Indeed, they bind to cytochrome P450 efficiently, but are extremely poor substrates of these enzymes. These xenobiotics, also called persistent organic pollutants (POP), are stored in the adipose tissue (Fig. 11.1).

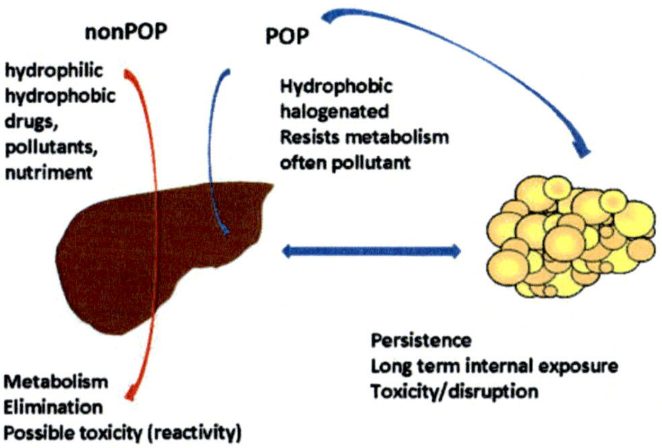

Fig. 11.1 Fate of persistent and non-persistent xenobiotics. Most xenobiotics are metabolized primarily by the liver and are thus detoxified. The detoxification system tends to render hydrophobic xenobiotics more hydrophilic which leads to their elimination in urine. Several halogenated xenobiotics are not metabolized and therefore tend to bind to liver proteins and to adipose mass. They can thus persist for years in the body and constitute a putative long-term threat

Evidence of a Protective Function of the Adipose Tissue

POPs cannot be metabolized by the metabolizing system and therefore tends to accumulate in ecosystems and in living organisms. These compounds are environmentally persistent and tend to accumulate along the food chain. The best studied are the POPs that were listed in the stockholm convention to limit their production and dissemination because of their possible long-term toxicity (Pelletier et al. 2003). POPs include certain organochlorine pesticides, dioxins, furans, polychlorobiphenyls, and polybrominated flame retardants. They do not readily undergo degradation by xenobiotic metabolizing enzymes (XMEs) most probably because they are halogenated. However, they do bind often with high affinity to certain xenobiotic receptors, as well as to certain XMEs such as CYP1A2 without undergoing catalytic transformation. The latter binding plays a significant role in their distribution as will be discussed below. Because of their hydrophobicity, POPs tend to distribute into fat mass such as the AT and milk. The AT is therefore a compartment which contains high amount of POPs, particularly in organisms that are at the top of the food chain. Such a bioaccumulation also leads to the age-dependent increase in POP content (Hue et al. 2007).

POPs are taken up by the adipocytes and probably localize within the lipid droplets. However, their precise location and their actual effects at the subcellular level are poorly understood. It is nevertheless believed that their accumulation within the AT decreases their availability for other cells and tissues thereby limiting their toxicity. Experimental evidence supports such a protective function for the AT. Indeed, studies carried in the 1980s and the 1990s showed that there was an inverse correlation between toxicity of POPs and fat mass of different animal species. Authors compared the 30 days toxicity of TCDD (2,3,7,8-TetraChloro-DibenzoDioxin) in approximately 20 terrestrial animal species and found a positive correlation between the BMI of these species and the LD50 (the dose that leads to 50 % death in the animal population) of dioxin (Geyer et al. 1997). They concluded that the species with the highest fat mass tended to display better resistance to dioxin in this particular acute exposure test. These conclusions were in line with studies showing that resistance of aquatic species to dioxin was also related to their fat mass content leading to the paradoxical notion of "survival of the fattest" (Lassiter and Hallam 1990). However, these observations should not be taken as evidence suggesting that the BMI is the only factor discriminating sensitive and resistant species. There is indeed strong evidence for a major contribution of the genetically determined arylhydrocarbon receptor affinity for dioxin.

It should be stressed that this protective function of the AT was revealed in acute or subacute exposure tests. These high dose treatments may allow the diffusion of the pollutants in all organism tissues unless an efficient filter or a buffer system captures them and thereby decreases exposure of the most sensitive tissues. This kinetic protective system does not only include the AT. Indeed, it has been established that proteins such as the dioxin-inducible liver CYP1A2 can bind this pollutant particularly during acute or subacute exposures and play an important role in its toxicokinetics (DeVito et al. 1998). It is now believed that POPs are first primarily captured by the liver-inducible protein compartment and then are redistributed to the AT. Obviously, these kinetic distribution mechanisms depend heavily on the treatment dose (Emond et al. 2006). In conclusion, organisms are protected against POP toxicity by a complex toxicokinetic system involving an inducible liver protein compartment and an AT compartment.

The Adipose Tissue as a Source of Endogenous Exposure

As mentioned earlier, POPs and other lipophilic contaminants distribute according to their affinity for proteins and lipids and are stored primarily in the liver and the AT. They are also found in blood from which they can contaminate other tissues. Blood POP content can be either related to their release from storage tissues or to recently absorbed pollutants. Several observations in both human and animals suggest that the release of pollutants from the AT is an important source of blood POPs.

In human, most of the evidence has been gathered from studies on drastic weight loss in obese individuals. Such a weight loss can be achieved through diet and bariatric surgery and could lead to a decrease of up to 30 kg of fat mass. Several independent studies have shown that there was an increase in blood POPs following fat mass loss elicited by either diet alone or diet coupled with bariatric surgery (Hue et al. 2006; Kim et al. 2011). If increased blood POP levels during weight loss are related to their release from AT, one would expect changes in POP content of this tissue. This has been addressed by Kim et al. (2011) who determined POP concentrations in both blood and AT and who also assessed the total amount of fat in the studied individuals. The data indicate that POP concentration in AT (expressed per gram lipid) increases with weight loss. While this may seem paradoxical, it is not particularly surprising since the total amount of fat mass decreases considerably thereby leading to an increased concentration of pollutants. It is also believed that released POPs can be taken up readily by the remaining fat. In line with these suggestions, we observed that POP concentrations in the AT of obese individuals is lower that that of lean individuals. However, the total amount of fat-stored POPs is two- to threefold higher in obese individuals as compared to lean controls. Furthermore, this total amount tends to decrease by 15 % following weight loss at least for certain POPs. This observation suggests that there is indeed some degree of POP release from AT during weight loss and that this release leads to a moderate decrease in total POP content.

Indirect evidence for POP release from fat storage tissue in human is provided by breastfeeding studies. Several POPs and other contaminants are found in breast milk because of its high lipid content. Because of the equilibrium between lipoid-associated POPs in AT, blood, and milk, it is likely that a significant part of breast milk POPs originates from the AT storage compartment, in addition to newly absorbed contaminants. In agreement with this model, it was observed that the half-life of dioxin in human is considerably reduced by breastfeeding. Consequently, the amount of POPs in breast milk reflects the POP body burden in an individual. In this context, it is interesting to note that the POP content in breast milk decreases continuously during the nursing period and is lower in the following nursing periods, in agreement with the decreased burden. Other studies also noted a decrease in breast milk POP content in multiparous mothers as compared to primiparous mothers. Taken together, the studies on breast milk contamination indicate that the pollutants stored in the AT can be released and eventually distributed to other fluids or tissues according to their lipid content (Iida et al. 1999). As a consequence, the AT is not only a sink for POPs but a real reservoir and an internal source of low grade contamination of other tissues.

Experimental evidence also suggests redistribution of POPs from their storage sites in the AT. Indeed, a study shows that in rodents pretreated with radiolabeled hexachlorobenzene, weight loss leads to a time-dependent increase in the brain content of this compound (Jandacek et al. 2005). Thus, decreased AT fat leads to a redirection of certain POPs toward sensitive and lipid-rich tissues. The study shows that weight loss alters the distribution of lipophilic pollutants, thus leading to enhanced localization in the brain with possible toxic outcome.

Observational studies were also carried out in northern elephant seals. These animals accumulate a large amount of fat in order to cope with extended fasting. Their fat is contaminated with PCBs. During the fasting period which could last several weeks, they lose a large amount of fat. Debier et al. have shown that fasting was accompanied by an increase in serum concentration of PCB which is likely due to their release from fat depots (Debier et al. 2006). Interestingly, the concentration of PCBs also increased in some of these depots (blubbers) because of the decreased fat content; however, different fat territories did not undergo similar changes, suggesting differences in the kinetics of POP exchange and release. It is suggested that the release of POPs during fasting may lead to toxic effects.

A critical issue is whether the release of POPs from AT observed during weight loss could lead to toxic outcomes in other organs and tissues. Indirect evidence was obtained in humans from several studies of weight loss triggered by either diet or diet associated with bariatric surgery. We have shown that the dynamic increase in serum POPs following drastic weight loss correlated with a delayed and reduced improvement of blood lipid parameters and liver toxicity biomarkers (Kim et al. 2011). Correlations between blood POP concentrations and other clinical parameters, in particular muscle parameters, were also observed by the group of Tremblay who conducted seminal studies in this field (Imbeault et al. 2002).

In conclusion, a number of human and animal studies suggest that the AT behaves as a toxicokinetic buffer for lipophilic pollutants. It is a privileged storage compartment for these pollutants. However, this is a dynamic situation and release from the AT occurs at a low basal level which can be magnified during weight loss. There is indirect evidence suggesting that released POPs are not inert and can exert some toxic effects. More direct evidence for this point is needed.

The Adipose Tissue as a Target of Pollutants

Several epidemiological studies carried out following industrial exposure of workers or accidental contamination by POPs indicated a relationship between serum concentration of certain POPs and markers of diabetes or of a prediabetic state. Such a correlation was also found in a large scale study carried in the general population (Lee et al. 2007). Because of the implication of the AT in metabolic diseases, it was hypothesized that this tissue could be a target of POPs and indeed, several effects were found. Thus, in addition to its function in the storage and release of POPs, the AT was shown to be altered by these compounds. Its vulnerability may be due to its ability to accumulate POPs; however, this pollutant localization remains unknown as well as the possible effects of pollutants on the structure and function of lipid droplets. Most of the studies were in vitro or ex vivo, but recently the effect of POPs on the AT of rodents was also assessed. Three main effects have been established by several labs including ours. POPs were shown to display anti-insulin effects in cellular models of adipocytes. For example, dioxin repressed the glucose transporter Glut4 expression and lipoprotein lipase in 3T3-F442a cells (Kern et al. 2002). This anti-insulin effect is not general and consistent for all genes. Indeed, while dioxin was found to antagonize insulin action on certain genes such as the IGFBP1 (insulin-like growth factor binding protein 1) gene in hepatocytes (Marchand et al. 2005), it displayed a different effect on other genes such as the liver PEPCK (phosphoenolpyruvate carboxykinase) gene, since it tended to inhibit gluconeogenesis in this tissue, similar to insulin (Stahl et al. 1993).

POPs have been shown to induce proinflammatory genes in rodent adipose cells (Kern et al. 2002). We found similar effects in human adipocytes (Kim et al. 2012). Importantly, in mice treated with dioxin, not only the gene expression of proinflammatory genes was increased, but also invasion of this tissue by macrophages and lymphocytes was observed (Kim et al. 2012). Finally, dioxin was shown to inhibit the differentiation of adipocyte precursor cells in certain model systems and to antagonize the effects of PPARγ; however the actual mechanisms remain elusive (Remillard and Bunce 2002).

In conclusion, preadipocytes and adipocytes are targeted by POPs which appear to disrupt certain signaling and differentiation pathways and to induce inflammation.

Conclusion and Hypothesis

The AT appears to play critical roles in the kinetics of POPs and in their pathogenic effects. It has a major role, together with the liver protein compartment in storing POPs and in preventing their distribution into more sensitive tissues. However, the AT storing capacity is constitutive and not inducible. This kinetic system acts as a buffer during acute or subacute exposure conditions. However, it translates an acute exposure into a long-term, low grade internal exposure (Fig. 11.2). It thus transforms an immediate threat into a latent chronic threat. This buffer system perfectly illustrates a previously developed hypothesis (Barouki and Morel 2001) which proposes that chronic toxicity of xenobiotics is in fact favored by the protective systems that are triggered upon repeated exposure to these compounds. In addition to these functions, the AT constitutes a target of POP toxicity. Indeed, the main toxic effect triggered by these compounds is inflammation which is a well-known risk factor for metabolic diseases. These observations support the contribution of POPs to metabolic diseases and suggest that AT alteration could at least partially mediate these effects.

Important points

- The AT plays an important role in the toxicokinetics of persistent organic pollutants.
- This tissue may have a protective function in case of acute or subacute exposure: it can take up these pollutants and thereby prevent their localization in other more sensitive tissues.

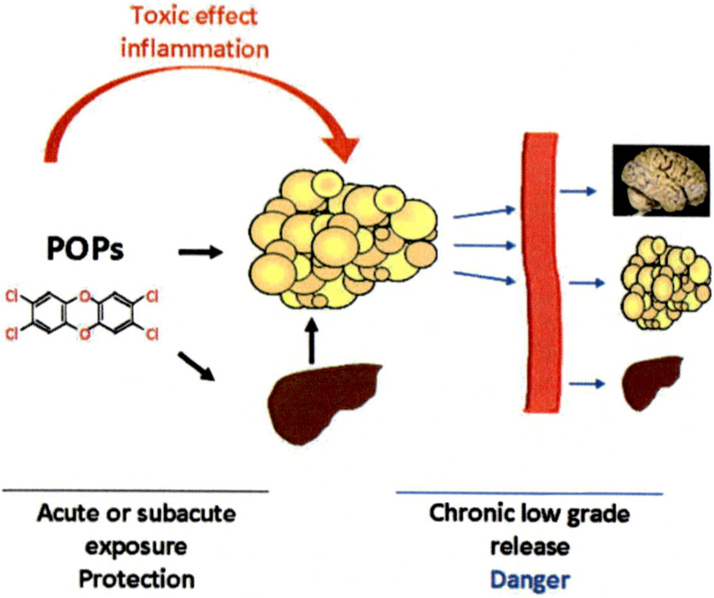

Fig. 11.2 Model representing the toxicological function of the adipose tissue. Upon acute of subacute exposure, POPs are first bound by liver proteins and adipose mass then tend to localize primarily in the adipose tissue. This constitute a protective function of the adipose tissue which diverts toxicants from other organs. However, in the long run, low grade diffusion into other organs takes place leading to a chronic exposure state. This property is revealed during drastic weight loss which leads to the redistribution of these pollutants. This chronic internal exposure could lead to toxic effects. Thus, adipose tissue translates an acute toxic threat into a chronic low grade toxicity. In addition, the adipose tissue is a direct target of POPs which can disrupt its functions and activate an inflammatory state

- In the long run, the AT constitutes an endogenous source of chronic low grade exposure. This is revealed in case of weight loss.
- The AT is also a target of toxicants which alter signaling pathways, differentiation programs and induce inflammation.

References

Barouki R (2010) Linking long-term toxicity of xeno-chemicals with short-term biological adaptation. Biochimie 92:1222–1226

Barouki R, Morel Y (2001) Repression of cytochrome P450 1A1 gene expression by oxidative stress: mechanisms and biological implications. Biochem Pharmacol 61:511–516

Debier C, Chalon C, Le Boeuf BJ et al (2006) Mobilization of PCBs from blubber to blood in northern elephant seals (*Mirounga angustirostris*) during the post-weaning fast. Aquat Toxicol 80:149–157

DeVito MJ, Ross DG, Dupuy AE Jr et al (1998) Dose-response relationships for disposition and hepatic sequestration of polyhalogenated dibenzo-p-dioxins, dibenzofurans, and biphenyls following subchronic treatment in mice. Toxicol Sci 46:223–234

Emond C, Birnbaum LS, DeVito MJ (2006) Use of a physiologically based pharmacokinetic model for rats to study the influence of body fat mass and induction of CYP1A2 on the pharmacokinetics of TCDD. Environ Health Perspect 114:1394–1400

Geyer HJ, Schramm KW, Scheunert I et al (1997) Considerations on genetic and environmental factors that contribute to resistance or sensitivity of mammals including humans to toxicity of 2,3,7,8-tetrachlorodibenzo-p-dioxin (TCDD) and related compounds. Part 1: genetic factors affecting the toxicity of TCDD. Ecotoxicol Environ Saf 36:213–230

Goldberg AL (2003) Protein degradation and protection against misfolded or damaged proteins. Nature 426:895–899

Hue O, Marcotte J, Berrigan F et al (2006) Increased plasma levels of toxic pollutants accompanying weight loss induced by hypocaloric diet or by bariatric surgery. Obes Surg 16:1145–1154

Hue O, Marcotte J, Berrigan F et al (2007) Plasma concentration of organochlorine compounds is associated with age and not obesity. Chemosphere 67:1463–1467

Iida T, Hirakawa H, Matsueda T et al (1999) Polychlorinated dibenzo-p-dioxins and related compounds in breast milk of Japanese primiparas and multiparas. Chemosphere 38:2461–2466

Imbeault P, Tremblay A, Simoneau JA, Joanisse DR (2002) Weight loss-induced rise in plasma pollutant is associated with reduced skeletal muscle oxidative capacity. Am J Physiol Endocrinol Metab 282:E574–E579

Jandacek RJ, Anderson N, Liu M et al (2005) Effects of yo-yo diet, caloric restriction, and olestra on tissue distribution of hexachlorobenzene. Am J Physiol Gastrointest Liver Physiol 288:G292–G299

Kern PA, Dicker-Brown A, Said ST et al (2002) The stimulation of tumor necrosis factor and inhibition of glucose transport and lipoprotein lipase in adipose cells by 2,3,7,8-tetrachlorodibenzo-p-dioxin. Metabolism 51:65–68

Kim MJ, Marchand P, Henegar C et al (2011) Fate and complex pathogenic effects of dioxins and polychlorinated biphenyls in obese subjects before and after drastic weight loss. Environ Health Perspect 119:377–383

Kim MJ, Pelloux V, Guyot E et al (2012) Inflammatory pathway genes belong to major targets of persistent organic pollutants in adipose cells. Environ Health Perspect, 19 Jan 2012 [Epub ahead of print]

Lassiter RR, Hallam TG (1990) Survival of the fattest: implications for acute effects of lipophilic chemicals on aquatic populations. Environ Toxicol Chem 9:585–595

Lee DH, Lee IK, Jin SH et al (2007) Association between serum concentrations of persistent organic pollutants and insulin resistance among nondiabetic adults: results from the National Health and Nutrition Examination Survey 1999–2002. Diabetes Care 30:622–628

Marchand A, Tomkiewicz C, Marchandeau JP et al (2005) 2, 3, 7,8-Tetrachlorodibenzo-p-dioxin induces insulin-like growth factor binding protein-1 gene expression and counteracts the negative effect of insulin. Mol Pharmacol 67:444–452

Morel Y, Barouki R (1999) Repression of gene expression by oxidative stress. Biochem J 342(Pt 3):481–496

Morel Y, Coumoul X, Nalpas A, Barouki R (2000) Nuclear factor I/CCAAT box transcription factor trans-activating domain is a negative sensor of cellular stress. Mol Pharmacol 58:1239–1246

Pelletier C, Imbeault P, Tremblay A (2003) Energy balance and pollution by organochlorines and polychlorinated biphenyls. Obes Rev 4:17–24

Remillard RB, Bunce NJ (2002) Linking dioxins to diabetes: epidemiology and biologic plausibility. Environ Health Perspect 110:853–858

Semenza GL (2001) HIF-1, O(2), and the 3 PHDs: how animal cells signal hypoxia to the nucleus. Cell 107:1 3

Stahl BU, Beer DG, Weber LW, Rozman K (1993) Reduction of hepatic phosphoenolpyruvate carboxykinase (PEPCK) activity by 2,3,7,8-tetrachlorodibenzo-p-dioxin (TCDD) is due to decreased mRNA levels. Toxicology 79:81–95

Part III
Endocrine Functions of Adipocyte

Chapter 12
Chatting Between the Brain and White Adipose Tissues

Luc Pénicaud and Anne Lorsignol

Introduction

Factors of neural origin play an important role in the control of energy homeostasis. Indeed, central and autonomic nervous systems are involved in the regulation of whole-body energy by regulating its different components: intake, expenditure, and storage. The metabolic or secretory activity of various tissues or organ is indeed under the control of the autonomic nervous system. This is the case for the liver, the pancreas, and the adrenal glands but there are data showing that this is true for muscles as well. The metabolic and secretory capacities of adipose tissues are also deeply controlled by autonomic nervous system.

In most mammals, two types of adipose tissue, white and brown, are present. Both are able to store energy in the form of triacylglycerols and to hydrolyze them into free fatty acids and glycerol. While white adipose tissue (WAT) provides lipids as substrates for other tissues such as muscles, brown adipose tissue uses fatty acids for heat production. Over a period of time, white fat mass reflects the balance between energy expenditure and energy intake. Remarkably, body fat mass remains relatively constant in adult suggesting that food intake and energy expenditure are linked. This has been supported by numerous studies that demonstrated the interdependency of these parameters and thus a feedback loop

L. Pénicaud (✉)
Centre des Sciences du Goût et de l'Alimentation, UMR 6265 CNRS, 1324
INRA-Université de Bourgogne, 15 rue Hugues Picardet, 21000 Dijon, France
e-mail: Luc.Penicaud@u-bourgogne.fr

A. Lorsignol
Equipe 1 "Plasticité des tissus adipeux", UMR UPS/CNRS/EFS 5273, U1031 Inserm,
BP 84 225, 31432 Toulouse, France

A. Lorsignol
Université de Toulouse, UPS; UMR 5241, BP 84 225, 31432, Toulouse, France

J.-P. Bastard and B. Fève (eds.), *Physiology and Physiopathology of Adipose Tissue*,
DOI: 10.1007/978-2-8178-0343-2_12, © Springer-Verlag France 2013

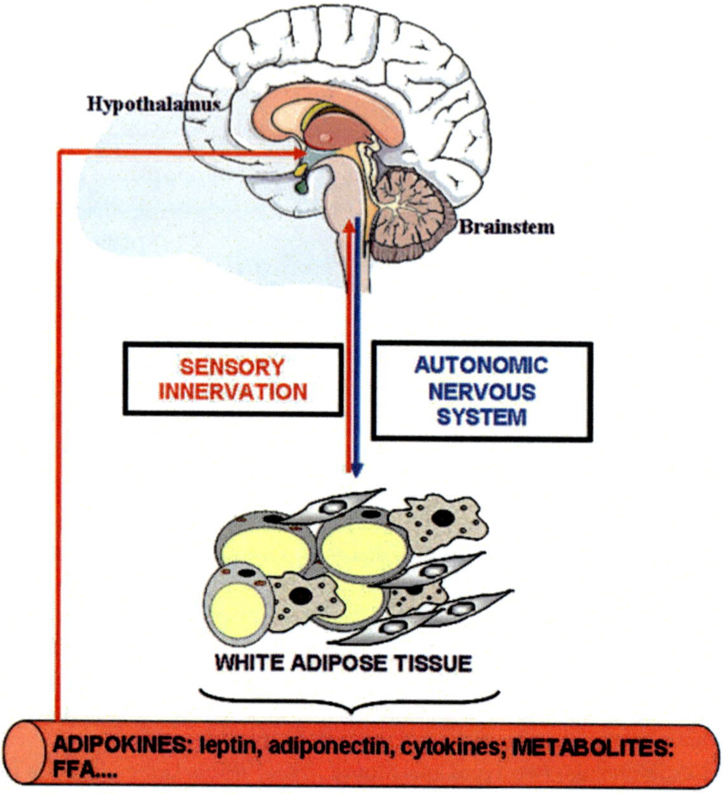

Fig. 12.1 The feedback loop between the brain and WAT

between the brain and adipose tissues with the involvement of the autonomic nervous system on one side and that of sensory fibers and metabolites or hormonal signals on the other (Fig. 12.1). This review will focus on WAT.

Efferent Innervation of White Adipose Tissue

A number of observations have clearly shown that WAT is innervated by sympathetic endings of the autonomic nervous system. The catecholaminergic fibers have initially been reported as closely associated with the blood vessels leading to implication of sympathetic nervous system (SNS) in WAT blood flow (Norman et al. 1988; Himms-Hagen 1990; Ballantyne and Raffery 1974; Slavin and Ballard 1978) (Fig. 12.2). While data questioned this in the late 1980s (Rebuffé-Scrive 1991), it was in the 1990s that Bartness and its group demonstrated direct neuroanatomical evidence for innervation of white adipocytes (Youngstrom and

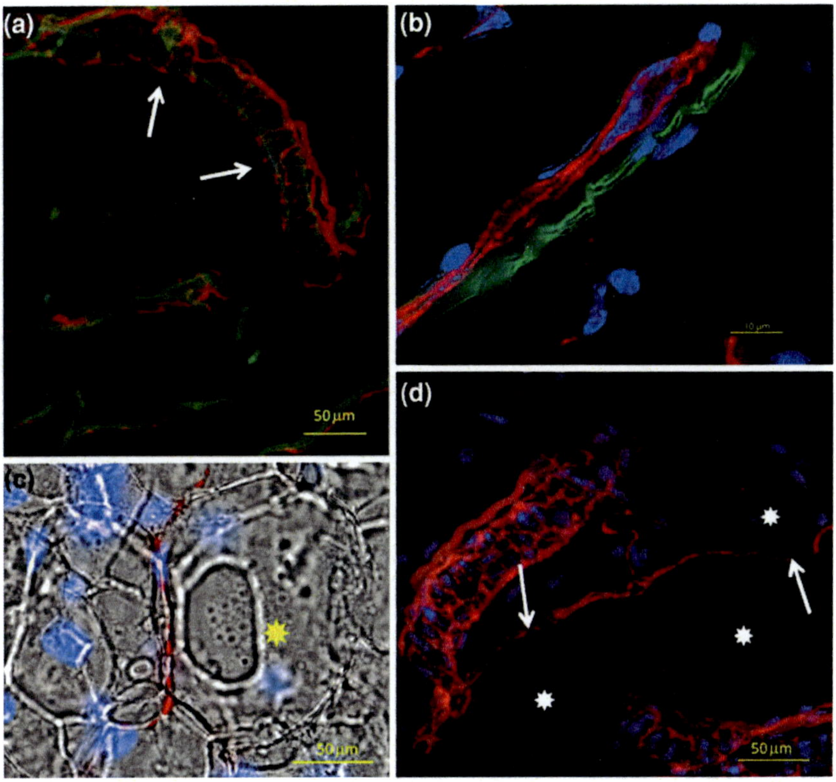

Fig. 12.2 White adipose tissue afferent innervation (Personal unpublished data). **a** Catecholaminergic innervation (*red*) of the vasculature (*green*) was observed to encapsulate the arteries and arterioles (*arrows*). **b** Nervous fibers (*green*) were also observed along capillaries (*red*). **c**, **d** Isolated sympathetic fibers (*red*) were also visualized surrounding (*arrows*) adipocytes (*asterisks*)

Bartness 1995). Although sparse, these sympathetic endings were of the "en passant" type thus allowing multiple sites of norepinephrine release. Using single neuron retrograde tracer and viral transsynaptic tracing methodologies, the sympathetic outflow from brain to WAT has been identified. Collectively, it appears that WAT receives input from central nervous system (CNS) cell groups that are part of the general SNS outflow from brain (hypothalamic nuclei, brainstem regions, intermediolateral cell groups of the spinal cord) (Bamshad et al. 1998; Bowers et al. 2004). More recently, Stanley et al. (2010) have elegantly demonstrated that most of neurons involved in the sympathetic input to WAT pads also projected to the liver, another key metabolically organ thus allowing a coordinated control of peripheral metabolism Stanley et al. (2010).

The main and classically investigated neurotransmitter of SNS is norepinephrine (NE) although these nerves contain and release other neurotransmitters among which is neuropeptide Y, as well as NE (Potter 1988; Giordano et al. 1996); these latter control lipolysis by activating different receptors subtypes present on adipocytes (Potter 1988; Giordano et al. 1996; Castan et al. 1992, 1994; Serradeil-Le Gal et al. 2000; Bradley et al. 2005; Lee et al. 2005). Adipose cells express different noradrenergic receptor subtypes (Lafontan and Berlan 1993). In WAT it has been demonstrated that the lipolytic activity of adipocytes depends on a balance between lipolysis-promoting ß-adrenergic receptor and lipolysis-inhibiting α2-adrenergic receptor (Grujic et al. 1997; Lafontan and Berlan 1995). Depending on this balance an increased sympathetic tone can lead to an increase or a decrease in lipolysis.

Most organs or tissues are innervated by both SNS and parasympathetic nervous system (PNS). For a long time it was thought that WATs did not receive parasympathetic nerves. Recent neuroanatomical studies in rats have reported parasympathetic innervation of WAT. A physiological role of such input was proposed since vagotomy was shown to reduce the insulin-dependent glucose and free fatty acid uptakes (Kreier et al. 2002). Such role of PNS can also be sustained by the demonstration of the presence of functional nicotinic receptor on white adipocytes as well as an increased insulin sensitivity of these cells under nicotine stimulation (Liu et al. 2004). However, the PNS innervation of WAT is highly controversial and remains a subject of debate (Berthoud et al. 2006; Giordano et al. 2006; Kreier and Buijs 2007).

Although adipose tissue is used as a general term, the fat pads are quite different in regard to their origin, anatomical characteristics, and functions so that one should rather speak about WATs. Indeed, both autonomic innervation (fibers density or sub-location) and the number and affinity of neurotransmitter receptors of fat depots are heterogeneous. First a relatively separated sympathetic innervation of inguinal and epididymal pads exists, since there are no overlapping patterns of labeled postganglionic cells within the sympathetic chain innervating these two deposits using fluorescent tracers (Youngstrom and Bartness 1995). In addition to this peripheral viscerotropic separation of sympathetic nerves, viscerotropy could also occur centrally, within spinal cord or brain (Bamshad et al. 1998; Kreier et al. 2002, 2006; Brito et al. 2007). Moreover, this heterogeneity of innervation may change according to nutritional status (Giordano et al. 2005). Second, taking NE turnover as an index of SNS activity, specific pattern has been delineated which might also depend of the stimulus considered (Brito et al. 2008; Bartnes et al. 2010). Altogether, these last data indicate a higher lipolysis in intra-abdominal fat pads as compared to subcutaneous one. Third, this is reinforced by the distribution of the different subclasses of receptors that depends on species, sex, and fat depot. Thus, in women, for example, it has been demonstrated that the number of α2 and ß1, ß2-adrenoceptors varies between omental, abdominal, and femoral adipose tissues and as a consequence the lipolytic response to epinephrine or NE (Lafontan and Berlan 1995; Mauriège et al. 1987, 1988; Pénicaud et al. 2000).

Effects of the Autonomic Nervous System on White Adipose Tissue Metabolism

The two main metabolic pathways of white adipocytes are on one hand the synthesis and accumulation of triglycerides and on the other their degradation into free fatty acid and glycerol (Wang et al. 2008). The increase in lipid store in adipocytes is performed by two ways. First by the direct uptake of triglycerides associated with lipoproteins coming from the circulation and which are hydrolyzed by lipoprotein lipase in non-esterified free fatty acids. These fatty acids are then transported into and in the cells by a family of fatty acid-binding protein (FABP, FAT, FATP, aP2,….). Second by the lipogenic pathways, i.e., the de novo synthesis from glucose. The last one is transported into the cell mainly via the insulin-sensitive glucose transporter isoform Glut 4. The glucose allows the synthesis of pyruvate and glycerol-3-phosphate, substrates, which will lead to the synthesis of triglycerides. Indeed, pyruvate will be utilized for the formation of acetyl-CoA and then its transformation into malonyl CoA under the control of acetyl-CoA carboxylase. The last step catalyzed by fatty acid synthase, a multi-enzyme complex, leads to the formation of long-chain fatty acids. These anabolic pathways are mainly under the control of insulin.

It is now recognized that lipolytic pathways are mainly under the dependency of three main players: adipose triglyceride lipase, hormone sensitive lipase, and perilipin A (Wang et al. 2008). In white adipocyte, both free fatty acids and glycerol are released into the adjacent blood vessels to provide fuel for other tissues. As already mentioned above catecholamines and particularly NE are the main hormones involved in the control of lipolysis. However, one has to underline that, the antilipolytic effect of insulin is predominant and thus catecholamines exert their effect when insulin level is low. From what is said above it is easy to conclude that the SNS is the main driver for adipose tissues lipolysis.

Apart from its well-known effect on lipolysis, SNS plays a role in regulating the anabolic pathways (Lafontan and Berlan 1993, 1995). Thus, it has been shown that stimulation of sympathetic nerves has no main effect on glucose uptake, utilization, and lipogenesis in WAT (Shimazu et al. 1991, 1999; Cousin et al. 1993). Whereas, as already mentioned, there are evidences that PNS innervation increases insulin sensitivity in WAT (Kreier et al. 2002).

Effects of the Autonomic Nervous System on White Adipose Tissue Secretion

Over the last 20 years the notion has emerged that WAT is not only involved in the storage and release of energy but could also be part of other physiological functions due to its capabilities in synthesis and secretion of numerous factors such as leptin, adiponectin, and many proteins involved in inflammation and immunity

(Halberg et al. 2008; Pénicaud et al. 2002). So that adipose tissue is now considered as a true endocrine organ.

The synthesis and secretion of some of these compounds are under the control of numerous factors among which the SNS via catecholamines plays a role. Leptin control has probably been the most studied. There are numerous evidences that stimulation of ß-adrenoceptor decreases the release of leptin. In human adipose tissue this occurs through a posttranslational mechanism, most likely secretion *per se*. In contrast, in rat adipose tissue, isoproterenol does not affect basal leptin secretion but has a short-term action to antagonize the insulin-stimulated leptin biosynthesis (Cammisotto and Bukowiecki 2002; Ricci et al. 2005). Also, an elegant study demonstrates a decrease leptin secretion when 3T3L1 adipocytes (a well-characterized white adipose cell line) are cultured in the presence of primary sympathetic neurons (Turtzo et al. 2001). It has then been proposed that catecholamines may mediate short-term decrease in plasma leptin that occurs within hours of fasting and cold exposure (Lee and Fried 2009).

Adiponectin is also negatively regulated by ß-adrenoceptor (Fu et al. 2007). By contrast the secretion of cytokines such as TNFα and IL6 is increased under ß-adrenergic stimulation (Mohamed-Ali et al. 2000; Vicennati et al. 2002). Overall, these data suggest that upregulation of proinflammatory cytokines and downregulation of adiponectin by ß-adrenoceptor activation may contribute to the pathogenesis of catecholamine-induced insulin resistance.

Effects of the Autonomic Nervous System on WAT Growth: Adipose Cells Proliferation, Differentiation and Angiogenesis

Fat mass is the result of two processes, i.e., the regulation of the size and the number of adipocytes. We have shown that the autonomic nervous system is indeed involved in the first one by regulating energy stores and degradation. There are also numerous evidence showing that the SNS is involved in the control of proliferation and differentiation and to a lesser extend in white adipocytes apoptosis.

Sympathetic activation would inhibit the development of WAT (Pénicaud et al. 2000; Cousin et al. 1996). Norepinephrine inhibits proliferation of adipocyte precursor cells in vitro and can be blocked by propranolol, a general ß-adrenoceptor antagonist (Jones et al. 1992). In vivo surgical or pharmacological (using 6OH-DA treatment) denervation of WAT triggers significant increase in the number of white preadipocytes and adipocytes both in rats and Siberian hamsters (Bowers et al. 2004; Cousin et al. 1993; Foster and Bartness 2006). We have been the first to demonstrate that 1 week after denervation of one retroperitoneal fat pad, DNA content was largely increased without change in the number of mature white adipocytes. Furthermore, the amount of A_2COL_6, an early marker of white adipocyte differentiation was enhanced in the denervated pad. One month later, the number of mature adipocytes was significantly increased in the denervated pad

(Cousin et al. 1993). According to the pad which has been denervated, the results were not always the same, reinforcing the idea of WAT pads heterogeneity principally depending on the innervation. A recent study using transgenic mice having a massive reduction of innervation due to the lack of Nscl-2, a neuronal-specific transcription factor, came in support of such observation (Ruschke et al. 2009). These mice present an increase in preadipocytes number and a bimodal distribution of the size of adipocytes indicating an increase in the number of small adipocytes. Moreover, recent data demonstrate that increase in sympathetic drive to WAT pads may induce emergence of brown (or brown-like) adipocytes within WAT depots, an effect likely due to $\beta 3$ receptors activation (Cao et al. 2011; Chao et al. 2011; Jimenez et al. 2003; Barbatelli et al. 2010). Altogether, these data agree with the major role of SNS in energy homeostasis, i.e., energy expenditure stimulation. Nevertheless, one must keep in mind that sympathetic efferent fibers synthesize and release other neurotransmitters such as NPY. Indeed, recent data demonstrated that sympathetic NPY release stimulates fat angiogenesis, proliferation and differentiation of new adipocytes, resulting in adipose tissue growth (Kuo et al. 2007; Ruohonen et al. 2008). These effects, which are mediated through Y1 and/or Y2 receptor subtypes can thus antagonize or minimize NE effects (Kuo et al. 2007; Yang et al. 2008). This duality would have to be analyzed in future studies on SNS outflow.

Although the importance of apoptosis in the biology of adipose tissues is still a controversial issue, there are different reports describing such process in white adipocytes. To our knowledge there is no direct demonstration of a role of the SNS in regulating the rate of apoptosis in adipose tissues; however, several observations are in support of such role. In brown adipocytes, it has been demonstrated that the proapoptotic effect of TNFα is abrogated by NE and that this neurotransmitter protects these cells from apoptosis (Nisoli et al. 1997; Navarro et al. 1998). Nothing is known concerning such effects on white adipose cells. Nevertheless, leptin is indeed known to induce a reduction in fat pads weight, this effect being observed under both peripheral and central injection. Furthermore, it has been reported that adipocyte apoptosis occurs after intracerebroventricular administration of leptin in rats (Qian et al. 1998; Hamrick et al. 2007; Gullicksen et al. 2003). On the other hand it is well demonstrated that leptin induces an increased SNS activity (Haque et al. 1999; Scarpace and Matheny 1998). From these data, it can be proposed that the signal that promotes apoptosis under leptin CNS activation is probably NE or another co-secreted neurotransmitter.

Altogether, these results demonstrate that in vivo SNS innervation acts as modulator of fat cell development.

From Adipose Tissues to the Brain

Circulating Signals

Energy balance is the result of ingestive behavior, energy expenditure, and energy storage in adipose tissue. To explain the precise overall regulation of these parameters it has been hypothetized, at first by Kennedy in the fifties, that signals generated in proportion to body fat stores will act in the brain to modulate food intake and/or energy expenditure (Kennedy 1953). Among these signals, the first to be proposed was insulin; since it was demonstrated that the pancreatic hormone increases proportional to body fat mass and acts in the CNS to reduce food intake (Wood et al. 1979; Porte et al. 2005). In 1994, Friedman and his colleagues identified and cloned the ob gene (Zhang et al. 1994; Campfield et al. 1996). They showed that it was predominantly expressed in adipose tissue and encodes a secreted peptide of 167 amino acids named leptin. Soon after, the receptor was cloned (Tartaglia et al. 1995). Several splices variants of this receptor have been identified. While the short form is expressed in various peripheral tissues, the long form (ob-Rb) is highly expressed in specific sites within the CNS (Elmquist et al. 1998). As expected, administration of leptin reduced food intake and body weight in ob/ob mice but did not in db/db mice (Halaas et al. 1995; Pelleymounter et al. 1995). Such an effect was also observed in other animal models of obesity as well as in humans. Thus, human obesity resulting from the lack of hormone can be reversed upon leptin treatment (Montague et al. 1997; Farooqi et al. 1999). Leptin is secreted in direct proportion to the amount of stored body fat (Considine et al. 1996). As a consequence, plasma leptin concentration increases while body fat rises whereas fasting and leanness lead to decreased leptin secretion.

Since the discovery of leptin other factors synthesized and released by adipocytes have been characterized and are grouped under the term adipokines. Among them, adiponectin, nesfatin, visfatin, as well as cytokines such as IL6, TNFα have been shown to be involved in energy homeostasis partly via their action in the brain (Schulz et al. 2010). Adiponectin is the most abundant protein secreted by WAT (Kadowaki and Yamauchi 2005). The protein is found in the CSF (Kubota et al. 2007; Kusminski et al. 2007). Its receptors are present in neurons of the hypothalamus known to control food intake and energy expenditure (Kubota et al. 2007; Guillod-Maximin et al. 2009). Adiponectin injected icv increases energy expenditure and reduces food intake (Kubota et al. 2007; Qi et al. 2004; Coope et al. 2008). Nesfatin (NEFA/nucleobinding2-encoded satiety- and fat-influencing protein) is an adipokine having strong anorectic effect by acting in the brain areas involved in energy homeostasis (different nuclei of the hypothalamus and brainstem) (Oh-I et al. 2006; Shimizu et al. 2009). It mainly interacts with the melanocortin system. It is synthesized by different tissues among which WAT is also present in the CNS (Oh-I et al. 2006). However, there is currently a lack of data to sustain nesfatin as a true signaling molecule since little is known on the control of its secretion by adipose cells and its transport through the blood–brain barrier.

Visfatin is mainly synthesized by visceral fat although its expression is not restricted to WAT (Fukuhara et al. 2005). It has an orexigenic effect at least in chicken and a positive correlation exists between plasma visfatin level and body fat mass and body weight in human (Cline et al. 2008; Hallschmid et al. 2009). Once again additional data are needed to definitely sustain a role of visfatin on body energy homeostasis via its action on the brain. TNFα and IL6 are secreted by adipose tissues but the main source is not the adipocyte itself but rather macrophages (Galic et al. 2010). Their release is proportional to the amount of fat and their anorectic effect has been demonstrated since many years (Galic et al. 2010; Buchanan and Johnson 2007; Langhans 2007; Dantzer et al. 1998).

Last, one also has to stress the role of nutrients of which the concentration might depend on the metabolic activity of adipose tissues such as glucose and particularly free fatty acids. Indeed both of these metabolites have been shown to play an important role as signals, reflecting energy homeostasis, to some part of the brain (Jordan et al. 2010; Lam 2010; Pénicaud et al. 2006; Wang et al. 2006). Glucose and lipids are detected by specialized fuel-sensing neurons that are incorporated in specific hypothalamic neuronal circuits. Hence, circulating nutrients cooperate with hormones (insulin) and adipokines (mainly leptin) to regulate the activity of distinct neuronal populations that control food intake, energy expenditure, and glucose homeostasis.

Sensory innervation

Apart from these circulating signals acting directly in the hypothalamus and other areas, adipose tissues sensory nerves may be part of this system. Indeed sensory innervation of WAT has been demonstrated by various facts. The identification of substance P and calcitonin gene-related peptide, markers of sensory neurons was a first demonstration (Giordano et al. 1996). Then a direct neuroanatomical demonstration was given by use of anterograde tracer (Fishman and Dark 1987). Finally, the sensory projection to different brain areas was studied by Bartness's and Kreier's groups. (Liu et al. 2004; Song et al. 2009; Shi et al. 2005). As stated by these authors, "labelling cells were found at all levels of the neuroaxis including both the nodose ganglia (visceral afferents) as well as the dorsal horn of the spinal cord (nociceptive and/or proprioceptive afferents) and in almost all the autonomic outputs areas in the brainstem and midbrain".

Although one does not know what (leptin, lipid molecules such as glycerol, free fatty acids, prostaglandins) these nerves "sense", data are in support of their role in informing the brain on lipid stores. First, when selective bilateral destruction of sensory fibers innervating epididymal fat pad was performed in hamster by injecting capsaicin; the weight of the other WAT pads (retroperitoneal and inguinal) was increased in a degree that approximated the lipid deficit if the pads had been removed by lipectomy (Shi and Bartness 2005). Second, leptin microinjection in WAT pad significantly increased electrical activity of sensory afferents emanating from the pad and elicited increase in sympathetic efferents in the contralateral pad suggestive of a reflex arc (Niijima 1998, 1999; Song et al. 2010).

Conclusions

This chapter underlines the progress made over the last years in delineating the crosstalk between WATs and the brain. Indeed data have clarified the neuroanatomical innervation of the different fat pads, the central localization of this innervation and its projections, the demonstration of sensory and parasympathetic innervation, and the brain's effects of adipokines. This represents a new field that merits further attention since they are, at least concerning the innervation by parasympathetic system, a matter of debate. The role of the autonomic nervous system in the control of adipose tissues functions has also received strong attention and led to new concept in regard not only to their metabolic and secretory activity but also to their development and plasticity even in the adults. Considering this last point, more data are needed to strongly sustain the influence of the SNS on apoptosis and also in the proliferation and differentiation of new adipose cells. To determine whether nervous control could play a role in the fate of the different progenitors (stem cells, multipotent cells), which have been demonstrated to be present in adipose tissues (Prunet-Marcassus et al. 2006; Crisan et al. 2008; Elabd et al. 2009; Casteilla and Dani 2006) and of which the role, the differentiation potential, and their regulation remain to be fully understood, appears crucial for a better view of adipose tissue biology and physiology.

Altogether this neural feedback loop between adipose tissues and the brain plays a crucial role in numerous physiological phenomenon, in particular the regulation of energy homeostasis and body fat mass and also reproduction and immune function. As it has been shown, it could be altered in numerous metabolic pathologies such as obesity, type II diabetes, and their complications. Thus, the comprehension of the biology and physiopathology of adipose tissues represents an important area of research with putative clinical implications.

Summary

Over the last decades, numerous papers have been published demonstrating the importance of the relationships between the brain and WATs in regard to body weight and metabolism regulation. Indeed the brain, mainly via the SNS, controls body fat mass both by regulating adipocytes metabolism (lipolysis and lipogenesis), secretory activity (leptin and other adipokines) as well as development. In turn fat mass will send informations to some brain areas via sensory nerves as well as via changes in metabolic and hormonal signals. Altogether these data are in support of a feedback loop between WATs and the brain. This crosstalk plays a major role in the regulation of energy homeostasis and has been shown to be altered according to physiological and pathological states.

References

Ballantyne B, Raffery AT (1974) The intrinsic autonomic innervation of white adipose tissue. Cytobios 10:187

Bamshad M, Aoki VT, Adkison MG et al (1998) Central nervous system origins of the sympathetic system outflow to white adipose tissue. Am J Physiol 276:R291–R299

Barbatelli G, Murano I, Madsen L et al (2010) The emergence of cold-induced brown adipocytes in mouse white fat depots is determined predominantly by white to brown adipocyte transdifferentiation. Am J Physiol Endocrinol Metab 298:E1244–E1253

Bartnes TJ, Shrestha Y, Vaughan CH et al (2010) Sensory and sympathetic nervous system control of white adipose tissue lipolysis. Mol Cell Endo 318:34–43

Berthoud HR, Fox EA, Neuhuber W (2006) Vagaries of adipose tissue innervation. Am J Physiol 291:R1240–R1242

Bowers RR, Festuccia WTL, Song CK et al (2004) Sympathetic innervation of adipose tissue and its regulation of fat cell number. Am J Physiol 286:R1167–R1175

Bradley RL, Mansfield JPR, Maratos-Flier E (2005) Neuropeptides, including neuropeptide Y and melanocortins, mediate lipolysis in murine adipocytes. Obseity Res 13:653–661

Brito MN, Brito NA, Baro DJ, Song CK, Bartness TJ (2007) Differential activation of the sympathetic innervation of adipose tissues by melanocortin receptor stimulation. Endocrinology 148:5339–5347

Brito NA, Brito MN, Bartness TJ (2008) Differntial sympathetic drive to adipose tissues after food deprivation, cold exposure or glucoprivation. Am J Physiol 294:R1445–R1452

Buchanan JB, Johnson RW (2007) Regulation of food intake by inflammatory cytokines in the brain. Neuroendocrinology 86:183–190

Cammisotto PG, Bukowiecki LJ (2002) Mechanisms of leptin secretion from white adipocytes. Am J Physiol 283:C244–C250

Campfield LA, Smith FJ, Burn P (1996) The ob protein (leptin) pathway—a link between adipose tissue mass and central neural networks. Horm Metab Res 28:619–632

Cao L, Choi EY, Liu X et al (2011) White to brown fat phenotypic switch induced by genetic and environmental activation of a hypothalamic-adipocyte axis. Cell Metab 14:324–338

Castan I, Valet P, Voisin T et al (1992) Identification and functional studies of a specific peptide YY-preferring receptor in dog adipocytes. Endocrinology 131:1970–1976

Castan I, Valet P, Quideau N et al (1994) Antilipolytic effects of alpha 2-adrenergic agonists, neuropeptide Y, adenosine, and PGE1 in mammal adipocytes. Am J Physiol 266:R1141–R1147

Casteilla L, Dani C (2006) Adipose tissue-derived cells: from physiology to regenerative medicine. Diabetes Metab 32:393–401

Chao PT, Yang L, Aja S et al (2011) Knockdown of NPY expression in the dorsomedial hypothalamus promotes development of brown adipocytes and prevents diet-induced obesity. Cell Metab 13:573–583

Cline MA, Nandar W, Prall BC et al (2008) Central visfatin causes orexigenic effects in chicks. Behav Brain Res 186:293–297

Considine RV, Sinha MK, Heiman ML et al (1996) Serum immunoreactive-leptin concentrations in normal- weight and obese humans. New Engl J Med 334:292–295

Coope A, Milanski M, Araujo EP et al (2008) AdipoR1 mediates the anorexigenic and insulin/leptin-like actions of adiponectin in the hypothalamus. FEBS Lett 82:1471–1476

Cousin B, Casteilla L, Lafontan M et al (1993) Local sympathetic denervation of white adipose tissue in rats induces preadipocyte proliferation without noticeable changes in metabolism. Endocrinology 33:2255–2262

Cousin B, Bascands-Viguerie N, Kassis N et al (1996) Cellular changes during cold acclimatation in adipose tissues. J Cell Physiol 167:285–289

Crisan M, Yap S, Casteilla L et al (2008) A perivascular origin for mesenchymal stem cells in multiple human organs. Stem Cells 3:301–313

Dantzer R, Bluthé RM, Gheusi G et al (1998) Molecular basis of sickness behavior. Ann N Y Acad Sci 856:132–138

Elabd C, Chiellini C, Carmona M et al (2009) Human multipotent adipose-derived stem cells differentiate into functional brown adipocytes. Stem Cells 27:2753–2760

Elmquist JK, Bjorbaek C, Ahima RS et al (1998) Distributions of leptin receptor mRNA isoforms in the rat brain. J Comp Neurol 395:535–547

Farooqi IS, Jebb SA, Langmac G et al (1999) Effects of recombinant leptin therapy in a child with congenital leptin deficiency. New Engl J Med 341:879–884

Fishman RB, Dark J (1987) Sensory innervation of white adipose tissue. Am J Physiol 253: R042–R044

Foster MT, Bartness TJ (2006) Sympathetic but not sensory denervation stimulates white adipocyte proliferation. Am J Physiol 291:1630–1637

Fu L, Isobe K, Zeng Q et al (2007) Beta-adrenoceptor agonists downregulate adiponectin, but upregulate adiponectin receptor 2 and tumor necrosis factor-alpha expression in adipocytes. Eur J Pharmacol 569:155–162

Fukuhara A, Matsuda M, Nishizawa M et al (2005) Visfatin: a protein secreted by visceral fat that mimics the effects of insulin. Science 307:426–430

Galic S, Oakhill JS, Steinberg GR (2010) Adipose tissue as ana endocrine organ. Mol Cell Endocrinol 316:129–139

Giordano A, Morroni M, Santone G et al (1996) Tyrosine hydroxylase, neuropeptide Y, substance P, calcitonin gene-related peptide and vasoactive intestinal peptide in nerves of rat periovarian adipose tissue: an immunohistochemical and ultrastructural investigation. J Neurocyto 25:125–136

Giordano A, Frontini A, Murano I et al (2005) Regional-dependent increase of sympathetic innervation in rat white adipose tissue during prolonged fasting. J Histochem Cytochem 53:679–687

Giordano A, Song CK, Bowers RR et al (2006) Hite adipose tissue lacks significant vagal innervation and immunohistochemical evidence of parasympathetic innervation. Am J Physiol 291:R1243–R1255

Grujic D, Susulic VS, Harper ME et al (1997) Beta3-adrenergic receptors on white and brown adipocytes mediate beta3-selective agonist-induced effects on energy expenditure, insulin secretion, and food intake. A study using transgenic and gene knockout mice. J Biol Chem 272:17686–17693

Guillod-Maximin E, Roy AF, Vacher CM et al (2009) Adiponectin receptors are expressed in hypothalamus and colocalized with proopiomelanocortin and neuropeptide Y in rodent arcuate neurons. J Endocrinol 200:93–105

Gullicksen PS, Della-Fera MA, Baile CA (2003) Leptin-induced adipose apoptosis: Implications for body weight regulation. Apoptosis 8:327

Halaas JL, Gajiwala KS, Maffei M et al (1995) Weight reducing effect of the plasma protein encoded by the obsess gene. Science 269:543–546

Halberg N, Wernstedt-Asterholm I, Scherer PE (2008) The adipocyte as an endocrine cell. Endocrinol Metab Clin N Am 37:753–768

Hallschmid M, Randeva H, Tan BK et al (2009) Relationship between cerebrospinal fluid visfatin (PBEF/Nampt) levels and adiposity in humans. Diabetes 58:637–640

Hamrick MW, Della Fera MA, Choi YH et al (2007) Injections of leptin into rat ventromedial hypothalamus increase adipocyte apoptosis in peripheral fat and in bone marrow. Cell Tissue Res 327:133
Haque MS, Minokoshi Y, Hamai M et al (1999) Role of the sympathetic nervous system and insulin in enhancing glucose uptake in peripheral tissues after intrahypothalamic injection of leptin in rats. Diabetes 48:1706
Himms-Hagen J (1990) Brown adipose thermogenesis: interdisciplinary studies. FASEB J 4:2890–2898
Jimenez M, Barbatelli G, Allevi R, Cinti S et al (2003) β3-adrenoceptor knockout in C57BL/6 J mice depresses the occurence of brown adipocytes in white fat. Eur J Biochem 270:699–705
Jones DD, Ramsay TG, Hausman GJ, Martin RJ (1992) Norepinephrine inhibits rat pre-adipocyte proliferation. Int J Obes 16:349–354
Jordan SD, Könner AC, Brüning JC (2010) Sensing the fuels: glucose and lipid signaling in the CNS controlling energy homeostasis. Cell Mol Life Sci 67:3255–3273
Kadowaki T, Yamauchi T (2005) Adiponectin and adiponectin receptors. Endocr Rev 26:439–451
Kennedy GC (1953) The role of depot fat in the hypothalamic control of food intake in the rat. Proc R Soc Lond 140:578–596
Kreier F, Buijs RM (2007) Evidence for parasympathetic innervation of white adipose tissue, clearing up some vagaries. Am J Physiol 293:R548–R549
Kreier F, Fliers E, Voshol PJ et al (2002) Selective parasympathetic innervation of subcutaneous and intra-abdominal fat-functional implications. J Clin Invest 110:1243–1250
Kreier F, Kap YS, Mettenleiter TC et al (2006) Tracing from fat tissue, liver, and pancreas: a neuroanatomical framework for the role of the brain in type 2 diabetes. Endocrinol 147:1140–1147
Kubota N, Yano W, Kubota T et al (2007) Adiponectin stimulates AMP-activated protein kinase in the hypothalamus and increases food intake. Cell Metab 6:55–68
Kuo LE, Kitlinska JB, Tilan JU et al (2007) Neuropeptide Y acts directly in the periphery on fat tissue and mediates stress-induced obesity and metabolic syndrome. Nature Med 13:803–811
Kusminski CM, McTernan PG, Schraw T et al (2007) Adiponectin complexes in human cerebrospinal fluid: distinct complex distribution from serum. Diabetologia 50:634–642
Lafontan M, Berlan M (1993) Fat cell adrenergic receptor and the control of white and brown fat cell function. J Lipid Res 34:1057–1091
Lafontan M, Berlan M (1995) Fat cell α2-adrenoceptors: the regulation of fat cell function and lipolysis. Endorine Rev 16:716–738
Lam TK (2010) Neuronal regulation of homeostasis by nutrient sensing. Nat Med16:392
Langhans W (2007) Signals generating anorexia during acute illness. Proc Nutr Soc 66:321–330
Lee MJ, Fried SK (2009) Integration of hormonal and nutrient signals that regulate leptin synthesis and secretion. Am J Physiol Endocrinol Metab 296:E1230–E1238
Lee H, Jun DJ, Suh BC, Choi BH et al (2005) Dual roles of purinergic receptors in insulin-stimulated leptin production and lipolysis in differentiated rat white adipocytes. J Biol Chem 280:28556–28563
Liu RH, Mizuta M, Matsukura S (2004) The expression and functional role of nicotinic acetylcholine receptors in rat adipocytes. JPET 310:52–58
Mauriège P, Galitzky J, Berlan M, Lafontan M (1987) Heterogeneous distribution of beta and alpha-2 adrenoceptor binding sites in human fat cells from various fat deposits: functional consequences. Eur J Clin Invest 17:156–165
Mauriège P, De Pergola G, Berlan M, Lafontan M (1988) Human fat cell beta-adrenergic receptors: beta-agonist-dependent lipolytic responses and characterization of beta-adrenergic binding sites on human fat cell membranes with highly selective beta 1-antagonists. J Lipid Res 29:587–601
Mohamed-Ali V, Bulmer K, Clarke D et al (2000) Beta-Adrenergic regulation of proinflammatory cytokines in humans. Int J Obes Relat Metab Disord 24(Suppl 2):S154–S155

Montague CT, Farooqi IS, Whitehead JP et al (1997) Congenital leptin deficiency is associated with severe early-onset obesity in humans. Nature 387:903–908

Navarro P, Valverde AM, Benito M, Lorenzo M (1998) Insulin/IGF-I rescues immortalized brown adipocytes from apoptosis down-regulating Bcl-xS expression, in a PI 3 kinase- and map kinase dependent manner. Exp Cell Res 15:213

Niijima A (1998) Afferent signals from leptin sensors in the white adipose tissue of the epididymis, and their reflex effect in the rat. J Auton Nerv Syst 73:19–25

Niijima A (1999) Reflex effects from leptin sensors in the white adipose tissue of the epididymis to the efferent activity of the sympathetic and vagus nerve in the rat. Neurosci Lett 262:125–128

Nisoli E, Briscini L, Tonello C et al (1997) Tumor necrosis factor-alpha induces apoptosis in rat brown adipocytes. Cell Death Differ 4:771–778

Norman D, Mukherjee S, Symons D et al (1988) Neuropeptides in interscapular and perirenal brown adipose tissue in the rat: a plurality of innervation. J Neurocytol 17:305–311

Oh-I S, Shimizu H, Satoh T et al (2006) Identification of nesfatin-1 as a satiety molecule in the hypothalamus. Nature 443:709–712

Pelleymounter MA, Cullen MJ, Baker MB, Hecht R, Winters D, Boone T, Collins F (1995) Effect of the obese gene product on body weight regulation in ob/ob mice. Science 269:540–543

Pénicaud L, Cousin B, Leloup C et al (2000) The autonomic nervous system, adipose tissue plasticity and energy balance. Nutrition 16:903–908

Pénicaud L, Cousin B, Laharrague P et al (2002) Adipose tissues as part of the immune system: role of leptin and cytokines. In: Kordon C (ed) Brain somatic cross-talk and the central control of metabolism. Springer, New York, p 81

Pénicaud L, Leloup C, Fioramonti X et al (2006) Brain glucose sensing: a subtle mechanism. Curr Opin Clin Nutr Metab Care 9:458–462

Porte D Jr, Baskin DG, Schwartz MW (2005) Insulin signaling in the central nervous system: a critical role in metabolic homeostasis and disease from *C. elegans* to humans. Diabetes 54:1264–1276

Potter K (1988) Neuropeptide Y as an autonomic neurotransmitter. Pharmacol Ther 37:251

Prunet-Marcassus B, Cousin B et al (2006) From heterogeneity to plasticity in adipose tissues: site-specific differences. Exp Cell Res 312:727–736

Qi Y, Takahashi N, Hileman SM, Patel HR et al (2004) Adiponectin acts in the brain to decrease body weight. Nat Med 10:524–529

Qian H, Azain MJ, Compton MM et al (1998) Brain administration of leptin causes deletion of adipocytes by apoptosis. Endocrinology 139:791

Rebuffé-Scrive M (1991) Neuroregulation of adipose tissue: molecular and hormonal mechanisms. Int J Obes 15:83–86

Ricci MR, Lee MJ, Russell CD et al (2005) Isoproterenol decreases leptin release from rat and human adipose tissue through posttranscriptional mechanisms. Am J Physiol 288:E798–E804

Ruohonen ST, Pesonen U, Moritz N et al (2008) Transgenic mice overexpressing neuropeptide Y in noradrenergic neurons. A novel model of increased adiposity and impaired glucose tolerance. Diabetes 57:1517–1525

Ruschke K, Ebelt H, Klöting N et al (2009) Defective peripheral nerve development is linked to abnormal architecture and metabolic activity of adipose tissue in Nscl-2 mutant mice. PLoS One 4:e5516

Scarpace PJ, Matheny M (1998) Leptin induction of UCP1 gene expression is dependent on sympathetic innervation. Am J Physiol 275:E259–E264

Schulz C, Paulus K, Lehnert H (2010) Adipocyte-brain: crosstalk. Results Probl Cell Differ 52:189–201

Serradeil-Le Gal C, Lafontan M, Raufaste D et al (2000) Characterization of NPY receptors controlling lipolysis and leptin secretion in human adipocytes. FEBS Lett 475:150–156

Shi H, Bartness TJ (2005) White adipose tissue sensory nerve denervation mimics lipectomy-induced compensatory increases in adiposity. Am J Physiol 289:R514–R520

Shi H, Song CK, Giordano A, Cinti S, Bartness TJ (2005) Sensory or sympathetic white adipose tissue denervation differentially affects depot growth and cellularity. Am J Physiol 288:R1028–R1037
Shimazu T, Sudo M, Minokoshi Y, Takahashi A (1991) Role of the hypothalamus in insulin dependent glucose uptake in peripheral tissues. Brain Res Bull 27:501–504
Shimizu Y, Nikami H, Saito M (1999) Sympathetic activation of glucose utilization in brown adipose tissue in rats. J Biochem 110:688–692
Shimizu H, Oh I, Hashimoto K et al (2009) Peripheral administration of nesfatin-1 reduces food intake in mice: the leptin-independent mechanism. Endocrinology 150:662–671
Slavin BG, Ballard KW (1978) Morphological studies of the adrenergic innervation of white adipose tissue. Anta Rec 191:377–389
Song CK, Schwartz GJ, Bartness TJ (2009) Anterograde transneuronal viral tract tracing reveals central sensory circuits from white adipose tissue. Am J Physiol 296:R501–R511
Song CK, Schwartz GJ, Lester B et al (2010) Leptin injected into white adipose tissue stimulates sensory nerves. In: Neuroscience meeting planner, Society for Neuroscience, San Diego
Stanley S, Pinto S, Segal J et al (2010) Identification of a neuronal subpopulations that project from hypothalamus to both liver and adipose tissue polysynaptically. Proc Natl Acad Sci U S A 107:7024–7029
Tartaglia IA, Dembski M, Weng X et al (1995) Identification and expression cloning of a leptin receptor OB-R. Cell 83:1263–1271
Turtzo LC, Marx R, Lane MD (2001) Cross-talk between sympathetic neurons and adipocytes in coculture. Proc Natl Acad Sci U S A 98:12385–12390
Vicennati V, Vottero A, Friedman C, Papanicolaou DA (2002) Hormonal regulation of interleukin-6 production in human adipocytes. Int J Obes Relat Metab Disord 6:905–911
Wang R, Cruciani-Guglielmacci C, Migrenne S et al (2006) Effects of oleic acid on disctint populations of neurons in the hypothalamus arcuate nucleux are dependent on extracellular glucose levels. J Neurophysiol 95:1491–1498
Wang S, Soni KG, Semache M et al (2008) Lipolysis and the integrated physiology of lipid energy metabolism. Mol Genet Metab 95:117–126
Wood S, Lotter E, Mc Kay L, Porte DJ (1979) Chronic intracerebroventricular infusion of insulin reduces food intake and body weight of baboons. Nature 282:503–505
Yang K, Guan H, Arany E et al (2008) Neuropeptide Y is produced in visceral adipose tissue and promotes proliferation of adipocyte precursor cells via the Y1 receptor. FASEB J 22:2452–2464
Youngstrom TG, Bartness TJ (1995) Cathecolaminergic innervation of white adipose tissue in the Siberian hamster. Am J Physiol 268:R744–R751
Zhang Y, Proenca R, Maffei M et al (1994) Positional cloning of the mouse obese gene and its human homologue. Nature 372:425–432

Chapter 13
Adiponectin: An Adipokine with Multiple Faces

Jean-Philippe Bastard, Camille Vatier and Bruno Fève

Introduction

Adipose tissue (AT) is now recognized as a secretory organ capable of synthesizing a wide variety of factors called adipokines (molecules synthesized and secreted by AT acting locally in an autocrine–paracrine way, or in an endocrine manner), which some are able to modulate insulin sensitivity and energy balance (Bastard et al. 2006; Antuna-Puente et al. 2008). These adipokines have pleiotropic physiological effects. It is now clear that they participate to the genesis of numerous metabolic disorders and their complications such as diabetes and

J.-P. Bastard (✉) · C. Vatier · B. Fève
Institut des maladies cardiométaboliques et de la nutrition (IHU ICAN), Paris, France
e-mail: jean-philippe.bastard@tnn.aphp.fr

B. Fève
Faculté de Médecine Pierre et Marie Curie site Saint-Antoine, INSERM, UPMC, UMR_S938, Paris, France
e-mail: bfeve@free.fr

J.-P. Bastard
Faculté de Médecine Pierre et Marie Curie site Saint-Antoine, INSERM, UPMC, UMR_S938, 27 rue Chaligny, 75012 Paris, France

B. Fève
Service d'Endocrinologie-Métabolisme, Hôpital Saint-Antoine, 27 rue Chaligny, 75571 Paris cedex 12, France

C. Vatier
Faculté de Médecine Pierre et Marie Curie site Saint-Antoine, INSERM, UPMC, UMR_S938, 27 rue Chaligny, 75012 Paris, France
e-mail: camille.vatier@inserm.fr

J.-P. Bastard
Service de Biochimie et Hormonologie, Hôpital Tenon, 4 rue de la Chine, 75970 Paris cedex 20, France

J.-P. Bastard and B. Fève (eds.), *Physiology and Physiopathology of Adipose Tissue*,
DOI: 10.1007/978-2-8178-0343-2_13,

cardiovascular diseases. AT is a heterogeneous tissue composed of adipocytes and multiple other cell types including preadipocytes, fibroblasts, endothelial, macrophagic cells, and the other cells of the immune system constituting the stroma vascular fraction (SVF). Some adipokines are mainly produced by adipocytes, whereas others are preferentially secreted by SVF cells, and in particular by macrophages. In this chapter, we shall present the properties of an adipokine mainly produced by adipocytes: adiponectin.

Discovery, Structural Data, Receptors, Regulation, and Metabolism of Adiponectin

Initially, the group of Lodish brought into light a new protein of 30-kDa secreted by the adipocyte, adipocyte complement-related protein of 30-kDa (Acrp30). Acrp30 was exclusively expressed in adipocytes and its mRNA levels were increased by a factor 100 during adipocyte differentiation (Scherer et al. 1995). At the same time as this discovery, Maeda et al. (1996) cloned the cDNA of a new AT factor related to collagen, which expression level was particularly high. For this purpose, it was named adiposity most abundant transcript gene 1 (apM1). By using a technique of differential display, the group of Spiegelman identified the cDNA of a new molecule of AT, and called it adipoQ. The peculiarity of this adipoQ was that its expression, which was very high, was also specific of AT but reduced in obesity (Hu et al. 1996). Finally, Nakano et al. (1996) described a new protein, gelatin-binding protein 28 (GBP 28), purified from human plasma. Thus, initially described between 1995 and 1996 under various names (A crp30 and adipoQ) for mouse (Scherer et al. 1995; Hu et al. 1996) and in human (GBP28 and ApM1) (Maeda et al. 1996; Nakano et al. 1996) by different and independent teams, this particular adipokine is now recognized as adiponectin. The circulating concentrations of adiponectin, which represent 0.01 % of plasma proteins, are around 5–30 mg/L in physiological conditions while those of leptin, for example, are only around 2–8 μg/L. As for leptin, there is a sex ratio for adiponectin levels with lower serum contents in men than in women. Hormonal regulation of adiponectin was suggested since testosterone inhibits synthesis and secretion of this adipokine in both rodent and human (Wang et al. 2008). The adiponectin gene consisted of three exons and two introns, and extends over 16 kb. Only two exons are translated into a protein of 244 amino acids in human. It consists of four domains with a signal sequence in the amino-terminal domain, followed by a variable region without specificity, a collagen domain, and a carboxy-terminal globular domain. Its mass is 30 kDa in mice and 28 kDa in human. The major part of the biological activity of adiponectin is attributable to the *C*-terminal domain. Further to its secretion, the monomer of adiponectin undergoes important translational modifications such as hydroxylation and glycation on specific proline and lysine residues. These translational modifications may play a crucial role in the formation of adiponectin oligomers and are essential for the inhibition of hepatic neoglucogenesis and for binding the ligand to

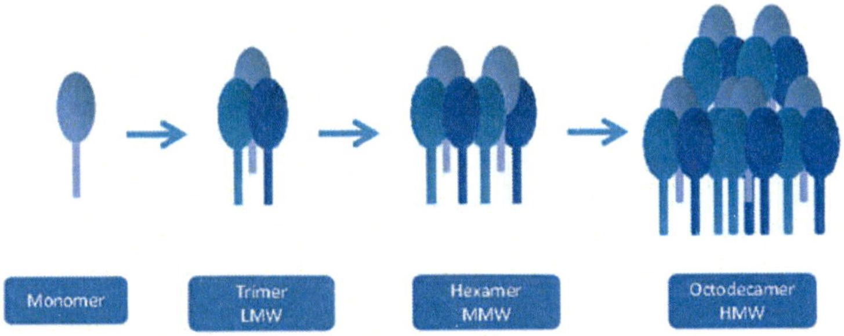

Fig. 13.1 Polymerization of adiponectin. *LMW* low molecular weight, *MMW* medium molecular weight, *HMW* high molecular weight

its membrane receptors (Wang et al. 2008). The secretion of adiponectin oligomers is controlled in the endoplasmic reticulum by chaperone proteins such as ERp44, which prevents the release of adiponectin oligomers, or Ero1-Lα, which on the contrary facilitates the release of high molecular mass complexes trapped by ERp44 (Wang et al. 2008). The assembly of the globular parts by hydrophobic interactions allows trimer formation while the interactions in the collagen domain level (disulfides bridges) allow hexamer formation and the high molecular mass forms (2–6 trimers). Thus, adiponectin can be found in the circulation under various molecular forms (Fig. 13.1), including low molecular weight corresponding to trimers, medium molecular weight corresponding to hexamers (assembly of two trimers) and high molecular weight (HMW), which corresponds to an assembly of three hexamers (Wang et al. 2008). HMW adiponectin is the main circulating form (>80 %) and would be the most active form of adiponectin (Wang et al. 2008). However, the physiological roles and the determinants that drive these various circulating forms of adiponectin are not still clearly clarified. Concerning the clearance and the degradation of adiponectin, its half-life would be of the order of 1 h and a half in human, while it would be shorter in mice. However, these data remain to be confirmed (Hoffstedt et al. 2004; Halberg et al. 2009). The liver would be involved in a major way in the clearance of adiponectin in mice (Halberg et al. 2009). This is in agreement with observations in cirrhotic patients, which report an increased circulating adiponectin concentration in these patients (Tietge et al. 2004). Nevertheless, a renal excretion is also suggested because adiponectin is found in urine, in particular in patients with renal damage (von Eynatten et al. 2009). Remarkably, adiponectinemia is increased, in adults (Stenvinkel et al. 2004; Beige et al. 2009) and in children with renal failure (Möller et al. 2012). Moreover, a decrease of adiponectinemia is observed in these patients after renal transplantation (Cui et al. 2011). Thus, the clearance of adiponectin is likely mainly hepatic, but a significant renal component must be also considered.

Two isoforms of adiponectin receptor (AdipoR1 and R2) have been cloned (Yamauchi et al. 2003). AdipoR1 expression is ubiquitous but higher in skeletal muscle. It displays a higher affinity for the globular form of adiponectin. AdipoR2 is predominantly expressed in the liver and possesses a higher affinity for the complete form of adiponectin (Yamauchi et al. 2003; Kadowaki et al. 2006). Invalidation of these receptors demonstrates their importance in the transmission of adiponectin signal (Yamauchi et al. 2007). Besides these two isoforms, the T-cadherin receptor expressed in endothelial and smooth muscle cells was also identified as a potential receptor of adiponectin (Hug et al. 2004).

Adiponectin has several physiological specificities as compared other adipokines. Although initially considered as an adipokine produced and secreted mainly by AT and in particular adipocytes (Bastard et al. 2006; Antuna-Puente et al. 2008), it was recently shown that other cell types as cardiomyocytes, skeletal muscle cells, osteoblasts, endothelial cells, or even pituitary cells could produce adiponectin.

Circulating concentrations of adiponectin are decreased in insulin resistant obese subjects, in type 2 diabetic patients or in patients with metabolic syndrome, and there is an inverse relationship between adiponectinemia and body mass index (BMI) in particular with the visceral AT (Arita et al. 1999; Staiger et al. 2003). A recently formulated hypothesis suggests that in a context of insulin resistance, the increased insulin plasma levels could inhibit the production of adiponectin (Cook and Semple 2010). Indeed, insulin can inhibit adipocyte adiponectin production and this property would not be altered in the case of "common" insulin resistance (Cook and Semple 2010). Nevertheless, it is only a hypothesis and numerous works are currently in progress to better understand the biological effects of adiponectin and its regulation. On the other hand, other studies performed in animal models showed that injection of recombinant adiponectin in insulin resistant obese or lipoatrophic mice dramatically improves their metabolic disturbances, in particular at the lipid metabolism level (Yamauchi et al. 2001).

Adiponectin, Insulin Sensitivity, and Insulin Resistance

The mechanism of action of adiponectin is still a source of important investigations. Nevertheless, although the signaling pathways of adiponectin receptors are not completely clarified, several molecules were shown to mediate its intracellular signaling and mechanism of action. The biological effects of adiponectin could partly involve the stimulation of fatty acids oxidation in both skeletal muscles and liver, the stimulation of glucose transport in muscle and a decrease of hepatic gluconeogenesis via the activation of AMP-activated protein kinase (AMPK) (Tomas et al. 2002; Yamauchi et al. 2002). Furthermore, recent data suggest that the signal transduction of adiponectin would imply an adaptor protein called APPL1 (adaptor protein, phospho-tyrosine interaction, PH domain, and leucine zipper containing 1), since the modulation of its expression into the muscle is associated with adiponectin capacities to stimulate AMPK activity (Mao et al. 2006). On the other hand, APPL2 isoform was reported to have a negative effect on adiponectin signaling (Wang et al. 2009). One of the roles of AMPK is to modulate cell concentrations of malonyl-CoA by inhibiting acetyl-CoA carboxylase that is the limiting enzyme of acetyl-CoA transformation into malonyl-CoA, and by stimulating malonyl-CoA decarboxylase, which occurs in the degradation of malonyl-CoA. This reduces the concentration of malonyl-CoA, leading to a decrease of lipogenesis and an increase of the mitochondrial carnitine-palmitoyl transferase-1 (CPT1) activity with a subsequent induction of fatty acid mitochondrial entry of and beta-oxidation that improves, *in fine*, insulin sensitivity. Furthermore, the increase in skeletal muscle lipid oxidation could be strengthened by the stimulation by adiponectin of Peroxisome proliferator-activated receptor-α (PPAR-α) activity (Yamauchi et al. 2001).

In agreement with these pathophysiological data, numerous studies reported an inverse relationship between insulin resistance and the circulating concentrations of adiponectin in several pathologies with a high cardiovascular risk such as obesity, metabolic syndrome, and type 2 diabetes (Bastard et al. 2006; Antuna-Puente et al. 2008). However, this association between adiponectin and insulin sensitivity seems less evident when considering studies in patients with severe insulin resistance or in patients with genetic lipodystrophies (Cook and Semple 2010). Thus, patients with mutations in the insulin receptor gene or with anti-insulin receptor antibodies, (type B insulin resistance) leading to a loss of function of the receptor, exhibit very high plasma levels, higher to those found in noninsulin resistant controls (Cook and Semple 2010; Semple et al. 2006, 2008; Antuna-Puente et al. 2010). Although the pathophysiological mechanism is not clarified, it is likely that the complete absence of insulin signaling pathway activation is involved in this major increase in circulating adiponectin levels. Indeed, in the context of type B insulin resistance, the circulating concentrations of adiponectin decrease in parallel with the reduction of anti-insulin receptor antibodies (Semple et al. 2008). Moreover, this is associated with the clinical improvement of the insulin resistance syndrome (Semple et al. 2008). Although the regulation of adiponectin secretion is not fully understood, this suggests that adiponectin measurement could help diagnosis in patients affected by extreme insulin resistance for searching mutations of the insulin receptor gene or the presence of anti-insulin receptor antibodies.

Studies carried out in patients affected by genetic lipodystrophies show that the concentrations of adiponectin, as those of leptin, were low according to the loss of fat mass (Semple et al. 2006; Antuna-Puente et al. 2010). Nevertheless, a peculiarity is present in patients presenting a generalized congenital lipoatrophy (Berardinelli-Seip syndrome) because, according to the involved gene, adiponectinemia is either not measurable or at values similar to those found in HIV lipoatrophic patients (Antuna-Puente et al. 2010). Indeed, adiponectin is undetectable in patients carrying a mutation of the gene coding for the enzyme 1-acylglycerol-3-phosphate-O-acyltransferase 2 (AGPAT2) involved in the metabolism of triglycerides. On the contrary, adiponectinemia is about 3 mg/L in patients with a mutation in the gene coding for seipin, which function is not still clearly clarified (Antuna-Puente et al. 2010). Once again, although the involved mechanisms are not identified, plasma adiponectin measurement allows to direct the genetic diagnosis to the genes of interest in patients with congenital generalized lipodystrophy.

Anti-Inflammatory and Anti-Atherogenic Effects of Adiponectin

Besides its insulin sensitizing effects, adiponectin also displays a vascular protective effect by acting at early steps in the process of atherogenesis through inhibiting the expression of adhesion molecules on the vascular endothelial cells, the transformation of macrophages into foam cells, and the proliferation of smooth muscle cells (Ouchi et al. 1999, 2001; Arita et al. 2002). Adiponectin also has anti-inflammatory properties that could modulate metabolic functions. Thus, several experiments support a role of adiponectin to regulate macrophage recruitment at the inflammatory site level. This could act via a modulation of reactive oxygen species production (Ouedraogo et al. 2006). In agreement with this hypothesis, there is a negative correlation between circulating concentrations of the oxidized low-density lipoprotein and those of adiponectin in type 2 diabetic patients or in subjects with a coronary disease (Lautamaki et al. 2007).

Numerous anti-inflammatory properties of adiponectin may depend on anti-Tumor necrosis factor-α (TNF-α) effects. This may contribute to its protective role in atherosclerosis and also to its insulin sensitive properties. Adiponectin can decrease the inflammatory response induced by TNF-α. Accordingly, in vitro studies show that adiponectin treatment induces a loss of macrophage activity and reduces macrophage production of TNF-α (Ouedraogo et al. 2006). Other in vitro experiments indicate that adiponectin counteracts the effects of hyperglycaemia and proinflammatory cytokines on endothelial cells, thus reducing the risk of atherosclerosis development (Goldstein et al. 2009). Thus, it was shown that adiponectin knockout mice were more prone to develop microvascular inflammation and cardiovascular disorders, in agreement with numerous epidemiological studies showing a higher cardiovascular risk in subjects with a low adiponectinemia (Ouedraogo et al. 2007). Furthermore, treatment with globular adiponectin improves the endothelial microinflammatory status in adiponectin deficient mice, suggesting a potential therapeutic role for this molecule (Ouedraogo et al. 2007). On the other hand, the proinflammatory cytokines TNF-α and interleukin-6 decrease adiponectin expression in human adipocytes (Lautamaki et al. 2007). Hence, it was reported that the neutralization of TNF-α increased adiponectinemia in patients affected by rheumatoid arthritis (Komai et al. 2007). Nevertheless, the inter-relationship between TNF-α and adiponectin is not fully clarified in inflammatory diseases and requires further studies.

Adiponectin and Renal Pathology

The presence of increased albuminuria, which is associated with obesity, high blood pressure and diabetes, is a risk factor of cardiovascular and renal diseases. Recently, adiponectin was involved in the pathophysiology of increased albuminuria. Thus, a negative correlation between adiponectinemia and albuminuria was shown in obese patients (Komaba et al. 2006; Lenghel et al. 2012). Furthermore, experiments in adiponectin-deficient mice demonstrate that increased albuminuria is associated with renal cell disorders characterized by a fusion of podocyte feet (Sharma et al. 2008). Cell culture studies of podocytes indicate that adiponectin, via the activation of AMPK, reduces the permeability of podocytes to albumin and podocyte's dysfunction (Sharma et al. 2008). These effects are associated with a lower oxidative stress that is related to a decrease of the NADPH oxidase NOX4 expression in podocytes (Sharma et al. 2008). Interestingly, following a treatment by adiponectin, a normalization of albuminuria was observed in association with improvement regression of the podocyte feet fusion. This suggests that in this animal model, adiponectin could have a major protective role in the pathophysiology of albuminuria (Sharma et al. 2008). However, the links between albuminuria and adiponectinemia remain to confirm in human. Indeed, it was reported that a treatment by thiazolidinediones, which increases the circulating concentrations of adiponectin, improves microalbuminuria in diabetic patients and suggest a beneficial role of adiponectin in renal pathophysiology (Pistrosch et al. 2005; Miyazaki et al. 2007). However, other studies underline that adiponectinemia is higher in patients with renal disorders exhibiting an increased cardiovascular risk, and that high adiponectin values are predictive of a higher risk of mortality (Stenvinkel 2011). For example, the MDRD study indicates that each one mg/L increase in adiponectin plasma levels is associated with a 6 % increase of cardiovascular mortality risk (Menon et al. 2006). Another study carried out in type 1 diabetic patients showed that high adiponectin plasma concentrations are associated with a higher risk of renal failure development and also of mortality (Jorsal et al. 2008). Although these results appear paradoxical considering the anti-atherogenic and anti-inflammatory properties of adiponectin, the increase of adiponectin in patients with chronic kidney disease could be the resultant of several associated mechanisms. For certain authors, the high rates of adiponectin are likely owed to a reduction of its renal clearance (Tentolouris et al. 2004; Komura et al. 2010). Alternatively, they could correspond to an adaptive response of the organism to counteract the development of vascular complications associated with chronic kidney disease. Finally, renal failure-associated hyperadiponectinemia may be due to an increase of energy expenditure, and in this context would reflect a higher risk of mortality. A recent study showed that when the patients with wasting syndrome were removed from the statistical analysis, the increase of adiponectinemia loses its predictive value of risk mortality in patients with renal dialysis (Ohashi et al. 2008). Collectively, these data illustrate the complexity of adiponectin regulation in human pathology and the difficulty for interpreting an adiponectin plasma values in term of patient's risk–profit, particularly in the case of chronic kidney disease.

Adiponectin and Cancer

Obesity is associated with a higher risk and a poorer prognosis of several cancers, but the mechanisms that underlie the links between obesity and tumor progression still remain unresolved. Besides oxidative stress, steroid hormones or insulin growth factor-1 (IGF-1), there are numerous clinical and experimental data suggesting that adipokines, or rather a disorder of the adipokine profile, could participate in tumor progression in a context of energy imbalance.

Low circulating concentrations of adiponectin are associated with a higher risk of breast cancer in postmenopausal women, independently of BMI, leptin, or IGF-1 concentrations (Körner et al. 2007; Wang et al. 2006). This suggests that adiponectin could play a role in the etiology of breast cancer, in particular in an environment devoid of oestrogens.

Likewise, low adiponectin plasma values are associated with the risk of endometrial cancer. Women with high BMI and a low adiponectinemia have a 6.5 times increased risk of developing an endometrial cancer (Dal Maso et al. 2004; Cust et al. 2007). This is independent from variations of IGF-1, leptin, and from the other classical risk factors for this type of cancer.

The risk of colorectal cancer is also negatively correlated with adiponectinemia in human (Wei et al. 2005; Guadagni et al. 2009), and always independently from BMI. Interestingly, colon cancers express high levels of adiponectin receptors, AdipoR1 and AdipoR2. Adiponectin gene variants could also influence the risk of colorectal cancer (Kaklamani et al. 2008).

Adiponectinemia is negatively correlated to the risk of prostate cancer (Michalakis et al. 2007), and the plasma values of this adipokine are lower in individuals affected by a prostatic adenocarcinoma as compared to patients suffering from a benign prostatic hypertrophy (Goktas et al. 2005). Adiponectinemia is negatively correlated to the histological rank and to the stage of the prostate cancer development.

Adiponectinemia is also lower in the context of gastric cancer, and is inversely related to both the tumor size and its invasivity (Ishikawa et al. 2005). The same observation is true for renal cancer (Spyridopoulos et al. 2007), but in this last case, the significance of the difference with healthy subjects disappears after adjustment to visceral AT, suggesting that adiponectin could relieve the effects of abdominal obesity.

Finally, disturbances of circulating adiponectin levels were also observed with leukemias, lymphomas, or myeloproliferative diseases (Avcu et al. 2006; Petridou et al. 2006).

Adiponectin could influence the evolution of these cancers by means of its insulin sensitive properties. However, numerous experimental works suggest that this adipokine could act directly on tumor cells, which express AdipoR1 and AdipoR2 receptors (Lang and Ratke 2009; Barb et al. 2007). Otherwise, adiponectin inhibits tumor growth in vivo in murine models, in particular in breast carcinogenesis (Wang et al. 2006), and perhaps partially via the control of tumor angiogenesis (Bråkenhielm et al. 2004). In vitro data on tumor cell lines are much more debated. In particular, on mammary, colic, or prostatic cancerous carcinoma cell lines, some authors observed a reduction of proliferation, whereas others noticed an increased proliferation (Lang and Ratke 2009; Barb et al. 2007). These controversial results could be due to differences in experimental conditions, in particular to the presence or the absence of serum in the culture medium. On the other hand, several works suggest that adiponectin could favor the migratory behavior of tumor cells, hereby promoting their capacity to metastasize (Chiu et al. 2009).

Some studies suggest that the anti-proliferative and anti-apoptotic effects of adiponectin could be relieved by the activation of AMPK. Indeed, AMPK regulates two major proteins involved in the control of growth stunting and apoptosis: p53 and p21 (Luo et al. 2005). Thus, adiponectin is capable of reducing mitogenic potential by inhibiting the mitogen-activated protein kinase (MAP-kinase) pathway in breast cancer cells (Dieudonne et al. 2006). The N-terminal c-Jun kinase (JNK) and signal transducer and activator of transcription 3 (STAT-3) pathways also mediate the effects of adiponectin on the control of tumor cells' proliferation (Lang and Ratke 2009).

Finally, it was also suggested that adiponectin could exert its anti-proliferative effects by reducing the bioavailability of growth factors. Thus, through a specific binding with platelet-derived growth factor-BB (PDGF-BB), heparin-binding epidermal growth factor (HB-EGF), and basic fibroblast growth factor (bFGF), adiponectin is capable to prevent the interaction of these growth factors with their specific membrane receptors, and therefore their mitogen action (Wang et al. 2005).

Adiponectin in Other Physiological and Pathological Contexts

Adiponectin is involved in other pathologies such as lung and bone diseases. Thus, adiponectinemia is negatively associated with bone mineral density in general population and seems to mediate a limiting effect on bone mass (Lenchik et al. 2003; Peng et al. 2008. Also, it was shown that high adiponectin plasma levels were associated with a decrease in bone mineral density in dialysed patients (Ealey et al. 2008). These data suggest that this adipokine could be involved in the regulation of bone mass.

Considering lung diseases, adiponectin was studied in asthma and chronic obstructive pulmonary diseases (Okuno et al. 2011), but its mechanisms of action in these conditions are poorly defined (Sood 2010). Several mouse models of these lung diseases indicate that adiponectin behaves as an anti-inflammatory agent, which is in balance with leptin as a "pro-inflammatory" adipokine (Sood 2010; Shore 2008). The studies performed in human are more controversial. Nevertheless, low circulating concentrations of adiponectin in addition to raised leptinemia

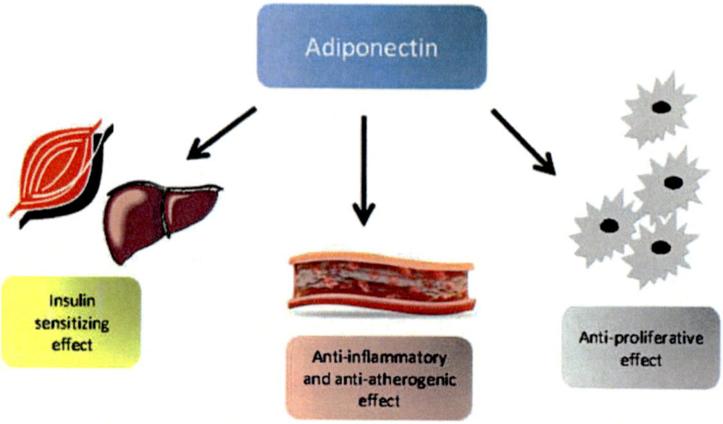

Fig. 13.2 Pleitropic functions of adiponectin

would be predictive for the development of asthma independently to the degree of obesity in targeted populations such as premenopausal women (Sood et al. 2008). By contrast, inverse variations of these adipokines both would be associated with chronic obstructive pulmonary disease (Shore 2008). In the latter case, the degree of obesity would be involved. Nevertheless, the role of adiponectin in the pathophysiology of lung disease is far to be established.

Finally, a recent review of the literature was focussed in the influence of dietary variations on the circulating concentrations of adiponectin in human (Silva et al. 2011). The daily consumption of fish or a complementation in omega 3 fatty acids is effective to increase adiponectinemia from 14 to 60 %. Weight loss induced by a hypocaloric diet associated with a physical activity also increased adiponectinemia from 18 to 48 %. In addition, two studies associating a complementation in dietary fiber to a diet program increased plasma levels of adiponectin from 60 to 115 %. Although low values of adiponectinemia are associated with several cardiovascular risk factors such as obesity, metabolic syndrome, and type 2 diabetes, the magnitude in the increase of adiponectinemia required to protect against these diseases is not known. In a more general manner, it is not currently established if the effects of these dietary interventions on the circulating concentrations of adiponectin have a clinical relevance.

Conclusion

The discovery of adipokines such as leptin, apelin, or adiponectin led us to revisit numerous metabolic, physiological, and physiopathological pathways. The functional duality of adiponectin including both beneficial or «malefic» effects shows all the complexity of its integration in a simple physiological model. Indeed, the role played by adiponectin in insulin resistance is very different when you consider insulin resistance of the metabolic syndrome, or that of the rare but severe insulin resistance syndromes, which alters either the insulin receptor functionality or AT development and triglyceride synthesis and storage. In the same way, adiponectin seems to protect against the occurrence and the progression of several types of cancer, but it also seems to promote the metastatic process. Likewise, the duality concerning the predictive power of adiponectin on cardiovascular complications is variable according to the renal function level of the subjects and can be very different in nondiabetic obese subjects or in patients with renal failure with wasting syndrome.

In spite of the still insufficient knowledge of the mechanisms of action and the physiological roles of adiponectin, several experimental, epidemiological, and clinical findings let us hope that the insulin sensitizing, anti-inflammatory, anti-atherogenic properties, and anti-proliferative effects of adiponectin (Fig. 13.2) could be of major interest for the development of new therapeutic opportunities. Nevertheless, human studies are still necessary to assess whether the modulation of circulating adiponectin levels has a real clinical relevance.

References

Antuna-Puente B, Fève B, Fellahi S, Bastard JP (2008) Adipokines: the missing link between insulin resistance and obesity. Diabetes Metab 34:2–11

Antuna-Puente B, Boutet E, Vigouroux C et al (2010) Higher adiponectin levels in patients with Berardinelli-Seip congenital lipodystrophy due to seipin as compared with 1-acylglycerol-3-phosphate-o-acyltransferase-2 deficiency. J Clin Endocrinol Metab 95:1463–1468

Arita Y, Kihara S, Ouchi N et al (1999) Paradoxical decrease of an adipose-specific protein, adiponectin, in obesity. Biochem Biophys Res Commun 257:79–83

Arita Y, Kihara S, Ouchi N et al (2002) Adipocyte-derived plasma protein adiponectin acts as a platelet-derived growth factor-BB-binding protein and regulates growth factor-induced common postreceptor signal in vascular smooth muscle cell. Circulation 105:2893–2898

Avcu F, Ural AU, Yilmaz MI et al (2006) Association of plasma adiponectin concentrations with chronic lymphocytic leukemia and myeloproliferative diseases. Int J Hematol 83:254–258

Barb D, Williams CJ, Neuwirth AK, Mantzoros CS (2007) Adiponectin in relation to malignancies: a review of existing basic research and clinical evidence. Am J Clin Nutr 86:s858–s866

Bastard JP, Maachi M, Lagathu C et al (2006) Recent advances in the relationship between obesity, inflammation and insulin resistance. Eur Cytokine Netw 17:4–12

Beige J, Heipmann K, Stumvoll M et al (2009) Paradoxical role for adiponectin in chronic renal diseases? An example of reverse epidemiology. Expert Opin Ther Targets 13:163–173

Bråkenhielm E, Veitonmäki N, Cao R et al (2004) Adiponectin-induced antiangiogenesis and antitumor activity involve caspase-mediated endothelial cell apoptosis. Proc Natl Acad Sci U S A 101:2476–2481

Chiu YC, Shieh DC, Tong KM et al (2009) Involvement of AdipoR receptor in adiponectin-induced motility and alpha2beta1 integrin upregulation in human chondrosarcoma cells. Carcinogenesis 30:1651–1659

Cook JR, Semple RK (2010) Hypoadiponectinmia—cause or consequence of human "insulin resistance"? J Clin Endocrinol Metab 95:1544–1554

Cui J, Panse S, Falkner B (2011) The role of adiponectin in metabolic and vascular disease: a review. Clin Nephrol 75:26–33

Cust AE, Kaaks R, Friedenreich C et al (2007) Plasma adiponectin levels and endometrial cancer risk in pre- and postmenopausal women. J Clin Endocrinol Metab 92:255–263
Dal Maso L, Augustin LS, Karalis A et al (2004) Circulating adiponectin and endometrial cancer risk. J Clin Endocrinol Metab 89:1160–1163
Dieudonne MN, Bussiere M, Dos Santos E et al (2006) Adiponectin mediates antiproliferative and apoptotic responses in human MCF7 breast cancer cells. Biochem Biophys Res Commun 345:271–279
Ealey KN, Kaludjerovic J, Archer MC, Ward WE (2008) Adiponectin is a negative regulator of bone mineral and bone strength in growing mice. Exp Biol Med (Maywood) 233:1546–1553
Goktas S, Yilmaz MI, Caglar K et al (2005) Prostate cancer and adiponectin. Urology 65:1168–1172
Goldstein BJ, Scalia RG, Ma XL (2009) Protective vascular and myocardial effects of adiponectin. Nat Clin Pract Cardiovasc Med 6:27–35
Guadagni F, Roselli M, Martini F et al (2009) Prognostic significance of serum adipokine levels in colorectal cancer patients. Anticancer Res 29:3321–3327
Halberg N, Schraw TD, Wang ZV et al (2009) Systemic fate of the adipocyte-derived factor adiponectin. Diabetes 58:1961–1970
Hoffstedt J, Arvidsson E, Sjolin E (2004) Adipose tissue adiponectin production and adiponectin serum concentration in human obesity and insulin resistance. J Clin Endocrinol Metab 89:1391–1396
Hu E, Liang P, Spiegelman BM (1996) AdipoQ is a novel adipose-specific gene dysregulated in obesity. J Biol Chem 271:10697–10703
Hug C, Wang J, Ahmad NS et al (2004) T-cadherin is a receptor for hexameric and high-molecular-weight forms of Acrp30/adiponectin. Proc Natl Acad Sci U S A 101:10308–10313
Ishikawa M, Kitayama J, Kazama S et al (2005) Plasma adiponectin and gastric cancer. Clin Cancer Res 11(2 Pt 1):466–472
Jorsal A, Tarnow L, Frystyk J et al (2008) Serum adiponectin predicts all-cause mortality and end stage renal disease in patients with type I diabetes and diabetic nephropathy. Kidney Int 74:649–654
Kadowaki T, Yamauchi T, Kubota N et al (2006) Adiponectin and adiponectin receptors in insulin resistance, diabetes, and the metabolic syndrome. J Clin Invest 116:1784–1792
Kaklamani VG, Wisinski KB, Sadim M et al (2008) Variants of the adiponectin (ADIPOQ) and adiponectin receptor 1 (ADIPOR1) genes and colorectal cancer risk. JAMA 300:1523–1531
Komaba H, Igaki N, Goto S et al (2006) Increased serum high-molecular-weight complex of adiponectin in type 2 diabetic patients with impaired renal function. Am J Nephrol 26:476–482
Komai N, Morita Y, Sakuta T et al (2007) Anti-tumor necrosis factor therapy increases serum adiponectin levels with the improvement of endothelial dysfunction in patients with rheumatoid arthritis. Mod Rheumatol 17:385–390
Komura N, Kihara S, Sonoda M et al (2010) Increment and impairment of adiponectin in renal failure. Cardiovasc Res 86:471–477
Körner A, Pazaitou-Panayiotou K, Kelesidis T, Kelesidis I, Williams CJ, Kaprara A, Bullen J, Neuwirth A, Tseleni S, Mitsiades N, Kiess W, Mantzoros CS (2007) Total and high-molecular-weight adiponectin in breast cancer: in vitro and in vivo studies. J Clin Endocrinol Metab 92(3):1041–1048
Lang K, Ratke J (2009) Leptin and Adiponectin: new players in the field of tumor cell and leukocyte migration. Cell Commun Signal 7:27
Lautamaki R, Ronnemaa T, Huupponen R et al (2007) Low serum adiponectin is associated with high circulating oxidized low-density lipoprotein in patients with type 2 diabetes mellitus and coronary artery disease. Metabolism 56:881–886
Lenchik L, Register TC, Hsu FC et al (2003) Adiponectin as a novel determinant of bone mineral density and visceral fat. Bone 33:646–651
Lenghel AR, Kacso IM, Bondor CI et al (2012) Intercellular adhesion molecule, plasma adiponectin and albuminuria in type 2 diabetic patients. Diabetes Res Clin Pract 95:55–61
Luo Z, Saha AK, Xiang X, Ruderman NB (2005) AMPK, the metabolic syndrome and cancer. Trends Pharmacol Sci 26:69–76

Maeda K, Okubo K, Shimomura I et al (1996) cDNA cloning and expression of a novel adipose specific collagen-like factor, apM1 (AdiPose Most abundant Gene transcript 1). Biochem Biophys Res Commun 221:286–289
Mao X, Kikani CK, Riojas RA et al (2006) APPL1 binds to adiponectin receptors and mediates adiponectin signalling and function. Nat Cell Biol 8:516–523
Menon V, Li L, Wang X et al (2006) Adiponectin and mortality in patients with chronic kidney disease. J Am Soc Nephrol 17:2599–2606
Michalakis K, Williams CJ, Mitsiades N et al (2007) Serum adiponectin concentrations and tissue expression of adiponectin receptors are reduced in patients with prostate cancer: a case control study. Cancer Epidemiol Biomarkers Prev 16:308–313
Miyazaki Y, Cersosimo E, Triplitt C, DeFronzo RA (2007) Rosiglitazone decreases albuminuria in type 2 diabetic patients. Kidney Int 72:1367–1373
Möller KF, Dieterman C, Herich L et al (2012) High serum adiponectin concentration in children with chronic kidney disease. Pediatr Nephrol 27:243–249
Nakano Y, Tobe T, Choi-Miura NH et al (1996) Isolation and characterization of GBP28, a novel gelatin-binding protein purified from human plasma. J Biochem 120:803–812
Ohashi N, Kato A, Misaki T et al (2008) Association of serum adiponectin levels with all-cause mortality in hemodialysis patients. Internal Med 47:485–491
Okuno S, Ishimura E, Norimine K et al (2011) Serum adiponectin and bone mineral density in male hemodialysis patients. Osteoporos Int 23:2027–2035
Ouchi N, Kihara S, Arita Y et al (1999) Novel modulator for endothelial adhesion molecules: adipocyte-derived plasma protein adiponectin. Circulation 100:2473–2476
Ouchi N, Kihara S, Arita Y et al (2001) Adipocyte-derived plasma protein, adiponectin, suppresses lipid accumulation and class A scavenger receptor expression in human monocyte-derived macrophages. Circulation 103:1057–1063
Ouedraogo R, Wu X, Xu SQ et al (2006) Adiponectin suppression of high-glucose-induced reactive oxygen species in vascular endothelial cells: evidence for involvement of a cAMP signaling pathway. Diabetes 55:1840–1846
Ouedraogo R, Gong Y, Berzins B et al (2007) Adiponectin deficiency increases leukocyte-endothelium interactions via upregulation of endothelial cell adhesion molecules in vivo. J Clin Invest 117:1718–1726
Peng XD, Xie H, Zhao Q et al (2008) Relationships between serum adiponectin, leptin, resistin, visfatin levels and bone mineral density, and bone biochemical markers in Chinese men. Clin Chim Acta 387:31–35
Petridou E, Mantzoros CS, Dessypris N et al (2006) Adiponectin in relation to childhood myeloblastic leukaemia. Br J Cancer 94:156–160
Pistrosch F, Herbrig K, Kindel B et al (2005) Rosiglitazone improves glomerular hyperfiltration, renal endothelial dysfunction, and microalbuminuria of incipient diabetic nephropathy in patients. Diabetes 54:2206–2211
Scherer PE, Williams S, Fogliano M et al (1995) A novel serum protein similar to C1q, produced exclusively in adipocytes. J Biol Chem 270:26746–26749
Semple RK, Soos MA, Luan J et al (2006) Elevated plasma adiponectin in humans with genetically defective insulin receptors. J Clin Endocrinol Metab 91:3219–3223
Semple RK, Cochran EK, Soos MA et al (2008) Plasma adiponectin as a marker of insulin receptor dysfunction: clinical utility in severe insulin resistance. Diabetes Care 31:977–979
Sharma K, Ramachandrarao S, Qiu G et al (2008) Adiponectin regulates albuminuria and podocyte function in mice. J Clin Invest 118:1645–1656
Shore SA (2008) Obesity and asthma: possible mechanisms. J Allergy Clin Immunol 121:1087–1093
Silva FM, de Almeida JC, Feoli AM (2011) Effect of diet on adiponectin levels in blood. Nutr Rev 69:599–612
Sood A (2010) Obesity, adipokines, and lung disease. J Appl Physiol 108:744–753
Sood A, Cui X, Qualls C et al (2008) Association between asthma and serum adiponectin concentration in women. Thorax 63:877–882

Spyridopoulos TN, Petridou ET, Skalkidou A et al (2007) Low adiponectin levels are associated with renal cell carcinoma: a case-control study. Int J Cancer 120:1573–1578
Staiger H, Tschritter O, Machann J et al (2003) Relationship of serum adiponectin and leptin concentrations with body fat distribution in humans. Obes Res 11:368–372
Stenvinkel P (2011) Adiponectin in chronic kidney disease: a complex and context sensitive clinical situation. J Ren Nutr 21:82–86
Stenvinkel P, Marchlewska A, Pecoits-Filho R et al (2004) Adiponectin in renal disease: relationship to phenotype and genetic variation in the gene encoding adiponectin. Kidney Int 65:274–281
Tentolouris N, Doulgerakis D, Moyssakis I et al (2004) Plasma adiponectin concentrations in patients with chronic renal failure: relationship with metabolic risk factors and ischemic heart disease. Horm Metab Res 36:721–727
Tietge UJ, Boker KH, Manns MP, Bahr MJ (2004) Elevated circulating adiponectin levels in liver cirrhosis are associated with reduced liver function and altered hepatic hemodynamics. Am J Physiol Endocrinol Metab 287:E82–E89
Tomas E, Tsao TS, Saha AK et al (2002) Enhanced muscle fat oxidation and glucose transport by ACRP30 globular domain: acetyl-CoA carboxylase inhibition and AMP-activated protein kinase activation. Proc Natl Acad Sci U S A 99:16309–16313
von Eynatten M, Liu D, Hock C et al (2009) Urinary adiponectin excretion: a novel marker for vascular damage in type 2 diabetes. Diabetes 58:2093–2099
Wang Y, Lam KS, Xu JY, Lu G, Xu LY, Cooper GJ, Xu A (2005) Adiponectin inhibits cell proliferation by interacting with several growth factors in an oligomerization-dependent manner. J Biol Chem 6;280(18):18341–18347. Epub 2005 Feb 25
Wang Y, Lam JB, Lam KS, Liu J, Lam MC, Hoo RL, Wu D, Cooper GJ, Xu A (2006) Adiponectin modulates the glycogen synthase kinase-3beta/beta-catenin signaling pathway and attenuates mammary tumorigenesis of MDA-MB-231 cells in nude mice. Cancer Res 66(23):11462–11470
Wang Y, Lam KS, Yau MH, Xu A (2008) Post-translational modifications of adiponectin: mechanisms and functional implications. Biochem J 409:623–633
Wang C, Xin X, Xiang R et al (2009) Yin-Yang regulation of adiponectin signaling by APPL isoforms in muscle cell. J Biol Chem 284:31608–31615
Wei EK, Giovannucci E, Fuchs CS et al (2005) Low plasma adiponectin levels and risk of colorectal cancer in men: a prospective study. J Natl Cancer Inst 97:1688–1694
Yamauchi T, Kamon J, Waki H et al (2001) The fat-derived hormone adiponectin reverses insulin resistance associated with both lipoatrophy and obesity. Nat Med 7:941–946
Yamauchi T, Kamon J, Minokoshi Y et al (2002) Adiponectin stimulates glucose utilization and fatty-acid oxidation by activating AMP-activated protein kinase. Nat Med 8:1288–1295
Yamauchi T, Kamon J, Ito Y et al (2003) Cloning of adiponectin receptors that mediate antidiabetic metabolic effects. Nature 423:762–769
Yamauchi T, Nio Y, Maki T et al (2007) Targeted disruption of AdipoR1 and AdipoR2 causes abrogation of adiponectin binding and metabolic actions. Nat Med 13:332–339

Chapter 14
Apelin Metabolic Functions

Isabelle Castan-laurell, Cédric Dray, Claude Knauf and Philippe Valet

Apelin

In 1998, Tatemoto and coworkers purified from bovine stomach extracts a protein that binds to the «orphan» APJ receptor (Tatemoto et al. 1998). The identified gene encodes a 77-amino acid polypeptide that includes a secretory signal sequence. The ligand of the orphan receptor APJ consisted in the C terminal part of this polypeptide and was called "apelin", for APJ Endogenous Ligand (Tatemoto et al. 1998). The APJ receptor is a G-protein-coupled receptor (GPCR) identified in 1993 in humans, displaying a close sequence homology to the angiotensin II receptor type 1 (O'Dowd et al. 1993). However, the receptor displayed no specific binding for angiotensin II. The APJ encoding gene was mapped to chromosome 11 and later sublocalized to the locus 11q12. Its transcripts were first detected in the brain but APJ is expressed in a wide range of tissues (Medhurst et al. 2003).

The human apelin gene has been localized on chromosome Xq25-q26.1 (Lee et al. 2000). Apelin gene is expressed in many peripheral tissues as well as in different brain areas particularly in the hypothalamus (Lee et al. 2000). So far, three active forms of apelin, consisting of 13, 17, or 36 amino acids and the pyroglutamated apelin-13 (Pyr(1)-apelin-13) have been identified. They originate

I. Castan-laurell · C. Dray · C. Knauf · P. Valet (✉)
INSERM U1048 I2MC, 1 Ave J. Poulhès BP 84 225,
31432 Toulouse Cedex 4, France
e-mail: philippe.valet@inserm.fr

I. Castan-laurell
e-mail: isabelle.castan@inserm.fr

C. Dray
e-mail: cedric.dray@inserm.fr

C. Knauf
e-mail: claude.knauf@inserm.fr

J.-P. Bastard and B. Fève (eds.), *Physiology and Physiopathology of Adipose Tissue*,
DOI: 10.1007/978-2-8178-0343-2_14, © Springer-Verlag France 2013

Human	LVQPRGSRNGPGPWQGGRRKFRRQRPRLSHKGPMPF
Rat	LVKPRTSRTGPGAWQGGRRKFRRQRPRLSHKGPMPF
Bovine	LVQPRGPRSGPGPWQGGRRKFRRQRPRLSHKGPMPF

Fig. 14.1 Sequence alignment of mammalian apelin-36. Amino acids that differ from the human sequence are highlighted in *red*

from a common 77-amino-acid pre-propeptide precursor (preproapelin) consisting in a dimer stabilized by disulfide bridges linking cysteine residues (Lee et al. 2005). The human, bovine, and rat preproapelin sequencing have revealed a high sequence homology among the three species and a perfect identity for the last 23 C-terminal amino acids (Fig. 14.1) (Tatemoto et al. 1998; Lee et al. 2000). The predominant molecular forms of endogenous hypothalamic and plasma apelin in rats were found to be apelin-36, apelin-17, and apelin-13 (Kawamata et al. 2001; De Mota et al. 2004). Pyr(1)-apelin-13 was found to be the predominant isoform in human cardiac tissue (Maguire et al. 2009).

Apelin Regulation in Obesity and Type 2 Diabetes

In Adipose Tissue

Apelin has been detected in adipose tissue by Tatemoto et al. (2001) and, later on, the work of Boucher et al. (2005) has demonstrated that apelin was not only produced but also secreted by adipocytes. Apelin has been then considered as a new adipokine. It has been shown that apelin expression increased during adipogenesis (Boucher et al. 2005; Hung et al. 2011) and recently, that the blockade of the renin-angiotensin system further augmented the apelin expression and secretion in 3T3-L1 adipocytes (Hung et al. 2011).

One of the main regulators of apelin production in adipocytes is insulin. There is a close relationship between apelin and insulin both in vivo and in vitro (Boucher et al. 2005). The expression of apelin in adipocytes is increased in various mouse models of obesity associated with hyperinsulinemia. During fasting and after re-feeding in mice, the pattern of apelin expression in adipocytes parallels the plasma levels of insulin. In cultured 3T3F442A adipocytes as well as in human isolated adipocytes, insulin treatment results in an increased expression and secretion of apelin (Boucher et al. 2005).

Other factors positively regulate apelin expression in adipocytes or adipose tissue (Fig. 14.2). Inflammatory factors such as TNFα (Daviaud et al. 2006) and lipopolysaccharide (Geurts et al. 2011) increase apelin expression but the physiological relevance of this upregulation of apelin has not yet been fully elucidated. In addition, hypoxia in adipose tissue can contribute to obesity. Different studies have shown that hypoxia induced apelin expression and secretion in both human

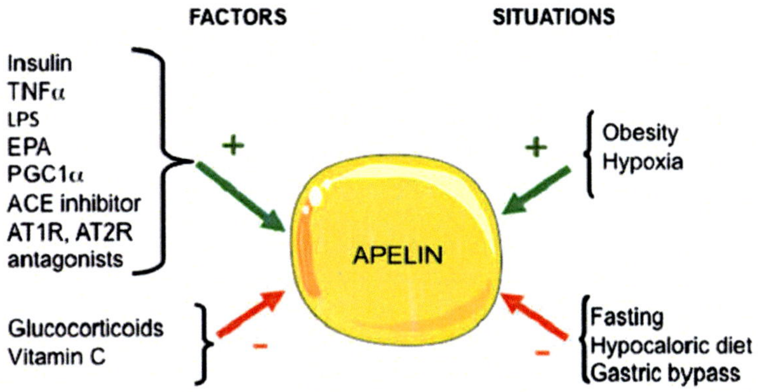

Fig. 14.2 Main factors and situations involved in apelin regulation in adipose tissue. *ACE* angiotensin-converting enzyme, *AT1R and AT2R* angiotensin type 1 and type 2 receptor, respectively, *EPA* eicosapentaenoic acid

and murine adipocytes (Geigr et al. 2011; Glassford et al. 2007; Kunduzova et al. 2008). Moreover, it has been demonstrated that induction of apelin under hypoxic conditions is mediated by a direct HIF-1 binding to the apelin gene. Since apelin is involved in angiogenesis, which is essential for adipose tissue expansion, apelin has been proposed to contribute to the development of new vasculature in expanding fat depot (Glassford et al. 2007; Kunduzova et al. 2008).

Overexpression of peroxisome proliferator-activated receptor gamma (PPARγ) coactivator-1α (PGC1α), a key regulator of cellular energy homeostasis in oxidative tissues, also induces apelin expression and secretion in human adipocytes (Mazzucotelli et al. 2008). Eicosapentaenoic acid, a polyunsaturated fatty acid from the omega-3 family, has been shown to increase both apelin gene expression and secretion in 3T3-L1 adipocytes (Lorente-Cebrian et al. 2010) and also in vivo in rat adipose tissue (Perez-Echarri et al. 2009). Curiously, negative modulators of apelin expression in adipocyte are not numerous and only glucocorticoids (dexamethasone) were described to decrease apelin mRNA levels in 3T3-L1 cells (Wei et al. 2005).

In Hypothalamus

Apelin mRNAs are present in different nuclei of the hypothalamus including the paraventricular, arcuate, and supraoptic nuclei involved in the control of behavioral, endocrine processes and energy homeostasis (Reaux et al. 2001). Apelin-positive nerve fibers in the hypothalamus imply the existence of apelinergic neurons and thus a dual action of apelin as a circulating peptide and a neurotransmitter. So far, it is not known whether peripheral plasma apelin can reach the hypothalamus and could modulate apelin levels in the hypothalamus. However hypothalamic apelin levels were found to be higher in HFD-fed C57Bl6/J and db/db mice (Reaux-Le Goazigo et al. 2011).

The role of central apelin in regulating feeding behavior has been mainly studied in rats by different groups but remains a matter of debate (for review, (Castan-Laurell et al. 2011)). Intracerebroventricular (icv) apelin injection has been shown to either stimulate or inhibit food intake depending of the nutritional status and whether apelin was injected during the feeding or fasting period. Clarke et al. showed that icv apelin injection decreased food and water intake and respiratory exchange ratio in control rats, but had no effect in high-fat fed rats (Clarke et al. 2009). Reaux-Le Goazigo et al. have recently demonstrated that apelin-immunoreactive neuronal cell bodies were distributed throughout the rostrocaudal extent of the Arc in rats and that apelin was weakly co-localized with NPY (an orexigenic peptide) but strongly co-localized with POMC known to decrease food intake (Reaux-Le Goazigo et al. 2011). Thus, increased hypothalamic apelin levels might be associated with reduced food intake and limited weight gain. However, with obesity, the beneficial effect of apelin is probably counteracted by the downregulation of apelin receptor by apelin itself as described by Clarke et al. in HFD-fed rats (Clarke et al. 2009).

In mice, only one study has been realized showing that icv apelin-13 infusion into the third ventricle during 10 days increased food intake (on day 3–7), adiposity, the locomotor activity especially during the nocturnal period (feeding period) and the body temperature only during the period of activity (Valle et al. 2008). Complementary investigations are necessary to delineate the role of the apelin/APJ system in mice feeding behavior and especially during obesity.

In Plasma

The first report of plasma apelin concentrations in humans was shown in obese and hyperinsulinemic subjects (Boucher et al. 2005) where plasma apelin levels are increased. Different groups also found increased plasma apelin levels in morbidly obese subject with or without type 2 diabetes or in patients without a severe obesity but exhibiting an impaired glucose tolerance or type 2 diabetes (for review, (Castan-Laurell et al. 2011)). However, reduced plasma apelin levels were described in obese subjects with untreated type 2 diabetes, compared to non-diabetic subjects (Erdem et al. 2008). These results could be consistent with the fact that 14 weeks of anti-diabetic treatment (rosiglitazone and metformin) promote an increase of plasma apelin levels and an improvement of the glycemic profile (Kadoglou et al. 2010). In women with gestational diabetes, no significant differences in plasma apelin levels were observed compared to women with normal glucose tolerance (Telejko et al. 2010), whereas in gestational diabetic lactating women, there was a trend to lower apelin concentrations in serum compared to lactating healthy women (Aydin 2010).

Plasma apelin concentrations were also measured in obese children. In pubertal obese children, apelin as well as adiponectin levels were lower when compared to non-obese children (Tapan et al. 2010). However, when comparing plasma apelin concentration in obese girls (between 14- and 18-year old) and in girls with either anorexia nervosa or with no otherwise specified eating disorders, apelin concentrations were significantly higher in obese compared to healthy control but lower in patients with both eating disorders compared to healthy control (Ziora et al. 2010). Recently, no difference in apelin levels and no significant correlations were found between apelin and weight status, body fat, insulin resistance, and cardiovascular risk factors associated with obesity between 80 obese and 40 lean children (Reinehr et al. 2011). In children with type 1 diabetes, plasma apelin levels were increased compared to non-diabetic subjects (Meral et al. 2010) suggesting that the lack of insulin is this situation has no impact on apelin levels.

Changes in apelin levels after weight loss or bariatric surgery in obese individuals were also investigated. Diet-induced weight loss decreases apelin levels in moderately obese women (Castan-Laurell et al. 2008) but not significantly in patients with the metabolic syndrome (Heinonen et al. 2009) or in obese children (Reinehr et al. 2011). Bariatric surgery resulted in a significant decrease in apelin levels only in morbidly obese patients exhibiting impaired fasting glucose or type 2 diabetes before surgery (Soriguer et al. 2009).

All together, these studies underline that obesity, *per se,* is probably not the main determinant of increased plasma apelin concentrations since circulating apelin levels are not necessarily correlated to the body mass index in all the published studies (Telejko et al. 2010; Aydin 2010; Reinehr et al. 2011; Soriguer et al. 2009). However, plasma apelin or changes in plasma apelin concentrations were found to correlate significantly with serum triglycerides, glucose (Soriguer et al. 2009), TNFα (Heinonen et al. 2009), Homeostasis Assessment Model of Insulin Resistance (HOMA-IR) (Castan-Laurell et al. 2008; Ercin et al. 2010) and HbA1c (Dray et al. 2010) suggesting also a link between apelin and the metabolic syndrome or type 2 diabetes.

Finally, genetic studies described a strong association between apelin gene polymorphisms and plasma levels of fasting glucose in the Chinese Han population (exhibiting type II diabetes) (Zhang et al. 2009) and recently with obesity in Chinese women (Liao et al. 2011).

Apelin Effects on Energy Metabolism

Glucose Metabolism

Intravenous apelin administration at low concentration (200 pmol/kg) decreased blood glucose in normoponderal mice and improved glucose tolerance (Dray et al. 2008). Furthermore, during a hyperinsulinemic-euglycemic clamp, when the hepatic glucose production is totally blunted, apelin increases glucose utilization throughout the whole organism mainly due to a rise in glucose uptake by skeletal muscles and adipose tissues.

In isolated skeletal muscle (soleus), apelin stimulates glucose transport and its effect is additive to that of insulin (Dray et al. 2008). The associated signaling pathway involved is dependent on AMP-activated protein kinase (AMPK) and endothelium NO synthase (eNOS) activation. AMPK is a key enzyme in energy metabolism, activated during ATP depletion in cells. It is involved in various metabolic processes stimulating the production of energy such as glucose transport. We demonstrated by both in vivo and in vitro experiments that AMPK was phosphorylated by apelin in soleus muscle and involved in apelin-stimulated glucose transport (Dray et al. 2008). More recently, in cultured C2C12 myotubes, apelin-induced glucose uptake was also shown to be dependent on AMPK activation (Yue et al. 2010). In addition, like insulin, apelin phosphorylates Akt and its activation is necessary for glucose transport both ex vivo in mouse soleus muscle and in C2C12 myotubes. Moreover, the activation of Akt is AMPK-dependent (Dray et al. 2008; Yue et al. 2010).

The role of central apelin on glucose metabolism has been recently studied in our group. Acute icv apelin has differential effect depending on the injected dose and the nutritional status (Duparc et al. 2011). Acute low-dose of icv apelin injection decreased peripheral-fed glycemia, and increased glucose and insulin tolerance in mice via a NO-dependent signaling pathway. However, acute high-dose of icv apelin injection in chow-fed and HFD mice increased fasted hyperglycemia. Thus, a rise in hypothalamic apelin levels, as described by Reaux-Le Goazigo et al. (Reaux-Le Goazigo et al. 2011), could be involved in the transition from normal to diabetic status. Both the abolished circadian plasma apelin regulation observed in HFD-treated mice and the chronic icv apelin treatment in normal mice triggering insulin intolerance are consistent with this hypothesis (Duparc et al. 2011).

Peripheral injection of apelin in obese and insulin-resistant mice has different consequences. Glucose tolerance is significantly improved in these mice receiving iv apelin bolus before an oral glucose tolerance test. In addition, the loss of insulin sensitivity observed during a hyperinsulinemic-euglycemic clamp in insulin-resistant mice was improved with an apelin perfusion during the clamp (Dray et al. 2008). Thus, peripheral apelin acute treatment remains efficient in obese insulin-resistant mice and improves the altered glucose metabolism especially by increasing glucose uptake in skeletal muscle.

Moreover, in insulin-resistant 3T3-L1 adipocytes (due to TNFα pretreatment), insulin-stimulated glucose uptake was reduced by 47 % whereas apelin treatment promoted an increased glucose uptake through the PI3 K/Akt pathway and improved insulin-stimulated glucose uptake (Zhu et al. 2011). Apelin was also shown to stimulate glucose transport in an AMPK-dependent manner in human adipose tissue (Attané et al. 2011).

Although plasma apelin levels are elevated in obese insulin-resistant mice, exogenous apelin is still efficient and thus apelin resistance is unexpected.

Recently, chronic apelin treatment in insulin-resistant young db/db mice was shown to improve insulin sensitivity (Yue et al. 2010). The role of apelin in glucose homeostasis was also confirmed by the phenotype of apelin null mice exhibiting hyperinsulinemia and insulin resistance. The loss of insulin sensitivity in apelin-/- mice was exacerbated by a high fat/high sucrose diet (Yue et al. 2010).

Lipid Metabolism

The first study that reported a role of apelin on lipid metabolism was related to chronic peripheral administration of apelin (during 2 weeks) in mice (Higuchi et al. 2007). Daily ip apelin injection in chow-fed mice was shown to decrease the triglycerides content in adipose tissue and the weight of different fat depots and also plasma triglycerides (Higuchi et al. 2007). The apelin treatment did not affect average food intake but increased rectal temperature and O_2 consumption. An increased expression of mitochondrial uncoupling protein 1 (UCP1) was observed in brown adipose tissue (BAT) (Higuchi et al. 2007). All together, the authors suggest that apelin increases energy expenditure through UCP1 activation.

Chronic apelin treatment also decreased adiposity in obese mice (Higuchi et al. 2007) and similar results were obtained in mice overexpressing apelin (apelin-Tg mice) fed a HFD (Yamamoto et al. 2011). Apelin-Tg mice exhibited a resistance against diet-induced obesity, increased oxygen consumption, and body temperature without modification of food intake (Yamamoto et al. 2011). Inversely, apelin-/- mice have increased abdominal adiposity and increased circulating FFA levels (Yue et al. 2011). After reintroduction of apelin (apelin infusion during 2 weeks) in apelin-/- mice, adiposity and fatty acids and also glycerol levels were decreased, suggesting a role of apelin in lipolysis regulation. In both isolated adipocytes and 3T3-L1 differentiated adipocytes, apelin was shown to inhibit isoproterenol- (β-adrenergic agonist) induced lipolysis through a pathway involving Gq, Gi, and AMPK (Yue et al. 2011). However, in human adipose tissue explants or human isolated adipocytes, apelin had no effect on basal or isoproterenol-stimulated lipolysis (Attané et al. 2011).

In addition to changes in adipose tissue metabolism, skeletal muscles have also been studied as a target of apelin. Increased UCP3 expression in skeletal muscle has been observed in different rodent models treated with apelin or overexpressing apelin (Higuchi et al. 2007; Yamamoto et al. 2011; Frier et al. 2009). This increase was associated with mitochondrial biogenesis. Accordingly, enzyme activities of β- hydroxyacyl CoA dehydrogenase (involved in the mitochondrial oxidative capacities), citrate synthase (involved in the citric acid cycle) and cytochrome C oxidase (involved in the respiratory chain) were increased in response to apelin treatment in rat triceps (Frier et al. 2009). However, the rise in mitochondrial markers was dependent on increased PGC-1β expression (Frier et al. 2009) but independent of PGC-1α, identified as a key player in the regulation of mitochondrial biogenesis. In apelin-Tg mice fed with HFD, increased mitochondrial

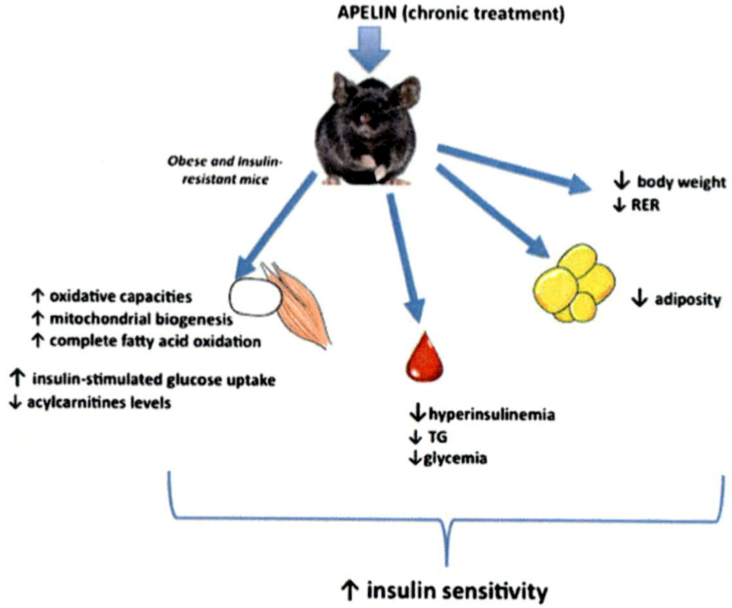

Fig. 14.3 Schematic representation of the different metabolic apelin effects in response to chronic apelin treatment in obese and insulin-resistant mice. *RER* respiratory exchange ratio, *TG* triglycerides

DNA was found in skeletal muscles without variation of PGC-1α expression (Yamamoto et al. 2011). Interestingly, diet-induced obesity resistance of apelin-Tg mice was correlated with vascular formation in skeletal muscle due to increased angiopoietin-1 and its receptor Tie 2 (Yamamoto et al. 2011).

Very recently, our group has also clearly demonstrated that chronic apelin treatment during 4 weeks, in obese and insulin-resistant mice, increases mitochondrial biogenesis and that the adverse alterations of mitochondria ultra-structure in muscle (reduced electron density of the matrix and loss of cristae) were decreased in both intramyofibrillar and subsarcolemmal mitochondria of soleus muscle of apelin-treated mice (Attané et al. 2012). Moreover, these effects were shown to be dependent on AMPK activation and increased PGC1α expression. We also pointed out that this higher mitochondrial biogenesis in muscle was associated to increased complete fatty acid oxidation (FAO) and mitochondrial respiratory capacity. Indeed, previous studies by using different mouse models have underlined the effect of apelin on adiposity and beneficial aspects in terms of increased oxygen consumption (Higuchi et al. 2007; Yamamoto et al. 2011) but the outcome of lipids was not clearly demonstrated. Our data show that the main whole-body substrates oxidized in vivo, measured by indirect calorimetry, were lipids in apelin-treated obese and insulin-resistant mice. Moreover, in skeletal muscles, apelin treatment increases the complete and not the incomplete FAO. The influx of lipid in mitochondria was associated with decreased acylcarnitines levels suggesting a tighter coupling between FAO and the tricarboxylic acid cycle. The overall improvement of insulin sensitivity observed at the end of the 4 weeks of apelin treatment appears to be the end point of increased FAO and mitochondrial biogenesis in muscle and decreased total adiposity (Attané et al. 2012) (Fig. 14.3).

Conclusion

Apelin exerts a wide range of effects and the newly described effects on energy metabolism, especially the fact that apelin treatment increases insulin sensitivity in obese and insulin-resistant mice, adds a new dimension, and enhances the beneficial role of apelin in obesity-associated type II diabetes. APJ thus represents an interesting target for pharmacological agent design. Different agonist and antagonist have been recently developed (for review see Castan-Laurell et al. 2011) and their use will help not only to better delineate the roles of the apelin/APJ system but also to avoid the issues of apelin bioavailability that deserves a better characterization. The apelin/APJ system, as a promising therapeutic target in obesity-associated diseases, now needs the input of preclinical studies and thus to be assessed in vivo in humans.

References

Attané C, Daviaud D, Dray C et al (2011) Apelin stimulates glucose uptake but not lipolysis in human adipose tissue ex vivo. J Mol Endocrinol 46:21–28

Attané C, Foussal C, Le Gonidec S et al (2012) Apelin treatment increases complete fatty acid oxidation, mitochondrial oxidative capacity and biogenesis in muscle of insulin-resistant mice. Diabetes 61:310–320

Aydin S (2010) The presence of the peptides apelin, ghrelin and nesfatin-1 in the human breast milk, and the lowering of their levels in patients with gestational diabetes mellitus. Peptides 31:2236–2240

Boucher J, Masri B, Daviaud D et al (2005) Apelin, a newly identified adipokine up-regulated by insulin and obesity. Endocrinology 146:1764–1771

Castan-Laurell I, Vitkova M, Daviaud D et al (2008) Effect of hypocaloric diet-induced weight loss in obese women on plasma apelin and adipose tissue expression of apelin and APJ. Eur J Endocrinol 158:905–910

Castan-Laurell I, Dray C, Attané C et al (2011) Apelin, diabetes and obesity. Endocrine 40(1):1–9

Clarke KJ, Whitaker KW, Reyes TM (2009) Diminished metabolic responses to centrally-administered apelin-13 in diet-induced obese rats fed a high-fat diet. J Neuroendocrinol 21:83–89

Daviaud D, Boucher J, Gesta S et al (2006) TNFalpha up-regulates apelin expression in human and mouse adipose tissue. FASEB J 20:1528–1530

De Mota N, Reaux-Le Goazigo A, El Messari S et al (2004) Apelin, a potent diuretic neuropeptide counteracting vasopressin actions through inhibition of vasopressin neuron activity and vasopressin release. Proc Natl Acad Sci U S A 101:10464–10469

Dray C, Knauf C, Daviaud D et al (2008) Apelin stimulates glucose utilization in normal and obese insulin-resistant mice. Cell Metab 8:437–445

Dray C, Debard C, Jager J et al (2010) Apelin and APJ regulation in adipose tissue and skeletal muscle of type 2 diabetic mice and humans. Am J Physiol Endocrinol Metab 298:E1161–E1169
Duparc T, Colom A, Cani PD et al (2011) Central Apelin Controls Glucose Homeostasis via a Nitric Oxide-Dependent Pathway in Mice. Antioxid Redox 15(6):1477–1496
Ercin CN, Dogru T, Tapan S et al (2010) Plasma apelin levels in subjects with nonalcoholic fatty liver disease. Metabolism 59:977–981
Erdem G, Dogru T, Tasci I et al (2008) Low plasma apelin levels in newly diagnosed type 2 diabetes mellitus. Exp Clin Endocrinol Diabetes 116:289–292
Frier BC, Williams DB, Wright DC (2009) The effects of apelin treatment on skeletal muscle mitochondrial content. Am J Physiol Regul Integr Comp Physiol 297:R1761–R1768
Geiger K, Muendlein A, Stark N et al (2011) Hypoxia induces apelin expression in human adipocytes. Horm Metab Res 43:380–385
Geurts L, Lazarevic V, Derrien M et al (2011) Altered gut microbiota and endocannabinoid system tone in obese and diabetic leptin-resistant mice: impact on apelin regulation in adipose tissue. Front Microbiol 2:149
Glassford AJ, Yue P, Sheikh AY (2007) HIF-1 regulates hypoxia- and insulin-induced expression of apelin in adipocytes. Am J Physiol Endocrinol Metab 293:E1590–E1596
Heinonen MV, Laaksonen DE, Karhu T et al (2009) Effect of diet-induced weight loss on plasma apelin and cytokine levels in individuals with the metabolic syndrome. Nutr Metab Cardiovasc Dis 19:626–633
Higuchi K, Masaki T, Gotoh K et al (2007) Apelin, an APJ receptor ligand, regulates body adiposity and favors the messenger ribonucleic acid expression of uncoupling proteins in mice. Endocrinology 148:2690–2697
Hung WW, Hsieh TJ, Lin T et al (2011) Blockade of the renin-angiotensin system ameliorates apelin production in 3T3-L1 adipocytes. Cardiovasc Drugs Ther 25:3–12
Kadoglou NP, Tsanikidis H, Kapelouzou A et al (2010) Effects of rosiglitazone and metformin treatment on apelin, visfatin, and ghrelin levels in patients with type 2 diabetes mellitus. Metabolism 59:373–379
Kawamata Y, Habata Y, Fukusumi S et al (2001) Molecular properties of apelin: tissue distribution and receptor binding. Biochim Biophys Acta 1538:162–171
Kunduzova O, Alet N, Delesque-Touchard N (2008) Apelin/APJ signaling system: a potential link between adipose tissue and endothelial angiogenic processes. FASEB J 22:4146–4153
Lee DK, Cheng R, Nguyen T et al (2000) Characterization of apelin, the ligand for the APJ receptor. J Neurochem 74:34–41
Lee DK, Saldivia VR, Nguyen T et al (2005) Modification of the terminal residue of apelin-13 antagonizes its hypotensive action. Endocrinology 146:231–236
Liao YC, Chou WW, Li YN et al (2011) Apelin gene polymorphism influences apelin expression and obesity phenotypes in Chinese women. Am J Clin Nutr 94(3):921–928
Lorente-Cebrian S, Bustos M, Marti A (2010) Eicosapentaenoic acid up-regulates apelin secretion and gene expression in 3T3-L1 adipocytes. Mol Nutr Food Res 54(Suppl 1):S104–S111
Maguire JJ, Kleinz MJ, Pitkin SL, Davenport AP (2009) [Pyr1]apelin-13 identified as the predominant apelin isoform in the human heart: vasoactive mechanisms and inotropic action in disease. Hypertension 54:598–604
Mazzucotelli A, Ribet C, Castan-Laurell I (2008) The transcriptional co-activator PGC-1alpha up regulates apelin in human and mouse adipocytes. Regul Pept 150:33–37
Medhurst AD, Jennings CA, Robbins MJ et al (2003) Pharmacological and immunohistochemical characterization of the APJ receptor and its endogenous ligand apelin. J Neurochem 84:1162–1172
Meral C, Tascilar E, Karademir F et al (2010) Elevated plasma levels of apelin in children with type 1 diabetes mellitus. J Pediatr Endocrinol Metab 23:497–502
O'Dowd BF, Heiber M, Chan A et al (1993) A human gene that shows identity with the gene encoding the angiotensin receptor is located on chromosome 11. Gene 136:355–360
Perez-Echarri N, Perez-Matute P, Marcos-Gomez B (2009) Effects of eicosapentaenoic acid ethyl ester on visfatin and apelin in lean and overweight (cafeteria diet-fed) rats. Br J Nutr 101:1059–1067

Reaux A, De Mota N, Skultetyova I et al (2001) Physiological role of a novel neuropeptide, apelin, and its receptor in the rat brain. J Neurochem 77:1085–1096
Reaux-Le Goazigo A, Bodineau L, Picco-DE Mota N et al. (2011) Apelin and proopiomelanocortin system: a new regulatory pathway of hypothalamic alpha-MSH release. Am J Physiol Endocrinol Metab (in press)
Reinehr T, Woelfle J, Roth CL (2011) Lack of association between apelin, insulin resistance, cardiovascular risk factors, and obesity in children: a longitudinal analysis. Metabolism 60(9):1349–1354
Soriguer F, Garrido-Sanchez L, Garcia-Serrano S et al (2009) Apelin levels are increased in morbidly obese subjects with type 2 diabetes mellitus. Obes Surg 19:1574–1580
Tapan S, Tascilar E, Abaci A et al (2010) Decreased plasma apelin levels in pubertal obese children. J Pediatr Endocrinol Metab 23:1039–1046
Tatemoto K, Hosoya M, Habata Y et al (1998) Isolation and characterization of a novel endogenous peptide ligand for the human APJ receptor. Biochem Biophys Res Commun 251:471–476
Tatemoto K, Takayama K, Zou MX et al (2001) The novel peptide apelin lowers blood pressure via a nitric oxide-dependent mechanism. Regul Pept 99(2–3):87–92
Telejko B, Kuzmicki M, Wawrusiewicz-Kurylonek N et al (2010) Plasma apelin levels and apelin/APJ mRNA expression in patients with gestational diabetes mellitus. Diabetes Res Clin Pract 87:176–183
Valle A, Hoggard N, Adams AC et al (2008) Chronic central administration of apelin-13 over 10 days increases food intake, body weight, locomotor activity and body temperature in C57BL/6 mice. J Neuroendocrinol 20:79–84
Wei L, Hou X, Tatemoto K (2005) Regulation of apelin mRNA expression by insulin and glucocorticoids in mouse 3T3-L1 adipocytes. Regul Pept 132:27–32
Yamamoto T, Habata Y, Matsumoto Y et al (2011) Apelin-transgenic mice exhibit a resistance against diet-induced obesity by increasing vascular mass and mitochondrial biogenesis in skeletal muscle. Biochim Biophys Acta 1810(9):853–862
Yue P, Jin H, Aillaud M et al (2010) Apelin is necessary for the maintenance of insulin sensitivity. Am J Physiol Endocrinol Metab 298:E59–E67
Yue P, Jin H, Xu S et al (2011) Apelin decreases lipolysis via G(q), G(i), and AMPK-Dependent Mechanisms. Endocrinology 152:59–68
Zhang Y, Shen C, Li X et al (2009) Association of apelin genetic variants with type 2 diabetes and related clinical features in Chinese Hans. Chin Med J 122:1273–1276
Zhu S, Sun F, Li W et al (2011) Apelin stimulates glucose uptake through the PI3 K/Akt pathway and improves insulin resistance in 3T3-L1 adipocytes. Mol Cell Biochem 353:305–313
Ziora K, Oswiecimska J, Swietochowska E et al (2010) Assessment of serum apelin levels in girls with anorexia nervosa. J Clin Endocrinol Metab 95:2935–2941

Chapter 15
Up-to-Date on Novel "Adipocrines"

Christian Carpéné and Jean-Sébastien Saulnier-Blache

Introduction

Alongside its function of storage and mobilization of fat, adipose tissue secretes numerous bioactive molecules (peptides, glycoproteins, fatty acids and their derivatives, phospholipids) that have been designed as "adipokines" and that are now strongly suspected to be involved in the regulation of energy balance and the etiology of pathologies associated with obesity (diabetes, hypertension, atherosclerosis, etc).

In our opinion, the use of the term "adipokine" should be debated. It comes from the contraction of adipocytokines, in which "-cyto" (cell) and "-kine" (movement) give cytokine, corresponding to a class of small molecules secreted by immune cells (lymphocytes, macrophages, glial cells, etc) and acting at distance. Cytokines have a messenger function, they are generally scarce in blood under normal conditions and their concentration increases sharply in traumatic conditions or during infection. Given these definitions, few adipokines may be considered as cytokines. Nevertheless, as for cytokines, the biological activities of "adipokines" are generally mediated by specific receptors to exert local (paracrine, autocrine) and/or systemic (endocrine) actions. "Adipokines" were initially considered as molecules exclusively secreted by adipose tissue (AT). This concept is obsolete because almost all the adipokines, except leptin and adiponectin, are also produced by other organs. Thus, for a given adipokine, it is crucial to examine the specific contribution of AT as

C. Carpéné · J.-S. Saulnier-Blache (✉)
Département Métabolisme et Obésité, Institut des maladies métaboliques et cardiovasculaires (I2MC), INSERM U1048/Université Paul Sabatier Toulouse Equipe n°3, BP 84225 31432 Toulouse Cedex 4, France
e-mail: jean-sebastien.saulnier-blache@inserm.fr

C. Carpéné
e-mail: christian.carpene@inserm.fr

J.-P. Bastard and B. Fève (eds.), *Physiology and Physiopathology of Adipose Tissue*,
DOI: 10.1007/978-2-8178-0343-2_15,

a source of the circulating levels. When AT substantially participates in the systemic biological activities of the candidate, we propose to replace the term "adipokine" by "adipocrine". We propose to use the term adipocrine only when AT (especially when extended in obesity) represents a substantial origin of the circulating levels of the considered molecule in the organism. This should replace the current nomenclature of adipokine which encompasses all the soluble molecules expressed/produced in adipose tissue, without taking into account whether such adipose-derived production is minor or plays a relevant role in the paracrine and endocrine influences of the secreted molecule.

Obviously, a complete survey of the novel adipokines not treated in the preceding chapters would have encompassed a lot of molecules since it has been estimated that AT physiologically releases more than one thousand products. When considering only peptides and proteins, the partners of ADAPT, an FP7 European project (Adipokines as Drug Targets to Combat Adverse Effects of Excess Adipose Tissue; coordinator: J. Eckel; website: http://www.adapt-eu.net/25.html), have detected more than 350 candidates in the conditioned media of differentiated human preadipocytes owing to an adipocyte secretome analysis. In the present review, we chose to focus attention on few recently proposed "adipocrines", which are still the subject of ongoing investigations, and that we considered original in their regulation and action: visfatin, autotaxin, PEDF, RBP4, resistin, chemerin.

Visfatin

The term "Visfatin" was first used in 2005 to characterize a protein secreted by visceral fat (Fukuhara et al. 2005). At this moment, the novel roles attributed to this protein were linked to obesity and insulin action. In fact, visfatin is identical to pre-B cell colony-enhancing factor (PBEF1), a protein already found in lymphoid and various other tissues (e.g., liver, placenta) (Samal et al. 1994), and known to modulate the production of inflammatory cytokines and chemokines by immune cells and their precursors. In man, Pbef1 gene is encoding a 491-residue long protein (Ognjanovic et al. 2001), which can be found as an extracellular product of 52-kDa, although lacking a signal sequence for secretion. The pioneering studies of the group of Shimomura indicated in 2005 that visfatin was able to reproduce insulin action in muscle or adipose cells. Then, in 2007, a novel insight was brought on PEBF1/visfatin by showing its identity with nicotinamide phosphoribosyltransferase (NAMPT), an enzyme participating in the synthesis of the coenzyme nicotinamide adenine dinucleotide (NAD) (Revollo et al. 2007). At this time, PEBF1/visfatin/NAMPT was shown to be secreted by cultured adipocytes (Revollo et al. 2007). Nowadays, PEBF1/visfatin/NAMPT is recognized as a multifunctional protein sharing at least three properties: (1) enzyme involved in NAD synthesis, (2) secreted adipokine, (3) pro-inflammatory cytokine. These three aspects will be evaluated after briefly stating on another proposed action of visfatin, which has been controversial for a long while: its direct insulin-mimicking action.

No Direct Insulin-Like Properties for Visfatin

It has been claimed that visfatin was able to activate the insulin receptor, and to stimulate glucose uptake in adipose and muscle cells (Fukuhara et al. 2005). Thereafter, only two reports confirmed that visfatin promotes glucose transport while various studies were unable to reproduce such promising effects of the adipokine, even in models that are highly responsive to insulin regarding glucose uptake activation: 3T3 L1 cells, cultured human preadipocytes (Revollo et al. 2007), mature adipocytes (Wanecq et al. 2009), or skeletal muscles (Harasim et al. 2011). Moreover, Shimomura and co-workers published a retraction of their previous findings on the insulin-like properties of visfatin (Fukuhara et al. 2007). Therefore, visfatin cannot be considered as beneficial adipokine on the sole basis of its putative insulin-mimicking effects in target tissues, albeit elusively quoted in various reviews dedicated to adipokines.

Visfatin as a NAD Biosynthetic Enzyme

Owing to its capacity to catalyze nicotinamide mononucleotide (NMN) formation from nicotinamide and 5-phosphoribosyl 1-pyrophosphate (PRPP), NAMPT/visfatin is a major component of NAD biosynthesis, together with nicotinamide/nicotinic acid mononucleotide adenylyltransferase, which converts NMN to NAD. Indeed, the multifunctional protein PEBF1/visfatin/NAMPT regulates the activity of NAD-consuming enzymes such as ADP ribosyl-transferases and sirtuins, known to be activated by calorie restriction and to influence a variety of biological effects, including metabolic regulation, inflammation, cancer, and aging.

The NAMPT-mediated NAD biosynthesis is useful for pancreatic β-cell function: it facilitates insulin secretion, and may participate in the hypoglycemic effect of injected visfatin (Revollo et al. 2007). Additionally, the in vivo administration of the NAMPT product NMN restores the impaired islet function in Nampt($\pm$) mice and in mice fed a deleterious fructose-rich diet (Caton et al. 2011). It has been proposed that cells that do not express enough endogenous intracellular NAMPT would depend on systemic NMN supply by extracellular NAMPT (noted hereafter as eNAMPT/visfatin) for their own NAD biosynthesis. Such systemic-dependent NAD synthesis has been called "NAD World" and has been applied to pancreatic endocrine cells. It also implies sirtuin activities and is altered in complications such as type 2 diabetes. However, major concerns have been recently raised regarding the relevance of eNAMPT/visfatin action in the extracellular space. First, kinetic studies show that NAMPT readily catalyzes the NMN formation from its substrates nicotinamide and PRPP only when ATP is present at millimolar levels. Second, the low concentrations of nicotinamide in plasma, together with the almost absence of circulating PRPP and ATP, suggest that eNAMPT/visfatin does not participate in NMN formation in blood. Accordingly, there is almost no NMN circulating in the plasma, indicating that this NAMPT product cannot accumulate upon activity of the adipokine (Hara et al. 2011). Thus, the systemic NAD synthesis is probably not regulated by circulating eNAMPT/visfatin.

The endocrine actions of visfatin involved in obesity, inflammation, and glucose handling are therefore far from being deciphered, in spite of the findings obtained by studying NAMPT activity/NAD supply in cell models or lysates. However, alternate mechanisms could explain the endocrine actions of the adipokine since eNAMPT/visfatin promotes the induction of nitric oxide synthase (iNOS) (Romacho et al. 2009) and the translocation of several subunits of NADPH oxidase (Malam et al. 2011), i.e., enzymes that are involved in immune responses and in adipocyte responses to hormonal activation.

eNAMPT/Visfatin as an Adipocrine Molecule

Various clinical studies have confirmed the link between circulating visfatin and body mass index (BMI) (Friebe et al. 2011) as originally proposed (Fukuhara et al. 2005). They have also demonstrated a positive correlation between visfatin mRNA abundance in omental AT and BMI or insulin resistance. The circulating eNAMPT/visfatin not only increases with obesity, but it also decreases in intervention studies lowering the body weight or treating the metabolic syndrome of obese patients, such as treatments with orlistat, rosiglitazone, statins, or after exercise training. However, eNAMPT/visfatin was not decreased after metformin treatment, low-calorie diet, or even after gastric surgery of obese patients. The existence of tight relationship between obesity and eNAMPT/visfatin levels is therefore still under debate (Stofkova 2010). We propose at least two major sources for such controversy:

- First, the plasma visfatin values vary a lot between clinical studies (from 1 to more than 20 ng/mL) and methodological causes may be as important as individual variations in such dispersion. Leucocytes themselves are rich in NAMPT activity. The relative proportions of the extracellular cytokine/adipokine (circulating under the form of a dimeric protein) and of the intracellular enzyme of white blood cells are not well established. Blood levels may vary in function of changes in these proportions, and actually, circulating eNAMPT/visfatin is strongly correlated with leukocyte counts (Friebe et al. 2011), reinforcing the visfatin role in inflammation and underlining a neglected source of variation.
- The second cause is that chronic inflammation (e.g., liver or kidney diseases), insulin resistance (e.g., gestational diabetes), immune diseases (e.g., rheumatoid

arthritis), polycystic ovary syndrome, cancer (breast, colorectal, or gastric), or cardiovascular diseases increase the circulating eNAMPT/visfatin values, independently from the obesity state of patients. This relies with the primary definition of PEBF1 as an immunomodulator.

Finally, it must be noted that models of genetic- or diet-induced obese rats (obese Zucker or cafeteria diet-fed) are clearly hyperleptinemic and hyperinsulinemic, but exhibit unchanged eNAMPT/visfatin serum levels and even reduced visfatin mRNA levels in WAT when compared to lean control (Mercader et al. 2008).

Regulation of Visfatin Production and Pharmacology

The reported regulators of eNAMPT/visfatin production are dexamethasone and glucose restriction, which upregulate gene expression in cultured adipocytes. Strikingly, tumor necrosis factor-alpha (TNFα) downregulates (Sommer et al. 2008) or is without influence (Lorente-Cebrián et al. 2009) on visfatin gene expression. Such findings do not easily deal with the in vivo increase in eNAMPT/visfatin in inflammatory states. On the opposite, the reduced release of visfatin by adipocytes upon insulin or GLP-1 treatment fits with the acute decrease of serum visfatin found in volunteers subjected to oral glucose tolerance test (Friebe et al. 2011). However, an increase of the adipokine production has been observed when volunteers are subjected to slow i.v. infusions of glucose (Haider et al. 2006) or after rosiglitazone treatment. Accordingly, macrophages, including those residing in AT, express visfatin in a manner positively regulated by PPARγ (Mayi et al. 2010) and repressed by LXR. Recent demonstrations performed in liver cells showed that Nampt gene expression is enhanced by FoxO transcription factors implicated in lipid metabolism, and by palmitate. At present, there is no doubt that the multifunctional protein eNAMPT/visfatin can be synthesized in other tissues than AT, therefore confirming by the high levels of NAMPT already found in skeletal muscles, placenta, cartilage, liver, heart (Revollo et al. 2007), and also in cancer cells and white blood cells (Friebe et al. 2011). Finally, the pharmacology of visfatin is emerging, the most representative of the NAMPT inhibitors, FK866 (also called APO866 or WK175), has been evidenced to inhibit tumor growth and leads to the development of other NAMPT inhibitors as candidates for novel treatments in oncology. Thus, the pharmacology of eNAMPT/visfatin is targeting other therapies than the mitigation of obesity.

Autotaxin

Autotaxin (ATX) is an enzyme secreted as a glycosylated protein of approximately 120 kDa containing several domains, including a catalytic site responsible for lysophospholipase D activity that generates lysophosphatidic acid (LPA), a phospholipid growth factor acting through specific receptors coupled with G proteins (LPA1R to LPA6R) (Okudaira et al. 2010). ATX was originally identified as a promigratory factor present in the conditioned medium of melanocytes in culture (Stracke et al. 1992). Since then it was found in many organs in animals and humans, especially in brain, lymph nodes, and AT. ATX is also present in the blood where it contributes mainly to circulating levels of LPA (van Meeteren et al. 2006).

ATX is abundantly secreted by AT where it is 2–3 times more expressed in the adipocytes than in the stromal-vascular cells (Ferry et al. 2003). In culture, ATX secretion is greatly increased during the adipogenesis process and is accompanied by a strong accumulation of LPA in the extracellular medium (Ferry et al. 2003). The expression of adipocyte ATX increases in obese individuals (mouse, human, rat), in correlation with the insulin resistance state rather than with their fat mass (Boucher et al. 2005).

The general knockout of ATX in mice is lethal as the result of an impaired neurogenesis and vasculogenesis of the embryo (van Meeteren et al. 2006). In contrast, specific invalidation of ATX in adipocytes (FATX-KO mice) are viable but exhibit a 40 % reduction in their plasma concentration of LPA (Dusaulcy et al. 2011). This demonstrates the important contribution of AT in the circulating level of this lipid mediator. Moreover, FATX-KO mice fed with a high fat diet exhibit higher fatness than wild mice with no change in food intake. This strongly suggests that, in the wild mice, ATX exerts an inhibitory effect on the expansion of fat mass. Adipocytes from high fat fed FATX-KO mice are larger but are similar in number (Dusaulcy et al. 2011) suggesting a hypertrophic rather than hyperplasic fattening. While adipocyte hyperplasia involves an increase in adipogenesis, adipocyte hypertrophy reflects a better ability of adipocytes to store triglycerides (lipogenesis), a metabolic event that is tightly dependent on insulin sensitivity. Interestingly, FATX-KO mice exhibit a better glucose tolerance and a stronger expression of the transcription factor PPARγ2 as well as of several of its target genes (adiponectin, FABP4, leptin, glut-1) when compared to wild-type (Dusaulcy et al. 2011). Consequently, AT from FATX-KO mice shows an apparent improvement of its sensitivity to insulin. Hence, the negative effect of ATX on fat mass expansion could at least in part be mediated by inhibition of AT sensitivity to insulin. Further experiments should now be conducted to validate this hypothesis in vivo.

LPA being the main mediator generated by ATX, its involvement in the negative regulation of fat mass is strongly suspected. The majority of LPA receptors in AT are belonging to the LPA1R subtype (Simon et al. 2005). The general knockout of LPA1R in mice (LPA1R-KO) increases postnatal mortality due to an alteration of the suckling behavior by the off-springs (Contos et al. 2000). Surviving animals (about 50 %) fed a normal diet have a reduced overall body development but exhibit a higher fat mass (Simon et al. 2005). As it was observed in FATX-KO mice, LPA1R-KO mice possess larger adipocytes (personal results) suggesting an hypertrophic fattening. On the other hand, LPA1R-KO mice have increased adipocyte expression of leptin and GLUT-4, and increased levels of plasma leptin (Dusaulcy et al. 2009). Over-fattening of LPA1R-KO mice is also observed after treatment with aurothioglucose (Jean Philippe Pradère Thesis, University Paul Sabatier, Toulouse 2007). In contrast, no over-fattening was observed when LPA1R-KO mice are fed a high fat diet. This has been attributed to a strong reduction of food intake likely related to the action of LPA on the satiety center (Dusaulcy et al. 2009). Nevertheless, the phenotype of LPA1R-KO mice partly corroborates the hypothesis that LPA (via LPA1R) would be involved in the inhibitory effect of ATX on body fat expansion. However, the cellular mechanisms responsible for that inhibitory effect still remain unclear. In culture, the LPA increases the proliferation of preadipocytes and inhibits their differentiation into adipocytes (Simon et al. 2005). But those effects unlikely explain in vivo observations since the obesity of both FATX-KO-KO and LPA1R is hypertrophic rather than hyperplasic. Preliminary results obtained in the team show that the LPA could partially oppose the stimulation of glucose transport induced by insulin in adipocytes in culture (unpublished results). This suggests that LPA may exert an inhibitory effect on insulin sensitivity of adipocytes and therefore reduce adipocyte hypertrophy. Hence, in vivo insulin sensitivity of FATX-KO and LPA1R mice AT should now be evaluated more precisely.

Pigment Epithelium-Derived Factor

Pigment epithelium-derived factor (PEDF) is a 50 kDa glycoprotein belonging to the serpin superfamily but does not possess anti-protease activity. PEDF was originally identified as a neurotrophic factor present in conditioned media prepared from epithelial cells pigment retina (Becerra et al. 1995). PEDF is expressed in many organs especially eyes, liver, and AT, and is also present in abundance in plasma. The PEDF is a multifunctional protein exerting antiagiogenic, antitumorigenic, antioxidant, and anti-inflammatory effects. Some of these biological responses are mediated via the binding of PEDF to extracellular matrix proteins (non receptor integrin to laminin, collagen, heparin, hyaluronan) (Rychli et al. 2009).

The expression of PEDF in AT is predominant in adipocytes. In cultured human preadipocytes, the expression and secretion of PEDF increase transiently during adipogenesis. In contrast, in murine preadipocyte line 3T3-L1, expression of PEDF decreases rapidly during adipogenesis (Kratchmarova et al. 2002)). Mass spectrometry identified PEDF as one the of the most abundant proteins secreted by the mouse and human adipocytes (Zvonic et al. 2007). In humans, circulating levels of PEDF are closely linked with visceral obesity and with the level of insulin resistance (Nakamura et al. 2009). Conversely, weight loss resulting from caloric restriction or bariatric surgery results in decreased expression adipocyte and of circulating levels of PEDF (Sabater et al. 2010).

The expression of PEDF in adipocytes increases significantly in genetic and nutritional obese mice. This increase is associated with increased plasma levels of

PEDF (Crowe et al. 2009). Mice invalidated for PEDF are valid and fertile, but their body fat was not analyzed. However, administration of PEDF in normal mice causes insulin resistance in skeletal muscle and liver, and neutralization of PEDF in obese mice increases insulin resistance (Crowe et al. 2009).

In vivo administration of PEDF is often associated with a pro-inflammatory reaction, the formation of ectopic fat deposition (mainly in liver and muscle) and the increase in adipocytes lipolysis (Chung et al. 2008). Interestingly, the effects of PEDF on lipid metabolism and insulin resistance are abolished in mice invalidated for ATGL (adipocyte triacylglycerol lipase also named phospholipase-linked plasma membrane protein) (Borg et al. 2011). Moreover, two hybrid experiments in yeast revealed the existence of a direct binding of PEDF to ATGL leading to an increase in PLA2 activity of the enzyme (Notari et al. 2006). This activation could be responsible for the generation of fatty acids or other lipid mediators that could mediate the biological effects of PEDF. However, there is a controversy over the fact that ATGL may actually represent a membrane receptor for PEDF. In addition, PEDF may also bind to the transcription factor PPARα (Chung et al. 2008) that plays an important role in lipid metabolism. However, these studies show the existence of a close relationship between PEDF and lipid metabolism, which may account for the involvement of this secreted factor in obesity and its involvement in the etiology of insulin resistance.

Retinol Binding Protein-4

The Retinol Binding Protein-4 (RBP4) is a liver protein of 21 kDa belonging to the lipocalin family. It enables the transport of small hydrophobic molecules in the blood from the liver to peripheral organs, in particular retinol (vitamin A). Circulating levels of RBP4 are positively correlated with the level of retinol (Zanotti and Berni 2004). AT is the second largest producer of RBP4 where it is predominantly expressed in adipocytes (Tsutsumi et al. 1992). In culture, its expression increases during the process of adipogenesis (Zovich et al. 1992). Adipocyte expression of RBP4 is correlated to changes in plasma RBP4 (Yang et al. 2005). Serum levels of RBP4 are increased in several animal models of obesity and insulin resistance. For example, RBP4 is highly expressed in AT of adipose-GLUT4 (-/-) mice that develops a strong insulin resistance in muscle and liver (Abel et al. 2001). Conversely, the overexpression of GLUT4 in AT reduces RBP4 expression (Yang et al. 2005). Pharmacological normalization of insulin resistance with rosiglitazone reduces the expression of RBP4 serum levels and adipocyte, but does not affect hepatic expression of the protein. Transgenic overexpression of RBP4 or injection of recombinant RBP4 in normal mice causes overall insulin resistance (Yang et al. 2005). Increased serum RBP4 induces the expression of gluconeogenic enzymes (phosphoenolpyruvate carboxykinase) in muscle and impairs its insulin sensitivity (Yang et al. 2005). Mice invalidated for RBP4 have an increased sensitivity to insulin. Finally, fenretinide, a synthetic retinoid that increases urine excretion of RBP4 and normalizes its serum concentration, improves insulin resistance and glucose tolerance in obese mice (Yang et al. 2005). In humans, serum RBP4 increases and is positively correlated with BMI and visceral AT in diabetic and non-diabetic obese patients (Yang et al. 2005). Weight reduction following a diet, exercise, or surgery reduces circulating levels of RBP4 as well as its expression in visceral AT. Moreover, in these subjects, the serum RBP4 is closely correlated with the degree of insulin resistance and this is independent of obesity (Graham et al. 2006). Serum RBP4 can also be correlated with indices of atherosclerosis and could negatively impact on the secretory function of pancreatic B cells (Janke et al. 2006). The expression of adipocyte RBP4 was higher in women than in men (Kos et al. 2011). This is partly attributed to estrogen and leptin, but could also be related to iron metabolism whose depletion reduces circulating RBP4 and increases insulin sensitivity (Fernandez-Real et al. 2008)). In peripheral tissues, RBP4 acts directly via cell surface receptors STRA6 (Stimulated by Retinoic Acid gene homolog 6) or indirectly through the action of retinoic acid receptor (Sivaprasadarao and Findlay 1988). The negative effects of RBP4 on insulin sensitivity is mediated, at least in part, by inhibition of IRS1 (insulin receptor substrate 1) phosphorylation, a key step in the signaling pathway of insulin (Ost et al. 2007).

In conclusion, RBP4 appears as an adipocrine associated with adiposity and insulin resistance. This was shown in animals and in a number of clinical studies. However, other clinical studies do not lead to the same conclusions, likely because of confounding parameters such as the status of retinol and iron patients.

Resistin

Resistin was identified in 2001 by two independent studies showing that this protein constitutes a potential link between obesity, inflammation, and insulin resistance (Steppan et al. 2001). Resistin is a protein of 12.5 kDa, which is able to counteract the stimulatory effects of insulin on glucose uptake in adipocytes (Steppan et al. 2001). Resistin also impairs the inhibition of glucose output in liver. Although resistin was first proposed to play a role in glucose homeostasis, it is also implicated in the regulation of innate immune response. Indeed, resistin is found in blood under a low molecular weight form (trimer) that appears to be biologically active, and also under a hexameric form. The latter appears to be a consequence of the richness in cysteine residues (approx 10 %), facilitating the formation of disulfide bonds (Patel et al. 2004) in resistin and other similar proteins from the same family, namely the "Found in inflammatory zone family" (FIZZ).

Resistin and Insulin Action

Resistin administration to laboratory rodents results in an impaired glucose tolerance and insulin action, mainly by impairing the insulin action on glucose metabolism. However, the exact mechanism of actions of resistin is not totally defined while various cellular targets involved in other processes have been reported. Among the mechanisms suspected to antagonize insulin action, the inhibition of AMPK by resistin could be the most important. The induction of suppressor of cytokine signaling (SOCS3) is also another major step allowing resistin to disturb proper insulin signaling. Based on these observations, resistin can be considered as a "deleterious" adipokine, exhibiting an influence totally opposed to the "beneficial" adipose-derived factors facilitating insulin action (such as adiponectin, apelin, omentin, chemerin, etc).

The actions reported so far for resistin are in agreement with an overall counter-regulatory function, aiming to reduce insulin influence when the pressure exerted by the pancreatic hormone is too high. Triggering resistance to insulin also appears an evident issue when considering several regulations of resistin expression in the adipose depots of animal models; decreased with fasting (and low insulin tonus) and increased with refeeding (or insulin treatment), or in the case of nutritional or genetic obesities.

Resistin and Inflammation

In humans, resistin has been reported to be not only expressed in adipocytes but also in monocytes, macrophages, and other immune cells circulating in blood and in other fluids (saliva, synovial, amniotic, or seminal fluid) or present in tissues such as placenta. In man, serum resistin increases with various chronic inflammations (liver diseases, arthritis, etc.), and in turn resistin increases the expression of IL-6 and TNFα in human peripheral blood mononuclear cells. Moreover, resistin expression is increased in such cells, once they are activated by LPS (lipopolysaccharide) or TNFα. The link between resistin and inflammation is therefore obvious, but the relative proportion of AT in the generation of resistin in inflammatory diseases is far from being stated.

A case of well-documented increase of resistin is that of pregnancy. The levels found in amniotic fluid from obese pregnant women are not more elevated than in normal weight pregnant women. Initial observations had suggested that resistin was increasing together with other insulin counter-regulatory factors at the moment of gestational diabetes mellitus, leading to insulin resistance, glucose intolerance, hyperglycemia, and fat deposition in late pregnancy. Nevertheless, it is currently proposed that, more than the AT enlargement, it is the increase of amniotic fluid white blood cell count that is involved in the increase of resistin in the amniotic fluid, transducing a local inflammation or infection, irrespective of the obesity state (Mazaki-Tovi et al. 2011).

Although resistin level increases in few obese conditions, it should be considered as an inflammatory marker rather than a mere indicator of AT accretion. Currently, it is proposed that increases in circulating resistin may have accompanied the AT low-grade inflammation and the systemic inflammation found in obese patients. Since there are numerous confounding factors that can elevate resistin, it should be consider as an inflammatory marker able to increase in chronic inflammatory diseases, irrespective of the obesity status. As far as we know the pharmacology of inhibiting resistin has not brought relevant anti-inflammatory approach to limit the links between obesity, AT low-grade inflammation, and insulin resistance. Nevertheless, the recent characterization of a cleaved form of decorin as the receptor for resistin especially expressed in adipose stromal cells leads to reconsider resistin as an adipokine having paracrine effects (Daquinag et al. 2011). The increased resistin expression in adipocytes from overdeveloped fat depots entering in inflammation process is therefore proposed to act on surrounding stroma cells or on committed preadipocytes to inhibit adipocyte differentiation or to maintain lipid droplets in a state of minimal size through activated lipolysis. In fact, the resistin generated in—or reaching—AT probably exerts actions distinct from those observed in other anatomical locations, owing to the presence on adipose progenitors a special form of deglycanated decorin (lacking 14 amino acids relative to the decorin commonly found in other tissues) that is the only isoform able to bind resistin. Moreover, as the pioneering definition resistin receptor has also screened a small peptide that specifically interacts with cells from the stroma vascular fraction of AT (Daquinag et al. 2011), the resistin pharmacology will shortly expand in a way that may be valuable to limit AT development.

Chemerin

Chemerin was recognized in 2003 as a chemo-attractant protein acting on immune cells and prompting their movement to the site of inflammation. Product of the gene RARRES2 located on human chromosome 7, chemerin is secreted as a precursor of low biological activity, which upon proteolysis gives a soluble protein of 14–18 kDa that acts as a selective agonist of its own receptor, a G protein-coupled receptor named chemokine-like receptor 1 (CCMKLR1, or ChemR23). Later, when chemerin was found to be expressed in human AT, together with its receptor, it was classified as an adipokine, owing to its known role in inflammation and because of its increasing expression and secretion during adipocyte differentiation (Bozaoglu et al. 2007; Roh et al. 2007).

The suspected role of macrophages and lymphocytes in linking chronic inflammation of expanding adipose tissues and insulin resistance prompted various research teams to investigate whether chemerin may be pivotal in the pathogenesis of obesity or metabolic syndrome. In fact, studies of mice deficient for chemerin recently revealed that chemerin plays an important role in glucose homeostasis, mainly by regulating insulin secretion by pancreatic islets rather than by directly influencing glucose utilization and lipid storage in adipose depots (Takahashi et al. 2011). Moreover, there is no change in fatness of the mice lacking chemerin, while in agreement with the recognized chemo-attractant properties of chemerin, the mice exhibit reduced macrophage accumulation in AT. However, this lesser grade of inflammation did not prevent the chemerin-deficient mouse to be glucose intolerant and did not help the insulin inhibition of hepatic glucose production.

Another concern raised by the unchanged adiposity of the chemerin-deficient mice is that it limits the relevance of the previously reported potentiation of insulin-stimulated glucose uptake and enhancement of insulin signaling by chemerin, performed on 3T3-L1 adipocytes (Takahashi et al. 2008).

Alternatively, chemerin invalidation results in an altered insulin secretion by pancreatic islets subjected to glucose challenge. Such unexpected observation relies with a notable expression of chemerin and its receptor in *β*-cells and with its role in maintaining the expression of MafA, a transcriptional factor crucial for B cell function (Takahashi et al. 2011). In spite of such evident interaction between chemerin and glucose homeostasis in the murine model, no change in the circulating chemerin was found when comparing control and type 2 diabetic patients, while there was a positive association with BMI (Bozaoglu et al. 2007).

The changes of plasma chemerin with obesity remain controversial, especially if one considers the increase of both chemerin and its receptor in AT of mice fed a high-fat diet (Borg et al. 2011) and in other models of diabetes (Bozaoglu et al. 2007) while serum chemerin concentration is decreased in db/db mice (Takahashi et al. 2008).

It remains to establish, for chemerin (as well as for many other adipokine candidates), the relative proportion of AT as a source of the signaling molecule.

References

Abel ED, Peroni O, Kim JK et al (2001) Adipose-selective targeting of the GLUT4 gene impairs insulin action in muscle and liver. Nature 409:729–733

Becerra SP, Sagasti A, Spinella P et al (1995) Pigment epithelium-derived factor behaves like a noninhibitory serpin. Neurotrophic activity does not require the serpin reactive loop. J Biol Chem 270:25992–25999

Borg ML, Andrews ZB, Duh EJ et al (2011) Pigment epithelium-derived factor regulates lipid metabolism via adipose triglyceride lipase. Diabetes 60:1458–1466

Boucher J, Quilliot D, Praderes JP et al (2005) Potential involvement of adipocyte insulin resistance in obesity-associated up-regulation of adipocyte lysophospholipase D/autotaxin expression. Diabetologia 48:569–577

Bozaoglu K, Bolton K, McMillan J et al (2007) Chemerin is a novel adipokine associated with obesity and metabolic syndrome. Endocrinology 148:4687–4694

Caton PW, Kieswich J, Yaqoob MM et al (2011) Nicotinamide mononucleotide protects against pro-inflammatory cytokine-mediated impairment of mouse islet function. Diabetologia 54:3083–3092

Chung C, Doll JA, Stellmach VM et al (2008) Pigment epithelium-derived factor is an angiogenesis and lipid regulator that activates peroxisome proliferator-activated receptor alpha. Adv Exp Med Biol 617:591–597

Contos JJ, Fukushima N, Weiner JA et al (2000) Requirement for the lpA1 lysophosphatidic acid receptor gene in normal suckling behavior. Proc Natl Acad Sci U S A 97:13384–13389

Crowe S, Wu LE, Economou C et al (2009) Pigment epithelium-derived factor contributes to insulin resistance in obesity. Cell Metab 10:40–47

Daquinag AC, Zhang Y, Amaya-Manzanares F et al (2011) An isoform of decorin is a resistin receptor on the surface of adipose progenitor cells. Cell Stem Cell 9:74–86

Dusaulcy R, Daviaud D, Pradere JP et al (2009) Altered food consumption in mice lacking lysophosphatidic acid receptor-1. J Physiol Biochem 65:345–350

Dusaulcy R, Rancoule C, Gres S et al (2011) Adipose-specific disruption of autotaxin enhances nutritional fattening and reduces plasma lysophosphatidic acid. J Lipid Res 52:1247–1255

Fernandez-Real JM, Moreno JM, Ricart W (2008) Circulating retinol-binding protein-4 concentration might reflect insulin resistance-associated iron overload. Diabetes 57:1918–1925

Ferry G, Tellier E, Try A et al (2003) Autotaxin is released from adipocytes, catalyzes lysophosphatidic acid synthesis, and activates preadipocyte proliferation. Up-regulated expression with adipocyte differentiation and obesity. J Biol Chem 278:18162–18169

Friebe D, Neef M, Kratzsch J et al (2011) Leucocytes are a major source of circulating nicotinamide phosphoribosyltransferase (NAMPT)/pre-B cell colony (PBEF)/visfatin linking obesity and inflammation in humans. Diabetologia 54:1200–1211

Fukuhara A, Matsuda M, Nishizawa M et al (2005) Visfatin: a protein secreted by visceral fat that mimics the effects of insulin. Science 307:426–430

Fukuhara A, Matsuda M, Nishizawa M et al (2007) Retraction. Science 318:565

Graham TE, Yang Q, Bluher M et al (2006) Retinol-binding protein 4 and insulin resistance in lean, obese, and diabetic subjects. N Engl J Med 354:2552–2563

Haider DG, Schaller G, Kapiotis S et al (2006) The release of the adipocytokine visfatin is regulated by glucose and insulin. Diabetologia 49:1909–1914

Hara N, Yamada K, Shibata T et al (2011) Nicotinamide phosphoribosyltransferase/visfatin does not catalyze nicotinamide mononucleotide formation in blood plasma. PLoS ONE 6:e22781

Harasim E, Chabowski A, Górski J (2011) Lack of downstream insulin-mimetic effects of visfatin/eNAMPT on glucose and fatty acid metabolism in skeletal muscles. Acta Physiol (Oxf) 202:21–28

Janke J, Engeli S, Boschmann M et al (2006) Retinol-binding protein 4 in human obesity. Diabetes 55:2805–2810

Kos K, Wong S, Tan BK et al (2011) Human RBP4 adipose tissue expression is gender specific and influenced by leptin. Clin Endocrinol (Oxf) 74:197–205

Kratchmarova I, Kalume DE, Blagoev B et al (2002) A proteomic approach for identification of secreted proteins during the differentiation of 3T3-L1 preadipocytes to adipocytes. Mol Cell Proteomics 1:213–222

Lorente-Cebrián S, Bustos M, Marti A et al (2009) Eicosapentaenoic acid stimulates AMP-activated protein kinase and increases visfatin secretion in cultured murine adipocytes. Clin Sci (Lond) 117:243–249

Malam Z, Parodo J, Waheed F et al (2011) Pre-B cell colony-enhancing factor (PBEF/Nampt/visfatin) primes neutrophils for augmented respiratory burst activity through partial assembly of the NADPH oxidase. J Immunol 186:6474–6484

Mayi TH, Duhem C, Copin C et al (2010) Visfatin is induced by peroxisome proliferator-activated receptor gamma in human macrophages. FEBS J 277:3308–3320

Mazaki-Tovi S, Kusanovic JP, Vaisbuch E et al (2011) Resistin in amniotic fluid. In: Preedy VR, Hunter RJ (eds) Adipokines. CRC Press, New York, pp 404–418

Mercader J, Granados N, Caimari A et al (2008) Retinol-binding protein 4 and nicotinamide phosphoribosyltransferase/visfatin in rat obesity models. Horm Metab Res 40:467–472

Nakamura K, Yamagishi S, Adachi H et al (2009) Serum levels of pigment epithelium-derived factor (PEDF) are positively associated with visceral adiposity in Japanese patients with type 2 diabetes. Diabetes Metab Res Rev 25:52–56

Notari L, Baladron V, Aroca-Aguilar JD et al (2006) Identification of a lipase-linked cell membrane receptor for pigment epithelium-derived factor. J Biol Chem 281:38022–38037

Ognjanovic S, Bao S, Yamamoto SY et al (2001) Genomic organization of the gene coding for human pre-B-cell colony enhancing factor and expression in human fetal membranes. J Mol Endocrinol 26:107–117

Okudaira S, Yukiura H, Aoki J (2010) Biological roles of lysophosphatidic acid signaling through its production by autotaxin. Biochimie 92:698–706

Ost A, Danielsson A, Liden M et al (2007) Retinol-binding protein-4 attenuates insulin-induced phosphorylation of IRS1 and ERK1/2 in primary human adipocytes. FASEB J 21:3696–3704

Patel SD, Rajala MW, Rossetti L et al (2004) Disulfide-dependent multimeric assembly of resistin family hormones. Science 304:1154–1158

Revollo JR, Körner A, Mills KF et al (2007) Nampt/PBEF/Visfatin regulates insulin secretion in beta cells as a systemic NAD biosynthetic enzyme. Cell Metab 6:363–375

Roh SG, Song SH, Choi KC et al (2007) Chemerin–a new adipokine that modulates adipogenesis via its own receptor. Biochem Biophys Res Commun 362:1013–1018

Romacho T, Azcutia V, Vázquez-Bella M et al (2009) Extracellular PBEF/NAMPT/visfatin activates pro-inflammatory signalling in human vascular smooth muscle cells through nicotinamide phosphoribosyltransferase activity. Diabetologia 52:2455–2463

Rychli K, Huber K, Wojta J (2009) Pigment epithelium-derived factor (PEDF) as a therapeutic target in cardiovascular disease. Expert Opin Ther Targets 13:1295–1302

Sabater M, Moreno-Navarrete JM, Ortega FJ et al (2010) Circulating pigment epithelium-derived factor levels are associated with insulin resistance and decrease after weight loss. J Clin Endocrinol Metab 95:4720–4728

Samal B, Sun Y, Stearns G et al (1994) Cloning and characterization of the cDNA encoding a novel human pre-B-cell colony-enhancing factor. Mol Cell Biol 14:1431–1437

Simon MF, Daviaud D, Pradere JP et al (2005) Lysophosphatidic acid inhibits adipocyte differentiation via lysophosphatidic acid 1 receptor-dependent down-regulation of peroxisome proliferator-activated receptor gamma2. J Biol Chem 280:14656–14662

Sivaprasadarao A, Findlay JB (1988) The interaction of retinol-binding protein with its plasma-membrane receptor. Biochem J 255:561–569

Sommer G, Garten A, Petzold S et al (2008) Visfatin/PBEF/Nampt: structure, regulation and potential function of a novel adipokine. Clin Sci (Lond) 115:13–23

Steppan CM, Bailey ST, Bhat S et al (2001) The hormone resistin links obesity to diabetes. Nature 409:307–312

Stofkova A (2010) Resistin and visfatin: regulators of insulin sensitivity, inflammation and immunity. Endocr Regul 44:25–36

Stracke ML, Krutzsch HC, Unsworth EJ et al (1992) Identification, purification, and partial sequence analysis of autotaxin, a novel motility-stimulating protein. J Biol Chem 267:2524–2529

Takahashi M, Takahashi Y, Takahashi K et al (2008) Chemerin enhances insulin signaling and potentiates insulin-stimulated glucose uptake in 3T3-L1 adipocytes. FEBS Lett 582:573–578

Takahashi M, Okimura Y, Iguchi G, et al (2011) Chemerin regulates β-cell function in mice. Scientific Reports 1

Tsutsumi C, Okuno M, Tannous L et al (1992) Retinoids and retinoid-binding protein expression in rat adipocytes. J Biol Chem 267:1805–1810

van Meeteren LA, Ruurs P, Stortelers C et al (2006) Autotaxin, a secreted lysophospholipase D, is essential for blood vessel formation during development. Mol Cell Biol 26:5015–5022

Wanecq E, Prévot D, Carpéné C (2009) Lack of direct insulin-like action of visfatin/Nampt/PBEF1 in human adipocytes. J Physiol Biochem 65:351–360

Yang Q, Graham TE, Mody N et al (2005) Serum retinol binding protein 4 contributes to insulin resistance in obesity and type 2 diabetes. Nature 436:356–362

Zanotti G, Berni R (2004) Plasma retinol-binding protein: structure and interactions with retinol, retinoids, and transthyretin. Vitam Horm 69:271–295

Zovich DC, Orologa A, Okuno M et al (1992) Differentiation-dependent expression of retinoid-binding proteins in BFC-1 beta adipocytes. J Biol Chem 267:13884–13889

Zvonic S, Lefevre M, Kilroy G et al (2007) Secretome of primary cultures of human adipose-derived stem cells: modulation of serpins by adipogenesis. Mol Cell Proteomics 6:18–28

Part IV
Pathology of Adipose Tissue

Chapter 16
Obesity: An Evolving Process

Arnaud Basdevant and Judith Aron-Wisnewsky

Introduction

The World Health Organization report published in 1998, "Obesity: preventing and managing the global epidemic", marks the emergence of obesity in the field of modern medicine (World Health Organization 1998). Although in recent years, physiopathological, clinical, and medico-economic data have led to the consideration of obesity as an authentic pathological situation, in any case pathogenic or even an authentic disease, the position of obesity in medical nosography still remains controversial.

The chronic and evolving nature of obesity raises a key question: why, during the evolution of the disease, does it become increasingly difficult to lose weight? In other words, why do most people manage to control their size, alternating short-term phases of gaining and losing weight, whereas others move into a process of chronic aggravation and find it very difficult to reverse this trend. The different

A. Basdevant (✉) · J. Aron-Wisnewsky
Institut des maladies cardiométaboliques et de la nutrition (IHU ICAN), Paris, France
e-mail: arnaud.basdevant@psl.aphp.fr

J. Aron-Wisnewsky
e-mail: judith.aron-wisnewsky@psl.aphp.fr

A. Basdevant · J. Aron-Wisnewsky
Nutriomique U872 équipe 7, Centre de recherche des Cordeliers, INSERM, UPMC, 15, rue de l'école de médecine, 75006 Paris, France

A. Basdevant · J. Aron-Wisnewsky
Centre de Recherche en Nutrition Humaine Ile-de-France (CRNH IdF), Paris, France

A. Basdevant · J. Aron-Wisnewsky
Assistance Publique Hôpitaux de Paris,Service de Nutrition,
Bâtiment Benjamin Delessert, Hôpital Pitié-Salpêtrière,
91 boulevard de L'hôpital, 75013 Paris, France

J.-P. Bastard and B. Fève (eds.), *Physiology and Physiopathology of Adipose Tissue*,
DOI: 10.1007/978-2-8178-0343-2_16,

chapters in this treaty concerning the "Physiology and physiopathology of adipose tissue" help to elucidate this question, each in its own way. Here, we propose to describe this evolution from a clinical point of view.

The Emergence of Obesity in the Field of Medicine

Although Hippocrates noted that obesity is linked to a "greater risk of sudden death" and Galen provided the first reasoned description of "*polysarkia*", it was the positivistic approach to the pathological condition as a *degradation of a perfect state* (on which the norm can be based) which gave the idea of "ideal weight" and hence obesity (Papavramidou et al. 2004). This idea of standard weight appeared in the early nineteenth century, particularly based on the work by Adolphe Quetelet. This Belgian scientist was in charge of analyzing the anthropometric characteristics of military conscripts and he defined "ideal weight" on the basis of a now-famous formula called body mass index, designed to express corpulence as a function of height (BMI is the ratio of weight in kg to height in m^2). This quantitative approach of the disease reached its height in the 1950–1960s, with the definition of risk factors and thresholds which insurance companies used to define new premiums. They identified smoking, hypercholesterolemia, high blood pressure, diabetes… and ideal weight, thus obesity. It was the *Metropolitan Life Insurance Company* which used body mass index thresholds (BMI) pragmatically, leading to higher premiums: obesity was defined as a BMI above 30 kg/m^2 and severe obesity as a BMI above 40 kg/m^2. The medico-economic perspective is at the very core of the medical definition of obesity (World Health Organization 1998; Wang et al. 2011; Gortmaker et al. 2011).

Obesity was initially marginalized among cardiovascular risk factors because of its low prevalence, and hence, relatively small contribution to morbimortality, compared with hypertension, smoking, or diabetes. The lack of medication and physiopathological basis also contributed to a certain lack of medical interest, all the more so in that the problem was, for a long time, confined to the United States. From the 1980s, epidemiology changed radically and the increase in prevalence reached the United Kingdom and the rest of Europe. Then from 1990 to 2000, it expanded to all countries, in particular emerging countries such as Brazil, Mexico, and China. The WHO estimates the prevalence of obesity in 7 % of the world's population, i.e., 400 million people and this figure should reach 12 % by 2020 if current evolutionary trends continue. It also seems that the medical and economic impact of obesity is becoming a concern, linked to the part played by obesity in chronic diseases, particularly type 2 diabetes, hypertension, and cardiovascular disease. More recently, a link has been made between obesity and certain cancers. There is also an important psychological and social impact (discrimination, no luck). These different consequences on health and well-being generate direct and indirect costs which are significant for individuals and the health system. The report published by the WHO had a considerable effect and marked the entry of

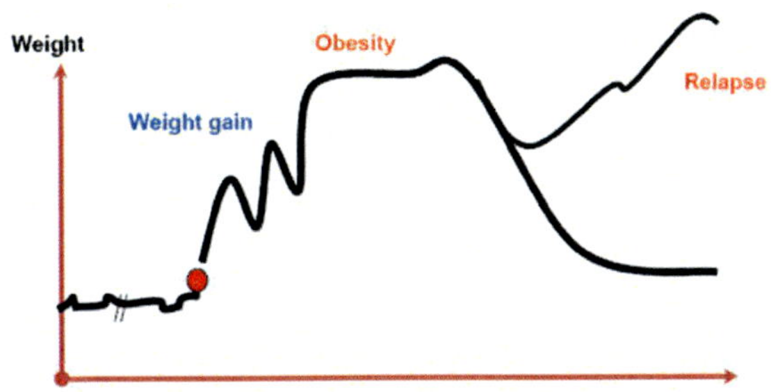

Fig. 16.1 Natural history of obesity

obesity into the field of modern medicine. Obesity is no longer confined to the *consumer* society, it is becoming a public health and medico-economic issue. Most importantly, medical and scientific credibility in the field have been reinforced by concomitant progress in biomedical and fundamental research into adipose tissue structure and function. Also, the role of environmental factors has been acknowledged (Kopelman 2000; Swinburn et al. 2011).

From Weight Gain to Obesity: An Evolving Process

Obesity is an evolving situation involving several phases (Fig. 16.1). It starts with a preclinical phase during which the person does not have any excess body fat according to standard clinical criteria. This phase, which goes from the intrauterine period to the first evidence of weight deviation, is silent, or almost apart from evocative signs such as the earliness of weight rebound. Following this preclinical phase is the initial period of weight gain and constitution of excess weight with still no pathological consequences. If the weight gain process continues and worsens over time, according to variable kinetics, it reaches a stage of defined obesity. The next phase is stabilization, where weight reaches a plateau. At this point, if there is no early intervention or if weight gain determinant persists, the disease becomes entrenched over time. During these phases of initiation, maintenance, and chronicity of the excess weight, there may be effects resulting from attempts to lose weight, either medically, surgically or by other methods. In a large number of cases, these attempts to lose weight, when successful, lead to further aggravation and progressively fail. This phase is described as recurrence, resistance to treatment, and eventually weight fluctuation. Although, this description of obesity evolution certainly appears schematic, it has the benefit of distinguishing phases marked by widely different physiopathology and medical issues.

Obesity can therefore be seen as an evolving chronic situation which results in the failure of systems regulating the level of energy reserves in adipose tissue. The accumulation of intra-adipose cell energy may, at first, be thought of as pathological inflation of an organ, the adipose tissue. This situation results from the interaction of many biological, behavioral, economic, social, and environmental determinants, the respective contributions of which vary according to the person concerned.

Evolution of the Energy Balance

The evolution of energy balance during the various phases of obesity has been remarkably well described by Sorensen (Sørensen 2009). In weight stable individuals, adipose reserves do not vary significantly, and energy intake (EI: energy intake) is balanced against expenditure (EE: energy expended); i.e., EI = EE. Constitution of excessive body fat reflects an imbalance between energy intake and expenditure: this is an unavoidable condition for constituting inflated energy reserves in the form of triglycerides in adipose tissue. During this phase of weight gain, EI > EE. The cost of energy storage must be taken into account when converting ingested energy into stored energy. According to Sorensen, to reflect this dynamic process, the equation becomes: $dES/dt = EI-(EE + EC)$, where EE corresponds to the resting metabolism (REE resting energy expenditure), corrected by a coefficient taking into account the contribution of activity (CAP) to energy expenditure: $EE = REE \times CAP$. Expenditure linked to metabolism and thermoregulation is added to this. Therefore, what is stored is less than what has been consumed in excess. It is difficult to define the exact cost of weight gain, because the evolution of body composition during weight gain, i.e., the proportion of "fat" to "lean" mass, is highly variable between individuals (Sørensen 2009). There is a key element which needs to be taken into account: any increase in body fat is accompanied by a simultaneous increase in lean mass (i.e., water, muscle, and different body components), and hence energy expenditure. Indeed, the REE largely depends on lean mass; therefore, the more a person grows fat, the greater the lean mass and energy expenditure. In other words, a person who gains weight gradually increases his energy expenditure: an obese person whose weight is stable therefore expends more energy than before he gained weight. During a weight stable period, an obese person is in a state of energy balance where intakes equal expenditures. The only difference is that the intakes and expenditures are greater than those of the non-obese period.

Now let us consider the phase of caloric restriction in an obese person. Since the energy expenditure is higher than during a non-obese situation, proposing to an obese person a so-called "normal" calorie intake, (i.e., the intake of a non-obese person), represents a drastic calorie restriction diet (Sørensen 2009).

Clinical research faces some major issues. First, it is virtually impossible to evaluate the deviations in energy balance that occur in real life. Studies on obese patients are generally done once the excess weight is confirmed, so there is almost never a prospective analysis of the initial period of weight gain. Second, the tools available to evaluate energy expenditure (calorimetry) are not able to measure energy expenditure and intake under real-life conditions with sufficient accuracy (Sørensen 2009). Therefore, it has to be said that owing to the lack of accurate tools, we cannot know what is the primary cause of obesity in a given individual: increased intake or reduced expenditure?

As Sorensen emphasizes, even if we could know whether there is a loss of equilibrium in a person's energy balance, we would not have the means of confirming that this is the primary event. Indeed, if it is correct to state that this energy imbalance contributes to the inflation of fat reserves, another hypothesis can also be proposed; the primary anomaly may be a modification in storage capacity. In other words, the initial phenomenon could be an increase in storage capacity through an increase in the number of cells or their capacity for lipogenesis, or a decrease in lipolysis or a combination of both (Sørensen 2009; Hall et al. 2011). Everything would happen as if the increase in intake was secondary to the "filling" of increased primary storage capacity (linked, for example, to adipose cells hyperplasia). In a way, the energy balance would then be at the level of the demand for storage capacity ("offer creates demand"). An increase in storage capacity could therefore be a central physiopathological mechanism in some cases of obesity.

We therefore face a difficult question: what is the *primum movens* of obesity? Is it an initial energy imbalance, a storage capacity anomaly, or both?

Evolution of Adipose Tissue

The cellular mechanisms involved in the inflation of fat mass are described in the first section of this book. Here, we approach this subject from a clinical angle. Adipose tissue is extremely plastic. Its volume may be multiplied by 5–10 in obese subjects. Physiologically, its development involves two periods of acceleration, one after birth, the other between the age of 9 and 13. Throughout life, it is capable of developing according to energy needs, hormonal situation, and temperature conditions. The increase in fat mass results from both an increase in the size of adipocytes (hypertrophy) and/or their number (hyperplasia). The relationship between body mass index and adipocyte volume follows a curve fit line, showing that the increase in volume is progressive according to the degree of corpulence, but soon reaches a plateau which also points out the role played by hyperplasia, particularly in cases of extreme obesity.

Hypertrophy due to the accumulation of triglycerides results from the imbalance between lipogenesis/lipolysis (see Chaps. 8, 9, 10) (Laharrague and Casteilla 2010; Arner et al. 2011; Langin 2011). The question of a primary anomaly in lipogenetic capacity in human obesity has been a question of debate for a long time. The activity of lipoprotein lipase has been found to be increased in many studies of obese subjects, but it is difficult to know whether this is a primary effect or consequence of weight gain. There are also close relations between lipolysis and cell size. Indeed the largest cells release more free fatty acid. This point is important in understanding certain physiopathological aspects. Hypertrophic obesities may have greater metabolic consequences than other types (Arner et al. 2011).

The number of adipocytes may increase in significant proportions. The increase in number is due to adipogenesis (see Chap. 1), i.e., the recruitment of a new adipocyte from a precursor (Arner et al. 2011). Many nutritional and non-nutritional factors can generate inappropriate adipogenesis, thus contributing to certain forms of obesity. This is the case of certain pollutants, endocrine disturbers, hormonal and nerve factors, and certain viruses (Hong et al. 2011; Pasarica et al. 2008). Penicaud et al. showed that denerving adipose tissue depots could lead to an increased number of adipocytes. The reduction in sympathetic tonus could promote the development of adipose tissue. These ideas are vital, when you study the physiopathology of certain human obesities, linked to stress with no evidence of changes in energy intake. A reduction in sympathetic tonus could contribute first to adipocyte recruitment, then lead to obesity (Penicaud et al. 2000; Woods and D'Alesio 2008; Kuo et al. 2007).

According to the so-called "critical size" hypothesis, differentiated adipose cells are loaded with triglycerides until they reach a critical size, beyond which they "recruit" a new pre-adipocyte. This is how more adipocytes can be produced, leading to hyperplasia. In case of prolonged positive energy balance, the number of adipose cells may continue to increase. This hypothesis is supported by overeating experiments; the initial response is an increase in size, then an increase in number. On the other hand, once differentiated, the cells do not return to the precursor stage. They remain available for more storage. Weight loss is linked to a reduction in the size of adipocytes, not the number which remains high. Hyperplasia seems not or only slightly reversible. This could partly explain a certain degree of resistance to weight loss noted in some people or why people put on more weight after a weight loss intervention. Cell size cannot be maintained below a certain value without triggering all the mechanisms for reconstituting fat mass; the minimum level of fat mass which can be reached is limited by the number of adipocytes. If this number is high, either constitutionally or following the recruitment of new cells during weight gain, it is difficult to reduce the volume of fat mass beyond a certain threshold (unless there is permanent dietary restriction).

Therefore, there are three potential components in the development of obesity to be considered: (1) energy intake, meaning dietary habits; (2) expenditure, thus the capacity for burning (oxidizing) energy nutrients; (3) the storage capacity related to the balance of lipogenesis/lipolysis on the one hand and adipocyte recruitment on the other.

However, the quantitative development of fat mass is neither the only cause of obesity nor its consequences. Indeed, fat distribution is a very important clinical factor. White adipose tissue is distributed in various subcutaneous, visceral, and ectopic depots. They have different characteristics and can evolve according to physiological or pathological situations. This inter-site variability is doubtlessly linked to differences in the expression of differentiation factors, levels of precursors, and the development of vascularization and innervation. One depot may undergo hyperplastic development while another is hypertrophic. Visceral depots are more vascularized and innervated by sympathetic fibers than subcutaneous. There is an inverse relationship between the development of sympathetic innervation and the capacity for developing hyperplasia. The differential effects of sex hormones are another factor of regional variation. The role of drugs (hormones or other) has also been suspected. These structural and developmental differences have metabolic consequences. For example, femoral lipid reserves are less easy to mobilize than abdominal adipocytes; the visceral tissue (omental) is particularly sensitive to glucocorticoids; breast adipose tissue physiology is radically different from that of other depots. It is therefore not surprising that, depending on the predominant topography of adipose inflation, the consequences on health and capacities for weight loss vary greatly (Unger and Scherer 2010).

These considerations have major consequences from a clinical point of view; obesity phenotyping must take into account the topography of adipose tissue depots and, an emerging question appears: could ectopic adipose depots with no notable consequences on weight be toxic even in the absence of obesity?

An interesting hypothesis is proposed by RH Unger to understand the difference between generalized obesity and localized adiposity as observed in metabolic syndrome (Unger and Scherer 2010). In common obesity, it can be considered that adipose tissue inflation continues in response to chronic energy imbalance, with insulin promoting storage as long as the tissues remain insulin-sensitive (and it remains the case long enough in morbid obesity). The increase in fat mass is linked to an increase in leptin, which is known to have antilipotoxic effects on peripheral tissues (leptin promotes lipid oxidation in none-adipocytic cells), so that excess in lipids accumulate in adipose tissues. In other words, the energy surplus moves toward general storage in adipose tissue. The metabolic disorders in this situation may develop with delay. Indeed, as long as adipocyte expansion involves small, well-vascularized cells which are not very inflammatory and free of fibrosis, the tissue can continue to grow. The process can therefore lead to morbid obesity. We speak of metabolically "healthy obesity" to indicate that lipid storage in adipose tissue can be well tolerated for as long as there is no inflammation or fibrosis. Furthermore, if experimental transgenic operations are done on animal models, adipogenesis can be increased (by overexpressing adiponectin for example) morbid obesity has been obtained without metabolic disorder. Similarly, if the constitution of fibrosis is prevented in experimental models, insulin-sensitivity persists whereas experimental models with excess of collagen lead to metabolic disturbances. Therefore, adipose tissue must reach the stage of hypertrophy and fibro-inflammatory alterations and leptin-resistance before the global expansion of fat mass becomes metabolically toxic.

The situation of metabolic syndrome linked to visceral lipid depots is different. If general expansion of adipose tissue is prevented in an obesogenic situation (high fat diet for example), metabolic disorders quickly appear. The hypothesis is that lipid accumulation in ectopic storage sites is metabolically deleterious. Leptin resistance appears to play an important part in reducing peripheral lipid oxidation, thus leading to lipotoxicity and pathological metabolic consequences. For example, ectopic fat depots lead to a loss of β-cell function and muscle alteration, particularly cardiac. The distribution of lipid depots also plays an important part in the development of metabolic syndrome, as is also seen in congenital lipodystrophies which are linked to the characteristics of metabolic syndrome very early on.

Organ Pathology with Systemic Impact

The inflation of fat mass is not only linked to an increase in the number and/or size of adipose cells, but also to a deep reorganization of its structure (see Chap. 20). As obesity develops, a real organ pathology develops which eventually evolves on its own. Thus, the behavioral and environmental factors, which cause obesity on a background of genetic or epigenetic predisposition, eventually induce an organic pathology (Mutch and Clément 2006). This is the model of chronic diseases linked to lifestyle, which are initially reversible if the behavioral and environmental determinants are controlled, but tend to become chronic and develop a certain degree of resistance over time. Then, this organ pathology causes systemic diseases since adipose tissue produces products such as adipokines (see Chaps. 11, 12, 13, 14, 15, 16 and 20), hormones, and different substances stored in adipose tissue (pollutants and others), which will act remotely from the adipose tissue to generate tissue or systemic functional damage, which is at the origin of these complications. Some of these secretions are involved in obesity-related disease. Systemic inflammation, to which adipose tissue and a high calorie diet both contribute, has multiple consequences (Keophiphath et al. 2009; Hotamisligil 2006; Horvath 2005; Lacasa et al. 2007). It is involved in the development of insulin resistance, as suggested by the relationship between high levels of inflammation biomarkers such as CRP, TNF, and IL6, and parameters such as insulin or HOMA-R. The expression of TNF in adipose tissue is increased in obese subjects and decreased after weight loss. Inflammation could also explain the gradual loss of hypothalamic sensitivity to signals emitted by adipose tissue, such as leptin, to inform the central nervous system on the status of energy reserves.

Production of PAI-1 (plasminogen activator inhibitor-1) contributes to a predisposition to thrombosis. This is how hepatic and vascular complications, and asthma associated with obesity can be accounted for by "low grade" inflammation (Hotamisligil 2006). One of the most intriguing new findings in the physiopathology of obesity is the revelation of local effects of adipose tissue production. A recently documented example concerns visceral adipose tissue and pericardial adipose tissue. We know that the regional distribution of adipose tissue is a major determinant of its metabolic consequences. The abundance of visceral tissue, also called abdominal obesity, is linked to a higher risk to develop type 2 diabetes and cardiovascular disease. There are also relationships between hepatic pathology (fibrosis and inflammation) and the degree of inflammation of visceral adipose tissue. Recent work in particular indicates that pericardial adipose tissue can contribute both mechanically and biologically to coronary anomalies, independent of other risk factors including abdominal adiposity. The hypothesis is that pericardial fat releases inflammatory substances (cytokines) and free fatty acids locally that may play a role in coronary alterations. It is known that this pericardial adipose tissue is more inflammatory in obese than in non-obese subjects (Lacasa et al. 2007; Clément et al. 2009).

To summarize, this cellular heterogeneity of obesities leads to the proposition for the future, definitions of composite phenotypes taking into account BMI, distribution of adipose tissue, ectopic fat depots, cellularity, cell volume, and the degree of inflammation, among others.

Failure of the System Regulating Energy Reserves

A person's body composition is remarkably stable physiologically, in the short and medium terms in any case, apart from minor fluctuations. We speak of "weight regulation", ponderostat, a concept which merits a more accurate definition because we must speak rather of the regulation of body compartments, including energy and hydraulic reserves as well as lean mass. The inflation of fat mass reflects the system's inability to regulate energy reserves under biological, behavioral, or environmental pressure (Morton et al. 2006).

The system which regulates energy reserves, and more generally, body composition, is generally very efficient: an increase in energy expenditure tends to be compensated by an increase in dietary intake. The value regulated, i.e., the level of energy reserves and hence fat mass, may change during the course of a lifetime under the influence of aging, hormonal events (e.g., pregnancy), and other factors, but overall body weight remains stable. These aspects are treated in Chap. 12.

Weight gain—and maintenance—reflects a failure of this regulatory system and, in any case, a change in its *set point*. The question is therefore to know why the counter-regulation system is deficient. Genetic, epigenetic factors, iatrogenic (drugs), nutritional (high-fat diets), metabolic, inflammatory, and lesional (hypothalamic tumors) alterations can modify the incorporation of messages providing information on the state of energy reserves and explain the regulatory system's inability (mainly hypothalamic) to maintain the stability of energy reserves (Morton et al. 2006).

Digestive Tract

The role of the digestive tract in obesity has long been underestimated for a long time. Experimental work and clinical research have revealed the role played by digestive hormones, in particular Ghrelin and glucagon-like peptide 1 (GLP1) on controlling food intake (Murphy and Bloom 2006). Intestinal neoglucogenesis could play a decisive part, not only in dietary intake, but also in the metabolic anomalies of obesity (Murphy and Bloom 2006). More recently, a focus has been made on gut microbiota which has different digestive, metabolic, and immune effects. Indeed, if you restore the gut microflora of a *germ free* mouse, she will develop weight gain, obesity, and insulin resistance, although she has the same amount of food intake, suggesting that gut microbiota induces greater metabolic

Fig. 16.2 Determinants of obesity over time

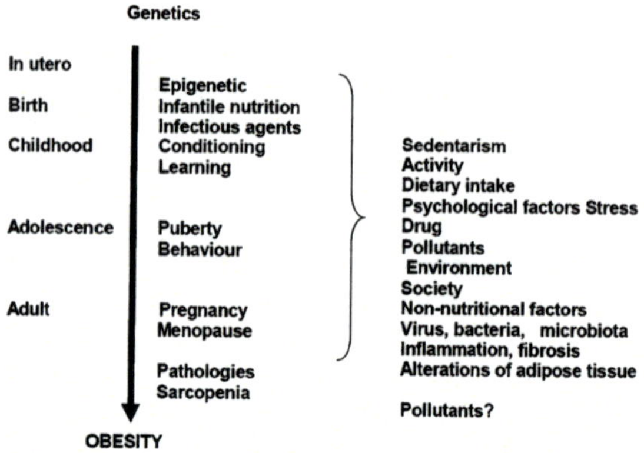

efficacy. Colonization of flora by that of an obese animal leads to weight gain in a non-obese animal. The microbiota contains a multitude of fermentation enzymes which can increase the digestion of complex glucides. Also, gut microbiota suppress the epithelial expression of *Fiaf* (fasting-induced adipocyte factor) which is a circulating inhibitor of the lipoprotein lipase, an enzyme involved in triglyceride storage. The absence of microbiota in *germ free* mice is linked to an increase in AMP-activated protein kinase (AMPK) in the liver and muscles. In humans, it has been suggested that obesity is associated with an increase in the *firmicutes/bacteroidetes* ratio and that this ratio is reduced with weight loss. Since these initial results, this has not been confirmed in all clinical studies, but has lead to more detailed study of bacterial populations using larger scale methods (Troy et al. 2008; Turnbaugh and Gordon 2009; Burcelin et al. 2009; Tilg et al. 2009; Turnbaugh et al. 2008).

The evolution of gut microbiota during the development of obesity is currently a major topic.

Thus, the physiopathology of energy reserves is not limited to the question of balance between input and output but must take into account cellular and anatomical storage capacity, the evolution of structure and function of adipose tissue, and the whole energy balance regulatory system.

Determinants of Obesity Over Time

Figure 16.2 summarizes the different determinants of obesity over time, in three main sections: (a) biological determinants: genetic, epigenetic, metabolic, hormonal, pharmacological, or others; (b) the behavioral determinants eventually linked to psychological or social factors; (c) environmental factors in the broadest sense. This distinction is artificial, because behavioral and environmental factors may lead to biological anomalies that are more or less reversible and reciprocally, biological anomalies can generate behavioral disorders.

From one extreme to the other, there are purely genetic forms of obesity, determined by rare mutations (e.g., mutation of the leptin gene or its receptor) or purely behavioral or environmental forms. Between these two extremes all situations may arise, but the rule is that there are complex interactions between environmental, behavioral, and biological factors.

Conclusion

Human obesity evolves in several phases of constitution (dynamic), maintenance (static), and resistance to weight loss. The constitution phase gives evidence of a positive energy balance no matter what the causes (excess of intake and/or reduction in energy expenditure). The phase of maintaining obesity results in a new energy balance and modifications to storage capacities.

Each of these phases has a corresponding physiopathological process and different etiological factors. The decisive elements in weight gain are often different from those that allow the overweight situation to persist and not regress easily.

These different factors of etiopathogenic interest should be distinguished. Anatomical, metabolic, neuroendocrine, psychological, and social factors intervene each in turn, some of which may be both genetically determined and/or acquired under environmental pressure and that of the obesity itself.

The whole clinical problem will be to try and recognize which factors and mechanisms predominate for each patient and which are accessible to treatment.

References

Arner P, Bernard S, Salehpour M et al (2011) Dynamics of human adipose lipid turnover in health and metabolic disease. Nature 478:110–113

Burcelin R, Luche E, Serino M, Amar J (2009) The gut microbiota ecology: a new opportunity for the treatment of metabolic diseases? Front Biosci 14:5107–5117

Clément K, Basdevant A, Dutour A (2009) Weight of pericardial fat on coronaropathy. Arterioscler Thromb Vasc Biol 29:615–616

Gortmaker SL, Swinburn BA, Levy D et al (2011) Changing the future of obesity: science, policy, and action. Lancet 378:838–847

Hall KD, Sacks G, Chandramohan D et al (2011) Quantification of the effect of energy imbalance on bodyweight. Lancet 378:826–837

Hong NS, Kim KS, Lee IK et al (2011) The association between obesity and mortality in the elderly differs by serum concentration of persistent organic pollutants: a possible exlplanation for the obesity paradox. Obes (Lond). doi: 10.1038/ijo.2011.187. [Epub ahead of print]

Horvath TL (2005) The hardship of obesity: a soft-wired hypothalamus. Nat Neurosci 8:561–565

Hotamisligil GS (2006) Inflammation and metabolic disorders. Nature 444:860–867

Keophiphath M, Achard V, Henegar C et al (2009) Macrophage-secreted factors promote a profibrotic phenotype in human preadipocytes. Mol Endocrinol 23:11–24

Kopelman PG (2000) Obesity as a medical problem. Nature 404:635–643

Kuo LE, Kintlinska JB, Tilan JU et al (2007) Neuropeptide Y acts directly in the periphery on fat tissue and mediates stress-induced obesity and metabolic syndrome. Nat Med 13:803–811

Lacasa D, Taleb S, Keophiphath M et al (2007) Macrophage-secreted factors impair human adipogenesis: involvement of proinflammatory state in preadipocytes. Endocrinology 148:868–877

Laharrague P, Casteilla L (2010) The emergence of adipocytes. Endocr Dev 19:21–30

Langin D (2011) In and out: adipose tissue lipid turnover in obesity and dyslipidemia. Cell Metab 14:569–570

Morton GJ, Cummings DE, Baskin DG et al (2006) Central nervous system control of food intake and body weight. Nature 443:289–295

Murphy KG, Bloom SR (2006) Gut hormones and the regulation of energy homeostasis. Nature 444:854–859

Mutch DM, Clément K (2006) Unraveling the genetics of human obesity. PLoS Genet 2:e188

Papavramidou NS, Papavramidis ST, Christopoulou-Aletra H (2004) Galen on obesity: etiology, effects, and treatment. World J Surg 28:631–635

Pasarica M, Mashtalir N, Mc Allister EJ et al (2008) Adipogenic human adenovirus Ad-36 induces commitment, differentiation, and lipid accumulation inhuman adipose-derived stem cells. Stem Cells 26:969–978

Penicaud L, Cousin B, Leloup C et al (2000) The autonomic nervous system, adipose tissue plasticity and energy balance. Nutrition 16:903–908

Sørensen TI (2009) Conference on "Multidisciplinary approaches to nutritional problems". Symposium on "Diabetes and health". Challenges in the study of causation of obesity. Proc Nutr Soc 68:43–54

Swinburn BA, Sacks G, Hall KD et al (2011) The global obesity pandemic: shaped by global drivers and local environments. Lancet 378:804–814

Tilg H, Moscehn AR, Kaser A (2009) Obesity and the microbiota. Gastroenterology 136:1476–1483

Troy S, Soty M, Ribeiro L et al (2008) Intestinal gluconeogenesis is a keyfactor for early metabolic changes after gastric bypass but not after gastric lap-band in mice. Cell Metab 8:201–211

Turnbaugh PJ, Gordon JI (2009) The core gut microbiome, energy balance and obesity. J Physiol 587(Pt 17):4153–4158

Turnbaugh PJ, Bäckhed F, Fulton L, Gordon JI (2008) Diet-induced obesity is linked to marked but reversible alterations in the mouse distal gut microbiome. Cell Host Microbe 3:213–223

Unger RH, Scherer PE (2010) Gluttony, sloth and the metabolic syndrome: a roadmap to lipotoxicity. Trends Endocrinol Metab 21:345–352

Wang YC, McPherson K, Marsh T et al (2011) Health and economic burden of the projected obesity trends in the USA and the UK. Lancet 378:815–825

Woods SC, D'Alesio DA (2008) Central control of body weight and appetite. J Clin Endocrinol Metab 93:S37–S50

World Health Organization (1998) Obesity: preventing and managing the global epidemic. Report of a WHO Consultation on Obesity, Geneva, 1997 (WHO/NUT/NCD/98.1)

Chapter 17
Obesity Phenotypes: Measures to Assess Adipose Tissue Mass in Humans

Jean-Michel Oppert

Introduction

Obesity is a chronic disease characterized by the fact it is multifactorial in origin and heterogeneous both in terms of determinants and phenotypes (WHO World Health Organization 2000). According to the World Health Organization, obesity is defined as "a condition of abnormal or excessive fat accumulation in adipose tissue, to the extent that health may be impaired" (WHO World Health Organization 2000). Body fat content and its relation to ill health are therefore central to the definition and understanding of obesity phenotypes.

Numerous obesity-related phenotypes can be identified (Table 17.1, (Oppert et al. 2008)). They can be divided into primary obesity phenotypes directly related to body composition and body fat amount, secondary obesity phenotypes related to the energy-balance effectors of body fat content (energy intake and energy expenditure) and multiple other obesity-related phenotypes related to adipose tissue morphology, metabolism, and secretory products as well as to the deleterious health consequences of increased adipose mass (obesity co-morbidities) including possible consequences of obesity treatments. In this chapter, we will focus on major body fat phenotypes that can be defined in adults and we will

J.-M. Oppert (✉)
Institut des maladies cardiométaboliques et de la nutrition (IHU ICAN), Paris, France
e-mail: jean-michel.oppert@psl.aphp.fr

J.-M. Oppert
Université Pierre et Marie Curie Paris 6, Paris, France

J.-M. Oppert
Centre de Recherche en Nutrition Humaine Ile-de-France (CRNH IdF), Paris, France

J.-M. Oppert
Assistance Publique Hôpitaux de Paris,Service de Nutrition, Bâtiment Benjamin Delessert, Hôpital Pitié-Salpêtrière, 91 boulevard de L'hôpital, 75013 Paris, France

J.-P. Bastard and B. Fève (eds.), *Physiology and Physiopathology of Adipose Tissue*,
DOI: 10.1007/978-2-8178-0343-2_17, © Springer-Verlag France 2013

Table 17.1 Overview of obesity-related phenotypes

Domain	Phenotype
Primary phenotypes	
Body fat/body composition	Overall corpulence (Body mass index, BMI)
	Total body fat
	Body fat distribution,
	Specific fat depots (abdominal visceral fat, hepatic fat, epicardial fat…)
	Fat-free mass relative to fat mass (sarcopenic obesity)
	Dynamics of weight and body composition changes over time
Secondary phenotypes	
Energy expenditure (EE)	Total (24-h) EE and components (resting EE, physical activity EE)
	Respiratory quotient (RQ), nutrient oxidation (lipid vs. carbohydrate)
	Physical activity level (PAL = TEE/REE)
	Non-exercise activity thermogenesis (NEAT)
Physical activity	"Dose" (frequency, intensity, duration), type
	Sedentary behavior (sitting time)
Food intake	Energy intake, macronutrient intake
	Eating patterns (meal frequency, snacking)
	Eating disorders
Other obesity-related phenotypes	
	Adipose tissue secretory products (adipokines)
	Inflammatory markers
	Hormones and markers modified by increased adipose tissue mass (e.g., insulin, insulin sensitivity)
	Structure and morphology of adipose tissue
	Obesity co-morbidities: metabolic and endocrine disorders; cardiovascular disease and HTA; respiratory disturbances; joint diseases; cancers; depression; and mood disorders

Adapted from (Oppert et al. 2008)

briefly describe the most important methods available for their assessment. More details can be found in recent reviews or statements on the topic (Oppert et al. 2008; Cornier et al. 2011).

Body Mass Index

BMI Cutoffs

There is an international consensus to use the Body Mass Index (BMI), defined as the ratio of weight (in kg) over height squared (in m^2), to assess weight status and define obesity (Table 17.2, (WHO World Health Organization 2000)). In adults of both genders, obesity is currently defined as a BMI $\geq$ 30 kg/m^2. The use of the BMI is based on two main reasons. First, on a group basis, the BMI was shown to be reasonably correlated with body fat content (r around 0.70–0.80) (Willett 1998).

Table 17.2 Classification of weight status in adults according to the World Health Organization

Classification	BMI (kg/m^2)
Underweight	<18.5
Normal range	18.5–24.9
Overweight	25.0–29.9
Obese	30.0–39.9
Morbidly obese	≥40.0

Source (WHO World Health Organization 2000)

Second, in a number of studies, a J- or U-shaped relationship was demonstrated between BMI and relative risk of mortality (all causes or cardiovascular). For example, in the Prospective Studies Collaboration, the progressive excess mortality above an apparent optimum BMI of about 22.5–25 kg/m^2 was so that, at 30–35 kg/m^2 median survival was reduced by 2–4 years and at 40–45 kg/m^2 it was reduced by 8–10 years (i.e comparable with the effects of smoking) (Prospective Studies Collaboration et al. 2009). Based on this type of relationship, current BMI cutoffs were defined, as shown in Table 17.2.

In subjects of Asian origin, lower cutoffs have been proposed (overweight: ≥23 kg/m^2, obesity: ≥25 kg/m^2) because a substantial proportion of Asian people was found at high risk of type 2 diabetes and cardiovascular disease at BMIs lower than 25 kg/m^2 (WHO Expert Consultation 2004). It is also important to emphasize the difference between the definitions of adult and childhood obesity. In adults, obesity is defined as a risk factor for morbidity and mortality, as noted above. In children, obesity is defined on a population distribution basis and/or the risk of being obese at age of 18 years, by using growth curves that describe, by sex, the evolution of BMI according to age (Cole et al. 2000).

Limitations of BMI

There are a number of limitations to the use of the BMI as a measure of obesity that need to be carefully considered (Prentice and Jebb 2001). At individual level, the BMI does not give precise indications about body composition, i.e., it does not distinguish between weight associated with lean or fat tissue. Commonly used BMI cutoff values to diagnose obesity have been shown to have high specificity, but low sensitivity to identify adiposity, as they apparently fail to identify half of the people with excess percent body fat (Okorodudu et al. 2010). BMI cutoffs do not allow to capture changes in body fat that may occur according to age, gender, ethnic groups, or with exercise training. That a given BMI may not correspond to the same degree of fatness across populations, is one justification for the different BMI thresholds in subjects of Asian origin mentioned above, as these individuals have a higher body fat percentage compared to Caucasians at an identical BMI (WHO Expert Consultation 2004). Thus, BMI can indeed be considered as the most useful although crude population-level indicator of obesity (WHO World

Health Organization 2000). However, it clearly does not account for the wide variation in obesity phenotypes between individuals and populations.

Weight History

Given the natural history of obesity, some body weight values represent important phenotypes when they occur at critical time periods for body weight gain or change (Dietz 1994). This means that there will be special interest in body weight at birth, at puberty, at age of 20 years (a time when individuals are thought to remember about their weight), during pregnancy (e.g., amount of weight retained after pregnancy) as well as the age when the adiposity rebound took place in children. Other phenotypes of interest include maximal weight during lifetime, minimal weight at adulthood, magnitude of weight change with indication of whether it corresponds to voluntary or involuntary weight loss. Maximum weight loss after bariatric surgery is often noted to analyze the trend in weight regain that occurs after some years of follow-up.

Total Body Fat

As the definition of obesity specifically refers to accumulation of fat, body fat content is to be considered as the primary phenotype. However, to assess accurately body fat remains difficult and expensive in the clinical setting and/or in large populations (Cornier et al. 2011; Snijder et al. 2006). Moreover, in contrast to BMI, there are no established reference data for body fat in adults or children (Prentice and Jebb 2001). Importantly, health risks specifically associated with variations in body compartments (fat, lean, or fat-free mass) still remain to be better defined (Baumgartner et al. 1995). It is usually assumed that increased risk of morbidity and mortality associated with lower BMI is explained by decreased fat-free mass whereas increased risk associated with higher BMI is related to increased fat mass. However, there is clearly a need for more research describing mortality as a function of increasing fat mass and decreasing lean mass (Heitmann et al. 2000; Oppert et al. 2002). Table 17.3 lists the main methods that can be used for measuring body composition and that we briefly describe below [for a recent review on body composition assessment see (Lee and Gallagher 2008)].

Anthropometry and Bio-Impedance

Anthropometric methods and bioelectrical impedance analysis (BIA) appear the most simple methods to use. Skinfold thicknesses measured at various locations

Table 17.3 Methods for estimation of total and regional body fat

Method	Precision	Measures total fat	Measures regional fat
Anthropometry			
Height, weight, BMI	High	Yes	No
Circumferences, ratios	Moderate	No	Yes
Skinfolds	Low	Yes	Yes
Impedancemetry			
Bio-electrical impedance (BIA)	High	Yes	No
Absorptiometry			
Dual-energy absorptiometry (DXA)	High	Yes	Yes
Densitometry			
Underwater weighing	High	Yes	No
Air-displacement plethysmography	High	Yes	No
Imaging techniques			
Computed tomography (CT)	High	No	Yes
Magnetic resonance imaging (MRI)	High	Yes	Yes
Others			
Isotope dilution (e.g., deuterium)	High	Yes	No
K isotope (^{40}K)	High	Yes	No

Adapted from (Oppert et al. 2008)

(e.g., tricipital, bicipital, subscapular, and suprailiac skinfolds) may be used to assess total fat using various equations. The sum of such skinfolds is also considered as an indicator of total subcutaneous fat. However, the inter-observer variability is high when recording skinfolds and in some obese subjects skinfolds may be too large to be measured. With BIA, the impedance (or resistance) of the body to a low intensity alternate current measured between specific locations at the extremities (arms, legs) is used to determine body water content. Assuming it has a constant hydration, fat-free mass is derived using specific equations and fat mass is calculated as body weight minus fat-free mass. The method is easy to perform, relatively cheap, repeatable, has a high precision (reproducibility) but rather moderate accuracy (validity compared to reference), especially during weight change in obese subjects (Verdich et al. 2011). New devices may, however, have improved properties even in severely obese patients (Linares et al. 2011).

Densitometry and DXA

Hydrostatic weighing (or hydrodensitometry), a method based on water displacement to assess total body density, was considered until recently as the reference method against which others were validated. Once body volume is determined with the subject submerged in water (using Archimedes' principle), density (ratio of body mass to volume) is calculated and fat mass is derived using conversion equations. Fat-free mass is calculated as body weight minus fat mass. Precision and accuracy are high, however there are practical limitations because subjects have to climb in a water tank and then to sit still under water for several seconds. The more recent air displacement method (plethysmography) seems a promising alternative, but would need validation in severe obese individuals. Dual-energy X-ray absorptiometry (DXA) is a three-compartment method that uses the differential attenuation of two low-energy X-ray beams to determine fat mass, lean body mass, and bone mineral content. DXA is increasingly considered as the gold standard in body composition studies. The method gives rise to only minimal radiation, provides information about whole-body as well as regional body composition, has a high accuracy though decreasing with increasing body weight and width. Many devices now accept subjects weighing up to 200 kg but there are limitations in the field-of-view due to increased body dimensions in most severely obese subjects. Regular cross-calibration, using the same phantom, is needed, especially in multicenter studies. Other methods using isotopes such as dilution methods or ^{40}K counting are performed only in few research centers in limited numbers of subjects.

Body Fat Distribution and Abdominal Obesity

Waist, Hip, and Waist–Hip Ratio

The concept of body fat distribution refers to the anatomic location of body fat. Since the pioneering work performed more than 60 years ago by the French physician Jean Vague in Marseille (Vague 1956), it is known that preferential accumulation of fat in the upper part of the body, that is at the level of trunk or abdomen, is associated with increased risk for cardiovascular and metabolic disease.

Table 17.3 indicates among the main body composition methods those that also allow to measure regional fat. Based on the work by Swedish investigators on cohorts of men and women from Göteborg, and published during the mid-1980s, the ratio of waist to hip circumference, or waist–hip ratio (WHR) has been extensively used in the epidemiologic literature as indicator of body fat distribution (Larsson et al. 1984). Waist circumference is measured, at the end of a gentle expiration, as the circumference midway between lower ribs and iliac crests on the midaxillary line; hip circumference is measured as the largest circumference at the trochanter level, in standing position (WHO World Health Organization 2000). An increase in WHR is interpreted as reflecting body fat accumulation in the region of the trunk and abdomen as opposed to the extremities (limbs). Other measures of body fat distribution have been used, though less frequently, such as the ratio of iliac to thigh circumference (e.g., in the Paris Prospective Study, (Ducimetiere and Richard 1989)) or the ratio of trunk (e.g., subscapular) to extremity (e.g., triceps) skinfolds (e.g., in the Framingham study, (Kannel et al. 1991)).

Results of numerous prospective studies consistently indicate that an increased WHR is associated with increased risk of cardiovascular disease, especially coronary heart disease, independent of the overall level of corpulence as assessed by the BMI (Bjorntorp 1993; Kissebah and Krakower 1994). Results from large-scale international studies such as Interheart, a case–control study on myocardial infarction that included 27,000 subjects in 52 countries, documented an odds ratio of 2.52 [95 % CI 2.31–2.74] when comparing the highest with the lowest WHR quintile, the latter considered as reference (Yusuf et al. 2005). The relation was consistent in men and women and persisted after adjustement for BMI and other risk factors. Interestingly, in that study, these relations were much stronger than that between BMI and myocardial infarction.

If the fact that the WHR is a strong risk marker for cardiovascular disease cannot be disputed, the biological significance of the WHR, what it means in terms of main body compartments, is not that straightforward. Indeed, as a ratio, the WHR can be elevated due to an increase in the numerator (waist circumference) and/or a decrease in the denominator (hip circumference). For waist circumference, it is likely that it reflects abdominal fat (without the possibility to distinguish between the subcutaneous and visceral compartments, see below). For hip circumference, bone and muscular elements at this level are likely to contribute in addition to fat. Independent of BMI and waist circumference, there is some evidence that enlarged hip circumferences as a marker of peripheral adiposity may confer protection toward risk of cardiovascular disease and mortality (Yusuf et al. 2005; Heitmann et al. 2004). Reports using DXA to assess regional body composition indicate that in contrast to and independently of total trunk fat mass, leg fat mass displays favorable associations with cardio-metabolic risk markers related to glucose tolerance (Snijder et al. 2004) or fatty liver disease (Perlemuter et al. 2008).

Although interest in the WHR as indicator of increased cardiovascular risk in epidemiological studies remains high (Yusuf et al. 2005), the WHR is not widely used in the clinical setting. One reason is that it is relatively cumbersome to measure two circumferences and compute their ratio, another is that there is no established cutoffs to denote increased values for this indicator.

Abdominal Visceral Fat

The development of imaging techniques (computed tomography, CT, and magnetic resonance imaging, MRI) applied to body fat assessment has led to emphasize the importance of the intra-abdominal fat compartment, or abdominal visceral fat, as opposed to abdominal subcutaneous fat (Cornier et al. 2011; Després and Lemieux 2006; van der Kooy and Seidell 1993). A large number of cross-sectional studies have documented stronger associations of abdominal visceral fat, compared to abdominal subcutaneous fat, with cardio-metabolic risk factors or cardiovascular events (Cornier et al. 2011; Bjorntorp 1993; Després and Lemieux 2006). There are much less prospective studies that have reported positive associations of abdominal visceral fat with cardiovascular endpoints such as incidence of coronary events, independently of other adipose tissue compartments (Nicklas et al. 2004). In addition, an important point when documenting relationships of abdominal visceral fat with health outcomes would be to take into account the total amount of fat mass which appears not to be the rule (Seidell and Bouchard 1997).

Anthropometric Indicators of Abdominal Visceral Fat

Measurement of abdominal visceral fat through imaging techniques represents the assessment of a specific adipose tissue depot, a notion which has to be differentiated from assessment of adipose tissue distribution through anthropometry. Also note that DXA allows assesment of regional body composition, e.g., fat mass in the trunk or abdominal region, but cannot differentiate between visceral and subcutaneous adiposity. For epidemiological and clinical studies, an important research question is to better define which easily obtained anthropometric indicator is best related to abdominal visceral fat content in order to identify subjects at increased cardiovascular and metabolic risk. Investigators in Quebec have shown that waist circumference (WC) alone showed stronger correlations with CT-assessed abdominal visceral fat than WHR in adult men and women (Pouliot et al. 1994). It has therefore been suggested that WC may represent a better indicator for assessing intrabdominal fat. In addition, WC is easy to assess and was shown to be independent of height (Han et al. 1995). Cutoffs for increased WC proposed by Lean et al. (Lean et al. 1995) have been endorsed by several organizations and major consensus conferences (WHO World Health Organization 2000; Executive Summary of The Third Report of The National Cholesterol Education Program (NCEP) 2001). These WC cutoffs were defined as those that would best identify subjects with increased BMI and/or increased WHR (Lean et al. 1995). Two WC levels were identified, separately for men and women: level 1 (increased risk) is ≥80 cm in women and 94 cm in men, level 2 (substantially increased risk) is ≥88 cm in women and 102 cm in men. Based on the relationships of WC with CT-measured abdominal visceral fat other cutoffs have been proposed (100 cm before and 90 cm after age 40, in both genders) (Lemieux et al. 1996).

Although WC appears better correlated to abdominal visceral fat than WHR, it has to be mentioned that WC is also correlated with abdominal subcutaneous fat (Pouliot et al. 1994). Therefore, WC is an important and useful anthropometric measure which is mainly an indicator of abdominal fatness. For clinical use, WC is of particular interest in subjects with a BMI between 25 and 35 kg/m^2. When BMI is over 35 kg/m^2, most subjects will have a WC over the cutoffs defined above.

Other anthropometric measures might be at least as well or better correlated with abdominal visceral fat that WC. One of these is the sagittal diameter, which corresponds to abdominal height (van der Kooy and Seidell 1993). For anatomic reasons, abdominal height might better reflect intrabdominal fat content than a circumference such as waist. Data from the Paris Prospective Study showed positive associations of this indicator with death from cardiac origin (Oppert et al. 2002). However, measurement of this indicator is not standardized yet and there are no published cutoffs to define values associated with increased health risks.

Ectopic Fat Depots in the Liver and Heart

The development of imaging techniques has not only benefitted the study of abdominal vsiceral fat but also the assessment of individual differences in liver fat content and epicardial fat and their relationship with ill health, especially cardiometabolic risk. Proton magnetic resonance spectroscopy (MRS) appears as a non-invasive method of choice to asses the intrahepatic triglyceride content (Szczepaniak et al. 2005). Measures of epicardial adipose tissue thickness, mass, and volume has mainly relied up to now on echography but assessment based on multidetector computed tomography (MDCT) or cardiac magnetic resonance imaging (MRI) allows a more precise although more expensive and cumbersome measurement (Iacobellis et al. 2011).

Conclusion

There is a need in human studies on adipose tissue to rely on more precise measures of total and regional body fat. From a clinical standpoint, the body composition perspective on patient evaluation and follow-up should not be overlooked. A major goal of the management of obese patients is indeed to decrease body fat mass while maintaining fat-free mass (Ciangura et al. 2010). Fat-free mass loss, or insufficiency, is a major concern with aging populations and the increasing occurence of 'sarcopenic obesity' (Stenholm et al. 2008). Altogether, this means that there is a strong need for easy to use but accurate methods to be developed, evaluated, and disseminated to better assess body fat content and repartition both in the research and clinical settings.

References

Baumgartner RN, Heymsfield SB, Roche AF (1995) Human body composition and the epidemiology of chronic disease. Obes Res 3:73–95

Bjorntorp P (1993) Visceral obesity: a "civilization syndrome". Obes Res 1:206–222

Ciangura C, Bouillot JL, Lloret-Linares C et al (2010) Dynamics of change in total and regional body composition after gastric bypass in obese patients. Obesity (Silver Spring) 18:760–765

Cole TJ, Bellizzi MC, Flegal KM et al (2000) Establishing a standard definition for child overweight and obesity worldwide: international survey. BMJ 320:1240–1243

Cornier MA, Després JP, Davis N et al (2011) Assessing adiposity: a scientific statement from the American heart association. Circulation 124:1996–2019

Després JP, Lemieux I (2006) Abdominal obesity and metabolic syndrome. Nature 444:881–887
Dietz WH (1994) Critical periods in childhood for the development of obesity. Am J Clin Nutr 59:955–959
Ducimetiere P, Richard JL (1989) The relationship between subsets of anthropometric upper versus lower body measurements and coronary heart disease risk in middle-aged men. The Paris Prospective Study. I. Int J Obes 13:111–121
Executive Summary of The Third Report of The National Cholesterol Education Program (NCEP) (2001) Expert Panel On Detection, Evaluation, And Treatment Of High Blood Cholesterol In Adults (Adult Treatment Panel III) JAMA 285:2486–2497
Han TS, van Leer EM, Seidell JC et al (1995) Waist circumference action levels in the identification of cardiovascular risk factors: prevalence study in a random sample. BMJ 311:1401–1405
Heitmann BL, Erikson H, Ellsinger BM et al (2000) Mortality associated with body fat, fat-free mass and body mass index among 60-year-old Swedish men-a 22-year follow-up. The study of men born in 1913. Int J Obes 24:33–37
Heitmann BL, Frederiksen P, Lissner L (2004) Hip circumference and cardiovascular morbidity and mortality in men and women. Obes Res 12:482–487
Iacobellis G, Malavazos AE, Corsi MM (2011) Epicardial fat: From the biomolecular aspects to the clinical practice. Int J Biochem Cell Biol 43:1651–1654
Kannel WB, Cupples LA, Ramaswami R et al (1991) Regional obesity and risk of cardiovascular disease; the Framingham Study. J Clin Epidemiol 44:183–190
Kissebah AH, Krakower GR (1994) Regional adiposity and morbidity. Physiol Rev 74:761–811
Larsson B, Svardsudd K, Welin L et al (1984) Abdominal adipose tissue distribution, obesity, and risk of cardiovascular disease and death: 13 year follow up of participants in the study of men born in 1913. BMJ 288:1401–1404
Lean ME, Han TS, Morrison CE (1995) Waist circumference as a measure for indicating need for weight management. BMJ 311:158–161
Lee SY, Gallagher D (2008) Assessment methods in human body composition. Curr Opin Clin Nutr Metab Care 11:566–572
Lemieux S, Prud'homme D, Bouchard C et al (1996) A single threshold value of waist girth identifies normal-weight and overweight subjects with excess visceral adipose tissue. Am J Clin Nutr 64:685–693
Linares CL, Ciangura C, Bouillot JL et al (2011) Validity of leg-to-leg bioelectrical impedance analysis to estimate body fat in obesity. Obes Surg 21:917–923
Nicklas BJ, Penninx BW, Cesari M et al (2004) Association of visceral adipose tissue with incident myocardial infarction in older men and women: the Health, Aging and Body Composition study. Am J Epidemiol 160:741–749
Okorodudu DO, Jumean MF, Montori VM et al (2010) Diagnostic performance of body mass index to identify obesity as defined by body adiposity: a systematic review and meta-analysis. Int J Obes (Lond) 34:791–799
Oppert JM, Charles AM, Thibult N et al (2002) Anthropometric estimates of muscle and fat mass in relation to cardiac and cancer mortality in men : the Paris Prospective Study. Am J Clin Nutr 75:1107–1113
Oppert JM, Laville M, Basdevant A (2008) Human phenotypes. In: Clément K, Sorensen TIA (eds) Obesity. Genomics and postgenomics. Informa Healthcare, New York, pp 1–18
Perlemuter G, Naveau S, Belle-Croix F et al (2008) Independent and opposite associations of trunk fat and leg fat with liver enzyme levels. Liver Int 28:1381–1388
Pouliot MC, Despres JP, Lemieux S et al (1994) Waist circumference and abdominal sagittal diameter: best simple anthropometric indexes of abdominal visceral adipose tissue accumulation and related cardiovascular risk in men and women. Am J Cardiol 73:460–468
Prentice AM, Jebb SA (2001) Beyond body mass index. Obes Rev 2:141–147
Prospective Studies Collaboration, Whitlock G, Lewington S, Sherliker P et al (2009) Body-mass index and cause-specific mortality in 900 000 adults: collaborative analyses of 57 prospective studies. Lancet 373:1083–1096

Seidell JC, Bouchard C (1997) Visceral fat in relation to health: is it a major culprit or simply an innocent bystander? Int J Obes Relat Metab Disord 21:626–631

Snijder MB, Dekker JM, Visser M et al (2004) Trunk fat and leg fat have independent and opposite associations with fasting and postload glucose levels: the Hoorn study. Diabetes Care 27:372–377

Snijder MB, van Dam RM, Visser M et al (2006) What aspects of body fat are particularly hazardous and how do we measure them? Int J Epidemiol 35:83–92

Stenholm S, Harris TB, Rantanen T et al (2008) Sarcopenic obesity: definition, cause and consequences. Curr Opin Clin Nutr Metab Care 11:693–700

Szczepaniak LS, Nurenberg P, Leonard D et al (2005) Magnetic resonance spectroscopy to measure hepatic triglyceride content: prevalence of hepatic steatosis in the general population. Am J Physiol Endocrinol Metab 288:E462–E468

Vague J (1956) The degree of masculine differentiation of obesities, a factor determining predisposition to diabetes, atherosclerosis, gout, and uric calculous disease. Am J Clin Nutr 4:20–34

van der Kooy K, Seidell JC (1993) Techniques for the measurement of visceral fat: a practical guide. Int J Obes Relat Metab Disord 17:187–196

Verdich C, Barbe P, Petersen M et al (2011) Changes in body composition during weight loss in obese subjects in the NUGENOB study: comparison of bioelectrical impedance vs. dual-energy X-ray absorptiometry. Diabetes Metab 37:222–229

WHO Expert Consultation (2004) Appropriate body-mass index for Asian populations and its implications for policy and intervention strategies. Lancet 363:157–163

WHO World Health Organization (2000) Obesity: preventing and managing the global epidemic. Report of a WHO Consultation on Obesity. WHO Technical Report Series n° 894, Geneva

Willett WC (1998). Anthropometric measures and body composition. In: Willett W (ed) Nutritional epidemiology, 2nd edn. Oxford University Press, Oxford, pp 244–272

Yusuf S, Hawken S, Ounpuu S et al (2005) Obesity and the risk of myocardial infarction in 27,000 participants from 52 countries: a case-control study. Lancet 366:1640–1649

Chapter 18
Animal Models of Obesity

Michèle Guerre-Millo

Introduction

Animal models have provided and still provide major contribution to our understanding of the physiological and genetic bases of obesity. Notwithstanding the usefulness of specific models such as dogs, pigs, and nonhuman primates, I will focus this chapter on laboratory rodents, mostly mice and rats. Indeed, these species represent the bulk of animals used for research due to their rapid and high reproduction rate, well-established breeding and caging conditions, and large availability of molecular tools for the genome cartography and various types of transgenic modification. Most rodent models of obesity have been investigated since the early 1950s, but it is only after an extending period of time that the mechanisms underlying their phenotype started to be identified. Despite the tremendous progress brought by the development of molecular biology, not all the components of obesity phenotypes have been deciphered and new animal models are still needed to unravel the complexity of energy balance regulation. Here, I will review various types of rodent models for obesity (Table 18.1), which each provides information related to specific aspects of human obesity.

M. Guerre-Millo (✉)
Institut des maladies cardiométaboliques et de la nutrition (IHU ICAN), Paris, France
e-mail: michele.guerre-millo@crc.jussieu.fr

M. Guerre-Millo
Nutriomique U872 équipe 7, Centre de recherche des Cordeliers, INSERM, UPMC,
15, rue de l'école de médecine, 75006 Paris, France

J.-P. Bastard and B. Fève (eds.), *Physiology and Physiopathology of Adipose Tissue*,
DOI: 10.1007/978-2-8178-0343-2_18,

Table 18.1 Different types of rodent models of obesity

Diet-induced obesity (High-fat diet)
Hypothalamic obesity (stereotaxic lesions)
Genetic obesity
Polygenic obesity (QTL)
Monogenic obesity (mutations *agouti, fat, tubby, Lep, and LepR*)
Obesity induced by mutagenesis or genetic manipulation (transgenic and knockout models)

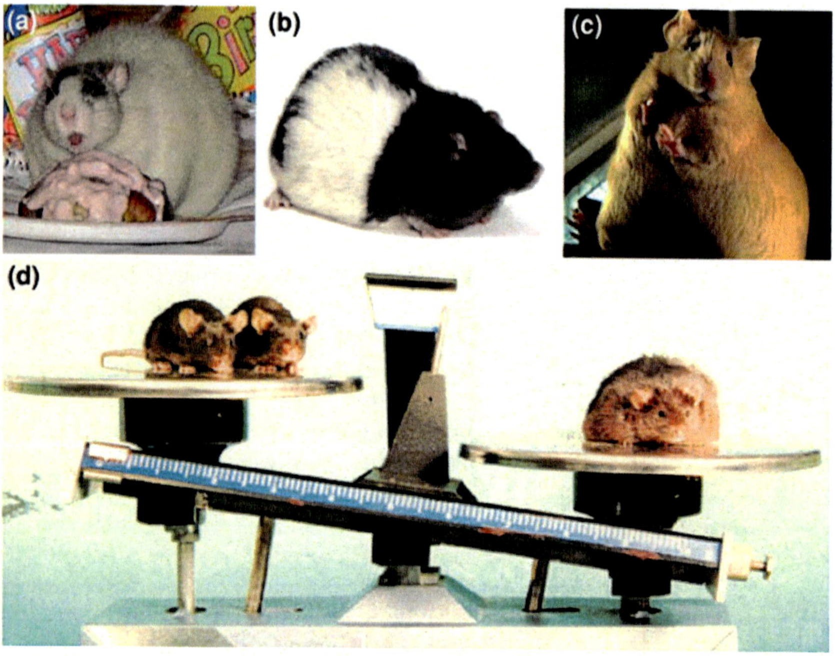

Fig. 18.1 Examples of rodents' models. **a** Nutritional obesity in rat; **b** Obese Zucker *fa/fa* rat; **c** Agouti mouse, obese and yellow; **d** Obese *Lep/Lep* mouse (*right*) and two non mutated control mice (*left*)

Rodent Models of Diet-Induced Obesity

Exploring the effects of the diet on the regulation of body weight was the driving force for the development of early rodent models of obesity. In this paradigm, mice or rats are fed calorically dense high-fat, high-fat/high-sugar, or cafeteria-type diets. Current commercial diets include 30, 45, or 60 % of calories from fat, while the control isocaloric diet contains around 10 % of calories from fat. Although rodents tend to reduce their food intake when fed a high-fat diet, they ingest more calories from fat, leading to increased adiposity and eventually obesity (Fig. 18.1). Not only the amount, but also the type of dietary fats can be changed to investigate responses to specific diets (Burcelin et al. 2002; Hariri and Thibault

2010). As in humans, genetic resistance or predisposition to obesity exists in rodents, with some strains gaining little weight on high fat compared to normal chow, whereas others rapidly progress to obesity. Studies comparing obesity-prone and obesity-resistant strains have been instrumental to identify the mechanisms and metabolic consequences of high-fat feeding (Surwit et al. 1995; Levin et al. 2004; Fearnside et al. 2008; Madsen et al. 2010).

Nowadays, rodent models are still frequently used in kinetic studies investigating systemic and organ-specific alterations associated with diet-induced obesity. One pathogenic component is leptin resistance, which develops gradually as evidenced by a progressive rise of circulating concentrations, but it is still debated whether it is secondary or causal to the diet-induced obesity (Scarpace and Zhang 2009). Insulin resistance is initially an adaptive response to high-fat diet that reduces hepatic glucose production and switches substrate utilization toward lipids, but becomes deleterious in the long run leading to type 2 diabetes. Diet-induced obese rodents represent valuable models for this life-threatening complication of human obesity. Importantly, these models have been crucial for the discovery of immune cells accumulation in the adipose tissue and its link to metabolic co-morbidities (see Chap. 20). The reversibility of high-fat diet-induced obesity has been addressed in obese rodents submitted to food restriction (Levin and Dunn-Meynell 2000; Kosteli et al. 2010) or to gastric bypass (Troy et al. 2008) after a period of high-fat feeding. In genetically modified mice, variations in the nature and amplitude of diet-induced responses are routinely checked to detect whether a specific gene is implicated or not in the regulation of body weight and obesity co-morbidities. Another area of research developed, to explore the still elusive mechanisms of early programming and epigenetic events, relies on diet-induced obese rodents (Ainge et al. 2011). Indeed, high-fat diet consumption during pregnancy and/or lactation, prepubertal high-fat feeding, as well as manipulation of milk intake by varying litter size, influence body weight, body fat content, and adipose tissue inflammation in adult offsprings (Guo and Jen 1995; Leibowitz et al. 2007; Boullu-Ciocca et al. 2008; Patterson et al. 2010). Finally, it is now established that gut bacteria interact with high-fat diet to promote obesity and insulin resistance (Cani et al. 2008; Ding et al. 2010), extending the use of diet in rodents to explore the complex relationship between energy balance and microbiota.

Rodent Models of Hypothalamic Obesity

A second model of obesity developed in the 1950s is the rat "VMH" (King 2006) that was produced by stereotaxic lesions of the ventro-medial hypothalamus (VMH). Rats with VMH lesions eat two to three times more food than normal, overeating starting almost immediately after the lesion. The weight gain that follows is rapid and quite dramatic as it is not uncommon to observe weight gains averaging 10 g per day, resulting in a doubling of body weight within 1 month. Eventually, daily food intake decreases and body weight usually stabilizes.

A similar morbid obese phenotype can be obtained by goldthioglucose injection in mice or rats (Marshall et al. 1955). Gold being toxic to neurons, it is linked to glucose by sulpfur (thio) to destroy the cells that take up glucose. When goldthioglucose is injected, there is an extensive damage in the VMH, which is enriched in glucose-responsive neurons. It is noteworthy that VMH lesion-induced hyperphagia and obesity have been reported in nonrodent species, including rabbits, cats, dogs, and monkeys [see Refs. in (King 2006)]. In humans, obesity is a severe sequel to tumors in the hypothalamic region or their surgical treatment (Pinkney et al. 2002).

Rodent Models of Genetic Obesity

Models of Polygenic Obesity

There are numerous evidences for a genetic influence on human obesity (see Chap. 24). In this field, both mice and rats provide highly relevant models for deciphering the genetic basis of obesity. Numerous genome-wide searches for obesity genes have been performed in rodents characterized by distinct body fat content or proneness to high-fat feeding. The experimental approach consists in correlation analyses between phenotypic traits related to obesity and the genotype using chromosomal markers that differ between groups of rodents under study. The chromosomal loci that statistically associate with the variation in the phenotypic character of interest have been named quantitative trait loci (QTL). Studies in two-strain backcrosses, F2 intercrosses, and recombinant inbred lines have led to the identification of hundreds of QTL influencing body weight, illustrating the polygenic nature of energy balance regulation (Brockmann and Bevova 2002; Rankinen et al. 2006). Unfortunately, identification of QTL at the gene level has proved mostly elusive. The dramatic improvement in genomic and bioinformatic resources holds promise to accelerate obesity gene discovery. Alternative rodent models, including heterogenous stocks of mice created from a limited number of founders (Churchill et al. 2004; Valdar et al. 2006) and new ways to exploit data, such as in silico mapping (Liu et al. 2007), the creation of genetic maps for gene expression (eQTL) (Schadt et al. 2005), or the application of systems biology to obesity genetics (Pomp et al. 2008) are expected to improved the identification of genes influencing energy balance and fat deposition.

Models of Monogenic Obesity

Genetic analysis of rodent models of monogenic obesity has been more successful, leading to the identification of five genes that, when mutated, result in obesity (Table 18.2). This significant progress has been achieved by applying the method

Table 18.2 Rodent models of monogenic obesity

Mutation	Chromosome	Type	Mutated gene, function	Phenotype
Mouse				
A^y, A^{vy}	2	Gain of function, dominant	Agouti, inhibition of α-MSH food intake reducing effect via MC4-R	Late-onset obesity, yellow fur
Fat	8	Loss of function, recessive	Carboxypeptidase E, hormones, and neuropeptides maturation	Late-onset obesity, diabetes, infertility
Tubby	7	Loss of function, recessive	Tub, unknown function	Late-onset obesity, blindness, deafness
Lep	6	Loss of function, recessive	Leptin, control of food intake	Early-onset obesity diabetes[a], infertility
LepR	4	Loss of function, recessive	Leptin receptor	Early-onset obesity, diabetes, infertility
Rat				
fa	5	Loss of function, recessive	Leptin receptor	Early-onset obesity infertility (females)
fa^k	5	Loss of function, recessive	Leptin receptor	Early-onset obesity diabetes, infertility

[a] Depends on the genetic background

of positional cloning in spontaneously obese mice detected in large breeding colonies. Most of the naturally occurring mutations being recessive, selective crosses were crucial to keep them in the progeny, thereby allowing identification of the mutated gene often decades after the discovery of the affected mice. Mutations in the same genes or members of their molecular pathways have been found (expect for the tubby mutation) in a very limited number of obese humans, but with remarkably similar phenotypes in humans and rodents (Clement 2006). In the case of leptin, identification of the mouse gene in 1994 was followed by the successful treatment of a leptin-deficient child only 5 years later (Farooqi et al. 1999). Leptin treatment was then successfully applied to 13 identified patients with leptin deficiency, reflecting a major therapeutic breakthrough based on experimental researches in obese mice started 50 years earlier (Table 18.3). Additionally, the study of the five mouse obesity genes has markedly increased our understanding of the mechanisms involved in energy regulation.

The *Agouti* Mutations

Several mutant alleles have been identified at the mouse "agouti" locus, given their easily detectable effect on coat color. Two of them, *lethal yellow* (A^y) and

Table 18.3 Leptin: from gene to therapeutic

1950	Description of monogenic obesity phenotypes in rodents (*ob/ob* and *db/db* mice)
1960	Hypothesis of a circulating satiety factor (parabiosis experiments)
1994 (December)	Cloning the leptin gene
1995 (July)	Successful treatment of *ob/ob* mice by recombinant leptin
1995 (December)	Cloning the leptin receptor gene
1997–1998	Discovery of rare mutations in the genes of leptin and its receptor in humans
1999–present	Successful treatment of 13 leptin deficient patients

viable yellow (A^{vy}), induce a phenotype of late-onset obesity and yellow fur in heterozygote mutants (Fig. 18.1). The mouse *agouti* gene was cloned independently by two groups in 1992 (Bultman et al. 1992; Miller et al. 1993). This breakthrough allowed showing that *agouti* transcripts, normally expressed exclusively in the skin of neonatal mice, were actually detectable in virtually all tissues in the obese yellow mutants. The structure of the A^y allele, where a 170-kb deletion brings the coding region of the *agouti* gene under the control of a ubiquitous promoter, accounts for this ectopic expression (Michaud et al. 1994). The double phenotype of the obese yellow mice has been attributed to a competitive antagonism between the agouti protein and the α-melanocyte stimulating hormone (α-MSH) on two melanocortin receptors, MC1-R, expressed in the skin, and MC4-R, in the hypothalamus [see Refs. in (Barsh et al. 1999; Moussa and Claycombe 1999)]. In the skin, agouti binding on MC1-R switches the production of black eumelanin pigment to yellow pheomelanin. In the hypothalamus, the protein inhibits the negative control of food intake exerted by MC4-R, leading to hyperphagia and obesity. An endogenous agouti-related protein (AGRP) was cloned subsequently and showed to act similarly to agouti as a potent antagonist for MC4-R. Deciphering the central mechanism of action of the agouti protein has led to the identification of a previously unknown pathway in the control of food intake, which comprises anorexigenic (α-MSH) and orexigenic (AGRP) components acting through MC4-R. The yellow obese mouse model is at the basis of the discovery of MC4-R mutations in humans (see Chap. 24), opening a whole area of pharmacological research targeting this receptor.

The *Fat* Mutation

In 1995, a single base mutation was found in the gene coding for the enzyme carboxypeptidase E (CPE) in the obese *fat/fat* mouse (Naggert et al. 1995). This was associated with loss of CPE activity, leading to hyperproinsulinemia, which is one of the earliest phenotypic characteristics caused by the fat mutation. Multiple defects in other peptide hormones or neuropeptides processing have been reported in CPE deleted mice (Cawley et al. 2010), but a causal relationship with the late-onset obesity observed in this model still remains unclear. Mice genetic manipulation yielding to an increased CPE expression specifically in proopiomelanocortin (POMC) neurons, shed light to this question (Plum et al. 2009). Indeed, this manipulation induces a reduction in food intake associated with alterations in the neuropeptide profile in the mediobasal hypothalamus, including increased amount of α-MSH, a product of CPE-dependent processing of POMC. These observations suggest that the obese phenotype of *fat/fat* mice relies, at least in part, on defect in POMC processing in hypothalamic neurons that alleviates the negative control of food intake by the α-MSH/MC4-R signaling pathway discovered in the agouti mouse model.

The *Tubby* Mutation

The mutated gene responsible for the *tubby* obesity was identified in 1996 (Kleyn et al. 1996; Noben-Trauth et al. 1996). The naturally occurring single base mutation in the *tub* gene abolishes a splice donor site, resulting in the replacement of 44-carboxy-terminal amino acids of the tub protein with 24 intron-encoded amino acids. The unusual phenotype of the *tubby* mouse combines vision and auditory deficits and late-onset obesity. The mutant mice begin to diverge in weight at about 12 weeks of age, and ultimately reach twice the weight of their wild-type littermates. It is now established that Tub is the founding member of the tub-like proteins family, composed of Tub and TULP1-3. These proteins are highly conserved among various vertebrate genomes and expressed primarily in nervous tissues. Unfortunately, little is known about their biochemical functions. Structure–function analyses and cell-based experiments have raised the possibility that Tub might function as heterotrimeric-G-protein-responsive intracellular signaling factor (Carroll et al. 2004). Although widespread *tub* gene expression in the hypothalamus suggests a potential control of food intake, the way Tub influences energy balance is far from being understood.

The *Lep* and *LepR* Mutations

As previously mentioned, the leptin gene was cloned in 1994 and two distinct mutations were identified in *Lep/Lep* mice, previously referred as *ob/ob* mice (Zhang et al. 1994). In the original strain, a single base mutation creates a premature stop codon, while the second is the result of the insertion of a retroviral-like transposon in the first intron of the gene leading to the absence of transcripts (Moon and Friedman 1997). Given the same phenotype in both strains, it was concluded that the single base mutation in the leptin gene was a loss-of-function mutation. *Lep/Lep* mice are visually obese at weaning demonstrating early onset obesity, and they can end up weighing more than 100 g, which is four times the weight of their littermate controls (Fig. 18.1). In this model, uncontrollable hyperphagia and reduced energy expenditure are constant characteristics, whereas the presence of type 2 diabetes is strain dependent. The absence of functional leptin is responsible for this extreme phenotype, as unambiguously demonstrated by the immediate and drastic reduction of food intake following recombinant leptin administration in *Lep/Lep* mice (Pelleymounter et al. 1995; Campfield et al. 1995; Halaas et al. 1995). These striking data helped to establish that leptin was, indeed, the unidentified circulating satiety factor proposed in 1973 on the basis of pioneer parabiotic experiments between wild type and *Lep/Lep* mice (reviewed in (Coleman 2010). Following the cloning of the leptin gene, a wealth of information became available, leading to the concept that leptin is an adipocyte secreted hormone that acts in the hypothalamus to regulate food intake in relation with the energy status. As a cytokine-like protein, leptin also exerts pleiotropic effects in a large variety of tissues and cell types, and is implicated in functions such as reproduction and immunity, besides energy homeostasis.

Cloning of the leptin receptor gene by virtue of leptin binding (Tartaglia et al. 1995) was followed by the demonstration that this gene is mutated in several strains of *db/db* mice, now referred as *LepR/LepR* mice (Lee et al. 1996, 1997; Chen et al. 1996; Li et al. 1998) and in two rats models of obesity, the *fa/fa* Zucker rat (Phillips et al. 1996) (Fig. 18.1) and the fa^k/fa^k Koletsky rat (Takaya et al. 1996). All mutations preclude leptin signaling, although through distinct mechanisms, including by creating truncated receptors that lack both the transmembrane and the intracellular domain required for leptin signaling. In all cases, the absence of functional leptin receptor results in massive early onset obesity as observed in leptin-deficient mice. The primary difference between the two models is that leptin receptor-deficient rodents have dramatic elevations in circulating leptin concentrations, reflecting their resistance to the hormone. Of note, leptin-resistant Zucker rats are able to reduce their food intake in response to a systemic load of calorie from glucose (Gilbert et al. 2003) or when placed in condition of hypobaric hypoxia (Simler et al. 2006), indicating that leptin independent anorexigenic pathways remain functional in this model.

Artificially Generated Mice Models of Obesity

Chemical Mutagenesis in Mice

To circumvent the arbitrary nature of spontaneous mutational events in critical obesity genes, attempts have been made to accelerate the process by increasing the mutation rate artificially. This can be performed by treating mice with mutagenic chemicals, such as ethylnitrosourea (Justice 2000). So far, a few mutant phenotypes differing in body weight or obesity-related traits have been selected. The confirmation of the mutation and its chromosomal localization requires time-consuming backcrossing to show that there is a single gene involved. Moreover, the final isolation of the mutated gene remains difficult. Currently, the obese mice issued from chemical mutagenesis are in the process of being crossed, but no gene has been yet identified using this strategy.

Transgenic Mouse Models

The 2005 update of the obesity gene map cited more than 200 genes that, when mutated or expressed as transgenes in mice, result in phenotypes that affect body weight and adiposity (Rankinen et al. 2006). These models support roles for a large array of genes and pathways in the regulation of energy homeostasis. Their phenotypic description is beyond the scope of this chapter, but it is interesting to note that two situations are informative: (1) when a gene targeted for a different purpose is unexpectedly found to influence body weight gain and (2) when a gene hypothesized to play a role in energy balance is targeted to confirm its implication. Examples of the latter case are worth mentioning. Following identification of the *tubby* mutation, it was unclear whether the Tub mutant protein retained any biological activity. A *tub* knockout was generated that recapitulated the full spectrum of the *tubby* mouse phenotype, thereby establishing the loss of function of the spontaneous mutant protein (Stubdal et al. 2000). Similarly, the generation of transgenic mice expressing the *agouti* gene from a ubiquitous promoter confirmed that the yellow and obese phenotype of agouti mice was directly related to ectopic expression of the protein (Klebig et al. 1995; Perry et al. 1995). Most importantly, after the demonstration that agouti antagonized α-MSH binding on MC4-R in reconstituted cell systems, the targeted deletion of this receptor in mice revealed its major role to inhibit food intake (Huszar et al. 1997). This knockout model provided crucial information to establish the molecular basis of the obese phenotype of agouti mice and, as a follow up, to promote the systematic screen for mutation in MC4-R in obese subjects.

Conclusion

Clearly, there is no perfect animal model of human obesity. Nevertheless, the use of rodent models to study the genetic, physiological, epigenetic, and environmental bases of obesity has provided an enormous amount of scientific knowledge, opening new avenues for urgent need of therapeutic targets in obesity. Although reduction in the use of animal models is ethically desirable, it is unlikely that in silico or cell-based experiments will overtake the use of live animals to explore the complexity and multifactorial nature of energy balance regulation in the near future. Moreover, rodent models of obesity represent necessary tools for testing innovatory interventions, such as treatment with chemical chaperone that reduces reticulum endoplasmic stress and improves leptin sensitivity in high-fat fed mice

(Ozcan et al. 2009) or transplantation of hypothalamic neurons that partly restore leptin signaling in *db/db* mice (Czupryn et al. 2011). Finally, new technical approaches, to obtain genetically engineered rodents more rapidly and at lower cost than before (Dow and Lowe 2012), are likely to provide new models in the field of obesity, increasing the chance of finding effective treatment strategies to curb the expanding progression of the disease worldwide.

References

Ainge H, Thompson C, Ozanne SE et al (2011) A systematic review on animal models of maternal high fat feeding and offspring glycaemic control. Int J Obes (Lond) 35:325–335

Barsh GS, Ollmann MM, Wilson BD et al (1999) Molecular pharmacology of Agouti protein in vitro and in vivo. Ann N Y Acad Sci 885:143–152

Boullu-Ciocca S, Achard V, Tassistro V et al (2008) Postnatal programming of glucocorticoid metabolism in rats modulates high-fat diet-induced regulation of visceral adipose tissue glucocorticoid exposure and sensitivity and adiponectin and proinflammatory adipokines gene expression in adulthood. Diabetes 57:669–677

Brockmann GA, Bevova MR (2002) Using mouse models to dissect the genetics of obesity. Trends Genet 18:367–376

Bultman SJ, Michaud EJ, Woychik RP (1992) Molecular characterization of the mouse agouti locus. Cell 71:1195–1204

Burcelin R, Crivelli V, Dacosta A et al (2002) Heterogeneous metabolic adaptation of C57BL/6 J mice to high-fat diet. Am J Physiol Endocrinol Metab 282:E834–E842

Campfield LA, Smith FJ, Guisez Y et al (1995) Recombinant mouse OB protein: evidence for a peripheral signal linking adiposity and central neural networks. Science 269:546–549

Cani PD, Delzenne NM, Amar J et al (2008) Role of gut microflora in the development of obesity and insulin resistance following high-fat diet feeding. Pathol Biol (Paris) 56:305–309

Carroll K, Gomez C, Shapiro L (2004) Tubby proteins: the plot thickens. Nat Rev Mol Cell Biol 5:55–63

Cawley NX, Yanik T, Woronowicz A et al (2010) Obese carboxypeptidase E knockout mice exhibit multiple defects in peptide hormone processing contributing to low bone mineral density. Am J Physiol Endocrinol Metab 299:E189–E197

Chen H, Charlat O, Tartaglia LA et al (1996) Evidence that the diabetes gene encodes the leptin receptor: identification of a mutation in the leptin receptor gene in db/db mice. Cell 84:491–495

Churchill GA, Airey DC, Allayee H et al (2004) The collaborative cross, a community resource for the genetic analysis of complex traits. Nat Genet 36:1133–1137

Clement K (2006) Genetics of human obesity. C R Biol 329:608–622

Coleman DL (2010) A historical perspective on leptin. Nat Med 16:1097–1099

Czupryn A, Zhou YD, Chen X et al (2011) Transplanted hypothalamic neurons restore leptin signaling and ameliorate obesity in db/db mice. Science 334:1133–1137

Ding S, Chi MM, Scull BP et al (2010) High-fat diet: bacteria interactions promote intestinal inflammation which precedes and correlates with obesity and insulin resistance in mouse. PLoS ONE 5:e12191

Dow LE, Lowe SW (2012) Life in the fast lane: mammalian disease models in the genomics era. Cell 148:1099–1109

Farooqi IS, Jebb SA, Langmack G et al (1999) Effects of recombinant leptin therapy in a child with congenital leptin deficiency. N Engl J Med 341:879–884

Fearnside JF, Dumas ME, Rothwell AR et al (2008) Phylometabonomic patterns of adaptation to high fat diet feeding in inbred mice. PLoS ONE 3:e1668

Gilbert M, Magnan C, Turban S et al (2003) Leptin receptor-deficient obese Zucker rats reduce their food intake in response to a systemic supply of calories from glucose. Diabetes 52:277–282

Guo F, Jen KL (1995) High-fat feeding during pregnancy and lactation affects offspring metabolism in rats. Physiol Behav 57:681–686

Halaas JL, Gajiwala KS, Maffei M et al (1995) Weight-reducing effects of the plasma protein encoded by the obese gene. Science 269:543–546

Hariri N, Thibault L (2010) High-fat diet-induced obesity in animal models. Nutr Res Rev 23:270–299

Huszar D, Lynch CA, Fairchild-Huntress V et al (1997) Targeted disruption of the melanocortin-4 receptor results in obesity in mice. Cell 88:131–141

Justice MJ (2000) Capitalizing on large-scale mouse mutagenesis screens. Nat Rev Genet 1:109–115

King BM (2006) The rise, fall, and resurrection of the ventromedial hypothalamus in the regulation of feeding behavior and body weight. Physiol Behav 87:221–244

Klebig ML, Wilkinson JE, Geisler JG et al (1995) Ectopic expression of the agouti gene in transgenic mice causes obesity, features of type II diabetes, and yellow fur. Proc Natl Acad Sci U S A 92:4728–4732

Kleyn PW, Fan W, Kovats SG et al (1996) Identification and characterization of the mouse obesity gene tubby: a member of a novel gene family. Cell 85:281–290

Kosteli A, Sugaru E, Haemmerle G et al (2010) Weight loss and lipolysis promote a dynamic immune response in murine adipose tissue. J Clin Invest 120:3466–3479

Lee GH, Proenca R, Montez JM et al (1996) Abnormal splicing of the leptin receptor in diabetic mice. Nature 379:632–635

Lee G, Li C, Montez J et al (1997) Leptin receptor mutations in 129 db3 J/db3 J mice and NIH facp/facp rats. Mamm Genome 8:445–447

Leibowitz KL, Chang GQ, Pamy PS et al (2007) Weight gain model in prepubertal rats: prediction and phenotyping of obesity-prone animals at normal body weight. Int J Obes (Lond) 31:1210–1221

Levin BE, Dunn-Meynell AA (2000) Defense of body weight against chronic caloric restriction in obesity-prone and -resistant rats. Am J Physiol Regul Integr Comp Physiol 278:R231–R237

Levin BE, Dunn-Meynell AA, Banks WA (2004) Obesity-prone rats have normal blood-brain barrier transport but defective central leptin signaling before obesity onset. Am J Physiol Regul Integr Comp Physiol 286:R143–R150

Li C, Ioffe E, Fidahusein N et al (1998) Absence of soluble leptin receptor in plasma from dbPas/dbPas and other db/db mice. J Biol Chem 273:10078–10082

Liu P, Vikis H, Lu Y et al (2007) Large-scale in silico mapping of complex quantitative traits in inbred mice. PLoS ONE 2:e651

Madsen AN, Hansen G, Paulsen SJ et al (2010) Long-term characterization of the diet-induced obese and diet-resistant rat model: a polygenetic rat model mimicking the human obesity syndrome. J Endocrinol 206:287–296

Marshall NB, Barrnett RJ, Mayer J (1955) Hypothalamic lesions in goldthioglucose injected mice. Proc Soc Exp Biol Med 90:240–244

Michaud EJ, Bultman SJ, Klebig ML et al (1994) A molecular model for the genetic and phenotypic characteristics of the mouse lethal yellow (Ay) mutation. Proc Natl Acad Sci U S A 91:2562–2566

Miller MW, Duhl DM, Vrieling H et al (1993) Cloning of the mouse agouti gene predicts a secreted protein ubiquitously expressed in mice carrying the lethal yellow mutation. Genes Dev 7:454–467

Moon BC, Friedman JM (1997) The molecular basis of the obese mutation in ob2 J mice. Genomics 42:152–156

Moussa NM, Claycombe KJ (1999) The yellow mouse obesity syndrome and mechanisms of agouti-induced obesity. Obes Res 7:506–514
Naggert JK, Fricker LD, Varlamov O et al (1995) Hyperproinsulinaemia in obese fat/fat mice associated with a carboxypeptidase E mutation which reduces enzyme activity. Nat Genet 10:135–142
Noben-Trauth K, Naggert JK, North MA et al (1996) A candidate gene for the mouse mutation tubby. Nature 380:534–538
Ozcan L, Ergin AS, Lu A et al (2009) Endoplasmic reticulum stress plays a central role in development of leptin resistance. Cell Metab 9:35–51
Patterson CM, Bouret SG, Park S et al (2010) Large litter rearing enhances leptin sensitivity and protects selectively bred diet-induced obese rats from becoming obese. Endocrinology 151:4270–4279
Pelleymounter MA, Cullen MJ, Baker MB et al (1995) Effects of the obese gene product on body weight regulation in ob/ob mice. Science 269:540–543
Perry WL, Hustad CM, Swing DA et al (1995) A transgenic mouse assay for agouti protein activity. Genetics 140:267–274
Phillips MS, Liu Q, Hammond HA et al (1996) Leptin receptor missense mutation in the fatty Zucker rat. Nat Genet 13:18–19
Pinkney J, Wilding J, Williams G et al (2002) Hypothalamic obesity in humans: what do we know and what can be done? Obes Rev 3:27–34
Plum L, Lin HV, Dutia R et al (2009) The obesity susceptibility gene Cpe links FoxO1 signaling in hypothalamic pro-opiomelanocortin neurons with regulation of food intake. Nat Med 15:1195–1201
Pomp D, Nehrenberg D, Estrada-Smith D (2008) Complex genetics of obesity in mouse models. Annu Rev Nutr 28:331–345
Rankinen T, Zuberi A, Chagnon YC et al (2006) The human obesity gene map: the 2005 update. Obesity (Silver Spring) 14:529–644
Scarpace PJ, Zhang Y (2009) Leptin resistance: a prediposing factor for diet-induced obesity. Am J Physiol Regul Integr Comp Physiol 296:R493–R500
Schadt EE, Lamb J, Yang X et al (2005) An integrative genomics approach to infer causal associations between gene expression and disease. Nat Genet 37:710–717
Simler N, Grosfeld A, Peinnequin A et al (2006) Leptin receptor-deficient obese Zucker rats reduce their food intake in response to hypobaric hypoxia. Am J Physiol Endocrinol Metab 290:E591–E597
Stubdal H, Lynch CA, Moriarty A et al (2000) Targeted deletion of the tub mouse obesity gene reveals that tubby is a loss-of-function mutation. Mol Cell Biol 20:878–882
Surwit RS, Feinglos MN, Rodin J et al (1995) Differential effects of fat and sucrose on the development of obesity and diabetes in C57BL/6 J and A/J mice. Metabolism 44:645–651
Takaya K, Ogawa Y, Hiraoka J et al (1996) Nonsense mutation of leptin receptor in the obese spontaneously hypertensive Koletsky rat. Nat Genet 14:130–131
Tartaglia LA, Dembski M, Weng X et al (1995) Identification and expression cloning of a leptin receptor, OB-R. Cell 83:1263–1271
Troy S, Soty M, Ribeiro L et al (2008) Intestinal gluconeogenesis is a key factor for early metabolic changes after gastric bypass but not after gastric lap-band in mice. Cell Metab 8:201–211
Valdar W, Solberg LC, Gauguier D et al (2006) Genome-wide genetic association of complex traits in heterogeneous stock mice. Nat Genet 38:879–887
Zhang Y, Proenca R, Maffei M et al (1994) Positional cloning of the mouse obese gene and its human homologue. Nature 372:425–432

Chapter 19
Contribution of "Omics" Approaches to Understand the Pathophysiology of Obesity

Nathalie Viguerie

Introduction

In the past 15 years, tremendous efforts have been made in the refinement of high-throughput "omics" technologies for the discovery of drug targets and molecular signatures associated with pathological states, disease subsets, or different responses to therapies. Before individual genome becomes one of the standards of medical investigation, the potential power of personalized medicine using molecular markers now appears more realistic since this approach is now close to in the cancer field (Auffray et al. 2011).

Such an approach needs the discovery of novel biomarkers which might be helped by defining new phenotyping classification using blood or tissue biomarkers (Curry 2008). Each molecular tissue biomarker (protein/peptide, lipid, RNA...) is likely to provide different information depending upon the consideration.

Significant progress has led to the generation of several new "omics" research fields: transcriptomics, proteomics, metabolomics, lipidomics, interactomics, etc. In contrast to traditional procedures, all experimental "omics" approaches can be considered to share three major features. First, "omics" are high-throughput, data-driven, holistic, and top-down methodologies. Second is the attempt to understand the cell or tissue metabolism as an integrated system. The last feature is that these high-throughput approaches generate large amounts of data and the analysis of

N. Viguerie (✉)
INSERM, UMR1048, Obesity Research Laboratory Team 4, I2MC,
Institute of Metabolic and Cardiovascular Diseases, 1 avenue Jean Poulhès,
BP 84225, 31432 Toulouse Cedex 4, France
e-mail: nathalie.viguerie@inserm.fr

N. Viguerie
University of Toulouse, UMR1048, Paul Sabatier University, Toulouse, France

J.-P. Bastard and B. Fève (eds.), *Physiology and Physiopathology of Adipose Tissue*,
DOI: 10.1007/978-2-8178-0343-2_19, © Springer-Verlag France 2013

these data often requires significant computational efforts and always specific statistical approaches.

To date, transcript profiling remains the most powerful method to comprehensively explore minute amounts of tissue or cells.

This chapter reviews some of the prominent data and concepts arising from the most commonly used experimental "omics" approaches to understand the physiopathology of obesity. Because of the difficulties to transpose concepts that arose from rodent studies to human physiopathology, especially regarding metabolism, we mainly focus on studies using human models and two tissues easily accessible using noninvasive sampling, blood, and adipose tissue (AT). Recent applications of the integrated "omics" for exploring metabolic and regulatory mechanisms are presented and advances in computational and statistical methodologies associated with integrated "omics" analyses are discussed.

Transcriptomics

Microarray-based transcriptome analysis may be considered as a mature genome-wide profiling technology. Consequently, it is widely used and applications of transcriptomics appear unlimited when applied to either cell culture systems, animal models, or clinical research. The various mRNA profiling platforms have the potential to easily identify specific transcript changes that respond to a given pharmacological treatment, nutritional, or exercise training challenge in a well-defined experimental setting. This does not mean that the changes in mRNA level can be taken as a causal marker, but rather can be used as a pattern of expressed mRNAs that changes in a characteristic and reproducible way. Since the technology has the feature of screening process covering thousands of potentially affected indicators of the metabolic status simultaneously it also reveals quite often totally unexpected findings.

Adipose Tissue

Understanding the biology of AT and its alteration in situation of obesity and insulin resistance is a major area of research. Despite many groups having pioneered AT as an active endocrine organ, there is limited knowledge of whether metabolic and energy signaling pathways as well as secretory factors may represent druggable targets. Gene expression profiling has shown promising potential for identifying undiscovered aspects of the metabolic and secretory aspects of various tissues including AT.

As a complement to genetics, microarrays have many applications in understanding pathologic AT. First, transcriptomic profiles of various stages of adipogenesis. Second, transcriptomic differences between different fat depots, AT from lean compared to obese individuals, glucose intolerance, versus diabetes or brown versus white AT. Third, effect of weight loss through dieting or surgery, cytokines, adipokines, hormones, environmental toxins, and drugs on transcriptomic profiles, and finally, influence of adipokines on transcriptomic profiles in skeletal muscle, liver, pancreatic β cells, etc.

Adipose Cells Differentiation

There are multiple advantages of using a microarray screening process to gain novel mechanistic insights into adipogenesis. First, novel characterized candidate genes could be identified based on their expression profiles and confirmed by further functional studies. Second, not yet characterized genes with modulated expression profile can be detected. Third, pangenomic screening allows a global view on biological processes and molecular networks during adipogenesis.

In mammals, adipogenesis includes distinct stages. After proliferation and determination phases, the cells reach confluence and then differentiation occurs. A re-entry into the cell cycle of growth-arrested cells at confluence involves several rounds of proliferation, referred to as mitotic clonal expansion. Growth arrest and induction of the transcription factors CCAAT-enhancer-binding protein (C/EBP) α and peroxisome proliferator-activated receptor (PPAR) γ mark the end of the mitotic clonal expansion phase and entry into terminal differentiation with transcriptional activation of genes defining the mature adipocyte phenotype.

In humans, studies have used primary preadipocyte and adipocyte cells, mesenchymal stem cells from bone marrow and AT, or the cellular model human multipotent AT-derived stem cells (hMADS), as in mice studies focused on primary immortalized embryonic fibroblasts and derived cell lines like the 3T3-L1 and 3T3-F442A. Most mouse cells undergo one or two rounds of clonal expansion during adipocyte differentiation. This event could not be observed in human adipocyte differentiation. Differences between rodent and human models were underscored in the expression profiles of genes known to be involved in the cell cycle with a sharp increase in 3T3-L1 cells at 24 h (Hackl et al. 2005) and only marginal changes in gene expression during hMADs adipocytes differentiation (Scheideler et al. 2008).

Preadipocytes remain in an undifferentiated state due to autocrine Wnt signaling, which blocks adipogenic conversion. Adipogenesis starts by inducing members of the nuclear receptor family of transcription factors, including the key regulators C/EBPs (β, δ, and α), PPARγ and their coactivators.

To identify additional factors capable of influencing adipocyte differentiation, expression profiling using microarrays has been used with different cell models and provided many genes potentially involved in the regulation of adipogenesis. The whole differentiation process was assessed using 3T3-L1 cells gene profiling to discover novel molecular players and also to obtain a global view on biological processes and molecular networks during adipogenesis.

In the early 2000s one of the first transcriptional programs associated with adipocyte differentiation in vitro and in vivo identified a panel of transcription factors that were expressed at higher level in AT, compared to mature adipocytes differentiated from 3T3-L1 preadipocytes (Soukas et al. 2001). Among these genes the zinc finger-containing transcription factor Krox20 also known as early growth response 2 (Egr2) (Soukas et al. 2001) is activated very early after induction and stimulates adipogenesis partially through and in cooperation with C/EBPβ.

Several members of the Krüppel-like zinc finger transcription factor family are known to be implicated in adipogenesis (Mori et al. 2005). On the basis of 3T3-L1 cells gene profiling, KLF9 appeared modulated during adipocyte differentiation (Henegar et al. 2008) and KLF4 was found as an immediate early regulator of adipogenesis by inducing C/EBPβ (Birsoy et al. 2008). The role of early B-cell factor EBF1, a helix-loop-helix transcription factor, was studied in adipocytes using Affymetrix microarrays to compare the adipogenic differentiation pathways of NIH-3T3 cells induced to undergo in vitro differentiation by ectopic expression of EBF1 or PPARγ2 (Akerblad et al. 2005). It was shown that the initial activation of genes associated with adipocyte development is independent of commitment to the adipogenic pathway and that EBF-1 and PPARγ2 induce adipocyte differentiation with comparable kinetics and efficiency. The xanthine oxidoreductase was also identified as a partner of PPAR γ and regulator of adipogenesis through PPARγ activity blockade (Cheung et al. 2007). Inhibition of its enzymatic activity or silencing blocks fat accumulation.

The effect of transcriptional coregulators has also been investigated using microarrays. As an example, the role of the corepressor RIP140 in adipogenesis and adipocyte function was studied using comparison of gene expression profiles from undifferentiated cells and adipocytes lacking and expressing RIP140 to identify derepressed and repressed genes that contribute to adipocyte function (Christian et al. 2005). Either ablation of RIP140 or depletion by silencing RNA in adipocytes altered expression of several clusters of genes involved in many metabolic pathways indicating that RIP140 is a global regulator of genes that control mitochondrial pathways and energy balance.

Besides the effects of Wnt signaling in maintaining preadipocytes in an undifferentiated state, a broader role of another nuclear hormone receptor involved in adipogenesis, NR1H3 (LXRα), was suggested in the regulation of adipocytes metabolism from a timed series microarray experiments of 3T3-L1 cells overexpressing Wnt1 (Ross et al. 2002). Using hMADS cells, LXRα was also shown to be transcriptionally activated in the adipocyte lineage among 65 candidate genes involved in the process of adipocyte/osteoblast commitment (Scheideler et al. 2008).

Comprehensive transcriptome analyses of 3T3-L1 cells also pointed out the role of lipid metabolism enzymes such as ELOVL6 involved in fatty acid elongation and the adipose triglyceride lipase ATGL which catalyzes the initial step in triglyceride hydrolysis (Hackl et al. 2005).

This also considerably extended upon previous microarray analyses of gene expression in fat cells as the analyses revealed the potential role of miRNAs in fat cell differentiation, since prediction of potential target genes for miRNAs was significant in 70 % of the genes with a unique 3′-UTR that were differentially expressed during adipocyte differentiation (Hackl et al. 2005). Some miRNAs have been proposed to be associated with adipogenesis using miRNA microarrays. Nakanishi and colleagues showed the upregulation of miR-335 in obesity (Nakanishi et al. 2009). The miR-335 levels were correlated with expression levels of adipocyte differentiation markers such as PPARγ, adipocyte-specific fatty acid-binding protein 4 (FABP4), and fatty acid synthase (FAS) in 3T3-L1 adipocytes. Several miRNAs, including let-7, were also shown upregulated during 3T3-L1 differentiation (Sun et al. 2009). Evidence was provided that let-7 contributes to adipogenesis by governing the transition from clonal expansion to terminal differentiation in part by targeting the transcription factor HMGA2 (Sun et al. 2009). These findings provide the first evidence that the regulated expressions of miRNAs might contribute to the pathophysiology of obesity.

Adipose Tissue Investigations

Adipose Tissue from Lean and Obese Subjects

In the past decade, a series of studies of the AT transcriptome reported the inflammatory status of adipose depots in rodent as well as human obesity. Weisberg and colleagues investigated the AT gene profiles correlated to adiposity by studying perigonadal depots from mice in which adiposity varied according to sex, diet, and obesity-related mutations. Among the top 100 genes with expression correlated to body mass, one-third encoded macrophage-related proteins (Weisberg et al. 2003). The moderate and chronic inflammatory status of AT from obese individuals was confirmed in humans with the first extensive analysis of inflammation-related genes in different clinically relevant conditions (Cancello et al. 2005, 2004). The separation of adipocytes from stromal vascular fraction revealed increased expression of inflammation-related genes in stromal vascular cells (Nair et al. 2005). Attempts have been made to characterize the immune cells' status in obesity. The AT macrophages secrete paracrine signals through a vast panel of pro-inflammatory cytokines. Increasing adiposity results in a shift in the inflammatory profile of AT macrophages as a whole from an alternatively activated M2 state to one in which classically activated M1 proinflammatory signals that impair insulin signaling in adipocytes predominate. In rodents, a novel alternatively activated and putatively beneficial macrophage population was shown in high fat diet-induced obesity with a typical chemokine and chemokine receptor pattern that revealed a complex network of attraction mechanisms (Zeyda et al. 2010). However, the precise phenotype of the immune cells infiltrating AT in humans remains unclear. Therefore, there is a crucial need for a precise molecular characterization of human AT immune cells upon the different obesity stages where gene profiling might be helpful for the development of therapeutic strategies to prevent obesity-related metabolic and cardiovascular complications.

Along with the immune cells accumulation in AT from obese individuals, evidence for fibrosis recently emerged as a novel component of obesity-related AT dysregulation. Microarray analyses identified an extracellular matrix remodeling signature in rodent models (O'Hara et al. 2009) and humans (Henegar et al. 2008). The modifications and occurrence of new adipose extracellular matrix components lead to an abnormal tissue fibrosis. Microarray analyses pertaining to studying the effect of macrophages conditioned medium on adipocytes showed that human preadipocytes exposed to molecules secreted by activated macrophages exhibited profound extracellular matrix remodeling (Keophiphath et al. 2009).

Despite the pathophysiological links between AT fibrosis and obesity-related complications have not yet been established, recent data suggests local tissue adaptations to fibrosis that could occur in an attempt to eliminate AT fibrosis and restore impaired adipogenesis and metabolic functions of hypertrophied AT.

Anatomical Localization

Due to its anatomical localization the accumulation of visceral AT (VAT) is related to the development of metabolic syndrome and increases the risk of co-morbidities such as type 2 diabetes (T2D) and cardiovascular diseases (CVD). It is not yet understood why visceral depot is more harmful than subcutaneous fat. Both depots show biological differences but all these differences may not explain the development of obesity complications. Coordinated depot-specific differences in expression of multiple genes involved in embryonic development and pattern specification has been shown in mice and in human subcutaneous AT (SAT) compared to VAT suggesting that genetically programmed developmental differences in adipocytes and their precursors in different depots play an important role in obesity, body fat distribution, and potential functional differences between visceral and subcutaneous AT (Gesta et al. 2006).

The macrophage content also is depot dependent. Omental AT of obese humans contains more macrophages than SAT. Omental fat macrophage infiltration correlates to fasting glucose, insulin, and hepatic fibro inflammatory lesions in obese individuals (Cancello et al. 2006). The relationships between obesity and metabolic syndrome and the expression of macrophage-specific markers were investigated in paired AT depot (subcutaneous and visceral) biopsies from women with a wide range of fat mass and metabolic complications (Klimcakova et al. 2010). Despite confirmation of differences between the two depots, an unexpected finding was that genes associated with obesity and metabolic complications were similar in subcutaneous and visceral fat. Both depots exhibited a coordinate decrease in expression of fat cell metabolic genes and an increase in expression of immune response genes with body mass index (BMI) and metabolic syndrome. This indicates that gene expression profiles from subcutaneous fat are as discriminating as those from visceral fat with respect to obesity and metabolic complications. It also suggests that impairment of fat cell metabolism may favor an inflammatory response in AT.

Weight Management

In obese individuals, weight loss is associated with improved insulin sensitivity and a reduction in obesity-related comorbidities.

As described above AT gene profiling has generated new insight into excess fat-related complications. Weight loss in obese subjects either through low calorie diet (Clement et al. 2004; Capel et al. 2009) or bariatric surgery (Cancello et al. 2005) decreased expression of inflammatory factors and increased expression of anti-inflammatory molecules, accompanied with decreased macrophage number and improved metabolic profile (Cancello et al. 2005; Clement et al. 2004; Capel et al. 2009). These data suggest that anti-inflammatory reagents or blockade of macrophage infiltration into AT may be beneficial for improving systemic insulin resistance in obese patients. Diet-induced weight loss in obese individuals also decreases the expression of genes associated with fatty acid metabolism (Capel et al. 2009) such as stearoyl-CoA desaturase-1 (SCD1) (Dahlman et al. 2005) and increased cell-death-inducing DFFA-like effector A (CIDEA), a key player in fat cell lipid droplet metabolism (Dahlman et al. 2005). The macronutrient content of the hypocaloric diet is of no or little importance for changes in gene expression in human AT (Dahlman et al. 2005; Capel et al. 2008). Although reduced caloric intake and increased physical activity favor a reduction in body fat mass and improve metabolic parameters, one of the greatest difficulties for obesity management is weight maintenance after successful weight loss. Understanding the molecular basis of weight maintenance following weight loss is an under-studied area of human physiology. Few studies focused on weight stabilization. Capel and colleagues found that the caloric restriction and weight maintenance phases of a weight management program were characterized by distinct patterns in SAT gene expression (Capel et al. 2009). The genes related to metabolism and adipose cells were downregulated during caloric restriction and mostly upregulated during weight stabilization. In contrast, macrophage genes involved in inflammatory pathways were unchanged or upregulated during caloric restriction, and down-regulated during weight stabilization. Marquez-Quinones and colleagues reported a study to identify differences in human SAT gene expression that may associate with weight regain and/or weight maintenance after the end of a low calorie diet (LCD)-induced weight loss program (Marquez-Quinones et al. 2010). Cellular growth and proliferation, cell death, cellular function, and maintenance were the main biological processes represented in SAT from subjects who regained weight. Mitochondrial oxidative phosphorylation was the major pattern associated with continued weight loss. This suggests that although cell proliferation may be detrimental, a greater mitochondrial energy gene expression may be beneficial for weight control.

These studies have begun to elucidate the AT gene expression networks underlying the physiological benefits reported with weight loss and weight maintenance.

Attempts have also been made to generate hypotheses that can potentially explain the role of AT in controlling weight changes during and after an LCD induced.

The AT gene expression profiles in slimmed individuals who regained weight after LCD compared to those who maintained the weight lost in response to LCD have been related to metabolic network analyses (Mutch et al. 2011). This study demonstrated that SAT from subjects predisposed to weight maintenance overexpress mitochondrial oxidative phosphorylation while those prone to weight regain overexpress cell growth and proliferation and apoptotic pathways.

Altogether, these studies reinforce that genetic and/or metabolic factors may help explain the interindividual variability in weight maintenance.

Peripheral Blood Cells

Peripheral blood is an easy biomaterial as an accessible biofluid that is easily sampled in the clinical routine. The continuous interaction between blood cells and body tissues raise the possibility that subtle changes occurring in the context of a disease may trigger specific changes in gene expression in blood cells that reflect the initiating stimulus. Circulating leukocytes may therefore contain informative transcriptional profiles as a first line of sentinels for disease of interest.

The first cross-sectional study in peripheral blood mononuclear cells within the diabetes field showed up regulation of JNK pathway genes and downregulation of mitochondrial oxidative phosphorylation pathway in T2D individuals (Takamura et al. 2007). The miR-126 was also shown downregulated in plasma circulating vesicles from T2D subjects (Zampetaki et al. 2010).

It can be argued that biopsy samples from AT, muscle, liver, or β-cells would be much more suitable than peripheral blood for transcriptomic studies to identify novel biomarkers of obesity-related complications. Data from deCODE Genetics are in line with this argument because they indicate that both the number of genes associated with BMI and the extent of correlation with BMI are considerably larger in AT than in peripheral blood (Emilsson et al. 2008). However, the scarcity of prospective studies and of studies that rigorously compare transcript signatures from different tissues with respect to their predictive value for T2D and CVD means that it is premature to come to any conclusion about the utility of gene expression studies in blood. Further studies are needed to address these issues.

Proteomics

As complementary to the gene profiling approach, large-scale proteomic studies serve as useful resource for fundamental biomedical research.

A comprehensive linear ion trap mass spectrometry (MS) analysis of 3T3-L1 cells' subcellular fractions (nuclei, mitochondria, membrane, and cytosol) identified up to 3000 proteins (Adachi et al. 2007). Compared to previous human cell lines and drosophila lipid droplet proteomic studies, a large proportion of the insulin signaling pathway and the ribosomal and proteasome complexes were comprehensively covered. Potentially secreted proteins were identified, indicating the secretory function of adipocytes.

The global approach of coupling microarray data with bioinformatics analysis (Mutch et al. 2009) has proved to be fruitful in gaining insights into adipocyte secretory potencies.

Transcriptomic studies of the glucose transporter GLUT4 -/- mice has identified the retinol binding protein 4 (RBP4) as new adipocyte-derived hormones suggested to link obesity with its comorbidities, especially insulin resistance, T2D, and certain components of the metabolic syndrome (Yang et al. 2005). To reach the objective of novel adipokine discovery, large-scale proteomic studies of adipocyte secretome may serve as a natural and useful resource for fundamental biomedical research.

Adipogenesis

By contrast to numerous transcriptomic studies designed to decipher the adipogenesis process, a few papers relate the use of proteomic approach, most of them being performed in 3T3-L1 cells.

Protein profiling during adipogenesis was performed with gel-based approaches using two-dimensional gel electrophoreses for separation and MS for protein identification as well as a liquid chromatography coupled with tandem MS (Adachi et al. 2007). The gel-based studies used either mouse 3T3-L1 models (Rahman et al. 2008) or human mesenchymal stem cells (Jeong et al. 2007) and reported up to 2000 protein species, whereas a non gel-based study (Adachi et al. 2007) identified approximately 3300 proteins. By comparing microarray data with the proteomic data, one-third could be mapped to the identified proteome. In another gel-based study using human subcutaneous adipose-derived stem cells as an in vitro model of adipogenesis, 170 individual proteins were detected, especially multiple novel serine protease inhibitors (serpins) (Zvonic et al. 2007).

Comparison of these data with reported secretomes showed varying molecular species reflecting the methodological and technical differences in proteomic studies.

The first large-scale quantitative proteomic study to analyze the changes in the secretome during the course of adipogenesis in humans (Zhong et al. 2010) used iTRAQ-based quantitative proteomics. Most of the 420 proteins identified showed differential expression during the course of differentiation with 40 % of the protein sequences containing signal peptides and 35 % with localization to extracellular compartment including numerous proteins whose dynamic expression had not been previously documented in the differentiation of human adipocytes (Zhong et al. 2010).

Secretion

The various bioactive molecules secreted by the AT are likely to play pivotal roles in the pathophysiology of obesity and its associated disorders. Many factors, including signaling metabolites, adipocytokines, and hormones have been proposed as a link to obesity-related complications such as metabolic syndrome, T2D, and CVD. Most of them are so far unknown. A recent challenge is therefore to identify novel or existing adipokines as drug targets that could be used to reverse adverse reactions related to excess fat.

A growing number of studies addressed this issue and profiled the adipocyte secretome.

The very first analyses dealt with gel electrophoresis combined with MS approaches to profile the secreted proteins from 3T3-L1 cells during adipocyte differentiation. Up to 41 proteins were for the first time identified as novel adipocyte secreted factors from 3T3-L1 adipocytes, including the pigment epithelium-derived factor (SERPINF1) secreted in preadipocytes, hippocampal cholinergic neurostimulating peptide (PEBP1), neutrophil gelatinase-associated lipocalin (LCN2), and haptoglobin (HP) in mature adipocytes in addition to a number of previously reported secreted factors such as adipsin or adiponectin (Wang et al. 2011). The secretion profiles indicated a dynamic environment including an active remodeling extracellular matrix and several factors involved in growth regulation (Wang et al. 2004).

A shotgun approach of secretome from primary human adipocytes differentiated in culture has led to the characterization of 44 novel adipokines including complement factor H (CFH), alphaB-crystallin (CRYAB), cartilage intermediate-layer protein (CILP), and heme oxygenase-1 (HMOX1) (Lehr et al. 2011). Using the same model and technology the incretin activating enzyme, dipeptidyl peptidase-4 (DPP4), was identified as a factor released by adipocytes with effects on skeletal muscle insulin signaling and smooth muscle cells proliferation, therefore providing a novel link between obesity and metabolic syndrome (Lamers et al. 2011).

Factors secreted by the stromal vascular fraction contribute to the secretome and modulate adipokine secretion by adipocytes. Therefore, Celis and colleagues utilized gel-based technology, MS, immunoblotting and antibody-array to identify extracellular and intracellular signaling molecules of interstitial fluid of fresh mammary AT samples and adipose cells from high-risk breast cancer patients (Celis et al. 2005). About 360 proteins were identified, including numerous signaling molecules, hormones, cytokines, and growth factors that are involved in a variety of biological processes, such as signal transduction and cell communication, energy metabolism, protein metabolism, cell growth and/or maintenance, immune response, transport, regulation of nucleic acid metabolism, and apoptosis. Among these proteins, the fatty acid binding protein FABP4 secretion was also shown using 3T3-L1 cells and demonstrated as positively correlated to BMI and metabolic syndrome (Xu et al. 2006).

Alvarez-Llamas and colleagues aimed at the characterization of the human VAT secretome rather than the adipocyte secretome (Alvarez-Llamas et al. 2007). Among 259 proteins 70 were considered secreted by AT and 40 % is involved in the modulation of the extracellular matrix.

A quantitative comparison of the VAT and SAT explants focused on isolated preadipocytes and endothelial cells and compared their secretomes to those from whole AT (Hocking et al. 2011). Although there were no discrete depot-specific differences in the secretomes from whole tissue, preadipocytes or endothelial cells, VAT exhibited an overall higher level of protein secretion than SAT. More proteins were secreted in greater abundance from VAT compared with SAT explants, preadipocytes, and endothelial cells. Almost 50 % of the VAT secretome was composed of factors with a role in angiogenesis suggesting that this depot may represent a more readily expandable tissue depot (Hocking et al. 2011).

Studies reported the composition of exosome-like vesicles (ELVs) released from AT, the vesicles by which adipocytes are supposed to perform part of their endocrine function with the secretion of adipokines. To address secretion of the adipocyte-derived microvesicles, namely adiposome, 3T3-L1 adipocytes were used (Aoki et al. 2007). Proteomic analysis revealed that 3T3-L1 ELVs contain some matrix metalloproteinases (MMP-2 and MMP-9) that colocalize with a novel angiogenic secreted protein, milk fat globule-epidermal growth factor 8 (MFG-E8), which is upregulated upon hypertrophy.

More recently, ELVs were proposed as a mode of communication between ATs and macrophages (Deng et al. 2009). ELVs released from mice AT were purified and labeled with green fluorescent dye, then injected into ob/ob or control mice fed a high-fat diet. Deng and colleagues found that AT ELVs are taken up by peripheral blood monocytes and stimulate the differentiation of the monocytes into activated macrophages. These exosomes are capable of a potent stimulatory effect which could mediate both the obesity-associated inflammatory responses and the development of insulin resistance. This finding suggests that the adiposome could play a role in activation of macrophages and subsequent development of insulin resistance.

The secreted proteins and vesicles identified consistently from different models of adipocytes and AT by different proteomics platforms better represent the characteristics of adipose secretome. Like transcriptomics, the use of proteomics has significantly enhanced our understanding of the proteins and pathways involved in the pathophysiology of obesity. Unfortunately, proteomic profiling of human AT is hindered by some of the similar shortcomings, as was seen with transcriptomics. Most prominent is the technology itself, as it is biased toward abundant proteins, that is, signal transduction molecules cannot be identified easily without extensive protein fractionation. Extensive validation of candidate proteins revealed by proteomics by an independent method, such as western blotting, remains of importance because proteins can be identified incorrectly due to a currently incomplete database.

Metabolomics

Despite a growing number of publications metabolomics is still in its infancy. Here, we provide a brief overview of studies that evaluated the clinical utility of results from serum or plasma samples.

Mutch and colleagues analyzed the metabolite profiles in the serum of morbidly obese individuals who underwent gastric surgery (Mutch et al. 2009). To identify metabolites associated with surrogates of steady-state insulin sensitivity the patients were subdivided into obese and T2D. The nervonic acid was negatively correlated with insulin resistance index. These results provided a new knowledge regarding the global alterations in serum metabolite profiles post bariatric surgery (Mutch et al. 2009). Nuclear magnetic resonance and MS have been used to define metabolic profiles that distinguish between patients with T2D or CVD and controls. Serum or plasma concentrations of sugars and sugar metabolites (e.g., glucose, mannose, deoxyhexose, and 1,5-anhydroglucitol), ketone bodies (α-hydroxybutyrate), lipids (e.g., phosphatidyl-cholines and nonesterified fatty acids), and branched-chain amino acids were found to be associated with insulin resistance or T2D (Suhre et al. 2010). In the first prospective metabolomics study, branched-chain and aromatic amino acids showed significant associations with incident T2D in the Framingham Offspring Study (Wang et al. 2011).

Additional studies with prospective designs, larger study samples, more advanced methods of metabolite quantification, and more sophisticated data processing and analysis strategies are needed to evaluate whether metabolite measurements would have a relevant impact on disease prediction and patient management.

Connecting Various "Omics"

While high-throughput "omics" approaches to analyze molecules at different cellular levels are rapidly becoming available, it is also becoming clear that any single "omics" approach may not be sufficient to characterize the complexity of biological systems.

Integrating data from various "omics" technologies may enable to draw a broader picture of a cell's behavior and of the impact of treatments or environmental signals. This will ultimately lead to systems biology, an emerging interdisciplinary study field that focuses on the complex interactions in biological systems. One goal of systems biology is to understand how genome-encoded parts interact to produce quantitative phenotypes. Systems biology has the power to transform the way biology and medicine have been viewed classically by way of dealing with biological entities on the systemic level rather than focusing on a system which is simplistically reduced to a small number of parts. However, the integration of diverse "omics" data sets posess major challenges to researchers (in particular bioinformaticians) and computational infrastructures. It also demands standards that make data sets from different sources (labs, platforms, technologies) reliable and comparable for integration on a broad scale. In this context it will be inevitable to apply mathematical modeling in order to interpret the flood of data.

All the aforementioned fields are under constant development, inevitably spawning major breakthroughs that make systems biology and its applications more and more palpable. As an example, next-generation sequencing technologies, also known as high-throughput sequencing or third-generation sequencing are about to shift "omics" strategies that rely on hybridization on microarrays to sequencing of RNA. In the case of transcriptomics several recent publications proved the utility of the RNA Seq technology to assess mRNA levels in different applications, highlighting its advantages over array-based methods, namely: higher signal-to-background-ratio, lower detection limit, unbiased measurements, unambiguous assignment of measured sequences, and quantitative linearity over a broader range. If throughput and quality increase and prices decrease as predicted, these sequencing technologies might become more widespread and, owing to their technical advantages over hybridization-based approaches might become the gold standard in measuring RNA specimen on a genome-wide scale.

In summary, "omics" technologies generated a plethora of data and provided novel mechanistic insights into adipogenesis and AT biology which can be exploited for developing novel drugs to prevent or cure obesity and/or related comorbidites. It also became evident that we are only at the beginning of drawing the complete picture of the complex cellular process of all the AT cells and that further integrative "omics" approaches will be necessary to elucidate the molecular network controlling the cell fate.

References

Adachi J, Kumar C, Zhang Y et al (2007) In-depth analysis of the adipocyte proteome by mass spectrometry and bioinformatics. Mol Cell Proteomics 6:1257–1273

Akerblad P, Mansson R, Lagergren A et al (2005) Gene expression analysis suggests that EBF-1 and PPARgamma2 induce adipogenesis of NIH-3T3 cells with similar efficiency and kinetics. Physiol Genomics 23:206–216

Alvarez-Llamas G, Szalowska E, de Vries MP et al (2007) Characterization of the human visceral adipose tissue secretome. Mol Cell Proteomics 6:589–600

Aoki N, Jin-no S, Nakagawa Y et al (2007) Identification and characterization of microvesicles secreted by 3T3-L1 adipocytes: redox- and hormone-dependent induction of milk fat globule-epidermal growth factor 8-associated microvesicles. Endocrinology 148:3850–3862

Auffray C, Caulfield T, Khoury MJ et al (2011) Genome medicine: past, present and future. Genome Med 4:9

Birsoy K, Chen Z, Friedman J (2008) Transcriptional regulation of adipogenesis by KLF4. Cell Metab 7:339–347

Cancello R, Henegar C, Viguerie N et al (2005) Reduction of macrophage infiltration and chemoattractant gene expression changes in white adipose tissue of morbidly obese subjects after surgery-induced weight loss. Diabetes 54:2277–2286

Cancello R, Tordjman J, Poitou C et al (2006) Increased infiltration of macrophages in omental adipose tissue is associated with marked hepatic lesions in morbid human obesity. Diabetes 55:1554–1561

Capel F, Viguerie N, Vega N et al (2008) Contribution of energy restriction and macronutrient composition to changes in adipose tissue gene expression during dietary weight-loss programs in obese women. J Clin Endocrinol Metab 93:4315–4322

Capel F, Klimcakova E, Viguerie N et al (2009) Macrophages and adipocytes in human obesity: adipose tissue gene expression and insulin sensitivity during calorie restriction and weight stabilization. Diabetes 58:1558–1567

Celis JE, Moreira JM, Cabezon T et al (2005) Identification of extracellular and intracellular signaling components of the mammary adipose tissue and its interstitial fluid in high risk breast cancer patients: toward dissecting the molecular circuitry of epithelial-adipocyte stromal cell interactions. Mol Cell Proteomics 4:492–522

Cheung KJ, Tzameli I, Pissios P et al (2007) Xanthine oxidoreductase is a regulator of adipogenesis and PPARgamma activity. Cell Metab 5:115–128

Christian M, Kiskinis E, Debevec D et al (2005) RIP140-targeted repression of gene expression in adipocytes. Mol Cell Biol 25:9383–9391

Clement K, Viguerie N, Poitou C et al (2004) Weight loss regulates inflammation-related genes in white adipose tissue of obese subjects. FASEB J 18:1657–1669

Curry SH (2008) Translational science: past, present, and future. Biotechniques 44: ii–viii

Dahlman I, Linder K, Arvidsson Nordstrom E (2005) Changes in adipose tissue gene expression with energy-restricted diets in obese women. Am J Clin Nutr 81:1275–1285

Deng ZB, Poliakov A, Hardy RW et al (2009) Adipose tissue exosome-like vesicles mediate activation of macrophage-induced insulin resistance. Diabetes 58:2498–2505

Emilsson V, Thorleifsson G, Zhang B et al (2008) Genetics of gene expression and its effect on disease. Nature 452:423–428

Gesta S, Bluher M, Yamamoto Y et al (2006) Evidence for a role of developmental genes in the origin of obesity and body fat distribution. Proc Natl Acad Sci U S A 103:6676–6681

Hackl H, Burkard TR, Sturn A et al (2005) Molecular processes during fat cell development revealed by gene expression profiling and functional annotation. Genome Biol 6:R108

Henegar C, Tordjman J, Achard V et al (2008) Adipose tissue transcriptomic signature highlights the pathological relevance of extracellular matrix in human obesity. Genome Biol 9:R14

Hocking SL, Wu LE, Guilhaus M et al (2011) Intrinsic depot-specific differences in the secretome of adipose tissue, preadipocytes, and adipose tissue-derived microvascular endothelial cells. Diabetes 59:3008–3016

Jeong JA, Ko KM, Park HS et al (2007) Membrane proteomic analysis of human mesenchymal stromal cells during adipogenesis. Proteomics 7:4181–4191

Keophiphath M, Achard V, Henegar C et al (2009) Macrophage-secreted factors promote a profibrotic phenotype in human preadipocytes. Mol Endocrinol 23:11–24

Klimcakova E, Roussel B, Marquez-Quinones A et al (2010) Worsening of obesity and metabolic status yields similar molecular adaptations in human subcutaneous and visceral adipose tissue: decreased metabolism and increased immune response. J Clin Endocrinol Metab 96:E73–E82

Lamers D, Famulla S, Wronkowitz N et al (2011) Dipeptidyl peptidase 4 is a novel adipokine potentially linking obesity to the metabolic syndrome. Diabetes 60:1917–1925

Lehr S, Hartwig S, Lamers D et al (2011) Identification and validation of novel adipokines released from primary human adipocytes. Mol Cell Proteomics 11(1):M111.010504. Epub 2011 Sep 26

Marquez-Quinones A, Mutch DM, Debard C et al (2010) Adipose tissue transcriptome reflects variations between subjects with continued weight loss and subjects regaining weight 6 mo after caloric restriction independent of energy intake. Am J Clin Nutr 92:975–984

Mori T, Sakaue H, Iguchi H et al (2005) Role of Kruppel-like factor 15 (KLF15) in transcriptional regulation of adipogenesis. J Biol Chem 280:12867–12875

Mutch DM, Rouault C, Keophiphath M et al (2009a) Using gene expression to predict the secretome of differentiating human preadipocytes. Int J Obes (Lond) (2005) 33:354–363

Mutch DM, Fuhrmann JC, Rein D et al (2009b) Metabolite profiling identifies candidate markers reflecting the clinical adaptations associated with Roux-en-Y gastric bypass surgery. PLoS One 4:e7905
Mutch DM, Pers TH, Temanni MR et al (2011) A distinct adipose tissue gene expression response to caloric restriction predicts 6-mo weight maintenance in obese subjects. Am J Clin Nutr 94:1399–1409
Nair S, Lee YH, Rousseau E et al (2005) Increased expression of inflammation-related genes in cultured preadipocytes/stromal vascular cells from obese compared with non-obese Pima Indians. Diabetologia 48:1784–1788
Nakanishi N, Nakagawa Y, Tokushige N et al (2009) The up-regulation of microRNA-335 is associated with lipid metabolism in liver and white adipose tissue of genetically obese mice. Biochem Biophys Res Commun 385:492–496
O'Hara A, Lim FL, Mazzatti DJ et al (2009) Microarray analysis identifies matrix metalloproteinases (MMPs) as key genes whose expression is up-regulated in human adipocytes by macrophage-conditioned medium. Pflugers Arch 458:1103–1114
Rahman A, Kumar SG, Kim SW et al (2008) Proteomic analysis for inhibitory effect of chitosan oligosaccharides on 3T3-L1 adipocyte differentiation. Proteomics 8:569–581
Ross SE, Erickson RL, Gerin I et al (2002) Microarray analyses during adipogenesis: understanding the effects of Wnt signaling on adipogenesis and the roles of liver X receptor alpha in adipocyte metabolism. Mol Cell Biol 22:5989–5999
Scheideler M, Elabd C, Zaragosi LE et al (2008) Comparative transcriptomics of human multipotent stem cells during adipogenesis and osteoblastogenesis. BMC Genomics 9:340
Soukas A, Socci ND, Saatkamp BD et al (2001) Distinct transcriptional profiles of adipogenesis in vivo and in vitro. J Biol Chem 276:34167–34174
Suhre K, Meisinger C, Doring A et al (2010) Metabolic footprint of diabetes: a multiplatform metabolomics study in an epidemiological setting. PLoS One 5:e13953
Sun T, Fu M, Bookout AL et al (2009) MicroRNA let-7 regulates 3T3-L1 adipogenesis. Mol Endocrinol 23:925–931
Takamura T, Honda M, Sakai Y et al (2007) Gene expression profiles in peripheral blood mononuclear cells reflect the pathophysiology of type 2 diabetes. Biochem Biophys Res Comm 361:379–384
Wang P, Mariman E, Keijer J et al (2004) Profiling of the secreted proteins during 3T3-L1 adipocyte differentiation leads to the identification of novel adipokines. Cell Mol Life Sci 61:2405–2417
Wang TJ, Larson MG, Vasan RS et al (2011) Metabolite profiles and the risk of developing diabetes. Nat Med 17:448–453
Weisberg SP, McCann D, Desai M et al (2003) Obesity is associated with macrophage accumulation in adipose tissue. J Clin Invest 112:1796–1808
Xu A, Wang Y, Xu JY et al (2006) Adipocyte fatty acid-binding protein is a plasma biomarker closely associated with obesity and metabolic syndrome. Clin Chem 52:405–413
Yang Q, Graham TE, Mody N et al (2005) Serum retinol binding protein 4 contributes to insulin resistance in obesity and type 2 diabetes. Nature 436:356–362
Zampetaki A, Kiechl S, Drozdov I et al (2010) Plasma microRNA profiling reveals loss of endothelial miR-126 and other microRNAs in type 2 diabetes. Circ Res 107:810–817
Zeyda M, Gollinger K, Kriehuber E et al (2010) Newly identified adipose tissue macrophage populations in obesity with distinct chemokine and chemokine receptor expression. Int J Obes (Lond) 34:1684–1694
Zhong J, Krawczyk SA, Chaerkady R et al (2010) Temporal profiling of the secretome during adipogenesis in humans. J Proteome Res 9:5228–5238
Zvonic S, Lefevre M, Kilroy G et al (2007) Secretome of primary cultures of human adipose-derived stem cells: modulation of serpins by adipogenesis. Mol Cell Proteomics 6:18–28

Chapter 20
Adipose Tissue Inflammation in Obesity

Christine Poitou, Elise Dalmas and Karine Clément

Introduction

Epidemiological studies show a concurrent increase of prevalence of obesity and associated conditions such as type 2 diabetes, cardiovascular diseases, liver damages, inflammatory joint diseases, asthma, and certain types of cancer. More and more data suggest that the inflammation associated to obesity would have led to the development of these pathological conditions. We will see what are the possible origins of this inflammation in obese people, especially the important role of adipose tissue and by which mechanisms the inflammation can have consequences in terms of comorbidities.

C. Poitou (✉) · E. Dalmas · K. Clément
Institut des maladies cardiométaboliques et de la nutrition (IHU ICAN), Paris, France
e-mail: christine.poitou-bernert@psl.aphp.fr

K. Clément
e-mail: karine.clement@psl.aphp.fr

C. Poitou · E. Dalmas · K. Clément
Nutriomique U872 équipe 7, Centre de recherche des Cordeliers, INSERM, UPMC,
15, rue de l'école de médecine, 75006 Paris, France

C. Poitou · K. Clément
Centre de Recherche en Nutrition Humaine Ile-de-France (CRNH IdF), Paris, France

C. Poitou · K. Clément
Assistance Publique Hôpitaux de Paris, Service de Nutrition, Bâtiment Benjamin Delessert,
Hôpital Pitié-Salpêtrière, 91 boulevard de L'hôpital, 75013 Paris, France

J.-P. Bastard and B. Fève (eds.), *Physiology and Physiopathology of Adipose Tissue*,
DOI: 10.1007/978-2-8178-0343-2_20, © Springer-Verlag France 2013

"Low-Grade" Systemic Inflammation and Obesity

Obesity is now considered as a chronic inflammatory state developing insidiously, commonly called *low-grade inflammation*. This concept is based on the fact that obese people present a moderate but chronic increase of the circulating rates of the inflammation's mediators, such as acute-phase proteins, pro-inflammatory cytokines, and adhesion molecules (Poitou et al. 2011; Dalmas et al. 2011). The number of molecules helping to characterise the low-grade inflammation has never stopped increasing since several years and are listed in Table 20.1.

The relative contribution from different tissues (liver, lymphoid, circulating immuno-inflammatory cells, subcutaneous and visceral adipose tissue, potentially muscle) in the production of circulating inflammatory molecules is hard to determine in human obesity, and especially during the various phases of its development. The liver is highly involved in low-grade inflammation by the increased synthesis of acute-phase molecules (CRP, SAA, orosomucoid, ferritin, etc…). The hepatic expression of pro-inflammatory cytokines such as IL-6, TNF-α, and IL-1β is increased with obesity probably by two complementary mechanisms. First, the steatosis, which is defined by an accumulation of lipids in the liver while putting on weight, can have a local inducing effect within the liver tissue of inflammatory reaction. Second, the pro-inflammatory cytokines and the fatty acids coming from adipose tissue via the portal circulation can also be involved. The circulating monocytes are also characterised by an inflammatory profile (Ghanim et al. 2004). If the liver and the circulating immuno-inflammatory cells clearly participate in low-grade inflammation in obese people, the adipose tissue has a major role when releasing secretory substances, thus modulating the inflammation of the parenchymal liver and participating also in the plasmatic increase of inflammatory factors.

Adipose Tissue Inflammation

The Origins of the Adipose Tissue Inflammation During Obesity

Many factors are evoked, but it is still complicated to clearly establish the events underlying the adipose tissue inflammation. Adipocytes are themselves capable to produce numerous pro- and anti-inflammatory biomolecules. The increase of the adipocytes' size, which characterises obesity, modifies their secretory profile and induces a pro-inflammatory status, without the involved signals being clearly defined. The characteristic distribution of macrophages as a crown around adipocytes presenting signs of cell death (Cinti et al. 2005) (cf Chap. 6) suggests the intervention of specific attracting factors. In adipose tissue, exogenous signals such as fatty acids or bacterial lipopolysaccharides (LPS) induce an inflammatory

Table 20.1 Systemic modifications of inflammatory molecules in obesity and weight loss

Name	Obesity	Weight loss
Acute-phase protein		
Haptoglobin	⇑	⇓
Serum amyloide A	⇑	⇓
C Raectiv protein (CRP)	⇑	⇓
Fibrinogen	⇑	⇓
Orosomucoïd	⇑	⇓
Cytokines/Chemokines		
Interleukin 6 (IL6)	⇑	⇓
Interleukin 8 (IL8)	⇑	⇓
Interleukine 9 (IL9)	⇑	⇓
Interleukin 18 (IL18)	⇑	⇓
Interleukin 10 (IL10)	⇑	⇓
Interleukin 1ra (IL1Ra)	⇑	⇓
Tumour necrosis factor (TNFα)	⇑	⇓ or →?
Monocyte chemoattractant protein MCP1	⇑	⇓
Monocyte chemoattractant protein MCP4	⇑	⇓
Macrophage migration inhibitory factor MIF	⇑	?
Macrophage inflammatory protein MIP1b	⇑	⇓
Interferonγ-inducible-protein 10 IP10	⇑	⇓ or→
EOTAXIN	⇑	⇓
RANTES	⇑	⇓
Monokine induced by IFNγ MIG	⇑	⇓
Other adipokines		
Leptin	⇑	⇓
Visfatin	⇑	⇓
Resistin	⇓	⇑ or →
Adiponectin	⇓	⇑
Omentin	⇓	?
Adhesion molecules/extracellular matrix prothrombotic		
Matrix metalloproteinase 9 (MMP9)	⇑	⇓
ICAM	⇑	⇓
VCAM	⇑	⇓
HGF	⇑	?
Plasminogen activator inhibitor PA-1	⇑	⇓
Cathepsin S	⇑	⇓
VEGF	⇑	⇓

response through the activation of the TLR4 (*toll like receptor* 4) and the associated intracellular signal pathways, mainly the NFKB pathway (Shi et al. 2006). Locally, the inflammatory adipocytes, which lipolytic activity is important, release fatty acids, which stimulate the TLR4/NFK B pathway in the macrophages, thus participating in the deleterious paracrine loop between both cell types (Suganami et al. 2009). In this context, the TLR4 inhibition by factors such as ATF3, transcription factor from the ATF/CREB family, or CTRP-3 (*C1q/TNF [tumour necrosis factor]- related protein-3*), member of the C1q/TNF family, could represent a therapeutic target to control adipose tissue inflammation (Suganami et al. 2009; Kopp et al. 2010). Among other factors, tissue hypoxia, classic cause of the macrophages' attraction in some tumours and in the atheromatous plaque, could also participate in triggering the adipose tissue inflammation in obese people (Ye 2009). The endoplasmic reticulum stress and the oxidative stress could also be able to intervene in the local inflammation.

The Cellular Actors

The Macrophages

In 2003, two American teams described an accumulation of macrophages in the adipose tissue of obese mice and in human, linked in this latter to the increase of the body mass index (Xu et al. 2003; Weisberg et al. 2003). The macrophages' infiltration in adipose tissue is only partly reversible with the weight loss induced by gastric surgery (Cancello et al. 2005; Aron-Wisnewsky et al. 2009). The macrophages in the adipose tissue are typically laid out as a crown around an adipocyte presenting call death signs such as the negativity of perilipin (Cinti et al. 2005; Cancello et al. 2005). This setting is specific to obese adipose tissue and more frequent in visceral adipose tissue (cf Chap. 6). These observations suggest that the macrophages play their "classic" role of phagocytosis and of eliminating the metabolically deficient adipocytes in obese subjects, especially those having reached a critical size provoking cell death.

If the increase of the number of macrophages is a major characteristic of the adipose tissue during obesity, what about their phenotype? The injection of a specific marker for monocytes in the mouse helped showing that obesity (Diet Induced Obesity (DIO) model) induces the recruitment of M1 macrophages, which are those setting up as a crown around adipocytes[13]. These macrophages are loaded with lipid droplets, which testify of their lipids' phagocytosis activity (Cinti et al. 2005; Xu et al. 2003). When there is no obesity, the "resident" macrophages are also detected in adipose tissue. They present a type M2 phenotype and their number is not or only a little modified in obesity (Lumeng et al. 2007; Fujisaka et al. 2009). A kinetic study in DIO model suggest a more complex situation: the accumulation of type M1 macrophages continues until 16 weeks of diet and then decreases in the advantage of type M2 macrophages, but we cannot know whether this is recruitment or a repolarisation of the in situ M1 macrophages (Strissel et al. 2007). Besides, a third population with an intermediary M1/M2 phenotype has been identified in the adipose tissue of DIO mice; it could form a group of cells undergoing repolarisation (Shaul et al. 2010). In human adipose tissue, macrophages are essentially of mixed phenotype intermediary between M1 and M2 polarisation states (Zeyda et al. 2007; Bourlier et al. 2008). In a population of

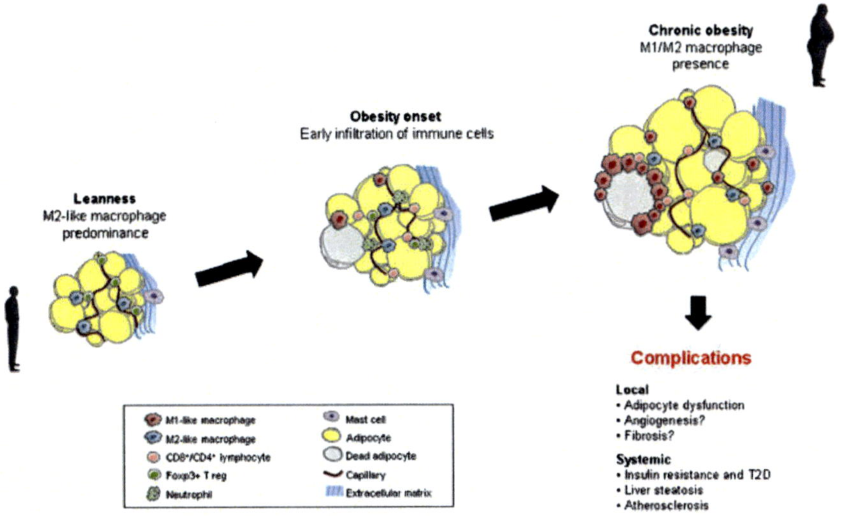

Fig. 20.1 Progress of the pathological change of adipose tissue during the obesity

massively obese people, we observed an increase of the number of positive cells for the CD40 marker and a steady number of cells expressing type M2 markers (CD206 [*mannose receptor*] and CD163 [member of the *scavenger receptors* family]) compared to healthy controls, suggesting that obesity promotes the accumulation of type M1 macrophages (Aron-Wisnewsky et al. 2009). A more precise characterisation shows that, as for the mouse, the macrophages laid out into a crown around adipocytes present a type M1 phenotype, whereas the interstitial macrophages have a polarised-M2 phenotype (Wentworth et al. 2010; Spencer et al. 2010). In human, however, it is impossible to distinguish the recruited macrophages from the resident macrophages. The lipids activating the TLR4/NFK B pathway (Ichioka et al. 2011) or the accumulation of lipids in the macrophages themselves (Prieur et al. 2011) favours the M1 polarisation in the adipose tissue of DIO mice. Surprisingly enough, the increase of the triglycerides' storage capacity in the macrophages via the overexpression of the diglyceride acyltransferase 1 (DGAT1) makes them resistant to an M1 activation (Koliwad et al. 2010). This phenotype can be explained by the fact that the triglycerides would be less pro-inflammatory than some other lipid species in macrophages. All these works on mice and humans reveal the complexity and the progressive aspect of the phenotype of adipose tissue macrophages in obesity (Fig. 20.1), a situation probably associated to functional aspects which should be clarified.

The Other Actors in Adipose Tissue Inflammation

Other cells from the innate and adaptive immunity are involved in the development of adipose tissue inflammation. The early and transitory infiltration of neutrophils suggests that an "acute" inflammation precedes the macrophages appearance in this model (Elgazar-Carmon et al. 2008). NKT cells (*natural killer T*), capable of recognising glycolipidic antigens, colonise also the adipose tissue in the first phases of obesity, a step which seems necessary to the macrophage infiltration (Ohmura et al. 2010). In a mouse model (*KitW-sh/W-sh* mouse), the absence of mast cells reduces the accumulation of macrophages in adipose tissue. How do these cells highly involved in allergy participate in the inflammatory process of adipose tissue remains to be clarified (Liu et al. 2009). Lymphocytes, especially T lymphocytes, have also been detected in adipose tissue in mice and humans. Kinetic studies in DIO model reveal that T lymphocytes precede the macrophages in the adipose tissue (Kintscher et al. 2008; Nishimura et al. 2009). Moreover, CD4+ T lymphocytes isolated from DIO mice's adipose tissue express a limited TCR repertoire, suggesting that they can recognise one—or several—antigens from fat origin (Winer et al. 2009). We distinguish several mobilisation profiles of T lymphocytes in adipose tissue in obesity: the number of cytotoxic T cells (CD8+) increases whereas the number of Th2 (CD4+ GATA3+) and Treg (Foxp3+) anti-inflammatory T cells decreases (Nishimura et al. 2009; Winer et al. 2009; Feuerer et al. 2009). The physiopathological importance of these changes is shown by the fact that the CD8+ T cells' inactivation (Nishimura et al. 2009) or the induction of Th2 or Treg lymphocytes (Winer et al. 2009; Feuerer et al. 2009) decreases the inflammation in DIO mice (Duffaut et al. 2009). In humans, a study shows the increase of CD8+ T cells and CD4+ lymphocytes in adipose tissue with the degree of adiposity (Duffaut et al. 2009). All these observations suggest there is a complex set of influence between the various cellular actors of the innate (neutrophiles, mast cells) or adaptive immunity, or at the interface of both systems (NKT). These different protagonists seem to act sequentially or altogether to favour *in fine* the accumulation of macrophages in the adipose tissue thanks to mechanisms not well known yet. A major stake in these studies is the identification of the factor(s) capable of triggering an immune response in the adipose tissue during the development phases of human obesity.

Mechanisms of the Macrophage Infiltration

Studies on bone marrow transplant in mice clearly showed that the circulating monocytes are the precursors of the majority of macrophages in adipose tissue (Weisberg et al. 2003), which was confirmed by marking experiments of in vivo monocytes (Lumeng et al. 2007). Several subpopulations of monocytes have been described in mice and humans. Their future depends on their phenotype, on one hand; and on the microenvironment in which they are going to differentiate into macrophages, on the other hand. In mice, the Grl + Ly-6Chigh monocytes, comparable to human CD14+CD16- classic monocytes, would be the precursors of pro-inflammatory macrophages of adipose tissue (Tsou et al. 2007; Westcott et al. 2009). In humans, early studies revealed that mononuclear circulating cells express a pro-inflammatory status in obese people (Ghanim et al. 2004). More recently, clinical studies indicate that the frequency of a subpopulation of CD14dimCD16+ monocytes recently identified is increased in some obese people (Poitou et al. 2011; Rogacev et al. 2010) and deceases with weight loss (Poitou et al. 2011). This subpopulation of monocytes could preferentially participate in the accumulation of macrophages in adipose tissue.

Another question concerns the cellular and molecular mechanisms involved in the recruitment of monocytes. Several chemoattractant molecules can intervene, and a major role of the MCP-1/CCR2 system (*monocyte chemotactic protein 1* and its receptor CCR2) in the macrophages accumulation in adipose tissue was suggested by studies on mice (Kanda et al. 2006; Weisberg et al. 2006; Inouye et al. 2007). Other chemokines, such as CXCL14 (Nara et al. 2007) or osteopontin (Nomiyama et al. 2007), have been described in mice. Adipocytes themselves produce soluble factors which increase the monocyte diapedesis and stimulate the production of adhesion proteins, which PECAM-1 (*platelet endothelial cell adhesion molecule*) and ICAM-1 (*intercellular adhesion molecule)* are expressed on endothelial cells (Curat et al. 2004). A work of our team shows that CCL5 cytokine (*alias* RANTES [*regulated upon activation, normal T cell expressed and secreted*]) could contribute to this process (Keophiphath et al. 2010). As a matter of fact, the endothelial cells are more inflammatory in the visceral adipose tissue than in the superficial adipose tissue, which could favour the recruitment of monocytes in deep tissues (Villaret et al. 2010). It is clear that other studies are necessary to define the full spectrum of the factors involved in monocytes diapedesis and in endothelial activation in adipose tissue during obesity.

The Consequences of Adipose Tissue Inflammation

Local Action on the Adipose Tissue's Structure and Functions

The presence of macrophages modifies locally the biology of adipocytes. These cells cultivated in environments conditioned with macrophages release pro-inflammatory factors and become insulin-resistant (Suganami et al. 2005). Moreover, the fatty acids released by the adipocytes made insensitive to the anti-lipolytic effect of insulin activate the macrophages towards an M1 phenotype via a deleterious paracrin loop between both cellular types. Another recognised effect of the macrophages is the inhibition of the pre-adipocytes' differentiation which become proliferative (Bourlier et al. 2008; Lacasa et al. 2007). Besides, some types of macrophages isolated from human adipose tissue have a pro-angiogenic effect in vitro, which could participate in the neovascularisation of the tissue during the weight gain (Bourlier et al. 2008).

The presence of macrophages could contribute to containing the adipocytes' hypertrophy. This eventuality is evoked to give account of the phenotype of deficient mice for the *CCR2* gene, which encodes for the MCP1 receptor: they present a reduction of the number of macrophages associated to an increase of adipocytes' diameter (Lumeng et al. 2007; Weisberg et al. 2006). It shall be noted that adipocytes being the only cells of the body capable of storing efficiently and without damaging the excessive caloric intake as triglycerides, limiting their expansion can be more noxious than beneficial if favouring the ectopic layers in other non-specialised organs such as the liver.

Finally, the macrophages are considered as major regulator of fibrosis in various tissues where they play pro- or antifibrotic roles according to their phenotype. In vitro, the conditioned environment of M1 macrophages activated by the LPS or isolated from the adipose tissue of obese people provides a profibrotic phenotype to human pre-adipocytes (Keophiphath et al. 2009). These observations suggest a rather profibrotic role of the adipose tissue's macrophages in obese people. Besides in vivo fibrotic layers were detected in human adipose tissue, which were more abundant in obese people than in controls (Henegar et al. 2008; Divoux et al. 2010). What are the consequences of fibrosis in adipose tissue? In our study on obese people, we observed that the lower the quantity of fibrosis, the more the obese patients lost fat in response to gastric surgery. The presence of excessive fibrosis could modify the adipose tissue's reshaping and potentially the metabolic adaptations that intervene over fast and drastic weight loss induced in this model (Divoux et al. 2010). Clarifying the cellular and molecular actors of fibrosis and its local or systemic consequences is a new challenge in the field of obesity.

Systemic Consequences

If it is suggested that immuno-inflammatory abnormalities present in obese people link obesity to cardiometabolic diseases (Lumeng and Saltiel 2011), asthma (Beuther et al. 2006), osteoarticular affections (Richette et al. 2011) or to the cognitive functions and the mood modifications (Capuron et al. 2011), the direct relations between adipose tissue inflammation and obesity complications are not all clearly established. In the mouse model, the abundance of macrophages can be modified in the adipose tissue by pharmacological means or via targeted genetic manipulations, which helped highlighting the importance of these cells in the induction of an insulin resistance. In humans, however, this relation remains controversial. Some studies report an opposite relationship between sensitivity to insulin and the expression of macrophages' markers in adipose tissue, especially in visceral adipose tissue (Harman-Boehm et al. 2007). The increased presence of macrophages in crown shape has also been associated to the modification of the sensitivity to insulin (Apovian et al. 2008). On the contrary, our works do not allow us to establish a link between the abundance of macrophages and the blood parameters of the glucose homoeostasis in very obese people (Tordjman et al. 2009). Moreover, we observed that an overfeeding induces insulin resistance without changes in macrophages detected in adipose tissue of non-obese subjects

(Tam et al. 2010). Associating abundance of a minor fraction of macrophages with a mixed phenotype and insulin resistance has been reported, suggesting that an accurate analysis of phenotypes could help distinguishing the types of macrophages more deleterious than others (Wentworth et al. 2010). Although highly suspected, a link between the macrophage accumulation and the increased cardiovascular diseases could not be established in humans. However, our clinical studies on a large population of massively obese people revealed that the seriousness of the hepatic histopathology (steatosis and inflammatory fibrosis) is related to the size of the macrophage accumulation in visceral adipose tissue, regardless of the glucose status (Tordjman et al. 2009).

Effect of Weight Loss on the Inflammation

Many studies indicate that the decrease of feed ration and sometimes the increase of physical activity are factors reducing the overall systemic inflammation. Thus, when losing weight, we observe a reduction of many molecules due to the inflammation and of adhesion molecules (Table 20.1) and an increase of adiponectin. We recently showed variations on a large scale of cytokines/chemokines in relation to the variations of caloric restriction and carbohydrate intakes after gastric surgery (Dalmas et al. 2011). The state of the inflammatory cells is modified by the weight variations (Poitou et al. 2011; Ghanim et al. 2004).

In the adipose tissue itself, we established a modification of the inflammatory genes' expression during a moderate weight reduction, due to a hypocaloric diet. The reduction of the inflammatory genes' expression in adipose tissue (Clement et al. 2004) is confirmed during a bigger weight reduction induced by surgery and is associated to a significant decrease of the macrophage infiltration (Cancello et al. 2005), which phenotype is modified (increase of the IL-10 expression).

However, a kinetic study on DIO mice's weight loss shows that, contrary to the initial hypothesis, a major accumulation of macrophages has been observed in adipose tissue during the early phases of caloric restriction, when the adipocyte lipolysis is stimulated (Kosteli et al. 2010). These macrophages act as phagocytes for lipids without triggering any inflammation, and participate in restoring locally the lipid homoeostasis. However, this observation remains to be reproduced in humans.

Conclusion

In obesity, adipose white tissue is the target of a major cellular and structural modification, which components are probably not all identified yet. Cell types and the signals involved are numerous and complex, creating locally a pro-inflammatory microenvironment, which favours macrophage recruitment and fibrosis formation. Although partly reversible, these changes remain after weight loss. The identification of the potential protective mechanisms, such as the immune neutralisation of some types of lymphocytes, could open onto new therapeutic perspectives aiming at containing the inflammation in adipose tissue. The discovery of this chronic inflammation also opens onto perspectives for research of the physiopathological mechanisms, which could explain the development of complications associated to obesity, their progression, and their preservation.

References

Apovian CM, Bigornia S, Mott M et al (2008) Adipose macrophage infiltration is associated with insulin resistance and vascular endothelial dysfunction in obese subjects. Arterioscler Thromb Vasc Biol 28(9):1654–1659

Aron-Wisnewsky J, Tordjman J, Poitou C et al (2009) Human adipose tissue macrophages: m1 and m2 cell surface markers in subcutaneous and omental depots and after weight loss. J Clin Endocrinol Metab 94(11):4619–4623

Beuther DA, Weiss ST, Sutherland ER (2006) Obesity and asthma. Am J Respir Crit Care Med 174(2):112–119

Bourlier V, Zakaroff-Girard A, Miranville A et al (2008a) Remodeling phenotype of human subcutaneous adipose tissue macrophages. Circulation 117(6):806–815

Bourlier V, Zakaroff-Girard A, Miranville A et al (2008b) Remodeling phenotype of human subcutaneous adipose tissue macrophages. Circulation 117(6):806–815

Cancello R, Henegar C, Viguerie N et al (2005) Reduction of macrophage infiltration and chemoattractant gene expression changes in white adipose tissue of morbidly obese subjects after surgery-induced weight loss. Diabetes 54(8):2277–2286

Capuron L, Poitou C, Machaux-Tholliez D et al (2011) Relationship between adiposity, emotional status and eating behaviour in obese women: role of inflammation. Psychol Med 41(7):1517–1528

Cinti S, Mitchell G, Barbatelli G et al (2005) Adipocyte death defines macrophage localization and function in adipose tissue of obese mice and humans. J Lipid Res 46(11):2347–2355

Clement K, Viguerie N, Poitou C et al (2004) Weight loss regulates inflammation-related genes in white adipose tissue of obese subjects. FASEB J 18(14):1657–1669

Curat CA, Miranville A, Sengenes C et al (2004) From blood monocytes to adipose tissue-resident macrophages: induction of diapedesis by human mature adipocytes. Diabetes 53(5):1285–1292

Dalmas E, Rouault C, Abdennour M et al (2011) Variations in circulating inflammatory factors are related to changes in calorie and carbohydrate intakes early in the course of surgery-induced weight reduction. Am J Clin Nutr 94(2):450–458

Divoux A, Tordjman J, Lacasa D et al (2010) Fibrosis in human adipose tissue: composition, distribution, and link with lipid metabolism and fat mass loss. Diabetes 59(11):2817–2825

Duffaut C, Galitzky J, Lafontan M et al (2009a) Unexpected trafficking of immune cells within the adipose tissue during the onset of obesity. Biochem Biophys Res Commun 384(4):482–485

Duffaut C, Zakaroff-Girard A, Bourlier V et al (2009b) Interplay between human adipocytes and T lymphocytes in obesity: CCL20 as an adipochemokine and T lymphocytes as lipogenic modulators. Arterioscler Thromb Vasc Biol 29(10):1608–1614

Elgazar-Carmon V, Rudich A, Hadad N et al (2008) Neutrophils transiently infiltrate intra-abdominal fat early in the course of high-fat feeding. J Lipid Res 49(9):1894–1903

Feuerer M, Herrero L, Cipolletta D et al (2009) Lean, but not obese, fat is enriched for a unique population of regulatory T cells that affect metabolic parameters. Nat Med 15(8):930–939

Fujisaka S, Usui I, Bukhari A et al (2009) Regulatory mechanisms for adipose tissue M1 and M2 macrophages in diet-induced obese mice. Diabetes 58(11):2574–2582
Ghanim H, Aljada A, Hofmeyer D et al (2004) Circulating mononuclear cells in the obese are in a proinflammatory state. Circulation 110(12):1564–1571
Harman-Boehm I, Bluher M, Redel H et al (2007) Macrophage infiltration into omental versus subcutaneous fat across different populations: effect of regional adiposity and the comorbidities of obesity. J Clin Endocrinol Metab 92(6):2240–2247
Henegar C, Tordjman J, Achard V et al (2008) Adipose tissue transcriptomic signature highlights the pathological relevance of extracellular matrix in human obesity. Genome Biol 9(1):R14
Ichioka M, Suganami T, Tsuda N et al (2011) Increased expression of macrophage-inducible C-type lectin in adipose tissue of obese mice and humans. Diabetes 60(3):819–826
Inouye KE, Shi H, Howard JK et al (2007) Absence of CC chemokine ligand 2 does not limit obesity-associated infiltration of macrophages into adipose tissue. Diabetes 56(9):2242–2250
Kanda H, Tateya S, Tamori Y et al (2006) MCP-1 contributes to macrophage infiltration into adipose tissue, insulin resistance, and hepatic steatosis in obesity. J Clin Invest 116(6):1494–1505
Keophiphath M, Achard V, Henegar C et al (2009) Macrophage-secreted factors promote a profibrotic phenotype in human preadipocytes. Mol Endocrinol 23(1):11–24
Keophiphath M, Rouault C, Divoux A et al (2010) CCL5 promotes macrophage recruitment and survival in human adipose tissue. Arterioscler Thromb Vasc Biol 30(1):39–45
Kintscher U, Hartge M, Hess K et al (2008) T-lymphocyte infiltration in visceral adipose tissue: a primary event in adipose tissue inflammation and the development of obesity-mediated insulin resistance. Arterioscler Thromb Vasc Biol 28(7):1304–1310
Koliwad SK, Streeper RS, Monetti M et al (2010) DGAT1-dependent triacylglycerol storage by macrophages protects mice from diet-induced insulin resistance and inflammation. J Clin Invest 120(3):756–767
Kopp A, Bala M, Buechler C et al (2010) C1q/TNF-related protein-3 represents a novel and endogenous lipopolysaccharide antagonist of the adipose tissue. Endocrinology 151(11):5267–5278
Kosteli A, Sugaru E, Haemmerle G et al (2010) Weight loss and lipolysis promote a dynamic immune response in murine adipose tissue. J Clin Invest 120(10):3466–3479
Lacasa D, Taleb S, Keophiphath M et al (2007) Macrophage-secreted factors impair human adipogenesis: involvement of proinflammatory state in preadipocytes. Endocrinology 148(2):868–877
Liu J, Divoux A, Sun J et al (2009) Genetic deficiency and pharmacological stabilization of mast cells reduce diet-induced obesity and diabetes in mice. Nat Med 15(8):940–945
Lumeng CN, Saltiel AR (2011) Inflammatory links between obesity and metabolic disease. J Clin Invest 121(6):2111–2117
Lumeng CN, Bodzin JL, Saltiel AR (2007a) Obesity induces a phenotypic switch in adipose tissue macrophage polarization. J Clin Invest 117(1):175–184
Lumeng CN, Deyoung SM, Bodzin JL et al (2007b) Increased inflammatory properties of adipose tissue macrophages recruited during diet-induced obesity. Diabetes 56(1):16–23
Nara N, Nakayama Y, Okamoto S et al (2007) Disruption of CXC motif chemokine ligand-14 in mice ameliorates obesity-induced insulin resistance. J Biol Chem 282(42):30794–30803
Nishimura S, Manabe I, Nagasaki M et al (2009) CD8+ effector T cells contribute to macrophage recruitment and adipose tissue inflammation in obesity. Nat Med 15(8):914–920
Nomiyama T, Perez-Tilve D, Ogawa D et al (2007) Osteopontin mediates obesity-induced adipose tissue macrophage infiltration and insulin resistance in mice. J Clin Invest 117(10):2877–2888
Ohmura K, Ishimori N, Ohmura Y et al (2010) Natural killer T cells are involved in adipose tissues inflammation and glucose intolerance in diet-induced obese mice. Arterioscler Thromb Vasc Biol 30(2):193–199

Poitou C, Dalmas E, Renovato M et al (2011) CD14dimCD16+ and CD14+ CD16+ monocytes in obesity and during weight loss: relationships with fat mass and subclinical atherosclerosis. Arterioscler Thromb Vasc Biol 31(10):2322–2330

Prieur X, Mok CY, Velagapudi VR et al (2011) Differential lipid partitioning between adipocytes and tissue macrophages modulates macrophage lipotoxicity and m2/m1 polarization in obese mice. Diabetes 60(3):797–809

Richette P, Poitou C, Garnero P, et al. (2011) Benefits of massive weight loss on symptoms, systemic inflammation and cartilage turnover in obese patients with knee osteoarthritis. Ann Rheum Dis 70(1):139–144

Rogacev KS, Ulrich C, Blomer L et al (2010) Monocyte heterogeneity in obesity and subclinical atherosclerosis. Eur Heart J 31(3):369–376

Shaul ME, Bennett G, Strissel KJ et al (2010) Dynamic, M2-like remodeling phenotypes of CD11c + adipose tissue macrophages during high-fat diet–induced obesity in mice. Diabetes 59(5):1171–1181

Shi H, Kokoeva MV, Inouye K et al (2006) TLR4 links innate immunity and fatty acid-induced insulin resistance. J Clin Invest 116(11):3015–3025

Spencer M, Yao-Borengasser A, Unal R et al (2010) Adipose tissue macrophages in insulin resistant subjects are associated with collagen VI, fibrosis and demonstrate alternative activation. Am J Physiol Endocrinol Metab 299:E1016–E1027

Strissel KJ, Stancheva Z, Miyoshi H et al (2007) Adipocyte death, adipose tissue remodeling, and obesity complications. Diabetes 56(12):2910–2918

Suganami T, Nishida J, Ogawa Y (2005) A paracrine loop between adipocytes and macrophages aggravates inflammatory changes: role of free fatty acids and tumor necrosis factor alpha. Arterioscler Thromb Vasc Biol 25(10):2062–2068

Suganami T, Yuan X, Shimoda Y et al (2009) Activating transcription factor 3 constitutes a negative feedback mechanism that attenuates saturated fatty acid/toll-like receptor 4 signaling and macrophage activation in obese adipose tissue. Circ Res 105(1):25–32

Tam CS, Viardot A, Clement K et al (2010) Short-term overfeeding may induce peripheral insulin resistance without altering subcutaneous adipose tissue macrophages in humans. Diabetes 59(9):2164–2170

Tordjman J, Poitou C, Hugol D et al (2009) Association between omental adipose tissue macrophages and liver histopathology in morbid obesity: influence of glycemic status. J Hepatol 51(2):354–362

Tsou CL, Peters W, Si Y et al (2007) Critical roles for CCR2 and MCP-3 in monocyte mobilization from bone marrow and recruitment to inflammatory sites. J Clin Invest 117(4):902–909

Villaret A, Galitzky J, Decaunes P et al (2010) Adipose tissue endothelial cells from obese human subjects: differences among depots in angiogenic, metabolic and inflammatory gene expression and cellular senescence. Diabetes 59(11):2755–2763

Weisberg SP, McCann D, Desai M et al (2003) Obesity is associated with macrophage accumulation in adipose tissue. J Clin Invest 112(12):1796–1808

Weisberg SP, Hunter D, Huber R et al (2006) CCR2 modulates inflammatory and metabolic effects of high-fat feeding. J Clin Invest 116(1):115–124

Wentworth JM, Naselli G, Brown WA et al (2010) Pro-inflammatory CD11c+ CD206+ adipose tissue macrophages are associated with insulin resistance in human obesity. Diabetes 59(7):1648–1656

Westcott DJ, Delproposto JB, Geletka LM et al (2009) MGL1 promotes adipose tissue inflammation and insulin resistance by regulating 7/4hi monocytes in obesity. J Exp Med 206(13):3143–3156

Winer S, Chan Y, Paltser G et al (2009) Normalization of obesity-associated insulin resistance through immunotherapy. Nat Med 15(8):921–929

Xu H, Barnes GT, Yang Q et al (2003) Chronic inflammation in fat plays a crucial role in the development of obesity-related insulin resistance. J Clin Invest 112(12):1821–1830

Ye J (2009) Emerging role of adipose tissue hypoxia in obesity and insulin resistance. Int J Obes (Lond) 33(1):54–66
Zeyda M, Farmer D, Todoric J et al (2007) Human adipose tissue macrophages are of an anti-inflammatory phenotype but capable of excessive pro-inflammatory mediator production. Int J Obes (Lond) 31(9):1420–1428

Chapter 21
Impact of Proinflammatory Cytokines on Adipocyte Insulin Signaling

Jean-François Tanti, Jennifer Jager and Yannick Le Marchand-Brustel

Introduction

During obesity, the metabolic inflammation which occurs in adipose tissue leads to the production of inflammatory cytokines (cf. Chap. 20) activating various signaling pathways, which interfere with insulin signaling and metabolic effects within adipocytes. This insulin resistance will contribute to adipose tissue dysfunctions and to the systemic insulin resistance which occurs during obesity and type 2 diabetes. In this chapter, we will describe the main molecular mechanisms, which are used by the inflammatory cytokines to decrease insulin signaling. We will also discuss how the blockade of these mechanisms could ameliorate insulin sensitivity.

J.-F. Tanti (✉) · Y. Le Marchand-Brustel
INSERM U1065, Centre Méditerranéen de Médecine Moléculaire, Université de Nice Sophia Antipolis, 151 route de St Antoine de Ginestière, BP 2 3194, 06204 Nice cedex 3, France
e-mail: tanti@unice.fr

Y. Le Marchand-Brustel
e-mail: lemarcha@unice.fr

J. Jager
School of Medicine, Division of Endocrinology Diabetes and Metabolism, University of Pennsylvania, 735 Clinical Research Building, 415 Curie Boulevard, Philadelphia, PA 19104, USA
e-mail: jagerj@mail.med.upenn.edu

J.-P. Bastard and B. Fève (eds.), *Physiology and Physiopathology of Adipose Tissue*,
DOI: 10.1007/978-2-8178-0343-2_21,

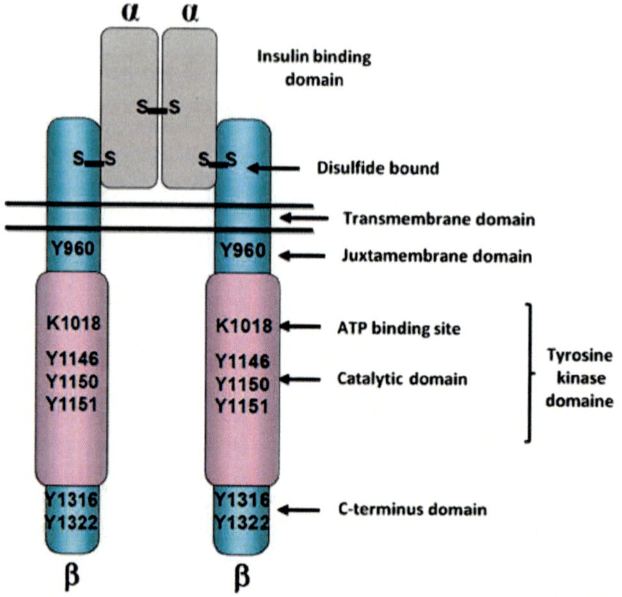

Fig. 21.1 Schematic representation of the insulin receptor. The insulin receptor is a heterotetrameric receptor which includes two extracellular α-subunits that bind insulin and two β-transmembrane-subunits that possess a tyrosine kinase activity. In the juxtamembrane region, the phosphorylation of the tyrosine^{960} allows the binding of the PTB (PhospoTyrosine Binding) domain of the insulin receptors substrate (IRS). In the kinase domain, the phosphorylation of the tyrosines$^{1146/1150/1151}$, located in the activation loop allows the modulation of the tyrosine kinase activity. In the C-terminus, the phosphorylation of the tyrosines$^{1316/1322}$ is important for the autophosphorylation of the receptor and could allow the binding of PI3 K. Y: tyrosine; K: lysine

Insulin Signaling

Insulin increases triglycerides synthesis and inhibits lipolysis in adipocytes. It also controls glucose transport by inducing the translocation of the glucose transporters Glut 4 from an intracellular compartment to the plasma membrane (cf. Chap. 7). Insulin effects are transmitted through a heterotetrameric receptor which includes two extracellular α-subunits and two β-transmembrane-subunits, linked by disulfide bound (Fig. 21.1). Insulin binding to the α-subunits induces their conformation change, leading to an increase in the tyrosine kinase activity of the β-subunits (Taniguchi et al. 2006). The β-subunits autophosphorylate on tyrosine residues Tyr^{1146}, Tyr^{1150}, and Tyr^{1151} located in the kinase domain, enhancing the kinase activity of the receptor. The phosphorylation of Tyr^{960} which is located in the NPXY motif within the juxtamembrane region is crucial for the binding of various insulin receptor substrates (Taniguchi et al. 2006).

The activated insulin receptor phosphorylates various substrates allowing the activation of two main signaling pathways. The Ras/mitogen-activated protein kinases (MAPK) ERK pathway is involved in the control of gene expression and cell growth and differentiation of various cell types including adipocytes. This signaling pathway is activated through the binding of the adaptor protein Shc to insulin receptors. Shc is then phosphorylated on tyrosine residues, and interacts with the SH2 (Src Homology -2) domain of the Grb2 protein. Grb2 then activates Sos (Son of Sevenless), a Ras GDP/GTP exchange factor thus allowing the activation of Ras and of the signaling cascade leading to the activation of the MAPK ERK (Fig. 21.2). The phosphatidylinositol 3-kinase (PI3 K) and Akt (also named PKB) are implicated in the metabolic effects of insulin and interacts with the Ras/MAPK pathway to control growth and differentiation. The IRS (Insulin receptor substrates) proteins are the insulin receptor substrates which, allow the insulin metabolic effects through the activation the PI3 K-Akt pathway (Taniguchi et al. 2006). The six IRS proteins are coded by six different genes. While IRS-1 and IRS-2 are ubiquitous, IRS-3 is the most abundant IRS protein in murine adipose tissue, but does not appear to be expressed in humans. IRS-4 is expressed in brain, thymus, and kidneys. More recently, IRS-5 has been identified, expressed in liver and kidney, as well as IRS-6 which is mostly expressed in skeletal muscles (White 2002). IRS-1 and -2 play a major role in the regulation of glucose homeostasis by insulin and their functions are altered in insulin resistant states (Taniguchi et al. 2006). They contain, in their N-terminal portion, a PH (pleckstrin homology) domain which recognizes the membrane phospholipids. This PH domain would allow the anchorage of the IRS proteins to the plasma membrane, in close proximity of insulin receptors. The PH domain is followed by a PTB (phosphotyrosine Binding) domain, which recognizes the phosphorylated tyrosine residues of the insulin receptor, and more specifically the NPXpY^{960} in the juxtamembrane domain. The PH and PTB domains, thus, allow a specific interaction of IRS with the insulin receptor and are necessary for their efficient phosphorylation. Further, IRS-2 contains a KRLB (kinase regulatory loop binding) domain, which interacts with the loop of the catalytic domain of the insulin receptor. In its C-terminal region, IRS proteins contain numerous tyrosine, serine, and threonine residues, which are potential phosphorylation sites. The phosphorylated tyrosine residues are binding sites for various proteins containing SH2 domains, among them is PI3 K. The binding of PI3 K with IRS 1/2 increases its lipid kinase activity leading to the phosphorylation of the phosphatidylinositol 4.5 phosphates (PIP2) at the position 3 of the inositol ring. The generated phosphatidylinositol 3.4.5 phosphates (PIP3) are intracellular second messengers which activate Akt and the atypical protein kinases ζ and λ through the activation of the PDK-1 protein (Taniguchi et al. 2006). (Fig. 21.2)

We will now describe the molecular mechanisms, which are used by inflammatory cytokines to diminish insulin signaling. These cytokines activate different signaling pathways which mainly interfere with IRS and Akt function, leading to a decrease in insulin metabolic effects in adipocytes.

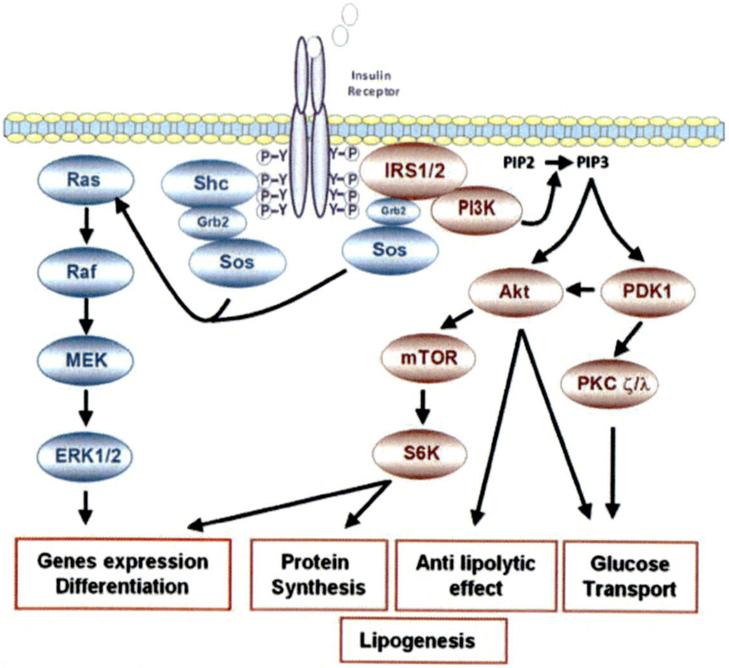

Fig. 21.2 Signaling pathways activated by insulin. The insulin receptor activates two main signaling pathways. The MAP kinase ERK pathway is involved in the control of gene expression and cell growth and differentiation. The PI3 K/Akt pathway is involved in the metabolic effect of insulin. The activation of this pathway requires the tyrosine phosphorylation of IRS-1 and -2 by the insulin receptor. In adipocytes, the PI3 K/Akt pathway is involved in insulin-induced glucose transport, lipogenesis, protein synthesis and anti-lipolysis

The Nitrosylation of the Proteins of the Insulin Signaling Pathway

Nitric oxide (NO) produced by the NO synthases is implicated in numerous physiological processes but is also involved in various pathological processes such as insulin resistance. Obesity is indeed associated with an increase in the activity of the inducible NO synthase (iNOS) in tissues which are insulin sensitive, in humans and rodents. Inflammatory cytokines, as well as other actors in insulin resistance, are involved in this increased expression. The large amounts of NO, which are produced are responsible for the nitrosylation of proteins of the insulin signaling pathway such as the insulin receptor, IRS-1 and Akt (Fig. 21.3), leading to IRS-1 degradation and to a decrease in Akt activity (de Luca and Olefsky 2008). In adipose tissue of obese patients or mice, the increase in the S-nitrosylation of insulin receptor, Akt and phosphodiesterase-3B (PDE3B) leads to a decrease in the insulin anti-lipolytic effect (Ovadia et al. 2011). The important role of the increase

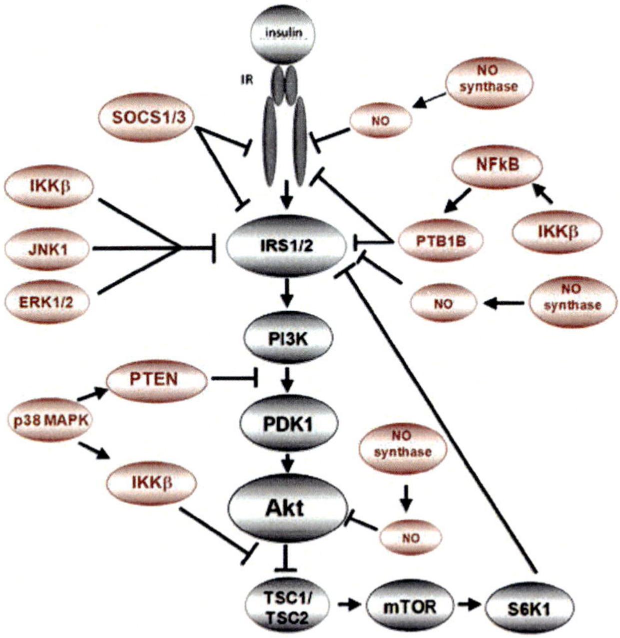

Fig. 21.3 Main proteins involved in the inhibition of the insulin signaling induced by the inflammatory cytokines. Inflammatory cytokines activate a network of signaling pathway that interferes with the activation of the insulin signaling pathway. The main proteins involved in this effect are: the tyrosine phosphatases PTP1B (protein tyrosine phosphatase-1B) that dephosphorylates the insulin receptor and/or the IRS; the inhibitory proteins SOCS1 and SOCS3 (Suppressor of Cytokine Signaling) that interact with the insulin receptor and the IRS; the serine/threonine kinases (JNK, ERK, IKK*β*, mTOR, S6K1) that phosphorylate the serine sites on the IRS; the NO synthase that increases NO production leading to the nitrosylation of the insulin receptor, IRS1/2 or Akt; the lipid phosphatase PTEN (phosphatase and tensin homolog) that dephosphorylates the PIP3 leading to an inhibition of Akt activation

in iNOS activity has been pointed out by the report that mice in which *Nos*2, the gene coding for iNOS, has been invalidated are protected against insulin resistance linked to obesity (de Luca and Olefsky 2008).

The SOCS proteins: negative regulators of insulin signaling, which are induced by inflammatory cytokines

The SOCS proteins (Suppressor of cytokine signaling) also named JAB (Janus family kinase-binding proteins) or SSI (signal transducer and activator of transcription induced Stat inhibitor) have been identified as negative regulators of cytokines action. The SOCS family includes eight members (SOCS1-7 and CIS), which possess a SH2 domain, and a SOCS-box domain, which controls the degradation of interacting proteins. SOCS1 and SOCS3 also contain in their N-terminal extremity a KIR (kinase inhibitory region), which allows to inhibit the kinase activity of other proteins such as the insulin receptor (Lebrun and Van Obberghen 2008). SOCS proteins are induced by various cytokines and are involved in a negative feedback pathway, thus allowing the fine tuning of the length and amplitude of their signal. At the molecular level, the SOCS proteins interact with the tyrosine kinases JAK (Janus activated kinases) or directly with the receptor of some cytokines, thus blocking the tyrosine phosphorylation of the transcription factors STAT (Polak et al. 2008). However, SOCS can also negatively regulate the signaling pathway induced by other hormones such as insulin. Further, insulin can induce the expression of some members of this family, such as SOCS3 (Lebrun and Van Obberghen 2008). Many cell studies have allowed the understanding of the mechanisms used by the SOCS to inhibit insulin signaling.

Molecular Mechanisms of the SOCS Induced Inhibition of Insulin Signaling

Among the various members of the SOCS family, SOCS1 and SOCS3 have been implicated in the negative regulation of insulin signaling (Fig. 21.3). Their expression, and more specifically that of SOCS3, is increased in adipose tissue and liver of obese mice (Lebrun and Van Obberghen 2008), as well as in adipose tissue and muscles of obese and/or diabetic patients (Rieusset et al. 2004). In adipose tissue, TNF-α largely contributes to the increase in SOCS3 (Lebrun and Van Obberghen 2008). At the molecular level, the SH2 domain of SOCS3 interacts with the phosphorylated Tyr960 of insulin receptor, thus preventing the binding of IRS1 and 2 to the receptor and inhibiting insulin signaling (Lebrun and Van Obberghen 2008). SOCS1 preferentially inhibits the interaction of IRS2 with the insulin receptor through its interaction with the kinase domain of insulin receptor necessary for the binding of the KRLB domain of IRS-2 (Lebrun and Van Obberghen 2008). SOCS1 can also inhibit the tyrosine kinase activity of insulin receptor. SOCS1 and SOCS3 also directly interact with the tyrosine phosphorylated IRS-1 and IRS-2. They can then recruit ubiquitin ligases through their SOCS-box domain leading to an increased ubiquitination and degradation of IRS proteins by the proteasome (Lebrun and Van Obberghen 2008).

The important role of SOCS3 in the negative regulation of insulin signaling in adipocytes has been pointed out using adipocytes differentiated from SOCS3$^{-/-}$ mouse embryo fibroblasts. When SOCS3 is absent, insulin-stimulated tyrosine phosphorylation of IRS-1 and IRS-2 and glucose transport are increased. Further, SOCS3 absence limits the inhibiting effect of TNF-α on insulin signaling through a partial prevention of the degradation of IRS.

This series of cellular studies demonstrates that SOCSs are important negative regulators of insulin signaling. To further evaluate their importance in the development of insulin resistance in vivo, many models of genetically modified mice have been used.

Role of SOCS in the Development of Insulin Resistance: Lessons from Animal Models

The importance of SOCS3 in the regulation of adipocyte insulin signaling has been confirmed in vivo by the characterization of mice with specific overexpression of SOCS3 in adipocytes. These mice develop adipose tissue insulin resistance due to a decreased expression and tyrosine phosphorylation of IRS in adipocytes (Shi et al. 2006). Other genetically modified mice are allowed to evaluate the importance of SOCS in other tissues in relation with the development of insulin resistance. The $SOCS3^{+/-}$ mice are partially resistant to a high fat diet (HFD) induced obesity and are more sensitive to leptin (Howard et al. 2004). This phenotype could result from an effect of SOCS3 in the central nervous system (SNC), since a specific invalidation of SOCS3 in SNC leads to the same phenotype (Mori et al. 2004). Hepatocyte overexpression of SOCS3 decreases IRS tyrosine phosphorylation and leads to glucose intolerance, systemic insulin resistance, and liver steatosis. Steatosis is due to STAT3 inhibition, which leads to an increased expression of SREBP-1c. Conversely, hepatic knockout of SOCS3 in obese diabetic mice slightly ameliorates their insulin sensitivity and normalizes SREBP-1c expression, hepatic steatosis and hypertriglyceridemia (Ueki et al. 2005). These results suggest that the inflammatory state in obese subjects could participate to the increased hepatic fatty acid synthesis through an increased expression of SOCS, which would lead to decrease the STAT3 inhibitory effect on SREBP1c expression. Another study confirms that SOCS3 is involved in liver insulin resistance but more inflammatory mediators are produced by the liver with aging leading to a systemic insulin resistance (Torisu et al. 2007). The role of muscle SOCS3 in the development of insulin resistance appears limited, but it is important for muscle integrity and locomotor activity (Lebrun et al. 2009).

The evaluation of SOCS1 contribution to the development of insulin resistance in vivo is less established. Indeed, $SOCS1^{-/-}$ mice die rapidly due to a massive inflammation. However, these mice are hypoglycemic and adipocyte differentiation of their embryo fibroblasts is increased (Lebrun and Van Obberghen 2008). These observations indirectly suggest that insulin signaling could be increased when SOCS1 is absent. However, a recent study using $SOCS1^{-/-}$ mice crossed with $RAG2^{-/-}$ mice, to prevent massive inflammation, demonstrates that the lack of SOCS1 is not sufficient to protect mice from HFD-induced insulin resistance (Emanuelli et al. 2008a). It has also been demonstrated that SOCS1 knockout in macrophages and T cells increases the inflammatory response induced by LPS or fatty acids. This is associated with glucose intolerance and hyperinsulinemia due to a decrease in liver insulin sensitivity (Sachithanandan et al. 2011). SOCS1 in immune cells could be protective through a limitation of metabolic inflammation in liver and perhaps adipose tissue. Conversely, SOCS1 could have a deleterious role in hepatocytes since its overexpression or knockout leads to a phenotype similar to that observed with SOCS3 (Ueki et al. 2005).

The Serine Phosphorylation of IRS Proteins: A Mechanism of the Negative Regulation of Insulin Signaling by Inflammatory Cytokines

Inflammatory cytokines activate numerous serine/threonine kinases in adipose tissue, which induce a decrease in insulin signaling through the serine phosphorylation of IRS-1 and IRS-2 proteins (Tanti and Jager 2009). IRS proteins contain a large number of phosphorylation sites, including tyrosine but also serine residues. In a pioneer work, we have demonstrated that IRS-1 serine phosphorylation is a molecular mechanism leading to a decrease in its tyrosine phosphorylation (Tanti et al. 1994). In normal conditions, insulin increases IRS-1 serine phosphorylation, which acts as a negative feedback on its own signaling pathway (Tanti and Jager 2009). In pathophysiologic conditions, such as obesity and diabetes, this mechanism is abnormally used by inflammatory cytokines and by various factors involved in insulin resistance, leading to desensitization of insulin signaling (Tanti and Jager 2009).

The recent development of mass spectrometry approaches has allowed to obtain a complete mapping of the phosphorylated IRS-1 serine residues, and to understand how this serine phosphorylation controls the tyrosine phosphorylation in response to insulin and how this process is abnormally activated in insulin resistance (Langlais et al. 2011). It is now clearly demonstrated that insulin induces a time-controlled and organized IRS-1 phosphorylation of numerous sites allowing a very precise control of the length of hormone action. The mechanism of IRS-1 serine phosphorylation is extremely complex, since a phosphorylation of one site can induces either positive or negative effects or both (Tanti and Jager 2009). According to the presently accepted scheme, insulin induces first the Akt phosphorylation of "positive" sites, which could prevent the phosphorylation of inhibitory sites or the association of IRS-1 with tyrosine phosphatases. Thus, this positive IRS-1 serine phosphorylation would allow a proper IRS-1 tyrosine phosphorylation in response to insulin, and thus the optimal cellular effects of the hormone (Tanti and Jager 2009). The phosphorylation of inhibitory serine residues (Fig. 21.4) is then occurring later and stops insulin signaling. Inhibitory sites are located in a cluster within or in close proximity of the IRS-1 PTB domain. Ser^{307} (in mouse) or Ser^{312} (in human) has been the most extensively studied serine residue. Its phosphorylation inhibits the interaction between IRS-1 and the insulin receptor leading to a conformation change of the PTB domain of IRS-1. It also favors IRS-1 degradation by the proteasome. Inhibitory sites, which are located in the C-terminal region (such as Ser^{612} and Ser^{632}) negatively regulate the IRS-1 interaction with PI3-kinase (Tanti and Jager 2009). Therefore, in physiological conditions, global insulin-induced IRS-1 phosphorylation on serine residues results from a finely controlled equilibrium between positive and inhibitory sites. During obesity and diabetes, inhibitory sites are phosphorylated in an uncontrolled manner in response to inflammatory cytokines, and also to other factors involved in insulin resistance, leading to desensitization of insulin signaling (Tanti and

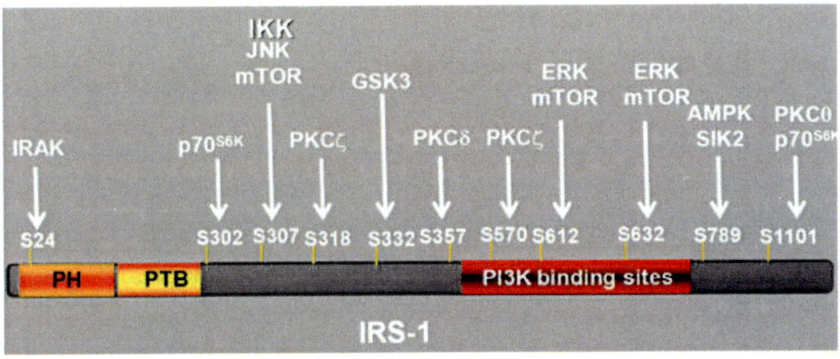

Fig. 21.4 - Main serine phosphorylation sites on IRS-1 involved in the desensitization of the insulin signaling. In physiological conditions, global insulin-induced IRS-1 phosphorylation on serine residues results from a finely controlled equilibrium between positive (not shown in the figure) and inhibitory sites. This process allows a fine tuning of the tyrosine phosphorylation of IRS-1 leading to an optimal propagation of the insulin signal. In insulin resistant state, the inhibitory sites are phosphorylated in an uncontrolled manner in response to inflammatory cytokines and to other mediators of the insulin resistance. The consequence is a desensitization of the insulin signaling. The main kinases involved in the phosphorylation of the inhibitory sites are represented. PH: Plekstrin Homology domain; PTB: PhosphoTyrosine Binding domain

Jager 2009). The implication of this abnormal IRS-1 serine phosphorylation in the development of insulin resistance has been recently confirmed by the use of genetically modified mice. Indeed, mice, which overexpress in muscles an IRS-1 protein deleted of three important regulatory sites ($Ser^{302/307/612}$) are partially protected against a HFD-induced insulin resistance (Morino et al. 2008). An increase in IRS-1 serine phosphorylation is now considered as a signature of an insulin resistant state in humans and animals. Although most published studies concerns IRS-1, IRS-2 is also phosphorylated on serine residues, which have been recently identified (Tanti and Jager 2009).

There are numerous kinases which are activated by the inflammatory cytokines and which are implicated in IRS-1 serine phosphorylation (Fig. 21.4). Some of them, such as mTOR or p70S6 kinase, are also involved in insulin-induced IRS-1 serine phosphorylation, but most of them are different. The MAP kinases ERK and JNK, as well as IKKβ, play a major role in this process (Tanti and Jager 2009).

We will now describe the role of these kinases in the development of insulin resistance and discuss the possibility of targeting these kinases as new therapeutic approaches of the metabolic disorders linked to obesity.

Implication of the Inflammatory Signaling Pathway IKKβ/NF-κB in Insulin Resistance

Cellular Studies

The IKKβ kinase plays a central role in the activation of the NF-κB transcription factor by various inflammatory stimuli. IKKβ activity is increased in adipose tissue of obese patients (Bashan et al. 2007). The inhibition of IKKβ in adipocytes or other insulin sensitive cells prevents the deleterious effects of inflammatory cytokines on insulin signaling (Donath and Shoelson 2011). At the molecular level, IKKβ interferes with insulin signaling through an increase in IRS-1 serine phosphorylation induced by inflammatory cytokines, such as TNF-α (Figs. 21.3 and 21.4). The phosphorylation of a cluster of sites inside the PTB domain, including Ser^{307}, is involved in the IKKβ-induced decreased interaction between the insulin receptor and IRS-1 (Tanti and Jager 2009). Although some studies have demonstrated a direct phosphorylation of IRS-1 by IKKβ, other studies have shown that IKKβ could also act indirectly, through the activation of other serine/threonine kinases. The activation of IKKβ would indeed be involved in insulin resistance induced by TNF-α via the deregulation of the TSC1/TSC2/mTOR complex (Lee et al. 2008). The IKKβ kinase allows for mTOR activation through the phosphorylation of TSC-1, leading to the activation of S6 Kinase-1 (S6K1) (Fig. 21.3). S6K1 and mTOR play a role in the IRS-1 phosphorylation of Ser^{307} and $Ser^{636/639}$ (Tanti and Jager 2009). Interestingly, $S6K1^{-/-}$ mice are protected against obesity and insulin resistance, as a consequence of increased energy expenditure and decreased S6K1-induced serine phosphorylation of IRS-1 (Pende et al. 2000). Further, the adipocyte knockout of raptor, a protein involved in the activation of the mTOR/S6K1 pathway mimics the phenotype of $S6K1^{-/-}$ mice (Polak et al. 2008). These results demonstrate the importance of the activation of the mTOR/S6K1 pathway in adipocyte in the control of insulin sensitivity and energy homeostasis.

Another recently described mechanism linking IKKβ activation and the decrease in IRS-1 tyrosine phosphorylation is the increased expression, via NFκB, of the PTB1B tyrosine phosphatase, which dephosphorylates the IRS-1 tyrosine residues (Fig. 21.3). This mechanism has been described in adipose tissue and other metabolic tissues in response to inflammatory cytokines (Zabolotny et al. 2008).

Implication of IKKβ in Insulin Resistance In Vivo: Lessons from Animal Models

The importance of the IKKβ/NFκB pathway in the development of insulin resistance in vivo has been pointed out by the use of various models of mice with a global or a tissue specific invalidation of the protein. The IKK$\beta^{+/-}$ mice are partially protected against HFD-induced insulin resistance (Arkan et al. 2005). Liver IKKβ knockout prevents the development of insulin resistance and liver inflammation, and partially protects the mice against the development of a systemic insulin resistance (Arkan et al. 2005). Conversely, the liver overexpression of a constitutively active IKKβ induces liver inflammation and insulin resistance (Cai et al. 2005). By contrast, muscle knockout of IKKβ does not prevent obesity associated insulin resistance, suggesting that this pathway is not important in muscle. The IKKβ/NFκB signaling pathway plays a central role in the production of inflammatory cytokines by myeloid cells and the specific IKKβ knockout in these cells prevents HFD-induced insulin resistance (Arkan et al. 2005). The hypothalamic activation of the IKKβ/NFκB signaling pathway is also implicated in the resistance to insulin and to leptin during obesity (Zhang et al. 2008). All those studies demonstrate that the abnormal activation of the IKKβ/NFκB signaling pathway during obesity results in an increase of the production of inflammatory cytokines, which act either locally, or systematically, to induce insulin resistance (Donath and Shoelson 2011).

IKKε, another protein kinase with sequence homology with IKKβ, is induced by inflammatory stimuli. Its expression has recently been shown to be increased in adipocyte, liver, and adipose tissue macrophages of HFD obese mice. IKK$\varepsilon^{-/-}$ mice are protected against HFD-induced obesity, through an increase in energy expenditure. Further, they display a decreased liver and adipose tissue inflammation, no steatosis and are protected against insulin resistance (Chiang et al. 2009).

In conclusion, the activation of the IKKβ/NFκB pathway is a central mechanism linking metabolic inflammation with the development of insulin resistance. A pharmacological targeting of this pathway could thus be useful in the treatment of insulin resistance and type 2 diabetes.

Is the Pharmacological Targeting of the IKKβ/NFκB Pathway a Novel Approach in the Treatment of Insulin Resistance?

Studies from the late nineteenth century and the early twentieth century reported that aspirin was able to decrease the hyperglycemia in diabetic patients (Goldfine et al. 2011). More recently, Shoelson et al. have shown that high dosages of aspirin and salicylate inhibit IKKβ and ameliorate insulin sensitivity in mice, either genetically obese, or rendered insulin resistant by a lipid infusion. Further, aspirin (7 g/day/2 weeks) given to a small group of diabetic patients ameliorates most of their metabolic parameters (29). However, such high doses preclude its use in clinic. By contrast, a non-acetylated dimer of salicylic acid (salsalate) induces less secondary effects such as bleedings of stomach injury. Proof of concept studies and clinical trials suggest that salsalate ameliorates glucose tolerance through a better insulin signaling and sensitivity, and/or via an increased insulin secretion (Goldfine et al. 2011). These encouraging observations suggest that the inhibition of the IKKβ/NFκB signaling pathway could be a novel therapeutic approach in diabetes and insulin resistance. However, since IKKβ/NFκB pathway is a central regulator of immunity, secondary unwanted effects could appear with chronic treatment.

Implication of MAP Kinases in the Alterations of Insulin Signaling and in the Development of Insulin Resistance

The Mitogen-Activated Protein kinases (MAP kinases) represent a large family of protein kinases, which are activated by various stimuli including inflammatory cytokines. Five groups of MAP kinases have been characterized in mammal cells: the Extracellular signal-Regulated Kinases 1 and 2 (ERK1 and ERK2, or p44 and p42); the c-jun amino-terminal kinases 1, 2, and 3 (JNK1, 2 and 3), also named Stress-Activated Protein kinases 1, 2 and 3 (SAPK1, 2 and 3); the p38 MAPK α, β, γ, and δ; ERK5; and the orphan MAP kinases ERK 3, 4, 6, 7 and 8. MAP kinases activation requires a cascade of phosphorylations which starts with the activation of a MAP kinase kinase kinase (MAP3 K, MAPKKK or MEKK). Once activated, the MAP3 K phosphorylates the MAP kinase kinase (MAP2 K, MAPKK, or MEK) on serine and/or threonine residues, located in their activation loop, leading to their activation. The MAP2 K then activates the MAPK through a double phosphorylation on threonine and tyrosine residues located in their activation loop (TXY motif). The activated MAPKs phosphorylate numerous substrates with serine or threonine residues close to a proline residue (Keshet and Seger 2010).

The total activity of JNK, ERK, and p38 MAPK is increased in adipose tissue, liver and muscles of obese patients or mice, an indication of the role of these pathways in insulin resistance (Tanti and Jager 2009). A large series of cellular studies have allowed the understanding of the molecular mechanisms linking MAP kinases activation to the desensitization of insulin signaling and the development of insulin resistance.

MAP Kinases and the Decrease in Insulin Signaling: Cellular Studies

MAP kinases play an important role in the abnormal serine phosphorylation of IRS proteins, leading to a decrease in insulin signaling (Figs. 21.3 and 21.4). JNK phosphorylates Ser^{307} of IRS-1 in adipocytes in response to inflammatory cytokines.

This increased in Ser^{307} phosphorylation is absent in $JNK1^{-/-}$ mice, submitted to HFD (Hirosumi et al. 2002). This observation strongly supports a major role of JNK, and more specifically of JNK1, in the phosphorylation of this site during obesity. However, some studies suggest that it is the phosphorylation of a cluster of sites including Ser^{307}, which allows the negative control of the interaction between IRS1 and the insulin receptor (Tanti and Jager 2009). IRS-2 is also phosphorylated by JNK on Thr^{348} and Ser^{488}. The phosphorylation of Ser^{488} permits the phosphorylation of Ser^{484} by GSK3β. These phosphorylations decrease the insulin signaling. These various sites are near the PTB domain of IRS-2, an indication that their phosphorylation could inhibit the interaction between IRS-2 and the insulin receptor (Tanti and Jager 2009). The ERK kinases phosphorylate Ser^{612} and Ser^{632} in mice (Ser^{616} and Ser^{636} in humans), which would decrease the interaction between IRS-1 and the PI3 K and would thus diminish the metabolic effects of insulin (Tanti and Jager 2009) (Figs. 21.3 and 21.4). Interestingly, basal ERK activity and IRS1-Ser^{636} phosphorylation is abnormally increased in primary muscle cells from type 2 diabetic patients (Bouzakri et al. 2003). Although the ERK kinases preferentially phosphorylate Ser^{612} and Ser^{632}, they are also able to phosphorylate Ser^{307} in response to TNF-α in 3T3-L1 adipocytes. Further, we have shown that the activation of the ERK pathway by the inflammatory cytokines in adipocytes also induces a decrease in the transcription of IRS-1 mRNA, leading to a decrease in insulin signaling and glucose transport (Jager et al. 2007). The pharmacological inhibition of p38MAPK also prevents the serine phosphorylation of IRS-1 induced by TNF-α. However, no study has been able to demonstrate a direct phosphorylation of IRS1 by p38MAPK. The p38MAPK could indirectly participate in the increased IRS-1 serine phosphorylation via IKKβ activation (Fig. 21.3) or through the transactivation of other receptors (Tanti and Jager 2009). Another mechanism to explain the p38MAPK inhibition of insulin signaling is the increased expression of the PTEN lipid phosphatase (Fig. 21.3), which decreases the Akt activation in response of insulin via a reduction of the PIP3 level (Liu et al. 2007).

The use of various genetically modified mice has highlighted the important role of the JNK or ERK pathways in the development of obesity and insulin resistance in vivo.

Implication of MAPK Kinases in the Development of Insulin Resistance In Vivo: Lessons from Animal Models

Consequences of JNK Invalidation on the Development of Insulin Resistance

Among the various JNK isoforms, the role of JNK1 in insulin resistance has been pointed out by genetic invalidation in mice. $Jnk1^{-/-}$ mice presents a lower body weight and adipose tissue mass than their wild-type counterparts, either on a normal or a HFD diet. This phenotype could be explained by a small increase in their energy expenditure. Insulin sensitivity of Jnk1$^{-/-}$ mice subjected to HFD is improved, probably through a decrease in the phosphorylation of IRS-1-Ser^{307} (Hirosumi et al. 2002). The mice, invalidated for JIP-1, a scaffold protein involved in the activation of JNK, are phenotypically very similar to the Jnk$^{-/-}$ mice (Jaeschke et al. 2004). Interestingly, a mutation of a human JIP-1 homolog has been found in some patients with rare genetic Type 2 diabetes. This mutation results in an increased activity of JNK. These results suggest that modifications of JNK activity could be involved in the development of Type 2 diabetes in humans (Waeber et al. 2000). Different from Jnk1$^{-/-}$ mice, Jnk2$^{-/-}$ mice are not protected against

HFD-induced obesity and insulin resistance (Hirosumi et al. 2002). This could be due to the fact that JNK1 is the predominant isoform expressed in liver, adipose tissue and skeletal muscle, and could compensate for the lack of the JNK2 isoform. In agreement with this hypothesis, genetic ablation of Jnk2 in Jnk1$^{+/-}$ mice induces a decrease in the body weight gain and insulin resistance under HFD (Tanti and Jager 2009). JNK2 could thus participate in the development of insulin resistance, although less extensively.

The observations that inflammatory parameters are decreased in Jnk1$^{-/-}$ mice and that JNK are implicated in the inflammatory response of adipose tissue macrophages suggest that JNK activity in myeloid cells could regulate the development of insulin resistance. Two studies using bone marrow transfer between wild-type and Jnk1$^{-/-}$ mice have shown that activation of JNK1 in myeloid cells plays an important role in HFD-induced inflammation. However, the conclusions of these two studies differ concerning the implication of JNK1 in myeloid cells for the insulin resistance development (Solinas et al. 2007; Vallerie et al. 2008). Vallerie et al. suggest that JNK1 activation in non-hematopoietic compartments is an important event in the development of insulin resistance. This hypothesis has been confirmed by studying mice with specific knockout of JNK1 in various tissues. Two studies have demonstrated that JNK1 invalidation in adipose tissue decreases its inflammation and ameliorates liver insulin sensitivity. However, they report contradictory results concerning the systemic insulin resistance (Zhang et al. 2011; Sabio and Davis 2010). The liver beneficial effect could be due to a decrease in IL-6 production by adipose tissue with a decreased SOCS3 expression in liver (Sabio and Davis 2010). Only IL-6 has been identified so far as being regulated by JNK1 in adipose tissue. However, in drosophila, JNK pathway activation by various stress control the lipocalin production by the fat body, an equivalent of liver and adipose tissue. This allows for a control of growth and metabolism in stress conditions via the regulation of insulin signaling (Hull-Thompson et al. 2009). Lipocalin family members are implicated in insulin resistance in humans, and a deregulation of JNK activity in adipose tissue could be responsible for their abnormal production. The sustained activation of JNK activity in adipose tissue could thus lead to abnormalities in the production of some adipokines leading to liver insulin resistance. Muscle specific knockout of JNK1 reduces muscle insulin resistance, likely through a decrease in IRS serine phosphorylation, but has small effect on systemic insulin resistance (Sabio and Davis 2010). By contrast, JNK1 appears to have a protective role in liver, since mice with a liver specific JNK1 knockout develops glucose intolerance, insulin resistance and liver steatosis. This is due to an increased expression of PGC-1β and of genes involved in hepatic lipogenesis and in lipoproteins secretion. On a similar fashion, a decreased JNK2 activity does not correct hepatic steatosis and even worsens hepatic lesions due to an elevated apoptosis (Sabio and Davis 2010). If these observations are also true in humans, a therapeutic strategy aiming at decreasing JNK hepatic activity could worsen hepatic steatosis and cardiovascular risks via an increased triglycerides secretion. However, it should be noted that one study describes a beneficial effect of hepatic inhibition of JNK on insulin sensitivity and steatosis (Nakatani et al. 2004).

Recent studies demonstrate that JNK1 invalidation in the central nervous system is sufficient to suppress HFD-induced obesity. This could be due to increased energy expenditure through the activation of the hypothalamo-pituitary-thyroid axis. These studies favor an unexpected role of the JNK1 signaling in hypothalamus and pituitary gland in the control of metabolic and endocrine homeostasis (Sabio and Davis 2010).

Consequences of ERK's Invalidation on the Development of Insulin Resistance

Our team has demonstrated that erk1$^{-/-}$ mice are resistant to HFD-induced obesity, because of a decrease in adipogenesis and an increase in postprandial energy expenditure. The lack of obesity is accompanied by a better glucose and insulin tolerance compared to wild-type mice (Bost et al. 2005). Conversely, mice with a genetic invalidation of p62, an inhibitor of the ERK signaling pathway, become obese and severely insulin resistant. This phenotype disappears when those mice are crossed with erk1$^{-/-}$ mice, demonstrating the role of ERK1 in this phenotype (Lee et al. 2010). These two studies show the implication of ERK in obesity development, but they do not allow to firmly conclude about the role of ERK in insulin resistance *per se*, since the correction of insulin resistance could be the consequence of the prevention of obesity. However, a direct implication of ERK in the development of insulin resistance is suggested by two other studies. First, the hepatic overexpression of MKP-4, a phosphatase involved in ERK and JNK inhibition, in obese mice leads to a decrease in the phosphorylation of ERK and JNK, and to the improvement of insulin signaling and glucose tolerance without modifying the weight of the mice (Emanuelli et al. 2008b). However, since MKP-4 inhibits both JNK and ERK, it is not possible to definitely conclude for a role of ERK in insulin resistance in vivo. A more direct proof of ERK involvement in insulin resistance, independently of obesity, has been obtained with our studies, in which ERK1 has been invalidated in the genetically obese (*ob/ob*) mice. Those erk1$^{-/-}$/*ob/ob* mice are as obese as *ob/ob* mice, but are more insulin sensitive and present a decrease in the adipose tissue inflammation (Jager et al. 2011).

These series of cellular and in vivo studies suggest that ERK or JNK inhibition could ameliorate insulin sensitivity. However, the direct targeting of these proteins could have a series of drawbacks since they are involved in numerous biological processes. A possible alternative choice would be to target the proteins which control ERK or JNK activity, specifically in response to inflammatory stress which develop during obesity. With this hypothesis in mind, we have recently identified Tpl2, a MAP kinase kinase kinase as a protein controlling ERK activity specifically in response to inflammatory cytokines in adipocytes. Tpl2 inhibition reduces TNF-α induced IRS-1 serine phosphorylation in adipocytes and decreases the lipolytic activity of these cytokines (Jager et al. 2010). A recent study showed that $Tpl2^{-/-}$ mice are protected against HFD-induced insulin resistance and adipose tissue inflammation (Perfield et al. 2011). Several studies in the domain of immunity indicate that Tpl2 is involved in inflammatory cytokines production by myeloid cells (Gantke et al. 2011). Targeting this protein could thus be beneficial to decrease the production of inflammatory mediators in adipose tissue immune cells and to inhibit their deleterious effects in adipocytes.

This series of studies suggest that the MAP kinases signaling pathways could be interesting targets for the treatment of insulin resistance and type 2 diabetes. Until now, only pharmacological JNK inhibitors have been tested in insulin resistance and type 2 diabetes.

Effect of Pharmacological Targeting of the JNK Pathway in Insulin Resistance

The first generation of competitive inhibitors of ATP directed against JNK displays protective effects against insulin resistance and ameliorates insulin secretion of obese mice. However, their specificity is limited, and it is difficult to insure that all the observed effects are the only consequence of JNK inhibition (Yang and Trevillyan 2008). More specific inhibitors have been obtained, and one of them prevents HFD-induced weight gain and insulin resistance. Further studies are required to confirm that the observed effects actually result from JNK inhibition. Other JNK inhibitors have also been developed, following the observation that peptides, which interact with the scaffold proteins JIP-1 could very selectively block JNK activation. One of those inhibitors prevents TNF-α deleterious effects in liver and ameliorates insulin sensitivity in *db/db* mice (Yang and Trevillyan 2008). All these results suggest that JNK inhibitors could be interesting therapeutic agents in insulin resistance and type 2 diabetes. However, as mentioned above, the implication of the JNK pathway in numerous physiologic processes could be an obstacle in their therapeutic use. Thus, to specifically inhibit one of the JNK isoform or some JNK signalosomes could be a better strategy in the future.

Conclusion

In adipose tissue of obese mice and patients, inflammatory cytokines activate a network of signaling pathway that targets key component of the insulin signaling pathway (Fig. 21.3). Several kinases including IKKβ, JNK, ERK, mTOR, and S6 K are activated by inflammatory cytokines and phosphorylated IRS proteins on several serine residues in an uncontrolled manner. Studies of genetically modified mice have uncovered an important role of these kinases in the control of insulin sensitivity and whole-body metabolism. These kinases are thus potential drug targets to fight insulin resistance and some studies suggest that therapy targeting the IKKβ/NF-κB or the JNK pathway may evolve into future diabetes medication. However, these kinases are involved in the control of several biologic processes and this can be problematic for the development of safe drugs for the treatment of a chronic disease. Identification of upstream regulators or downstream substrates of these kinases activated specifically in response to metabolic or inflammatory stresses could be a strategy to circumvent this problem. Future studies should tell us whether the targeting of the proteins belonging to the inflammatory network activated in obese adipose tissue may evolve into the development of new therapeutic agents against insulin resistance and diabetes.

References

Arkan MC, Hevener AL, Greten FR et al (2005) IKK-beta links inflammation to obesity-induced insulin resistance. Nat Med 11:191–198

Bashan N, Dorfman K, Tarnovscki T et al (2007) Mitogen-activated protein kinases, inhibitory-kappaB kinase, and insulin signaling in human omental versus subcutaneous adipose tissue in obesity. Endocrinology 148:2955–2962

Bost F, Aouadi M, Caron L et al (2005) The extracellular signal-regulated kinase isoform ERK1 is specifically required for in vitro and in vivo adipogenesis. Diabetes 54:402–411

Bouzakri K, Roques M, Gual P et al (2003) Reduced activation of phosphatidylinositol-3 kinase and increased serine 636 phosphorylation of insulin receptor substrate-1 in primary culture of skeletal muscle cells from patients with type 2 diabetes. Diabetes 52:1319–1325

Cai D, Yuan M, Frantz DF et al (2005) Local and systemic insulin resistance resulting from hepatic activation of IKK-beta and NF-kappaB. Nat Med 11:183–190

Chiang SH, Bazuine M, Lumeng CN et al (2009) The protein kinase IKKepsilon regulates energy balance in obese mice. Cell 138:961–975

de Luca C, Olefsky JM (2008) Inflammation and insulin resistance. FEBS Lett 582:97–105

Donath MY, Shoelson SE (2011) Type 2 diabetes as an inflammatory disease. Nat Rev Immunol 11:98–107

Emanuelli B, Macotela Y, Boucher J, Ronald Kahn C (2008a) SOCS-1 deficiency does not prevent diet-induced insulin resistance. Biochem Biophys Res Commun 377:447–452

Emanuelli B, Eberle D, Suzuki R, Kahn CR (2008b) Overexpression of the dual-specificity phosphatase MKP-4/DUSP-9 protects against stress-induced insulin resistance. Proc Natl Acad Sci U S A 105:3545–3550

Gantke T, Sriskantharajah S, Ley SC (2011) Regulation and function of TPL-2, an IkappaB kinase-regulated MAP kinase kinase kinase. Cell Res Dec 7 [Epub ahead of print]: 1–15

Goldfine AB, Fonseca V, Shoelson SE (2011) Therapeutic approaches to target inflammation in type 2 diabetes. Clin Chem 57:162–167

Hirosumi J, Tuncman G, Chang L et al (2002) A central role for JNK in obesity and insulin resistance. Nature 420:333–336

Howard JK, Cave BJ, Oksanen LJ et al (2004) Enhanced leptin sensitivity and attenuation of diet-induced obesity in mice with haploinsufficiency of Socs3. Nat Med 10:734–738

Hull-Thompson J, Muffat J, Sanchez D et al (2009) Control of metabolic homeostasis by stress signaling is mediated by the lipocalin NLaz. PLoS Genet 5:e1000460

Jaeschke A, Czech MP, Davis RJ (2004) An essential role of the JIP1 scaffold protein for JNK activation in adipose tissue. Genes Dev 18:1976–1980

Jager J, Gremeaux T, Cormont M et al (2007) Interleukin-1beta-induced insulin resistance in adipocytes through down-regulation of insulin receptor substrate-1 expression. Endocrinology 148:241–251

Jager J, Gremeaux T, Gonzalez T et al (2010) The Tpl2 kinase is up-regulated in adipose tissue in obesity and may mediate IL-1{beta} and TNF-{alpha} effects on ERK activation and lipolysis. Diabetes 59:61–70

Jager J, Corcelle V, Gremeaux T et al (2011) Deficiency in the extracellular signal-regulated kinase 1 (ERK1) protects leptin-deficient mice from insulin resistance without affecting obesity. Diabetologia 54:180–189

Keshet Y, Seger R (2010) The MAP kinase signaling cascades: a system of hundreds of components regulates a diverse array of physiological functions. Methods Mol Biol 661:3–38

Langlais P, Yi Z, Finlayson J et al (2011) Global IRS-1 phosphorylation analysis in insulin resistance. Diabetologia 54:2878–2889

Lebrun P, Van Obberghen E (2008) SOCS proteins causing trouble in insulin action. Acta Physiol (Oxf) 192:29–36

Lebrun P, Cognard E, Bellon-Paul R et al (2009) Constitutive expression of suppressor of cytokine signalling-3 in skeletal muscle leads to reduced mobility and overweight in mice. Diabetologia 52:2201–2212

Lee DF, Kuo HP, Chen CT et al (2008) IKKbeta suppression of TSC1 function links the mTOR pathway with insulin resistance. Int J Mol Med 22:633–638

Lee SJ, Pfluger PT, Kim JY et al (2010) A functional role for the p62-ERK1 axis in the control of energy homeostasis and adipogenesis. EMBO Rep 11:226–232

Liu HY, Collins QF, Xiong Y et al (2007) Prolonged treatment of primary hepatocytes with oleate induces insulin resistance through p38 mitogen-activated protein kinase. J Biol Chem 282:14205–14212

Mori H, Hanada R, Hanada T et al (2004) Socs3 deficiency in the brain elevates leptin sensitivity and confers resistance to diet-induced obesity. Nat Med 10:739–743

Morino K, Neschen S, Bilz S et al (2008) Muscle-specific IRS-1 Ser->Ala transgenic mice are protected from fat-induced insulin resistance in skeletal muscle. Diabetes 57:2644–2651

Nakatani Y, Kaneto H, Kawamori D et al (2004) Modulation of the JNK pathway in liver affects insulin resistance status. J Biol Chem 279:45803–45809

Ovadia H, Haim Y, Nov O et al (2011) Increased adipocyte S-nitrosylation targets anti-lipolytic action of insulin: relevance to adipose tissue dysfunction in obesity. J Biol Chem 286:30433–30443

Pende M, Kozma SC, Jaquet M et al (2000) Hypoinsulinaemia, glucose intolerance and diminished ß-cell size in S6K1-deficient mice. Nature 408:994–997

Perfield JW 2nd, Lee Y, Shulman GI et al (2011) Tumor progression locus 2 (TPL2) regulates obesity-associated inflammation and insulin resistance. Diabetes 60:1168–1176

Polak P, Cybulski N, Feige JN et al (2008) Adipose-specific knockout of raptor results in lean mice with enhanced mitochondrial respiration. Cell Metab 8:399–410

Rieusset J, Bouzakri K, Chevillotte E et al (2004) Suppressor of cytokine signaling 3 expression and insulin resistance in skeletal muscle of obese and type 2 diabetic patients. Diabetes 53:2232–2241

Sabio G, Davis RJ (2010) cJun NH2-terminal kinase 1 (JNK1): roles in metabolic regulation of insulin resistance. Trends Biochem Sci 35:490–496

Sachithanandan N, Graham KL, Galic S et al (2011) Macrophage deletion of SOCS1 increases sensitivity to LPS and palmitic acid and results in systemic inflammation and hepatic insulin resistance. Diabetes 60:2023–2031

Shi H, Cave B, Inouye K et al (2006) Overexpression of suppressor of cytokine signaling 3 in adipose tissue causes local but not systemic insulin resistance. Diabetes 55:699–707

Solinas G, Vilcu C, Neels JG et al (2007) JNK1 in hematopoietically derived cells contributes to diet-induced inflammation and insulin resistance without affecting obesity. Cell Metab 6:386–397

Taniguchi CM, Emanuelli B, Kahn CR (2006) Critical nodes in signalling pathways: insights into insulin action. Nat Rev Mol Cell Biol 7:85–96

Tanti JF, Jager J (2009) Cellular mechanisms of insulin resistance: role of stress-regulated serine kinases and insulin receptor substrates (IRS) serine phosphorylation. Curr Opin Pharmacol 9:753–762

Tanti JF, Gremeaux T, van Obberghen E, Le Marchand-Brustel Y (1994) Serine/threonine phosphorylation of insulin receptor substrate 1 modulates insulin receptor signaling. J Biol Chem 269:6051–6057

Torisu T, Sato N, Yoshiga D et al (2007) The dual function of hepatic SOCS3 in insulin resistance in vivo. Genes Cells 12:143–154

Ueki K, Kadowaki T, Kahn CR (2005) Role of suppressors of cytokine signaling SOCS-1 and SOCS-3 in hepatic steatosis and the metabolic syndrome. Hepatol Res 33:185–192

Vallerie SN, Furuhashi M, Fucho R, Hotamisligil GS (2008) A predominant role for parenchymal c-Jun amino terminal kinase (JNK) in the regulation of systemic insulin sensitivity. PLoS One 3:e3151

Waeber G, Delplanque J, Bonny C et al (2000) The gene MAPK8IP1, encoding islet-brain-1, is a candidate for type 2 diabetes. Nat Genet 24:291–295

White MF (2002) IRS proteins and the common path to diabetes. Am J Physiol Endocrinol Metab 283:E413–E422

Yang R, Trevillyan JM (2008) c-Jun N-terminal kinase pathways in diabetes. Int J Biochem Cell Biol 40:2702–2706

Zabolotny JM, Kim YB, Welsh LA et al (2008) Protein-tyrosine phosphatase 1B expression is induced by inflammation in vivo. J Biol Chem 283:14230–14241

Zhang X, Zhang G, Zhang H et al (2008) Hypothalamic IKKbeta/NF-kappaB and ER stress link overnutrition to energy imbalance and obesity. Cell 135:61–73

Zhang X, Xu A, Chung SK et al (2011) Selective inactivation of c-Jun NH2-terminal kinase in adipose tissue protects against diet-induced obesity and improves insulin sensitivity in both liver and skeletal muscle in mice. Diabetes 60:486–495

Chapter 22
Adaptive Changes in Human Adipose Tissue During Weight Gain

Maud Alligier, Emmanuel Disse, Hubert Vidal and Martine Laville

Introduction

Weight gain refers to the gradual development of adipose tissue (AT), which may ultimately leads to obesity if too excessive and prolonged. The expansion of AT was long thought to be a passive phenomenon by which adipocytes store excess energy in the form of triglycerides. Currently, AT is regarded as a complex and heterogeneous endocrine organ whose role is not limited to storage. Recent studies of AT development during weight gain have revealed complex multiple mechanisms, both structural and functional, which accompany AT expansion (Strissel et al. 2007). These mechanisms correspond to the adaptation of AT to the nutritional stress created by positive energy balance.

However, the lipid storage function of adipocytes and the resulting AT expansion are vitally important processes, not only for controlling energy homeostasis, but also because any impairment of these processes could expose

M. Alligier · E. Disse · H. Vidal (✉) · M. Laville
Unité INSERM U1060 - INRA 1235 - INSA-Lyon - Université-Lyon1,
Laboratoire CarMeN et Centre Européen Nutrition Santé.
Centre de Recherche en Nutrition Humaine Rhône-Alpes, Lyon, France
e-mail: hubert.vidal@univ-lyon1.fr

M. Alligier
e-mail: maud.alligier@chu-lyon.fr

E. Disse
e-mail: emmanuel.disse@chu-lyon.fr

M. Laville
e-mail: martine.laville@univ-lyon1.fr

M. Alligier · E. Disse · H. Vidal · M. Laville
Faculté de Médecine Lyon Sud-BP 12, 165 Chemin du Grand Revoyet,
69921 Oullins Cedex, France

J.-P. Bastard and B. Fève (eds.), *Physiology and Physiopathology of Adipose Tissue*, DOI: 10.1007/978-2-8178-0343-2_22, © Springer-Verlag France 2013 other organs, such as the liver, muscles, pancreas, and heart to toxicity related to ectopic depots of lipids (i.e. lipotoxicity) (Lionetti et al. 2009; Tan and Vidal-Puig 2008; Virtue and Vidal-Puig 2010).

Overfeeding: A Model to Study Adipose Tissue Expandability

To identify the factors involved in AT development, along with their respective roles, scientists rapidly used experimental overfeeding in both animals and man. Thanks to this model, which perturbs energy balance, it was possible to study adaptations leading to excess AT development and its metabolic effects. This approach made it possible to study the early phases of excess AT formation that lead to obesity. Animal overfeeding experiments are often extreme. In contrast, because of methodological and ethical considerations, they are conducted only rarely in man and are usually of short duration and moderate intensity. It is important to note, however, that extrapolating the results of animal studies to man must be done carefully.

Most overfeeding studies in healthy volunteers have reported weight gain related to increased lean and fat body mass (Bouchard et al. 1990; Samocha-Bonet et al. 2010) with an accumulation of body fat in omental and subcutaneous AT (Bouchard et al. 1990; Tchoukalova et al. 2010). Some studies have also reported the induction of excess lipid deposition in the hepatic parenchyma (Samocha-Bonet et al. 2010) and metabolic reorientation toward storing triglycerides in muscle (Meugnier et al. 2007). Most human studies have also revealed an impairment of insulin sensitivity, as measured by hyperinsulinemic-euglycemic clamp or HOMA index (Samocha-Bonet et al. 2010; Forbes et al. 1989; Tam et al. 2010; Bisschop et al. 2001; Alligier et al. 2011). Overfeeding is also generally associated with an increase in circulating inflammatory markers (CRP, MCP1) and in leptin (Tam et al. 2010; Alligier et al. 2011).

Decreased plasma free fatty acid levels are often seen with overfeeding (Bisschop et al. 2001), probably due to lipolysis inhibition associated with reduced lipid oxidation, as measured by indirect calorimetry (Samocha-Bonet et al. 2010; Brons et al. 2009; Ngo Sock et al. 2010). As for the AT itself, several overfeeding studies, in both animals and man, have showed very early induction of the metabolic pathways involved in the lipid storage process (Alligier et al. 2011; Franck et al. 2011; Shea et al. 2009; Grant 2011). Among the most highly regulated genes are those encoding enzymes involved in the re-esterification pathways that transform fatty acids into triglycerides, particularly diacylglycerol acyltransferase (DGAT2), together with fatty acid conversion enzymes such as elongase and desaturase, which promote fatty acid storage. These studies have showed that the induction of lipid storage processes is a key early event in the response of subcutaneous AT to overfeeding (Alligier et al. 2011).

Adipose Tissue Expandability: Hyperplasia and Hypertrophy

AT grows by two distinct mechanisms: hypertrophy and hyperplasia. It is logical that in response to an excess uptake, lipids should be stored initially within existing adipocytes, thus increasing cell size: this is known as hypertrophy. When adipocytes can no longer expand, new cells are formed from precursors: this is known as hyperplasia. Hyperplasia during weight gain has been well documented in animals on high energy diets (Faust et al. 1978). However, in man, adipocyte hyperplasia is a matter of debate. According to Arner et al. (Spalding et al. 2008), the number of adipocytes is determined during childhood and adolescence and remains stable throughout adult life. Hence, any weight gain or obesity in adults is accompanied by an increase in the size of existing adipocytes, rather than by their proliferation (Spalding et al. 2008). This hypothesis is supported by the results of human overfeeding studies which generally reveal no hyperplasia, but rather hypertrophy (Salans et al. 1971). However, the moderate weight gains occurring during these experiments may be insufficient to trigger recruitment of new adipocytes. Of note however, a recent study by Jensen et al. (Tchoukalova et al. 2010) has demonstrated the mobilization of new cells in the subcutaneous AT of the lower part of the body in healthy volunteers overfed for 8 weeks.

Since the AT of obese subjects is composed essentially of enlarged adipocytes, it might be thought that there would be no, or only little, recruitment of new adipocytes during AT expansion, particularly in obese subjects with metabolic complications (Arner and Spalding 2010). The signal for triggering adipocyte recruitment has not yet been clearly identified, but enlarged adipocytes could secrete growth factors that promote pre-adipocyte proliferation and differentiation (Marques et al. 1998). It is known that the adipocytes of obese subjects are dysfunctional, particularly with regard to secretory functions (Skurk et al. 2007). Unlike small adipocytes, enlarged adipocytes secrete more pro-inflammatory molecules, such as IL6 and TNFα, which contribute to adipocyte dysfunction and the low-grade inflammation seen in obese subjects. A possible pathway for regulating new adipocyte recruitment in AT could be Wnt signaling. MacDougald's team was the first to demonstrate that in vitro inhibition of the Wnt canonical pathway stimulated adipocyte differentiation (Ross et al. 2000). During a recent human overfeeding trial, we found evidence of a modification in the expression of several genes involved in regulating the Wnt signaling pathway in AT, without an increase in the number of adipocytes (Alligier M et al. 2011). Currently, despite the availability of data from man (Alligier et al. 2011; Christodoulides et al. 2006a, b), the exact contribution of the Wnt signaling pathway in vivo during AT development remains to be clarified.

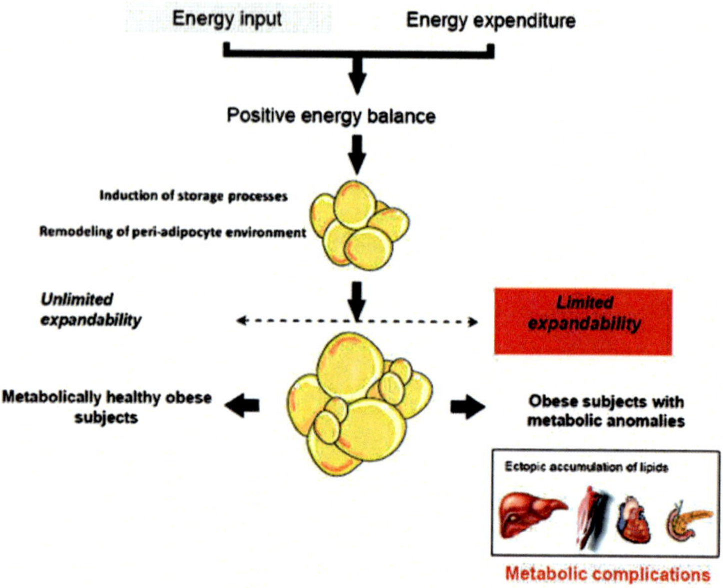

Fig. 22.1 Hypothesis of the limit of expandability of AT

Obesity: Critical Expandability of Adipose Tissue?

AT is the tissue with the greatest potential to change size and was long thought to have no limit to its expansion in a person with a chronic excess energy intake. It seems this is true for some people, and in that case, the development of AT, even when great, is not associated with classic metabolic impairment related to obesity, such as insulin resistance. These subjects are metabolically healthy and could account for 10–25 % of the obese population (Bluher 2010). For the remainder, there is a limit to tissue expandability contributing to defective lipid storage, which could favor ectopic fat depots and lipotoxicity (Lionetti et al. 2009; Tan and Vidal-Puig 2008; Virtue and Vidal-Puig 2010). This hypothesis, proposed by Vidal-Puig et al, remains to be tested in man (Fig. 22.1).

According to this hypothesis, under chronic positive energy balance, the adipocyte storage capacity reaches its limit and the tissue is unable to recruit new adipocytes to meet the excess of lipids. A limited expansion results in numerous enlarged adipocytes responsible for two factors contributing to the metabolic anomalies associated with weight gain:

- *Impairment of lipid storage capacity*: enlarged adipocytes store less lipid, resulting in release of lipids into the circulation (referred to in the literature as spill-over) (Fielding 2011). In this case, other organs, such as the liver, muscle or the pancreas, are submitted to prolonged lipid exposure that lead to ectopic depots.

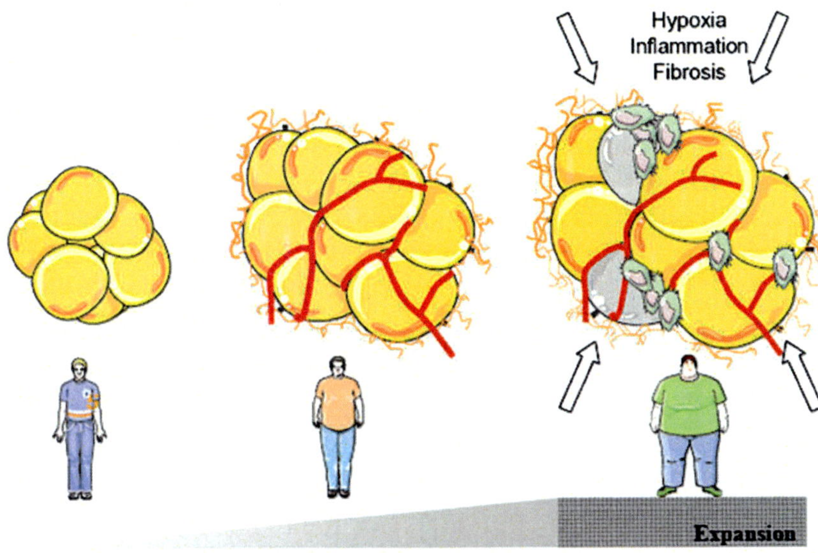

Fig. 22.2 Mechanisms associated with AT expansion limit (Adapted from ref (Sun et al. 2011))

- *Modified pro-inflammatory adipokine and/or cytokine secretion*: secretions from adipocytes and other AT cells such as immune cells could be modified in the event of adipocyte hypertrophy. The wide range of action of these molecules on both glucose and lipid metabolism could maintain or exacerbate adipocyte dysfunction and the associated metabolic anomalies.

The underlying causes of defective AT expansion are not yet known, but studies suggest that, under certain conditions, the adipocyte environment is not suited to harmonious tissue expansion. During weight gain, AT is remodeled (Alligier et al. 2011), a process that, in the face of any defect, could limit expansion. Remodeling involves angiogenesis, changes of the extracellular matrix and infiltration with inflammatory cells (Fig. 22.2) (Sun et al. 2011; Lee et al. 2010; Rutkowski et al. 2009).

Hypoxia and Angiogenesis

Enlarged adipocytes can reach impressive sizes (>100 μm) exceeding the maximum oxygen diffusion distance and creating areas of hypoxia. This hypoxia could explain the death of enlarged adipocytes and also limits lipid storage and stimulates the infiltration of immune cells.

AT development during weight gain requires a constant input of oxygen and nutrients. The tissue's vascular network must adapt to meet these needs. The relationship between the vascular network and AT development has been well documented in vivo in rodents (Cao 2007). By using angiogenesis inhibitors, researchers have demonstrated weight loss and reversibility of obesity-related disease (Rupnick et al. 2002; Brakenhielm et al. 2004). Adipogenesis and angiogenesis are closely related mechanisms with the ability to stimulate each other.

However, recent in vivo data from mice expressing the HIF1α transcription factor (Halberg et al. 2009) show that the hypoxia signal cannot induce the complete angiogenic sequence, even though the AT of these mice appears to develop normally. In this model, HIF1α appears to be associated with the induction of several genes associated with tissue remodeling and with the extracellular matrix, leading to fibrosis (Halberg et al. 2009). This suggests that hypoxia could be the cause of the fibrotic areas found in the AT of obese subjects (Divoux et al. 2011).

Extracellular Matrix

The extracellular matrix forms an environment for AT cells. It is not an inert milieu and many interactions take place there. It undergoes constant remodeling via complex systems of synthesis and degradation (Mariman and Wang 2010). These modifications could impair AT function.

Each adipocyte is surrounded by a thin layer of extracellular matrix, known as the basal lamina, which is composed principally of type IV collagen. It has been suggested that the basal lamina may play a role in cell survival (Mariman and Wang 2010). Adipocytes can grow very large, and the mechanical forces exerted by increasing lipid droplets could weaken the cell membrane.

It has been shown in several animal models, and in man, that weight gain is accompanied by remodeling of the extracellular matrix (Strissel et al. 2007; Alligier M et al. 2011; Pasarica et al. 2009). One study focussed more specifically on type VI collagen (Pasarica et al. 2009). It revealed the presence of type VI collagen in human AT and reported that this collagen type was correlated to the body mass index (BMI) and fat mass. Moreover, subjects with the highest expression of type VI collagen were those with the largest omental AT and highest inflammatory marker level, suggesting that the accumulation of type VI collagen in subcutaneous AT may have deleterious effect in man (Pasarica et al. 2009).

Until now, few studies have been conducted on the role of the extracellular matrix in human AT. Prof K. Clément has pioneered research on this topic (Divoux and Clement 2011), and her team has demonstrated fibrosis in subcutaneous and omental AT of obese subjects (Henegar et al. 2008). Histology has revealed different types of depots: large collagen fibers diffused throughout the tissue and others more localized: depots surrounding adipocytes (Divoux et al. 2011). A comparison of the AT transcriptome of lean and obese subjects revealed marked differences in the genes related to the extracellular matrix, whose expression was strongly correlated to BMI and to several inflammatory markers (Henegar et al. 2008). In this study, it was also reported that weight loss was accompanied by the modified expression of over 200 genes related to the extracellular matrix and its remodeling. There was also a negative correlation between the amount of subcutaneous AT fibrosis and weight loss after bariatric surgery (Divoux et al. 2011).

In responding to nutritional input changes, AT may modify the synthesis of extracellular matrix components that participate in fibrosis. Fibrosis may affect adipocyte function and lead to cell stress. Although the extracellular matrix enables adipocytes to maintain their integrity in the face of lipid droplet pressure, depots of excess matrix would cause rigidity in the peri-adipocyte environment. The extracellular matrix would then exert compression force on the adipocyte, resulting in mechanical stress. This stress could be responsible, at least in part, for the defective development of AT during positive energy balance. Several in vitro and in vivo models have validated these hypotheses. Subjecting 3T3-L1 pre-adipocyte cultures to mechanical stretching limited adipocyte differentiation and reduced the expression of PPARγ2, the nuclear receptor essential for differentiation (Tanabe et al. 2004). The matrix could impose a mechanical stress on pre-adipocytes, thus reducing the development and storage capacity of AT. In vivo, obese collagen VIa3-null mice present greatly enlarged adipocytes, which seem to grow with uninhibited expansion (Khan et al. 2009). Despite major hypertrophy of adipocytes, these animals present no impairment of lipid or carbohydrate metabolism. On the contrary, their sensitivity to insulin is increased in comparison to non-mutated obese mice. Adipocytes appeared to have become "hyper-active" with regard to storage, thus avoiding ectopic fat depots. According to the authors, collagen VIa3, when present, could induce a mechanical stress on adipocytes, stimulating signaling pathways specific to cell stress and contributing to inflammation and insulin resistance of adipocytes (Khan et al. 2009).

As well as type VI collagen, which has been studied because of its specificity of expression in AT, there are other proteins involved in the synthesis and degradation of extracellular matrix, which may be involved in AT biology. SPARC (secreted protein, acidic, and rich in cysteine) is a protein capable of regulating the synthesis of extracellular matrix components and SPARC-null mice are resistant to weight gain when submitted to a high-fat diet (Bradshaw et al. 2003). It has also been noted that obese subjects have higher plasma concentrations of MMP2 and MMP9 (matrix metalloproteinase) than lean subjects (Derosa et al. 2008). Other associations with matrix synthesis and degradation have been found, such as plasma concentrations of TIMP-1 (tissue inhibitor of metalloproteinases-1) which are positively correlated to human adiposity (Kralisch et al. 2007).

Reports in the literature suggest that remodeling of the peri-adipocyte environment during weight gain, which increases in obesity, might reduce the elasticity of AT, thus inhibiting expandability and reducing lipid storage capacity.

Infiltration by Inflammatory Cells

Another factor confirming the distressed state of enlarged adipocytes is the presence of macrophages arranged in characteristic crown-like structures around the adipocytes of obese subjects. These macrophages infiltrated into the AT of obese subjects seem to eliminate dead adipocytes (Weisberg et al. 2003). A study in a transgenic mice model of lipoatrophy confirmed that massive death of adipocytes was accompanied by rapid macrophage accumulation (Pajvani et al. 2005). However, once active in AT, macrophages stimulate, via their secretions, the recruitment of new inflammatory cells, resulting in their accumulation (Lumeng et al. 2007). The presence of large numbers of immune cells in the AT of obese subjects is associated with the main metabolic anomalies found in obesity, and with systemic inflammation (Cancello et al. 2006).

Interestingly, however, human overfeeding studies have not revealed macrophage infiltration of AT during moderate weight gain (Tam et al. 2010; Alligier et al. 2011), or any change in the numbers of other immune cell types. Since these studies concern the early phases of AT development, it would appear that infiltration by immune cells, particularly macrophages, is of relatively late-onset process during weight gain and probably not a causal or initiating factor of fat mass development (Alligier et al. 2011).

Conclusion

Episodes of positive energy balance are periods of nutritional stress which require the body to adapt. AT plays a major role in this adaption, since it manages energy reserves. The adaptive response of AT is primarily the storage of excess energy in the form of lipids, and it results in AT expansion. The tissue remodeling that accompanies this expansion is a highly coordinated and finely regulated process that aims to create a suitable environment for the development of adipocytes. However, during obesity, these mechanisms may reach their limit and, in the event of excessive input, AT expansion may be insufficient. This may result from a defective recruitment of new adipocytes and be associated with an accumulation of enlarged adipocytes producing pro-inflammatory cytokines that promote the infiltration of AT with immune cells. Another important factor seems to be the extracellular matrix and the architecture of AT which could play a decisive role in tissue plasticity and remodeling. It is necessary to understand the different phases of these adaptive processes and any impairment occurring during pathological situations if we are to propose new preventive and curative strategies for obesity in the near future.

References

Alligier M, Meugnier E, Debard C, Lambert-Porcheron S, Chanseaume E, Sothier M, Loizon E, Ait Hssain A, Brozek J, Scoazec J, Morio B, Vidal H, Laville M (2011) Subcutaneous adipose tissue remodeling during the initial phase of weight gain induced by overfeeding in humans. Article in press, J Clin Endocrinol Metab 66(6):765–774

Arner P, Spalding KL (2010) Fat cell turnover in humans. Biochem Biophys Res Commun 396:101–104

Bisschop PH, de Metz J, Ackermans MT, Endert E, Pijl H, Kuipers F, Meijer AJ, Sauerwein HP, Romijn JA (2001) Dietary fat content alters insulin-mediated glucose metabolism in healthy men. Am J Clin Nutr 73:554–559

Bluher M (2010) The distinction of metabolically 'healthy' from 'unhealthy' obese individuals. Curr Opin Lipidol 21:38–43

Bouchard C, Tremblay A, Despres JP, Nadeau A, Lupien PJ, Theriault G, Dussault J, Moorjani S, Pinault S, Fournier G (1990) The response to long-term overfeeding in identical twins. N Engl J Med 322:1477–1482

Bradshaw AD, Graves DC, Motamed K, Sage EH (2003) SPARC-null mice exhibit increased adiposity without significant differences in overall body weight. Proc Natl Acad Sci U S A 100:6045–6050

Brakenhielm E, Cao R, Gao B, Angelin B, Cannon B, Parini P, Cao Y (2004) Angiogenesis inhibitor, TNP-470, prevents diet-induced and genetic obesity in mice. Circ Res 94:1579–1588

Brons C, Jensen CB, Storgaard H, Hiscock NJ, White A, Appel JS, Jacobsen S, Nilsson E, Larsen CM, Astrup A, Quistorff B, Vaag A (2009) Impact of short-term high-fat feeding on glucose and insulin metabolism in young healthy men. J Physiol 587:2387–2397

Cancello R, Tordjman J, Poitou C, Guilhem G, Bouillot JL, Hugol D, Coussieu C, Basdevant A, Bar Hen A, Bedossa P, Guerre-Millo M, Clement K (2006) Increased infiltration of macrophages in omental adipose tissue is associated with marked hepatic lesions in morbid human obesity. Diabetes 55:1554–1561

Cao Y (2007) Angiogenesis modulates adipogenesis and obesity. J Clin Invest 117:2362–2368

Christodoulides C, Laudes M, Cawthorn WP, Schinner S, Soos M, O'Rahilly S, Sethi JK, Vidal-Puig A (2006a) The Wnt antagonist Dickkopf-1 and its receptors are coordinately regulated during early human adipogenesis. J Cell Sci 119:2613–2620

Christodoulides C, Scarda A, Granzotto M, Milan G, Dalla Nora E, Keogh J, De Pergola G, Stirling H, Pannacciulli N, Sethi JK, Federspil G, Vidal-Puig A, Farooqi IS, O'Rahilly S, Vettor R (2006b) WNT10B mutations in human obesity. Diabetologia 49:678–684

Derosa G, Ferrari I, D'Angelo A, Tinelli C, Salvadeo SA, Ciccarelli L, Piccinni MN, Gravina A, Ramondetti F, Maffioli P, Cicero AF (2008) Matrix metalloproteinase-2 and -9 levels in obese patients. Endothelium 15:219–224

Divoux A, Clement K (2011) Architecture and the extracellular matrix: the still unappreciated components of the adipose tissue. Obes Rev 12:e494–e503

Divoux A, Tordjman J, Lacasa D, Veyrie N, Hugol D, Aissat A, Basdevant A, Guerre-Millo M, Poitou C, Zucker JD, Bedossa P, Clement K (2011) Fibrosis in human adipose tissue: composition, distribution, and link with lipid metabolism and fat mass loss. Diabetes 59:2817–2825

Faust IM, Johnson PR, Stern JS, Hirsch J (1978) Diet-induced adipocyte number increase in adult rats: a new model of obesity. Am J Physiol 235:E279–E286

Fielding B (2011) Tracing the fate of dietary fatty acids: metabolic studies of postprandial lipaemia in human subjects. Proc Nutr Soc 70:342–350

Forbes GB, Brown MR, Welle SL, Underwood LE (1989) Hormonal response to overfeeding. Am J Clin Nutr 49:608–611

Franck N, Gummesson A, Jernas M, Glad C, Svensson PA, Guillot G, Rudemo M, Nystrom FH, Carlsson LM, Olsson B (2011) Identification of adipocyte genes regulated by caloric intake. J Clin Endocrinol Metab 96:E413–E418

Grant RW (2011) Vester Boler BM, Ridge TK, Graves TK, Swanson KS: Adipose tissue transcriptome changes during obesity development in female dogs. Physiol Genomics 43:295–307

Halberg N, Khan T, Trujillo ME, Wernstedt-Asterholm I, Attie AD, Sherwani S, Wang ZV, Landskroner-Eiger S, Dineen S, Magalang UJ, Brekken RA, Scherer PE (2009) Hypoxia-inducible factor 1alpha induces fibrosis and insulin resistance in white adipose tissue. Mol Cell Biol 29:4467–4483

Henegar C, Tordjman J, Achard V, Lacasa D, Cremer I, Guerre-Millo M, Poitou C, Basdevant A, Stich V, Viguerie N, Langin D, Bedossa P (2008) Zucker JD, Clement K: Adipose tissue transcriptomic signature highlights the pathological relevance of extracellular matrix in human obesity. Genome Biol 9:R14

Khan T, Muise ES, Iyengar P, Wang ZV, Chandalia M, Abate N, Zhang BB, Bonaldo P, Chua S, Scherer PE (2009) Metabolic dysregulation and adipose tissue fibrosis: role of collagen VI. Mol Cell Biol 29:1575–1591

Kralisch S, Bluher M, Tonjes A, Lossner U, Paschke R, Stumvoll M, Fasshauer M (2007) Tissue inhibitor of metalloproteinase-1 predicts adiposity in humans. Eur J Endocrinol 156:257–261

Lee MJ, Wu Y, Fried SK (2010) Adipose tissue remodeling in pathophysiology of obesity. Curr Opin Clin Nutr Metab Care 13:371–376

Lionetti L, Mollica MP, Lombardi A, Cavaliere G, Gifuni G, Barletta A (2009) From chronic overnutrition to insulin resistance: the role of fat-storing capacity and inflammation. Nutr Metab Cardiovasc Dis 19:146–152

Lumeng CN, Deyoung SM, Bodzin JL, Saltiel AR (2007) Increased inflammatory properties of adipose tissue macrophages recruited during diet-induced obesity. Diabetes 56:16–23

Mariman EC, Wang P (2010) Adipocyte extracellular matrix composition, dynamics and role in obesity. Cell Mol Life Sci 67:1277–1292

Marques BG, Hausman DB, Martin RJ (1998) Association of fat cell size and paracrine growth factors in development of hyperplastic obesity. Am J Physiol 275:R1898–R1908

Meugnier E, Bossu C, Oliel M, Jeanne S, Michaut A, Sothier M, Brozek J, Rome S, Laville M, Vidal H (2007) Changes in gene expression in skeletal muscle in response to fat overfeeding in lean men. Obesity (Silver Spring) 15:2583–2594

Ngo Sock ET, Le KA, Ith M, Kreis R, Boesch C, Tappy L (2010) Effects of a short-term overfeeding with fructose or glucose in healthy young males. Br J Nutr 103:939–943

Pajvani UB, Trujillo ME, Combs TP, Iyengar P, Jelicks L, Roth KA, Kitsis RN, Scherer PE (2005) Fat apoptosis through targeted activation of caspase 8: a new mouse model of inducible and reversible lipoatrophy. Nat Med 11:797–803

Pasarica M, Gowronska-Kozak B, Burk D, Remedios I, Hymel D, Gimble J, Ravussin E, Bray GA, Smith SR (2009) Adipose tissue collagen VI in obesity. J Clin Endocrinol Metab 94:5155–5162

Ross SE, Hemati N, Longo KA, Bennett CN, Lucas PC, Erickson RL, MacDougald OA (2000) Inhibition of adipogenesis by Wnt signaling. Science 289:950–953

Rupnick MA, Panigrahy D, Zhang CY, Dallabrida SM, Lowell BB, Langer R, Folkman MJ (2002) Adipose tissue mass can be regulated through the vasculature. Proc Natl Acad Sci U S A 99:10730–10735

Rutkowski JM, Davis KE, Scherer PE (2009) Mechanisms of obesity and related pathologies: the macro- and microcirculation of adipose tissue. FEBS J 276:5738–5746

Salans LB, Horton ES, Sims EA (1971) Experimental obesity in man: cellular character of the adipose tissue. J Clin Invest 50:1005–1011

Samocha-Bonet D, Campbell LV, Viardot A, Freund J, Tam CS, Greenfield JR, Heilbronn LK (2010) A family history of type 2 diabetes increases risk factors associated with overfeeding. Diabetologia 53:1700–1708

Shea J, French CR, Bishop J, Martin G, Roebothan B, Pace D, Fitzpatrick D, Sun G (2009) Changes in the transcriptome of abdominal subcutaneous adipose tissue in response to short-term overfeeding in lean and obese men. Am J Clin Nutr 89:407–415

Skurk T, Alberti-Huber C, Herder C, Hauner H (2007) Relationship between adipocyte size and adipokine expression and secretion. J Clin Endocrinol Metab 92:1023–1033

Spalding KL, Arner E, Westermark PO, Bernard S, Buchholz BA, Bergmann O, Blomqvist L, Hoffstedt J, Naslund E, Britton T, Concha H, Hassan M, Ryden M, Frisen J, Arner P (2008) Dynamics of fat cell turnover in humans. Nature 453:783–787

Strissel KJ, Stancheva Z, Miyoshi H, Perfield JW 2nd, DeFuria J, Jick Z, Greenberg AS, Obin MS (2007) Adipocyte death, adipose tissue remodeling, and obesity complications. Diabetes 56:2910–2918

Sun K, Kusminski CM, Scherer PE (2011) Adipose tissue remodeling and obesity. J Clin Invest 121:2094–2101

Tam CS, Viardot A, Clement K, Tordjman J, Tonks K, Greenfield JR, Campbell LV, Samocha-Bonet D, Heilbronn LK (2010) Short-term overfeeding may induce peripheral insulin resistance without altering subcutaneous adipose tissue macrophages in humans. Diabetes 59:2164–2170

Tan CY, Vidal-Puig A (2008) Adipose tissue expandability: the metabolic problems of obesity may arise from the inability to become more obese. Biochem Soc Trans 36:935–940

Tanabe Y, Koga M, Saito M, Matsunaga Y, Nakayama K (2004) Inhibition of adipocyte differentiation by mechanical stretching through ERK-mediated downregulation of PPAR-gamma2. J Cell Sci 117:3605–3614

Tchoukalova YD, Votruba SB, Tchkonia T, Giorgadze N, Kirkland JL, Jensen MD (2010) Regional differences in cellular mechanisms of adipose tissue gain with overfeeding. Proc Natl Acad Sci U S A 107:18226–18231

Virtue S, Vidal-Puig A (2010) Adipose tissue expandability, lipotoxicity and the Metabolic Syndrome–an allostatic perspective. Biochim Biophys Acta 1801:338–349

Weisberg SP, McCann D, Desai M, Rosenbaum M, Leibel RL, Ferrante AW Jr (2003) Obesity is associated with macrophage accumulation in adipose tissue. J Clin Invest 112:1796–1808

Chapter 23
Differences Between Subcutaneous and Visceral Adipose Tissues

Max Lafontan

Introduction

Humans have wide variability in the distribution of body fat, which has important implications for metabolic health. Obese individuals with an upper body fat distribution have increased health complications. Obesity plays a causative role in the pathogenesis of a cluster of several abnormalities including insulin resistance, type 2 diabetes, dyslipidemia, hypertension, and cardiovascular disease, which lead to increased morbidity and mortality risk. However, all obese subjects are not equal in terms of the metabolic and cardiovascular risks in that striking differences exist between individuals with upper body fat and those with lower body fat distribution (Votruba and Jensen 2007). A number of epidemiologic reports have focused attention on the major role of body fat distribution in the appearance of metabolic abnormalities. As opposed to the extent of subcutaneous adipose tissue (scAT), the increase in visceral adipose tissue (vAT) (i.e., visceral obesity), which is easily measured by the expansion of waist circumference, is associated with increased metabolic disturbances and cardiovascular diseases (CVD). These epidemiological observations have raised a number of questions that will be discussed in this chapter. Why do obese individuals with upper body fat distribution have more health complications compared with obese individuals with lower body fat distribution? Why does the accumulation of vAT exert stronger deleterious effects than scAT accumulation? Is vAT expansion a causal factor or only a marker of an altered metabolic status? Are visceral adipocytes and the other

M. Lafontan (✉)
INSERM 1048, Team 1, I2 MC, Batiment L4, CHU Rangueil, 1 Avenue Jean poulhès, BP 84225 31432 Toulouse cedex 4, France
e-mail: max.lafontan@inserm.fr

J.-P. Bastard and B. Fève (eds.), *Physiology and Physiopathology of Adipose Tissue*,
DOI: 10.1007/978-2-8178-0343-2_23,

cells of the stroma-vascular fraction (SVF) of adipose tissue involved in the initiation of the metabolic disturbances observed in visceral obesity? Is the role of visceral fat correctly evaluated or overestimated due to its modest extent compared with total fat mass and scAT expansion in obese patients? Some results obtained with animal models that have positively contributed to the improved understanding of the results of clinical investigations will be briefly evoked.

Heterogeneity of Adipose Tissue Distribution: The Major Contribution of Imaging Techniques

The measurement of the thickness of fat layers and of body fat distribution rapidly improved in the mid 1970s due to computed tomography (CT), followed soon thereafter by magnetic resonance imaging (MRI). The complete reconstruction in vivo of the anatomy of all the major body compartments and tissues became possible, thus providing major new research opportunities for the evaluation of fat mass and various fat depots (Heymsfield 2008). scAT is composed of two distinct anatomical layers where deep and superficial layers are separated by the *fascia superficialis* (Scarpa's facia). Subcutaneous fat depots represent 80 % of the total fat mass in normal weight subjects. Intra-peritoneal adipose tissue, defined as vAT, is mainly composed of the omental and mesenteric fat depots. vAT only represents 10–20 % of the total fat mass in lean and obese subjects respectively and 5–10 % in women. For an equivalent body mass index (BMI) or equivalent fat mass, vAT accumulation is higher in men than women (after correction for the differences in total adiposity). An individual with a normal BMI but with an increased vAT is at higher risk of developing metabolic disturbances than an obese person with less vAT. The extent of vAT is associated with the prevalence of CVD. New imaging methods have brought to attention neglected fat depots such as intra-thoracic, pericardiac, perivascular, intramuscular, and fat lying around lymph nodes (i.e., perinodal adipose tissue). This tissue may not only provide a specific lipid resource but also fatty acids, dendritic cells, and soluble mediators that modulate local immunity (Knight 2008). Attention has been particularly focused on the expansion of pericardial adipose tissues. These fat depots include epicardiac fat (localized more or less deeply in pericardium) and pericardial fat (around the pericardium surface). Only pericardial fat was associated with prevalent myocardial infarction after adjusting for conventional measures of adiposity. Nevertheless, the correlations between vAT expansion and CVD risk are stronger than those for pericardial fat.

Morphological Characteristics, Adipogenic and Angiogenic Differences Between Fat Depots

Both sex and site differences in regional fat storage have been described. Gender-related and depot-specific differences exist in the expansion of the adipose tissue mass. The increase in fat cell size (hypertrophy of adipocytes), the replication of fat cell precursors, and the processes of preadipocytes differentiation (fat cell hyperplasia) are regulated in a tissue-specific manner (Votruba and Jensen 2007). Morphologically, visceral adipocytes are smaller than subcutaneous adipocytes in normal weight individuals, both in men and women, but these depot-specific differences decrease in obese women in parallel with an increase in BMI. Omental, but not subcutaneous, adipocyte hypertrophy is associated with an altered lipid profile (Veilleux et al. 2011). Recent studies, on gene expression profiling and miRNA identification have revealed that the expression profile of miRNAs (i.e., known to modulate stem cells and adipocyte differentiation) in vAT and scAT, suggest a common development in both depots although 16 different miRNAs exhibited depot-specific differences. In human adipose tissue, differences in the expression profile of developmental genes (among them: Shox2, En1, Tbx15, Hoxa5, Hoxc8, and Hoxc9) have been reported between vAT and scAT (Gesta et al. 2006; Yamamoto et al. 2010). In addition to mature adipocytes, the visceral and subcutaneous fat stroma vascular fraction (SVF) contains a heterogeneous population of cells (i.e., macrophages, lymphocytes, preadipocytes, microvascular endothelial cells (EC)...etc.), the activity of which could interfere with development and metabolic functions inside the fat depots. Macrophage density is higher in vAT than in scAT (Curat et al. 2004; Cancello et al. 2006), particularly in obese patients. Factors involved in the control of adipose tissue cellularity have been explored in both fat depots. Differences have been reported in the dynamics of progenitor expansion and in the level of commitment/filling of differentiating preadipocytes. In vitro studies have revealed that replication capacities of fat cell precursors are lower in vAT than in scAT. A difference was also found in the proliferation of fat cell precursors in the adipose tissue of obese individuals of both sexes. In the case of differentiation, the proportion of early differentiated adipocytes in the subcutaneous adipose tissue SVF of women was greater than in men, especially in the femoral depot. Nevertheless, in vitro adipogenesis, as assessed by the expression of peroxisome proliferator-activated receptor gamma (PPARγ), was not increased in femoral (lower body) preadipocytes cultured from women compared with men. However, baseline PPARγ2 and C/EBPα mRNA were higher in abdominal (upper body) than femoral subcutaneous preadipocytes, consistent with the ability of abdominal subcutaneous adipocytes to attain a larger size. Preadipocytes undergoing differentiation are more numerous in the scAT of women than men, and the differentiation of femoral adipocytes (i.e., lower body fat) is reduced compared with that of abdominal adipocytes (i.e., upper body fat). Moreover, femoral preadipocytes are more resistant to the pro-apoptotic effect of TNF-α than those in abdominal scAT. The turnover and utilization of the preadipocyte pool may be reduced in lower versus the upper body fat in women (Tchoukalova et al. 2010). The apparent number of preadipocytes in the abdominal scAT that can undergo differentiation is reduced in obesity with enlarged fat cells.

Several hormones are involved in the control of progenitor replication. The gender-specific differences in body fat distribution suggest that sex steroids play an important role in regulating body fat distribution. Estradiol enhances the proliferation of omental and subcutaneous fat cell precursors in both sexes, but the effect is greater in the adipose tissue of women. Both estrone and dehydrotestosterone (i.e., an androgen) had no significant gender- or site-specific effect on the rate of proliferation. In conclusion, estradiol may act as an important local factor influencing the proliferation of preadipocytes that may affect the fat cell number in a depot- and gender-specific manner in human abdominal scAT and omental adipose tissue (Anderson et al. 2001). With age, the replication of fat cell precursors declines in subcutaneous but not in omental depots while the differentiation of preadipocytes does not differ in either depot. These observations could explain the preferential expansion of vAT observed with aging, particularly if the proliferative and repletion capacities of scAT are saturated.

Is it possible to explain such depot-specific differences? The proportion of various cell types in the SVF is variable according to the anatomical location of the adipose tissue. These cells secrete cytokines and growth factors that will interfere with the proliferation, differentiation, and metabolism of fat cell precursors. Thus, important limitations exist in using in vitro results to interpret biological events occurring in the various fat depots in vivo. During in vitro studies, survival of the various cell types of the SVF largely depends upon cell culture conditions and co-cultures are difficult. A recent study has shown that fat cell progenitors are included in a cell subpopulation of the SVF identified by specific plasma membrane markers (i.e., CD34 +/CD31−). These cells differ from adult mesenchymal cells and from hematopoietic stem cells (Sengenes et al. 2005). These adipose-derived stromal cells were found to be scattered in adipose tissue stroma, and although some of them were lying along vessels, they did not express markers of pericytes. Their spontaneous commitment toward an adipocyte lineage was enhanced in adipose tissue from obese individuals (Maumus et al. 2011).

Close relationships exist between adipocytes and the vascular bed. Adipose tissue has long been known for its pro-angiogenic potencies and has been used by surgeons for plastic surgery. Recent studies have clearly established that adipogenesis depends on the formation of new vessels in the adipose tissue. The coupling of adipogenesis and angiogenesis is essential for the terminal differentiation of adipocytes. Living tissue imaging techniques have provided evidence for the dynamic interactions existing between differentiating adipocytes, stromal cells, and angiogenesis (Nishimura et al. 2007). Mature adipocyte secretions, such as vascular endothelial growth factor (VEGF-A or VEGF), leptin, and apelin, are considered to be the major factors for the control of angiogenesis inside adipose tissue. Investigations on obese rodent models have shown that chronic anti-angiogenic treatments reduced fat mass development. In humans, according to a recent study focused on the role of angiogenesis in the expandability of human adipose tissue depots, scAT was shown to possess a greater angiogenic capacity than vAT, even after normalization for its higher initial capillary density. The capillary density of scAT and its angiogenic capacity decreased with morbid obesity. The angiogenic capacity of scAT, but not vAT, correlated negatively with insulin sensitivity. It was concluded that impairment of angiogenesis in scAT may contribute to the pathogenesis of metabolic diseases (Gealekman et al. 2011). A study defining the properties of EC, isolated from scAT and vAT biopsied in parallel from obese subjects, has produced different conclusions. vAT EC were markedly more angiogenic and pro-inflammatory than scAT cells. This phenotype may be related to premature EC senescence. vAT EC may contribute to hypoxia and inflammation in vAT (Villaret et al. 2010). This research field is just beginning in human adipose tissue biology; it is a neglected area in the exploration of fat mass development according to the anatomical location and one, which requires deeper investigation.

Triglyceride Synthesis: Differences Between Lipoprotein Lipase and Glucose and Fatty Acid Uptake Between Visceral and Subcutaneous Adipose Tissue

At sexual maturity, both sex and site differences in regional fat storage have been described. Females have more body fat, a greater proportion of fat in their lower body, and much less visceral fat than do lean males of the same BMI. Regional differences in triglyceride storage capacity and lipolysis could be the causes of regional fat accumulation. Triglyceride accumulation could be related to various regulatory steps such as uptake of triglyceride precursors (i.e., glucose and fatty acids), efficiency of triglyceride synthesis pathways, and the magnitude of lipolytic responses to hormonal factors.

Triglyceride synthesis in adipocytes plays a major role in the turnover and maintenance of the size of fat depots. Triglyceride synthesis involves glucose uptake on one hand, and on the other hand, uptake of nonesterified fatty acids (NEFAs), after their release from circulating triglycerides by adipose tissue lipoprotein lipase (LPL) located in the capillary lumen. In women, LPL activity is lower in vAT when compared with various scAT depots (i.e., abdominal, gluteal, and femoral); these differences decrease after the menopause. Site-specific differences in LPL activity are less evident in men although LPL activity is positively correlated with fat cell size in both sexes. Translocation/activation of LPL from the adipocyte to the capillary lumen seems important in the induction of differential fat storage. Recent studies have revealed striking differences in NEFA uptake after a meal according to gender and fat depots. Greater thigh adipose tissue in women is associated with greater efficiency of fat storage after a meal under conditions of energy balance, whereas the opposite is seen with vAT (Votruba et al. 2007). Insulin and glucocorticoids are major regulators of biosynthetic processes in the adipocyte. Insulin stimulates LPL activity, glucose, and NEFA uptake and triglyceride synthesis; its effects on LPL activity are greater in scAT than in vAT. Higher LPL activities were found in the lower body subcutaneous sites with enlarged fat cells in women (Fried and Kral 1987). Fat cell enlargement is associated with insulin resistance in nondiabetic individuals independently of BMI. Glucocorticoids increase LPL expression in vAT but are without noticeable effect in scAT.

In the case of glucose uptake by adipocytes, the basal uptake level and insulin-stimulated glucose transport are higher in omental than in subcutaneous adipocytes. Round ligament adipose cells showed a significantly greater responsiveness to insulin when compared with subcutaneous and omental adipocytes. Insulin-stimulated glucose uptake is doubled in omental adipocytes compared with subcutaneous adipocytes removed from the same patient. The expression level of mRNA of the glucose transporter GLUT-4 is higher in mature adipocytes of vAT than in scAT (i.e., four times higher than in scAT). Similar differences between vAT versus scAT were described for the expression of proteins involved in insulin signaling. Insulin receptor phosphorylation was more intense and rapid and the insulin receptor protein content was higher in omental than in subcutaneous adipose tissue. Insulin-induced phosphorylation of the Akt enzyme also occurred to a greater extent and earlier in omental than in subcutaneous fat (Laviola et al. 2006). In vitro results have been supported by in vivo measurement of [18F]-2-fluoro-2-deoxy-D-glucose uptake using positron emission tomography (PET). Insulin-stimulated glucose uptake per kilogram of fat is higher in vAT than in scAT (Virtanen et al. 2002). The increased uptake of glucose observed in omental adipocytes could facilitate triglyceride synthesis and counteract the increased lipolytic responsiveness observed in these cells (Fig. 23.1). Based on the increased lipolytic responsiveness (discussed later) and the effects of insulin on glucose uptake, triglyceride turnover is higher in vAT than in scAT.

In addition to glucose utilization, NEFA uptake and the activity of Acyl-CoA synthases also contribute to the triglyceride synthesis in adipocytes. Older studies have revealed that the uptake of labeled triolein is higher in vAT than in scAT in both lean and obese individuals. Several plasma membrane proteins (i.e., plasma membrane fatty acid binding protein (FABPm), fatty acid translocase (FAT/CD36), and fatty acid transporters (FATP-1 and FATP-4) facilitate NEFA uptake. Depot-specific differences in the expression of fatty acid binding proteins have been reported. The decrease in CD36 protein content parallels the increase in fat cell size in the omentum; tissue CD36 content was not correlated with CD36 mRNA (Allred et al. 2011). The expression of fatty acid binding protein FABP-4, also known as aP2, is higher in subcutaneous adipocytes than in the omental fat cells. Contradictory results regarding lipogenic enzymes in scAT and vAT have been reported. Recent results have shown that gene expression of the main lipogenic enzymes (i.e., fatty acid synthase and acetyl-CoA carboxylase) was down regulated in the vAT of obese subjects. In another study, expression of these genes was higher in scAT than in vAT from obese subjects. A reduction in the genes

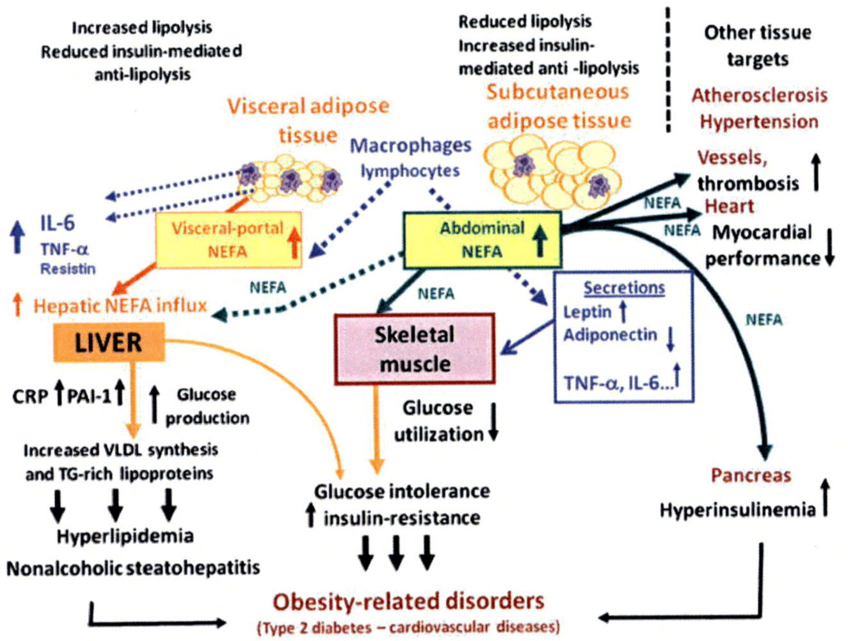

Fig. 23.1. Working hypothesis for the mechanism(s) related to adipose tissue distribution and the appearance of obesity-related diseases. Major agents involved in the metabolic control of the adipocyte and adipocyte secretions are depicted with their putative effects on diseases. Visceral deposits represent 5-20% of body fat mass. In the obese state, macrophages infiltrate visceral and subcutaneous adipose tissues and contribute, in addition to adipocytes, to cytokine secretions. Visceral adipocytes are characterized by an increased response to catecholamines and reduced insulin responsiveness (see Table 23.1). Non-esterified fatty acids (NEFAs) and cytokines [interleukin-6 (IL-6), tumor necrosis factor-α (TNF-α) and resistin] are delivered to liver by visceral fat but upper-body subcutaneous fat deposit also contributes to disturbed NEFA availability. An increase in the levels of NEFAs causes a decrease of glucose utilization in skeletal muscle, leading to glucose intolerance and insulin resistance which can result in type 2 diabetes-related disorders. An increase of the levels of NEFAs can cause similar consequences by increasing the production of glucose and the synthesis of very-low-density-lipoproteins (VLDLs) in the liver. This process could be at the origin of non alcoholic fatty liver disease (NAFLD). The action of NEFAs on insulin release by pancreas and on other target tissues is mentioned but not detailed. Subcutaneous fat deposits represent a major site of fat storage (see mechanisms in Table 23.1). Due to their widespread distribution, subcutaneous fat deposits also play a major role in the production of various adipokines (e.g., IL-6, TNF-α, resistin, leptin and adiponectin) also produced by visceral fat. Adipokines affect NEFA and glucose utilization by skeletal muscle and glucose and VLDL production by liver as well as insulin secretion by pancreas. Abbreviations: ↑ increase, ↓ decrease, *PAI-1* plasminogen activator inhibitor-1, *CRP* C-Reactive Protein

involved in de novo synthesis of fatty acids and an upregulation of genes facilitating triglyceride/fatty acid cycling has been reported in morbid obesity.

Acylation stimulating protein (ASP), also known as C3adesArg, is a circulating protein obtained from adipsin synthesized in the adipocyte. The protein stimulates triglyceride synthesis and glucose transport in preadipocytes/adipocytes through a G-protein-coupled receptor, C5L2, via stimulation of diacylglycerol-acyltransferase. Effects on preadipocyte differentiation have also been reported (Maslowska et al. 2005). ASP is secreted after a meal and plays a role in the post-prandial clearance of triglycerides. In rodents, ASP deficiency protects against the development of obesity. Altered binding of ASP to its receptor could reduce the efficiency of fatty acid trapping. The binding capacity and affinity of ASP for its receptor is lower in vAT than in scAT. It was suggested that ASP might be an important factor in maintaining regional adipose tissue mass (Saleh et al. 1999). The physiological role of this protein remains a subject of debate.

Adipogenesis and the expression of various enzymes involved in the control of adipocyte metabolism are under the coordinated control of transcription factors such as strerol regulatory element binding protein-1c/adipocyte differentiation and determination factor-1 (SREBP-1c/ADD-I), peroxisome proliferator-activated receptors (PPARα,β/γ and PPARγ1/γ2), and CCAT-enhancer binding protein-α and β(C/EBPα and β). Comparisons of the tissue-specific expression levels of these proteins are limited in humans. SREBP-1c mRNA levels are lower in vAT compared with abdominal and femoral scAT and there is no striking gender difference. According to some studies, SREBP-1c mRNA levels are lower in obese individuals and increase after weight loss. Contradictory results have been published concerning SREBP-1c and PPARγ2 mRNA changes after weight loss. A genetic predisposition for type 2 diabetes is associated with an impaired ability to recruit new adipose cells to store excess lipid in scAT. Several studies have shown that PPARγ-regulated genes are commonly reduced in the scAT of patients with a genetic predisposition for type 2 diabetes. This research area requires deeper investigations into both fat depots in physiological and pathological conditions.

Lipolytic Activity of Adipocytes and Mobilization of Triglycerides Stored in Visceral and Subcutaneous Fat Depots

In addition to lipogenic pathways, differences among lipolytic responses to catecholamines, natriuretic peptides, and insulin between visceral and subcutaneous adipocytes have been reported (Lafontan and Langin 2009). By controlling the differential recruitment of the various fat depots, these hormones exert a potent action on NEFA turnover. Growth hormone, interleukin-6 (IL-6) and tumor necrosis factor-α (TNF-α) also stimulate lipolysis; the effects are delayed and involve complex mechanisms. TNF-α stimulates triglyceride hydrolysis via multiple intracellular pathways acting on insulin signaling, on G proteins and on perilipin, the protein located at the surface of lipid droplets and involved in the regulation of lipolysis (Lafontan and Langin 2009; Langin and Arner 2006).

Lipolytic activity in vitro and in vivo depends on the balanced functional efficiency between lipolytic and antilipolytic regulators and the level of expression of adipocyte lipases (i.e., adipocyte triglyceride lipase, ATGL; hormone sensitive lipase, HSL, and monoacylglycerol lipase, MGL) and their regulators (i.e., perilipin and ABHD5 (for *α/β-hydrolase domain-containing protein 5*), a regulator of ATGL) (Chap. 10). Variations in perilipin levels according to adipocyte size may contribute to differences in basal lipolysis between fat depots and may regulate lipid accumulation in adipocytes. Expression of the perilipin gene is up-regulated in vAT versus scAT in isolated adipocytes and in the whole tissue. Reduced perilipin expression was observed in vAT and scAT in obese women. With regard to the responsiveness of adipocytes, basal lipolytic rates (e.g., spontaneous lipolysis) are lower in vAT than in scAT adipocytes in normal weight or obese individuals. The increased basal lipolytic rate, observed in scAT adipocytes could be related to the expression level of the fatty acid binding protein FABP-4, which controls the efflux of NEFAs released by triglyceride lipolysis. Expression levels of ATGL and its activity are reduced in the scAT of obese patients. Both mRNA and the protein levels of ATGL and HSL are also reduced in the scAT of insulin-resistant subjects compared with insulin-sensitive ones (Jocken et al. 2007).

Effect of Catecholamines

The sympathetic nervous system (SNS) and epinephrine secreted by the adrenal medulla play a major role in the control of lipolysis and lipid mobilization in humans. Catecholamines act both on the control of lipolysis in adipocytes and on the local blood flow in the vascular bed of adipose tissue. They also control the release of the major antilipolytic hormone, insulin. The mechanisms of action of catecholamines are well-known and have been recently reviewed (Lafontan and Langin 2009). In normal weight individuals, catecholamines exert the most potent lipolytic effects in visceral adipocytes when compared with scAT adipocytes (i.e., abdominal, gluteal and femoral); this difference persists in adipocytes differentiated in vitro. However, although various in vitro studies have shown that beta-adrenergic receptor agonists exert similar effects in scAT and vAT adipocytes, a recent study has shown that the response of omental adipocytes to lipolytic stimuli was greater than in subcutaneous adipocytes (Tchernof et al. 2006). Detailed study of intracellular signaling pathways has not revealed major differences in post-receptor events. In vitro results on the presence of putative functional beta3-adrenergic receptors in human fat cells have not been followed by convincing in vivo results. In fact, discrepancies in lipolytic responses could be explained by differences in the expression of plasma membrane beta- (mainly beta1 and beta2) and alpha2A-adrenergic receptors in the adipocytes of various fat depots (Table 23.1). In normal weight obese women, gluteal and femoral adipocytes are less responsive to the lipolytic action of catecholamines. They are characterized by a potent alpha2-adrenergic receptor-mediated antilipolytic responsiveness (and a higher density of alpha2A-adrenergic receptors) linked to a reduced density of beta-adrenergic receptors (Mauriège et al. 1987). Subcutaneous abdominal adipocytes of obese men have functional characteristics very similar to those of femoral and gluteal adipocytes from women. On the contrary, visceral adipocytes (particularly the omental ones) have reduced alpha2A-adrenergic responsiveness with a low density of alpha2A-adrenergic receptors, which explains the greater lipolytic efficiency of catecholamines in vAT adipocytes.

Independently of gender and age, the functional equilibrium between beta1/2- and alpha2A-adrenergic responsiveness correlates with fat mass expansion and adipocyte hypertrophy. When compared to smaller adipocytes of vAT, hypertrophic adipocytes (abdominal, gluteal, and femoral), are characterized by a reduced lipolytic responsiveness to catecholamines, a high density of alpha2A-adrenergic receptors and a low density of beta1/2-adrenergic receptors. The increased expression of alpha2A-adrenergic receptors, concomitant with the reduced expression of beta-adrenergic receptors, could represent a physiological adaptation of the fat cell to its hypertrophy. Since the subcutaneous hypertrophic adipocyte is a weaker responder to catecholamines than the smaller visceral adipocyte, the reduction of lipolysis could represent an adaptation to limit excessive NEFA release from hypertrophied scAT adipocytes and ease the deleterious effects of NEFA overflow on health. By their limited lipolytic responsiveness, scAT adipocytes exert an important role for NEFA sequestration under the form of triglycerides in the adipocytes. These results underline the role of scAT and particularly, gluteofemoral fat, as a determinant of health by the long-term entrapment of excess fatty acids, thus protecting from the adverse effects associated with ectopic fat deposition (Manolopoulos et al. 2010). With aging and depending on the extent of obesity, the functional balance between adrenergic receptor distribution and activity could be associated to other dysfunctions affecting downstream lipolytic pathways such as lipase expression, perilipins, and subunits of protein-kinase-A (Mantovani et al. 2009).

Physiological studies have confirmed most of the results obtained from the isolated fat cell. In situ, microdialysis studies have shown that exercise-induced lipolysis is reduced in the scAT of obese subjects and that the stimulation of alpha2A-adrenergic receptors of scAT adipocytes is responsible for this lipolytic defect (Stich et al. 2000). The extent of NEFA release into the portal vein by visceral lipolysis correlates with the extent of vAT. The contribution of splanchnic lipolysis to hepatic NEFA delivery ranged from less than 10 % to almost 50 % and increased as a function of visceral fat in men and women. Noticeable differences have been reported concerning the relative amount of NEFAs provided by visceral and subcutaneous fat depots (Nielsen et al. 2004). Fatty acid release from vAT triglycerides is substantial in some obese subjects and could be an important factor in developing hepatic insulin resistance. Nevertheless, excessive fatty acid release from visceral fat is unlikely to be a major factor in the pathogenesis of insulin resistance in skeletal muscle, because it represents a very small percentage of total NEFAs delivered to muscle tissues (Klein 2004). In the face of the complexity of experiments, additional investigations must be carried out to reach convincing conclusions. vAT adipocytes, which are characterized by increased lipolytic responsiveness and reduced antilipolytic effects of insulin (see later) are probably

Table 23.1. Differences in lipolytic, anti-lipolytic and other biochemical pathways between visceral and subcutaneous adipocyte

Effects and factors	Regional differences
Lipolytic pathway	
Hormone sensitive lipase protein level and activity	scAT > vAT
Basal rate of lipolysis (spontaneous glycerol release)	scAT > vAT
Lipolytic response to catecholamines	vAT > scAT
Isoproterenol-stimulated adenylyl cyclase activity	vAT > scAT
β-adrenergic receptor-mediated lipolysis	vAT > scAT
$\beta_{1,2}$-adrenergic receptor number	vAT > scAT
α_2-adrenergic receptor-mediated anti-lipolysis	scAT > vAT
α_2-adrenergic receptor number	scAT > vAT
Fatty acid binding protein-4 (FABP-4)	scAT > vAT
Anti-lipolytic pathway	
Anti-lipolytic action of insulin	scAT > vAT
Insulin receptor substrate (IRS-1) protein and expression	scAT > vAT
Insulin-induced insulin receptor tyrosine phosphorylation	scAT > vAT
Insulin-induced IRS-1 tyrosine phosphorylation	scAT > vAT
Insulin-induced PI3-kinase pathways	scAT > vAT
Protein tyrosine phosphatase 1B activity	vAT > scAT
Insulin receptor (exon 1 deleted)	vAT > scAT
Lipogenesis, glucose, and fatty acid uptake	
Lipoprotein lipase (LPL)	scAT > vAT
Fatty acid uptake by human adipocytes and preadipocytes	scAT > vAT
Acylation stimulating protein mRNA	scAT > vAT
High-density lipoprotein D fractions binding	scAT > vAT
Glucose transporter (GLUT-4) protein and expression	vAT > scAT
Miscellaneous factors, secretions and hormone receptors	
Mean fat cell size in normal weight individuals	scAT > vAT
Leptin mRNA and protein secretion	scAT > vAT
Plasminogen activator inhibitor-1 (PAI-1)	scAT > vAT
Angiotensinogen and angiotensin 2 expression	vAT > scAT
Adiponectin (Acrp30, AdipoQ)	vAT > scAT
interleukin-6 (IL-6) and interleukin-8 (IL-8) secretion	vAT > scAT
Retinol binding protein-4 (RBP-4)	scAT > vAT
Inhibitor of apoptosis 2 (cIAP2) mRNA	vAT > scAT
Monocyte chemoattractant protein 1 (MCP-1)	vAT > scAT
Insulin growth factor-1 (IGF-1) and IGF-binding protein-3	vAT > scAT
Tumor necrosis factor-α (TNF-α)	vAT > scAT
11β-hydroxysteroid dehydrogenase type 1 activity (11β-HSD-1)	vAT > scAT
Endothelial nitric oxide synthase mRNA and protein	vAT > scAT
Glucocorticoid receptor mRNA	vAT > scAT
Peroxisome proliferator-activated receptor- γ mRNA	vAT > scAT
Estrogen receptor ER-β	scAT > vAT
Androgen receptor	vAT > scAT
Thyroid hormone receptor-α1 (TR-α1)	scAT > vAT
Thiazolidinedione-induced preadipocyte differentiation	scAT > vAT

Some of these differences can be increased or reduced according to the extent of the fat mass and adipocyte hypertrophy. Moreover, some sex-related differences have not been considered. Expression profiling studies of micro RNAs (miRNA) and of developmental genes have revealed disparities between visceral and subcutaneous adipose-tissues which are not detailed here

scAT subcutaneous adipose tissue, *vAT* visceral adipose tissue the adipocytes that will be responsible for the portal NEFA flux to the liver and liver dysfunctions. Studies in dogs support the role of vAT in the development of hepatic insulin resistance observed with visceral obesity (Kabir et al. 2005; Bergman et al. 2006). By using PET imaging to measure fatty acid metabolism in the liver, it was found that obese individuals have increased fatty acid flux from visceral fat and increased hepatic oxidation of fatty acids with an increased production of reactive oxygen species (ROS), in the context of adipose tissue insulin resistance. Fatty acid flux from visceral fat is proportional to the mass of the corresponding depot (Iozzo et al. 2010).

In the case of scAT depots, population studies have shown that an increased gluteofemoral fat mass is independently associated with a protective lipid and glucose profile, as well as a decrease in cardiovascular and metabolic risk (Manolopoulos et al. 2010). The distinct properties of femoral fat characterized by a reduced lipolytic activity and an increased fatty acid uptake suggest that it is more inert than abdominal scAT or vAT. Its protective properties seem related to its ability to control long-term fatty acid storage, and thus limit increases in plasma NEFA levels and avoid adverse effects associated with ectopic fat deposition. Deficient NEFA storage in scAT and vAT will initiate ectopic storage of lipids in liver, skeletal muscle, and even pancreas, followed by the appearance of insulin resistance and type 2 diabetes (Ravussin and Smith 2002; Unger 2003).

Antilipolytic Effects of Insulin

Insulin plays a major role in the control of NEFA disposal via inhibition of lipolysis and NEFA efflux from adipocytes but also by its ability to control the stimulation of triglyceride synthesis (i.e., re-esterification process) and fat storage. Insulin also controls LPL activity and glucose and NEFA uptake. The effects of insulin on lipolysis and NEFA re-esterification are lower in omental than in abdominal subcutaneous fat cells. In vitro studies have shown that various steps of insulin transduction pathways are negatively affected (Zierath et al. 1998). Reduced expression of genes involved in insulin signaling has been reported in omental fat cells from insulin resistant individuals (Table 23.1). Conversely, an in vivo study in nonobese individuals receiving insulin treatment has revealed an increased expression of proteins involved in insulin signaling with a rapid and increased insulin effect in vAT adipocytes. The adipocyte insulin signaling system of omental fat shows greater and earlier responses to insulin than that of subcutaneous fat (Laviola et al. 2006). Although these results fit with previously reported insulin effects on glucose transport in vAT adipocytes they contradict the results obtained when studying lipolysis (i.e., reduced antilipolytic effect). In the absence of further details on the expression level of proteins and enzymes involved in the lipolytic pathways (i.e., adrenergic receptors, protein kinase-A, phosphodiesterase-3B, perilipin, and lipases), it is impossible to interpret such discrepancies. The results of in vitro studies on lipolysis have been confirmed by in vivo investigations.

The rate of appearance of systemic, leg and splanchnic palmitate (i.e., [9,10-3H]palmitate), was measured in healthy adults using the hyperinsulinemic-euglycemic clamp technique to achieve a physiological range of plasma insulin concentrations. vAT lipolysis was shown to be more resistant to insulin suppression than in scAT (i.e., leg lipolysis) in humans (Meek et al. 1999).

Production of Hormones, Cytokines, and Proinflammatory Molecules in Visceral and Subcutaneous Adipose Tissue

In addition to its essential function in the storage of lipids and in the control of the bioavailability of NEFAs according to the "buffering role" of the adipocyte, adipose tissue has obtained the status of an endocrine organ. The adipose tissue undergoes a continuous process of remodeling that could be enhanced in obesity. The composition and the stiffness of the extracellular matrix (ECM) has been shown to regulate adipogenesis and adipose tissue expansion. Obesity has been linked to increased mRNA and protein expression of adhesion molecules such as intercellular adhesion molecule-1 (ICAM-1 or CD54) and vascular cell adhesion molecule-1 (VCAM-1 or CD106) in vAT but not in scAT. Expression of ICAM-1 and VCAM-1 positively correlated with the expression of CD68 (i.e., microsialin, a macrophage marker) in both adipose depots. Protein expression was correlated positively with BMI. Some molecules produced by adipose tissue (i.e., adipokines) have their origin in the adipocyte, while others are secreted by some cells of the stroma-vascular fraction. LPL was the first identified adipocyte-secreted enzyme. A summary of the molecules produced by adipose tissue and of their major biological effects is given in Table 23.2. Various cell types of the SVF are the object of intense investigations. Differential gene expression profiling methods and proteomic approaches aimed at describing the secretome are currently being used on rodent and human adipose tissues and SVF fractions. The respective contribution of the different cell types of adipose tissue to adipokine secretions needs to be evaluated. Macrophages infiltrating the adipose tissue of the obese probably impact, via their secretions, on preadipocyte/adipocyte functions and on other cells of the SVF. Macrophages contribute to TNF-α secretion and to the production of various cytokines and interleukins (IL-6, IL-8, IL-1β, resistin, and visfatin) and other agents identified in the vAT secretome. The macrophage phenotype is modified by the extent of the fat mass: macrophages that accumulate with fat mass development exhibit a particular M2 remodeling phenotype in humans (Bourlier et al. 2008). The expression of genes related to inflammation is altered and parallels immune cell infiltration. Recruitment of macrophages is not always associated with inflammation: macrophages in adipose tissue in lean mice express IL-10, which has a protective effect against insulin resistance.

Adipocytes secrete a variety of lipid metabolites and bioactive peptides that play important roles in cell growth, inflammation, energy homeostasis, and insulin

Table 23.2. Adipose tissue productions

Lipid and lipoprotein metabolism
Lipoprotein lipase (LPL)
Acylation stimulating protein (ASP/C3desArg))
Prostaglandin E2 (PGE_2), prostaglandins $F_{2\alpha}$ ($PGF_{2\alpha}$), prostacyclin (PGI_2)
Autotaxin (lysophospholipase D) and production of lysophosphatidic acid (LPA)
Cholesterol ester transport protein (CETP)
Retinol binding protein-4 (RBP-4)
Metabolism and energy homeostasis
Leptin
Adiponectin
Resistin
Interleukin-6 and -8 (IL-6 and IL-8)
Retinol binding protein-4 (RBP-4)
Insulin sensitivity of muscle, hepatocyte and adipocyte
Adipsin/acylation stimulating protein (ASP)
Leptin
Adiponectin
Resistin
Visfatin
Omentin-1
Vaspin
Interleukin-6 (IL-6)
Apelin
Growth factors influencing adipose tissue development
Insulin growth factor-1 (IGF-1)
Nerve growth factor (NGF)
Vascular endothelial growth factor (VEGF)
Thrombopoietin
Food intake and activation of sympathetic nervous system
Leptin
Metabolism of extracellular matrix
Type 6 collagen
Plasminogen activator inhibitor-1 (PAI-1)
Metalloproteases (gelatinases MMP-2 and MMP-9)
Tissue inhibitors of metalloproteases (TIMP -1 to TIMP-3)
Immune system, acute phase reactants, and inflammation
Tumor necrosis factor-α (TNF-α)
Interleukins-1β, -6, -8, -10 (IL-1β,IL-6,IL-8, IL-10)
Interleukin-1 receptor antagonist (IL-1Ra), macrophage inflammatory protein-1β (MIP-1β)
Regulated on activation, normal T cell expressed, and secreted (RANTES)
Adipsin, factors C3, B, and D of alternate complement system
Monocyte chemoattractant protein-1 (MCP-1)
α1-acid glycoprotein
Serum amylo A 3 (SAA-3)
Haptoglobin, pentraxin-3, lipocalin 24p3
Metallothionein

(continued)

Table 23.2. (continued)

Cathepsin S and L
Vessels and angiogenesis
Vascular endothelial growth factor (VEGF), thrombopoietin
Monobutyrin
Leptin, apelin
Fasting-induced adipose factor (FIAF) or peroxisome proliferator-activated receptor γ angiopoietin-related gene) (PGAR)
Angiopoietin-2, -angiotensinogen/angiotensin-2, adrenomedullin

This list of productions/secretions originating from adipocyte and/or various cells of the stroma-vascular fraction is non exhaustive. The factors are grouped according to their contribution in the control of major functions. Some productions possess pleiotropic actions and are found in different groups.

Diversity of fat cell secretions has been considered several times. Overviews can be found in various recent review papers (Halberg et al. 2008; Galic et al. 2010; Poulos et al. 2010)

action. Among the bioactive peptides secreted by adipocytes, some of them have hormonal status and act on distal organs, while a number of other secreted factors are probably limited to autocrine and paracrine action inside the adipose tissue mass. They contribute to the development and remodeling of adipose tissue (i.e., remodeling of ECM elements, control of fat cell precursors proliferation/differentiation, modulation of the functional activity of microvascular EC, and angiogenesis). Among the most commonly studied factors, leptin, adiponectin (also identified as *Acrp30* encoded by the *apM1* gene), and apelin possess well-established hormonal activities and a number of pleiotropic effects. In addition to hormones, various factors of the alternate complement system (adipsin, factors C3, B, and D, C3adesArg, known as ASP, which results from a conversion of complement factor C3) and acute phase reactants (serum amyloid A3, haptoglobin, α1-acid glycoprotein, pentraxin-3, lipocalin 24p, cathepsin S) are secreted by adipocytes. Biologically, active pro-coagulant molecules, such as plasminogen activator inhibitor-1 (PAI-1), factor VII and the tissue factor (TF), involved in the TF pathway, are also produced in direct proportion to adiposity. PAI-1, an important factor in the control of fibrinolysis via plasmin production is not exclusively produced by adipocytes. Various elements of the renin-angiotensin system have been identified in adipose tissue. Angiotensinogen and enzymes of angiotensin production are secreted by the adipocyte. A number of other molecules are not secreted strictly by adipocytes. Among them, it has been possible to identify a number of pro-inflammatory cytokines (TNF-α, IL-1β, IL-6, IL-8, IL-18, and interferon γ), chemokines such as monocyte chemoattractant protein-1 (MCP-1), macrophage inflammatory protein-1α (MIP-1α), and various matrix metalloproteinases (MMP-2, MMP-9, MMP-11…etc.) and the tissue inhibitors of matrix metalloproteinases (TIMPs); all of them are supposed to mainly act as paracrine regulators. A strong, positive relationship was found among TNF-α secretion and BMI, total body fat, and adipocyte volume (a measure of morphologic characteristics of adipose tissue) in scAT while BMI, fat mass, and adipocyte volume were not related to the rates of secretion of interleukin-6, leptin, or adiponectin. TNF-α secretion was increased in patients with adipose hypertrophy (in which adipose tissue consists of a few large adipocytes) and decreased in adipose hyperplasia (in which adipose tissue consists of many small adipocytes) (Arner et al. 2010). TNF-α, but not IL-6 or resistin, completely inhibited the normal differentiation of the preadipocytes and induced a proinflammatory and macrophage-like phenotype of the cells. The list of adipocyte and SVF cell productions is increasing regularly. The nonexhaustive list of recently identified adipokines includes omentin, visfatin, chemerin, vaspin, and dipeptidyl peptidase 4, although their physiological roles remain to be fully delineated. The respective productions of the different cell types identified in SVF and the cross-talk between the various cell types remains an open field of investigation. A recent study to characterize the rodent secretome in vAT and scAT opens new perspectives. vAT had a higher secretory capacity than scAT. This difference seems to be an intrinsic feature of its cellular components (Hocking et al. 2010). These results open experimental perspectives for future investigations on the human adipose tissue secretome; it is an area of research that will probably expand rapidly in the near future. The major results on secreted factors and hormonal receptors identified in vAT and scAT comparative studies are summarized in Table 23.2. Some recent reviews summarize the main aspects of adipose tissue secretions but they will not be detailed here (Halberg et al. 2008; Galic et al. 2010). The development and endocrine functions of adipose tissue have been recently reviewed (Poulos et al. 2010).

Hypothesis to Explain the Impact of vAT Expansion on Endocrine and Metabolic Deterioration: The Debated Question of Visceral Obesity

Epidemiological studies have clearly revealed that a subgroup of individuals with central visceral obesity is particularly prone to developing cardiovascular disease, stroke, and type 2 diabetes. Visceral obesity is associated with multiple central endocrine aberrations. The oldest hypothesis, "the portal theory", was initially proposed by Per Björntorp. Several other hypotheses have been postulated and pursued to account for the deleterious effect of vAT accumulation on glucose utilization and insulin resistance. According to the first hypothesis, because of its anatomical location, vAT partly drains into the liver via the portal vein. Portal NEFAs, coming from the highly lipolytic visceral depots may affect hepatic metabolism to increase hepatic gluconeogenesis and the production of very low density lipoproteins, as well as inhibiting insulin clearance and promoting insulin resistance. There is data from both humans and rodents that argue against this "portal hypothesis" since in humans some studies have shown a stronger association between abdominal scAT and insulin resistance. However, recent results support the original hypothesis. Using [11C]-palmitate imaging by PET with compartmental modeling to measure fatty acid metabolism in adipose tissue and the liver, an increase in the fatty acid flux from vAT toward the liver has been described. Fatty acid flux from vAT is proportional to its mass. Obese individuals had increased hepatic oxidation of fatty acids with increased production of ROS, which induce insulin resistance (Iozzo et al. 2010). According to the portal theory, the liver plays an important role in the initiation of insulin resistance commonly observed in visceral obesity. Investigations using dogs submitted to a high fat diet have clarified some mechanisms (Kabir et al. 2005). It was shown that NEFAs coming from vAT altered liver functions, promoted hepatic insulin resistance and increased glucose production. This hepatic insulin resistance was also linked to a reduction in apolipoprotein B degradation, increased production of triglyceride-rich lipoproteins and accumulation of liver lipids (Bergman et al. 2006). Nevertheless, the situation could be more complex since vAT is not the unique source of systemic NEFAs. The contribution of NEFAs emanating from other nonvisceral depots and abdominal scAT must be considered in the initiation of insulin resistance in humans (Klein 2004; Jensen 2006). A recent original study in rodents, designed to mimic visceral fat accumulation by transplanting fat pads from a donor mouse to the mesenterium of a recipient littermate, has shown that these fat pads were drained by the portal vein only. The delivery of cytokines to liver was clearly established, and hence this study supports the portal theory and also proposes a role for IL-6 as previously (Rytka et al. 2011) shown in humans (Fontana et al. 2007).

The discovery of the secretory activity of adipocytes and the other cells of the SVF has promoted a reconsideration of the NEFA-centered "portal hypothesis". In fact, various bioactive peptides, such as TNF-α, IL-6, IL-1β, IL-8, leptin, resistin, and adiponectin coming from vAT, are supposed to be delivered to the portal vein. These proinflammatory factors are known to exert multiple biological effects as discussed previously and in other chapters of this book (see Chaps. 3, 12, 13, 14, 15, 20, and 21). A recent study of these factors in the portal vein of morbidly obese patients has shown that vAT was a privileged site of IL-6 production. It was suggested that visceral fat is an important site for IL-6 secretion and provides a potential mechanistic link between visceral fat and hepatic and systemic inflammation in people with abdominal obesity (Fontana et al. 2007). Obesity is now considered as a low-level inflammatory state leading to a chronic activation of the immune system. Obese individuals are known to have elevated plasma levels of inflammation markers. The evolution of inflammatory processes will be considered in Chap. 20.

Issues and Trends

It is unquestionable that vAT expansion precedes the appearance of various metabolic and endocrine disorders, such as glucose intolerance, insulin resistance, dyslipidemia, and hypertension, which will progress toward associated pathologies such as type 2 diabetes and CVD. The importance of visceral obesity is recognized by endocrine, diabetes, and cardiology scientific societies. Theoretically, from a therapeutic point of view, some practices could limit the health-related risks related to increased vAT. For example, a patient with visceral obesity performing regular physical activity could noticeably reduce vAT with a subsequent improvement in metabolic parameters (Despres et al. 2008). There are still some debated questions about the relative importance of vAT and scAT depots in the initiation of metabolic (i.e., via NEFAs) or humoral changes (i.e., via adipokine production by adipocytes and immune cells infiltrated in the SVF of adipose tissues). A number of components of the human adipose tissue secretome remain unknown or their actions are poorly understood. The secretory activity of adipocytes and of the other cells of the SVF will probably be characterized in the near future and it might be anticipated that the complexity of the paracrine interactions between the various cell types will open new perspectives.

For the time being, although some questions persist based on the results of epidemiological studies, an expanded "portal hypothesis" (i.e., including NEFAs and all the adipokines secreted by vAT in the concept) cannot be totally refuted. The role of other intra-abdominal nonvisceral depots remains poorly studied in humans; it is suspected that perivascular and pericardial fat depots could contribute. Consideration of various scAT depots must be continued to delineate the determinants of insulin resistance in skeletal muscle, which cannot be explained by vAT-related problems (Klein 2004; Jensen 2006). One innovative aspect of adipose tissue biology is the discovery of the infiltration of the adipose tissue of the obese by a number of immune cells such as macrophages and various lymphocyte subtypes. The mechanisms explaining this phenomenon remain debated although various hypotheses have been proposed. The incidence of this infiltration on the loss of insulin sensitivity remains to be fully validated.

Finally, it is clear that adipogenesis and angiogenesis are highly implicated in the regulation of the expansion of fat mass. Although some progress exists, the cellular and physiological determinants of vAT and scAT depots expansion remain to be evaluated more thoroughly, and in humans the task is not easy. The existence of putative tissue-specific regulatory mechanisms related to the anatomical site must be explored and understood. Moreover, when considering physiological relevance, the role of the vascular bed and the putative impact of SNS innervation on these events still remains a black box. The understanding of the function of the microvascular bed of adipose tissue, which plays a major role in the exchanges between adipocytes and the vascular lumen, is quite limited. Moreover, the biology of the microvascular EC of capillary and lymphatic beds must be clarified in the near future. They contribute to the functions of the vessels and to the partitioning of adipocyte secretions between capillary and lymphatic transport. Selected animal species and transgenic animal models are of interest for the investigation of cellular mechanisms and can offer useful opportunities to solve some problems. However, the results cannot be fully extrapolated to human-related problems in the absence of true anatomical correspondence in the location of some fat depots.

References

Allred CC, Krennmayr T, Koutsari C et al (2011) A novel elisa for measuring cd36 protein in human adipose tissue. J Lipid Res 52:408–415

Anderson LA, McTernan PG, Barnett AH et al (2001) The effects of androgens and estrogens on preadipocyte proliferation in human adipose tissue: influence of gender and site. J Clin Endocrinol Metab 86:5045–5051

Arner E, Ryden M, Arner P (2010) Tumor necrosis factor alpha and regulation of adipose tissue. N Engl J Med 362:1151–1153

Bergman RN, Kim SP, Catalano KJ et al (2006) Why visceral fat is bad: mechanisms of the metabolic syndrome. Obesity (Silver Spring) 14(Suppl 1):16S–19S

Bourlier V, Zakaroff-Girard A, Miranville A et al (2008) Remodeling phenotype of human subcutaneous adipose tissue macrophages. Circulation 117:806–815

Cancello R, Tordjman J, Poitou C et al (2006) Increased infiltration of macrophages in omental adipose tissue is associated with marked hepatic lesions in morbid human obesity. Diabetes 55:1554–1561

Curat CA, Miranville A, Sengenes C et al (2004) From blood monocytes to adipose-tissue-resident macrophages: Induction of diapedesis by human mature adipocytes. Diabetes 53:1285–1292

Despres JP, Lemieux I, Bergeron J et al (2008) Abdominal obesity and the metabolic syndrome: contribution to global cardiometabolic risk. Arterioscler Thromb Vasc Biol 28:1039–1049

Fontana L, Eagon JC, Trujillo ME et al (2007) Visceral fat adipokine secretion is associated with systemic inflammation in obese humans. Diabetes 56:1010–1013

Fried SK, Kral JG (1987) Sex differences in regional distribution of fat cell size and lipoprotein lipase activity in morbidly obese patients. Int J Obes 11:129–140

Galic S, Oakhill JS, Steinberg GR (2010) Adipose tissue as an endocrine organ. Mol Cell Endocrinol 316:129–139

Gealekman O, Guseva N, Hartigan C et al (2011) Depot-specific differences and insufficient subcutaneous adipose tissue angiogenesis in human obesity. Circulation 123:186–194

Gesta S, Bluher M, Yamamoto Y et al (2006) Evidence for a role of developmental genes in the origin of obesity and body fat distribution. Proc Natl Acad Sci U S A 103:6676–6681

Halberg N, Wernstedt-Asterholm I, Scherer PE (2008) The adipocyte as an endocrine cell. Endocrinol Metab Clin North Am 37:753–768

Heymsfield SB (2008) Development of imaging methods to assess adiposity and metabolism. Int J Obes (Lond) 32(Suppl 7):S76–S82

Hocking SL, Wu LE, Guilhaus M et al (2010) Intrinsic depot-specific differences in the secretome of adipose tissue, preadipocytes, and adipose tissue-derived microvascular endothelial cells. Diabetes 59:3008–3016

Iozzo P, Bucci M, Roivainen A et al (2010) Fatty acid metabolism in the liver, measured by positron emission tomography, is increased in obese individuals. Gastroenterology 139(846–856):e841–e846

Jensen MD (2006) Is visceral fat involved in the pathogenesis of the metabolic syndrome? Human model. Obesity (Silver Spring) 14(Suppl 1):20S–24S

Jocken JW, Langin D, Smit E et al (2007) Adipose triglyceride lipase and hormone-sensitive lipase protein expression is decreased in the obese insulin-resistant state. J Clin Endocrinol Metab 92:2292–2299

Kabir M, Catalano KJ, Ananthnarayan S et al (2005) Molecular evidence supporting the portal theory: A causative link between visceral adiposity and hepatic insulin resistance. Am J Physiol Endocrinol Metab 288:E454–E461

Klein S (2004) The case of visceral fat: argument for the defense. J Clin Invest 113:1530–1532

Knight SC (2008) Specialized perinodal fat fuels and fashions immunity. Immunity 28:135–138

Lafontan M, Langin D (2009) Lipolysis and lipid mobilization in human adipose tissue. Prog Lipid Res 48:275–297

Langin D, Arner P (2006) Importance of TNFalpha and neutral lipases in human adipose tissue lipolysis. Trends Endocrinol Metab 17:314–320
Laviola L, Perrini S, Cignarelli A et al (2006) Insulin signaling in human visceral and subcutaneous adipose tissue in vivo. Diabetes 55:952–961
Manolopoulos KN, Karpe F, Frayn KN (2010) Gluteofemoral body fat as a determinant of metabolic health. Int J Obes (Lond) 34:949–959
Mantovani G, Bondioni S, Alberti L et al (2009) Protein kinase a regulatory subunits in human adipose tissue: decreased r2b expression and activity in adipocytes from obese subjects. Diabetes 58:620–626
Maslowska M, Wang HW, Cianflone K (2005) Novel roles for acylation stimulating protein/c3adesarg: a review of recent in vitro and in vivo evidence. Vitam Horm 70:309–332
Maumus M, Peyrafitte JA, D'Angelo R et al (2011) Native human adipose stromal cells: localization, morphology and phenotype. Int J Obes 35:1141–1153
Mauriège P, Galitzky J, Berlan M et al (1987) Heterogeneous distribution of beta- and alpha2-adrenoceptor binding sites in human fat cells from various fat deposits: Functional consequences. Eur J Clin Invest 17:156–165
Meek SE, Nair KS, Jensen MD (1999) Insulin regulation of regional free fatty acid metabolism. Diabetes 48:10–14
Nielsen S, Guo Z, Johnson CM et al (2004) Splanchnic lipolysis in human obesity. J Clin Invest 113:1582–1588
Nishimura S, Manabe I, Nagasaki M et al (2007) Adipogenesis in obesity requires close interplay between differentiating adipocytes, stromal cells, and blood vessels. Diabetes 56:1517–1526
Poulos SP, Hausman DB, Hausman GJ (2010) The development and endocrine functions of adipose tissue. Mol Cell Endocrinol 323:20–34
Ravussin E, Smith SR (2002) Increased fat intake, impaired fat oxidation, and failure of fat cell proloferation results in ectopic fat storage, insulin resistance and type 2 diabetes mellitus. Ann NY Acad Sci 967:363–378
Rytka JM, Wueest S, Schoenle EJ et al (2011) The portal theory supported by venous drainage-selective fat transplantation. Diabetes 60:56–63
Saleh J, Christou N, Cianflone K (1999) Regional specificity of asp binding in human adipose tissue. Am J Physiol 276:E815–E821
Sengenes C, Lolmède K, Zakaroff-Girard A et al (2005) Preadipocytes in the human subcutaneous adipose tissue display distinct features from the adult mesenchymal and hematopoietic stem cells. J Cell Physiol 205:114–122
Stich V, deGlisezinski I, Crampes F et al (2000) Activation of alpha2-adrenergic receptors impairs exercise-induced lipolysis in scAT of obese subjects. Am J Physiol 279:R499–R504
Tchernof A, Belanger C, Morisset AS et al (2006) Regional differences in adipose tissue metabolism in women: minor effect of obesity and body fat distribution. Diabetes 55:1353–1360
Tchoukalova YD, Koutsari C, Votruba SB et al (2010) Sex- and depot-dependent differences in adipogenesis in normal-weight humans. Obesity 18:1875–1880
Unger R (2003) Lipid overload and overflow: metabolic trauma and the metabolic syndrome. Trends Endocrinol Metab 14:398–403
Veilleux A, Caron-Jobin M, Noel S et al (2011) Visceral adipocyte hypertrophy is associated with dyslipidemia independent of body composition and fat distribution in women. Diabetes 60:1504–1511
Villaret A, Galitzky J, Decaunes P et al (2010) Adipose tissue endothelial cells from obese human subjects: Differences among depots in angiogenic, metabolic, and inflammatory gene expression and cellular senescence. Diabetes 59:2755–2763
Virtanen KA, Lonnroth P, Parkkola R et al (2002) Glucose uptake and perfusion in subcutaneous and visceral adipose tissue during insulin stimulation in nonobese and obese humans. J Clin Endocrinol Metab 87:3902–3910
Votruba SB, Jensen MD (2007) Regional fat deposition as a factor in ffa metabolism. Annu Rev Nutr 27:149–163

Votruba SB, Mattison RS, Dumesic DA et al (2007) Meal fatty acid uptake in visceral fat in women. Diabetes 56:2589–2597

Yamamoto Y, Gesta S, Lee KY et al (2010) Adipose depots possess unique developmental gene signatures. Obesity (Silver Spring) 18:872–878

Zierath J, Livingston J, Thorne A et al (1998) Regional difference in insulin inhibition of non-esterified fatty acid release from human adipocytes: relation to insulin receptor phosphorylation and intracellular signalling through the insulin receptor substrate-1 pathway. Diabetologia 41:1343–1354

Chapter 24
Genetics of the Human Obesities

Frédéric Fumeron

Introduction

Before looking at the DNA, it has long been admitted that obesity ran in families. Familial correlations studies have shown a high degree of resemblance concerning body mass index (BMI) and a high degree of heritability of obesity. However, as once stated by Claude Bouchard, one of the best specialists in the genetics of the energy balance, in obese families, even cats and dogs are obese! Actually, the shared environment in families is an important issue while studying the genetics of obesity. Most of the time, obesity is the result of complex interactions between genetics and environment, as most of the diseases important in terms of Public Health, the multifactorial and polygenic diseases. Nevertheless, some rare cases of obesities are due to defects in one gene. Although only a very small percentage of subjects carry these defects, their study has provided a large part of our knowledge concerning the physiology of the regulation of energy balance and feeding behavior. The genes of which major defects cause monogenic obesities can be involved in the susceptibility to common multifactorial obesities as well through common genetic variations (polymorphisms). For many years, genetic variants of candidate genes involved in different pathways (food intake, energy expenditure, adipocyte biology, lipid metabolism, etc.) as well as markers of different regions of the genome, have been tested for the susceptibility to obesity in association studies in populations of unrelated subjects or by linkage analyses in families. In the most recent studies allowed by the progress in genetic techniques, hundreds of thousands genetic markers located in the whole genome have been tested in large populations of

F. Fumeron (✉)
INSERM U695 Déterminants génétiques du diabète de type 2 et
de ses complications vasculaires UFR de Médecine Paris Diderot—Paris 7,
site Bichat, 16, rue Henri Huchard, BP 416 75870 Paris Cedex 18, France
e-mail: frederic.fumeron@inserm.fr

J.-P. Bastard and B. Fève (eds.), *Physiology and Physiopathology of Adipose Tissue*,
DOI: 10.1007/978-2-8178-0343-2_24, © Springer-Verlag France 2013

unrelated subjects with various degrees of corpulence. Such studies evidenced a few already known candidate genes but also novel genes of which role has yet to be defined. In all these studies, the relative risks conferred by the individual variants are low, with a very poor predictive value. Therefore, the combinations of alleles at risk (genetic scores) as well as the gene–environment interactions must be considered. The recent epidemic of obesity is an argument in favor of the interactions with environment since in a few decades, the genetic structure of the population cannot have been changed, contrary to the lifestyle factors.

Heritability of Obesity

The odds ratio of being obese when one first degree relative is obese is 3 when compared to the general population (after elimination of identical twins from the statistics) (Allison et al. 1996). It has sometimes been claimed that the risk was higher when the mother is obese than when the father is obese. This might be the consequence of the prenatal or postnatal nutritional environment effect, and/or sex dependent genetic effects. Studies of twins and nuclear families have repeatedly indicated that heritability was high for BMI, with estimates between 40 and 90 %. One big issue is to separate familial effects due to common environment to true genetic effects. To this regard, the most important evidence came from studies of adoptees and twins reared apart. In Denmark, there is a high correlation between the BMI of 540 adopted individuals and the corpulence of their biological parents, without any relationship with that of their foster parents (Stunkard et al. 1986). In Sweden, the intrapair correlation for the BMI was much higher in monozygotic twins than in dizygotic twins, and was very similar between those reared together and those reared apart (Stunkard et al. 1990). Although many of these studies show a highly significant heritability, it has been proposed by using a model based on data aggregated from twins, nuclear families, foster parents with adopted offspring, that true heritability was not so high, with only 5 % of the variance in BMI explained by genetic transmission (Bouchard et al. 1988). This has to be kept in mind when discussing the results of the last huge genetic studies which explain only a small part of the supposed heritability.

The most elegant study concerning twins demonstrates the importance of genes in response to nutritional environment, in other words the importance of the gene X nutrition interaction (Bouchard et al. 1990). A total of 12 pairs of monozygotic twins have been overfed by 1,000 kcal/day during a period of 100 days. The intrapair correlation of the response to overfeeding was very high, although between twin pairs, the response had a very high range of variation (between 3 and 14 kgs for the body weight, for a mean of 8.4).

Gene Identification

The studies described above have yielded the evidence for the genetic susceptibility to obesity and to gain weight in response to the environment. Different approaches were used during the last 30 years to discover the molecular basis for this susceptibility (Bouchard 2010). First, single-gene disorders in animal models (mainly rodents) and in humans allowed to identify important metabolic pathways in the regulation of energy balance, although the single-gene defects are not common in human obesities (nevertheless, it has sometimes been estimated that it could account for 5 % of severe early-onset obesities).

Monogenic Obesities

These obesities are usually very severe, with beginnings in childhood (O'Rahilly 2009; Ramachandrappa and Farooqi 2011). The genetic models of obesity in rodents are very close to this kind of obesity (see Chap. 18). The discovery of the mutations responsible for the genetic obesities in rodent led to the identification of most of the defects in human monogenic obesities. The first mutation described was that of the *ob/ob* mice and the gene in which the mutation occurred as well as its human homolog was cloned in 1994 (Zhang et al. 1994). This conducted to the description of a novel hormone, the leptin (see Chap. 12). In humans, the first deficit in leptin was reported in 1997 in a consanguineous family with a very severe form of recessive obesity (Montague et al. 1997). Affected subjects develop an early-onset massive obesity with hyperphagia, hypogonadism and delayed puberty, and increased frequency of infections and immunodeficiency. These defects are corrected by leptin injection (Farooqi et al. 1999). Actually, the few individuals which carry this defect are the only ones in the world who can benefit from a leptin treatment. Subjects heterozygous for the mutation do not suffer of massive obesity but have a higher measured fat mass than predicted from the usual determinants (age, sex, body weight, height) and a lower increase in leptin levels with increasing BMI when compared to controls non-carriers of the mutation (Farooqi et al. 2001) (Table 24.1).

Following the leptin deficiency, other mutations were described in families with evidence of monogenic form of obesities. Only a few genes are described, which explain a very small part of human obesities (Blakemore and Froguel 2010; Farooqi 2011). The mutated genes are, as well as the leptin gene (*LEP*) (Montague et al. 1997; Ozata et al. 1999), the leptin receptor (*LEPR*) (Clément et al. 1998), the pro-opiomelanocortin (*POMC*) (Krude et al. 1998), the proconvertase 1 (*PCSK1*) (Jackson et al. 1997), the hypothalamic receptor to melanocortins 4 (*MC4R*) (Vaisse et al. 1998; Yeo et al. 1998; Hinney et al. 1999; Sina et al. 1999; Gu et al. 1999; Vaisse et al. 2000; Farooqi et al. 2000; Mergen et al. 2001; Dubern et al. 2001), and a gene involved in the hypothalamus development, *SIM1* (Holder et al. 2000).

Table 24.1 Genetic obesity syndromes. In these disorders, obesity is largely driven by increased appetite and/or diminished satiety. Adapted from (O'Rahilly 2009; Ramachandrappa and Farooqi 2011)

Obesity syndrome	Gene name	Chromosome location	Gene function	Main supplementary distinguishing features
LEP deficiency	*LEP*	7q31.3	Adipocyte-derived satiety hormone, leptin	Hypogonadism, frequent infections, undetectable serum leptin
LEPR deficiency	*LEPR*	1p31	Receptor for leptin	Hypogonadism
POMC deficiency	*POMC*	2p23.3	Hypothalamic neuropeptide	Hypopigmentation, isolated ACTH deficiency
PCSK1/3 deficiency	*PCSK1*	5q15	Processes pro-peptides (including POMC) to active moieties	Postprandial hypoglycemia, hypogonadism, elevated plasma proinsulin
MC4R deficiency	*MC4R*	18q22	Receptor for POMC products αMSH and βMSH	Accelerated growth, increased final height
SIM1 deficiency	*SIM1*	6q16.3-q21	Transcription factor necessary for hypothalamic development	Spectrum of developmental delay
BDNF deficiency	*BDNF*	11p12-p14	Deletion of BDNF (brain-derived neurotrophic factor) in a subset of patients with WAGR syndrome is associated with obesity	Developmental delay, hyperactivity, impaired memory, impaired pain sensation
TrkB deficiency	*NTRK2*	9q22.1	Receptor for BDNF and neurotrophin 5 (NTF5)	Developmental delay, hyperactivity, impaired memory, impaired pain sensation
Bardet-Biedl syndrome	*BBS1–16*	Many chromosomal locations	At least 12 genes identified, most of which affect the function of the primary cilium, involved in leptin receptor signaling	Polydactyly, retinal dystrophy, hypogonadism, renal abnormalities, developmental delay

(continued)

Table 24.1 (continued)

Obesity syndrome	Gene name	Chromosome location	Gene function	Main supplementary distinguishing features
Prader–Willi syndrome	–	15q11-q13	Imprinted locus on chromosome 15q including snoRNA cluster	Hypotonia, short stature, hypogonadotropic hypogonadism
Albright's hereditary osteodystrophy (pseudohypoparathyroidism)	*GNAS1*	20q13.32	Encodes α-subunit of the stimulatory G protein, maternal transmission of mutations for obesity	Short stature, skeletal defects, hormone resistance
Alström syndrome	*ALMS1*	2p13.1	Disorder of the ciliary function	Photophobia, nystagmus, visual impairment, deafness, severe insulin resistance
Chromosome 16p11 deletion	*SH2B1*	16p11.2	16p11.2 deletion encompasses multiple genes but always include SH2B1. SH2B adaptator protein 1, neuronal role in energy homestasis, leptin and insulin signang	Cognitive deficits, severe insulin resistance

The discovery of these genes allowed describing the leptin–melanocortin pathway. Proopiomelanocortin, which is cleaved by PCSK1, is the precursor of different peptides including ACTH, β-endorphin, and α-MSH (melanocyte-stimulating hormone). The latter peptide binds different receptors (melanocortin receptors 1–4) according to cell type or organ. When binding to the MC4R in the hypothalamus, it results in an anorexigenic effect, but when binding to the skin MC1R, it is implicated in melanin synthesis. To summarize, leptin, satiety factor synthesized and secreted by the adipose tissue, acts in hypothalamus through a specific receptor, which released the secretion of a POMC derived protein, α-MSH, acting itself on a the MC4R leading to the anorexigenic effect (Cummings and Schwartz 2003). In this pathway, a normal development of the hypothalamus is needed, which explains the obesity associated to mutations in *SIM1*, *BDNF* (brain-derived neurotrophic factor), and in its receptor TrkB (*NTRK2*). These mutations cause severe early-onset obesity, with hyperphagia, and severe developmental delay.

LEP and LEPR defects are phenotypically very similar, i.e., early-onset obesity with hyperphagia, hypogonadism and delayed puberty, alterations in immune functions (less important in *LEPR* mutations). In the *POMC* defect, the obesity is associated with adrenal insufficiency and red hair. The defect in *PCSK1* is characterized by an extreme childhood obesity, hypogonadism, hypocortisolism, abnormal glucose homeostasis, low plasma insulin with high plasma proinsulin, indicating a defect in prohormone processing. The defects in *MC4R* are usually characterized by severe obesity in childhood, increased lean mass and growth rate, hyperinsulinemia.

The *MC4R* defects seem to be the most frequent cause of monogenic obesity: up to 6 % of childhood and 2 % of adult severe obesities might be due to them (Blakemore and Froguel 2010; Farooqi 2011). More than 130 mutations of this gene had been described in 2010 (Hinney et al. 2010). Actually, it is the only form of monogenic obesity where most of the cases are heterozygotes, indicating a dominant or codominant inheritance, since homozygotes are more severely affected. In heterozygous carriers, the expression and the penetrance of the mutation are variable. Some of the mutations do not completely suppress the signaling ability of MC4R, leading to less severe forms of obesity (Farooqi et al. 2003). A systematic research of loss-of-function mutations in normal weight ($n = 2257$) and severely obese adult and children ($n = 2677$) has shown both a relatively high frequency of these in obese subjects (1.72 %) as well as a variable penetrance since 0.15 % of non-obese carry these mutations (Stutzmann et al. 2008). Abnormal eating behavior was more frequent in obese subjects carrying the mutations than in the other obese subjects. The penetrance was age dependent. In the cross-sectional study, the penetrance increased with the generations: from 40 % in grandparents but 78 % in children, showing the change in obesogenic environment in the last decades. In the longitudinal study, the penetrance increased with age, showing that those born with these mutations will develop the disease early or lately. The demonstration of the variable penetrance for the carriers of such mutations is a very strong argument in favor of gene–environment interaction, even in the context of monogenic obesity.

Deletions in the region of *SH2B1* on chromosome 16p11 are associated with severe early-onset obesity, severe insulin resistance and cognitive deficits in heterozygous subjects (Bochukova et al. 2010; Walters et al. 2010). Such deletions have been found at relatively high frequencies (0.8 %) in cohorts of children with severe obesity (BMI $\geq$ 40) (Walters et al. 2010), most of the time but not always associated with mental disorder or retardation. Other genes are located in this region but *SH2B1* is involved in leptin and insulin signaling. Mice with a specific deletion of this gene develop hyperphagia and obesity. Remarkably, it has been found that if a duplication of the region occurs instead of a deletion, the reciprocal phenotype is observed, i.e., underweight (Jacquemont et al. 2011). The carriers of the duplication show reduced weight at birth, a failure to thrive when younger than 5 years. In adults, the duplication carriers have an 8.3 relative risk of being clinically underweight, associated with a high frequency of restrictive eating behavior. The observed phenotypes are the converse of those reported in carriers of deletions at this locus.

Other complex genetic disorders such as Prader–Willi and Bardet–Biedl syndrome include obesity with hyperphagia, but also mental retardation and various dysmorphisms (Table 24.1). The study of these syndromic obesities raised some hypothesis for understanding the physiopathology of common obesity like the importance of gene imprinting (epigenetics) or how the structure and function of the primary cilium can affect hypothalamic neurons involved in the control of energy balance.

Other defects have been reported possibly involved in monogenic obesities but, which need confirmation (Rankinen et al. 2006): mutations in the genes of *GPR24,* receptor MCHR1 to MCH, orexigen neuropeptide in rodents, of *CRHR1* and *CRHR2,* receptors to CRH (corticotropin releasing hormone) involved in glucocorticoids (cortisol) secretion. Rare mutations in *MC3R* have been found in subjects with severe obesity but their role in monogenic obesity is still a matter of debate. They might be a susceptibility factor.

Common Obesities: Candidate Gene Studies

Monogenic obesities represent less than a few percents of all obesity cases. In these cases, the genetic defect is directly causal, either without or with few interactions with environment. In common, multifactorial polygenic obesities, each of the genetic factors will have a small effect, depending on the environment. The recent increase in obesity, what is called "the obesity epidemics", cannot be attributed to genetics but rather to changes in the environmental conditions, and to the response of genes to these changes (as in the overfeeding experiment in twins). The observation of the Pima Indians is a striking example of this interaction. In the Pimas of Arizona, lifestyle changes of the late fifties—high-fat diet and sedentarity—has resulted in an epidemic of obesity and diabetes, with one of the highest prevalence in the world. By contrast, the Pimas who live in the mountains of

Mexico have been isolated from Western influence and kept their traditional low-fat diet and are physically active as farmers and sawmill workers. The Pimas of Mexico have a much lower incidence of obesity and diabetes despite the same genetic background as the Pimas in Arizona (mean BMI in Mexico: 24.9 *vs.* 33.4 in Arizona) (Ravussin et al. 1994).

The candidate gene approach is based on the current understanding of the biology and physiopathology of obesity. Variants of genes coding for proteins that are thought to be involved in metabolisms or behaviors related to obesity, based on data derived from animal models, cellular systems, or extreme/monogenic forms of obesity, are tested at a population level, or less frequently in families. Obesity, at least in its constituting phase, results from an excess of energy intake when compared to energy expenditure, the excess calories being stored as body fat. Candidate genes may be involved in food intake (signals, or neurotransmitters with orexigenic or anorexigenic effects: leptin and its receptor, neuropeptide Y, serotonin receptors and transporter etc.), energy expenditure (uncoupling proteins, adrenergic receptors, etc.), adipocyte biology (PPARγ, adipokines, etc.). This led to a classification of obesity as either a disorder of the energy balance, or a disorder of the adipocyte, or a neurobehavioral disorder (Walley et al. 2009). Nevertheless, these pathways are not mutually exclusive and the genetic susceptibility is a mix of variants belonging to these pathways interacting with environment.

The variants associated (in populations of unrelated subjects) or linked (in familial studies) to common obesity are genetic polymorphisms, defined by a frequency at least equal to 1 % in the general population. These polymorphisms may be functional by modifying the protein sequence, or the regulation of its levels by affecting transcriptional sites, mRNA splicing or mRNA stability. These polymorphisms may as well be non-functional, being intronic or intergenic. In this case, a significant result may be indirect, due to linkage disequilibrium (alleles at multiple loci appearing together [= haplotype] more often than would be expected by chance) with a functional mutation or variant not known yet.

Till 2007, the candidate gene approach has been the most widely used. Many candidate genes have been studied because of their involvement in the pathways indicated above, the first "natural" candidate genes being those involved in monogenic obesity. Many of these studies failed to replicate. The recent development in genetic research with large number of SNPs in large series of subjects evidenced that the individual effects for each gene were small, indicating that most of the classical candidate gene studies were underpowered. Nevertheless, in some cases, the evidence is strong enough to suggest that these candidate genes have a true effect on obesity, although modest.

Common variants of *LEP* and *LEPR* have been associated with adult and childhood obesity, and sometimes with leptin levels (Mammès et al. 1998; Mammès et al. 2000; Mammès et al. 2001; Li et al. 1999; Chagnon et al. 2000). Differences in response to restrictive diet have been reported according to these variants (Mammès et al. 1998; Mammès et al. 2001). A promoter variant of the *LEP* gene, which has been shown to modulate the transcriptional expression of the gene (Hoffstedt et al. 2002), modifies the relationship between fat mass and plasma leptin levels (Le Stunff et al. 2000).

Some point mutations in *POMC* significantly increase obesity risk, without being always associated with obesity (Ramachandrappa and Farooqi 2011). Common non-synonymous variants in *PCSK1* have been associated with multifactorial obesity in some European cohorts of adults and children (Benzinou et al. 2008). Besides the mutations of *MC4R* which cause monogenic obesities with variable penetrance, two rare coding variants (allelic frequencies ≈ 0.5–2 %) have been associated with a protective effect against obesity (Walley et al. 2009). Polymorphisms in coding and in flanking regions of *MC4R* have been associated with adiposity, glucose metabolism, energy expenditure, and food intake (Cole et al. 2010). Some polymorphisms of *BBS* genes have also been involved in the susceptibility to common obesities (Benzinou et al. 2006).

The common Pro12Ala variant of PPARγ is expressed only in the adipocyte-specific γ2 isoform and results from a C to G transversion in the exon B of the PPARγ encoding gene (*PPARG*). The minor Ala allele has consistently been negatively associated with insulin resistance and susceptibility to type 2 diabetes. This variant has also been associated with obesity and body weight changes, with conflicting results. In meta-analyses, the Ala–Ala homozygous genotype seems to be associated with slightly higher BMI. Actually, the association of this genotype with high BMI might be present only when consuming a high total or saturated fat diet, indicating a gene–nutrition interaction (Lamri et al. 2011). This is consistent with the metabolic role of PPARγ through binding of fatty acids.

Some variants in the adiponectin gene (*ADIPOQ*) have been associated with childhood or adult obesity (Bouatia-Naji et al. 2006) or with higher weight gain at follow-up in a French cohort (Fumeron et al. 2004). Surprisingly, although low levels of adiponectin are usually found in obesity, the risk alleles were those associated with high adiponectin levels, including the -11391A (rs17300539) situated in the promoter and associated with increased transcriptional activity.

Common Obesities: Pangenomic Studies

Genome-wide linkage studies are useful to localize a gene in a chromosomal region ("positional cloning"). In families, the transmission of the phenotypic trait of interest (or disease) is studied jointly ("cosegregation") with that of a set of genetic markers encompassing the whole genome, with known localization. The regions with peaks of linkage are then submitted for fine mapping to evidence candidate genes, since they may contain hundreds of different genes. This approach has been successfully employed to localize the genes responsible for the monogenic diseases. In complex diseases, linkage analyses succeed when the linkage peak is due to few common variants with a strong effect in one gene (Walley et al. 2009). For example, this approach has been successful for the

PCSK1 variants indicated above. However, many positive results of more than 60 genome scans (Rankinen et al. 2006) failed to replicate (Saunders et al. 2007).

Genome wide association studies (GWAS)

This approach has been made possible by recent major advances: the considerable increase in our knowledge of common genetic variation with the Human Genome project and the International HapMap, and the progress in high-throughput genotyping. Now, it is possible to genotype up to 1 million genetic variants in a single analysis on single nucleotide polymorphism (SNP) chips, which can capture more than 80 % of the common genetic variation reported in the HapMap. This hypothesis-free approach has been very successful in a few years to detect common variants involved in the susceptibility to polygenic diseases (Table 24.2).

FTO

The SNPs of the *FTO* gene (Fat Mass and Obesity associated) have been the most strongly associated with BMI and/or obesity. Initially, this gene was identified in a GWAS screening for the genetic susceptibility to type 2 diabetes (Frayling et al. 2007). When adjusted for BMI, the association with diabetes completely disappeared. In the same time, other GWAS evidenced the association with BMI/obesity (Dina et al. 2007). The effect on BMI has since been replicated in many cohorts, in adult and in childhood obesity. The homozygotes for the allele at risk (16 % of the general population of European origin) weighed 3 kg more and have a 67 % increased risk of obesity when compared to the homozygotes TT (36 % of the general population). The genetic variation at the *FTO* locus would account for 22 % of the common obesities (Dina et al. 2007).

At that time, the function of the *FTO* gene was unknown. These results led to research concerning the role of this gene and how it is implicated in the physiopathology of obesity.

This gene has been found widely expressed, in particular in hypothalamus, adipose tissue, liver, skeletal muscle. The expression in hypothalamus is influenced by the nutritional state. *FTO* mRNA in the arcuate nucleus of mice is up-regulated by feeding and down-regulated by fasting but opposite expression patterns are observed in rats. In the hypothalamic arcuate nucleus of rats but not in the paraventricular nucleus, knockdown of the *FTO* gene increased food intake while overexpression decreased it. Mice carrying additional copies of *FTO* showed a wide variation in tissue expression levels compared to wild type (Church et al. 2010). They were obese and had an increased food intake. They also had a reduction in leptin levels although usually, obesity is associated with hyperleptinemia, suggesting that hypoleptinemia might mediate the effects of *FTO* genetic variation on obesity.

FTO belongs to a family of genes coding for an acid nucleic demethylase, indicating that it might be involved in epigenetic regulation of other genes unknown for the moment.

Table 24.2 Gene variants associated with obesity related phenotypes in genome wide association studies (GWAS). From (Dina et al. 2007; Loos et al. 2008; Thorleifsson et al. 2009; Willer et al. 2009; Meyre et al. 2009; Speliotes et al. 2010; Frayling 2007; Cho et al. 2009; Lindgren et al. 2009; Scherag et al. 2010; Heid et al. 2010). Other reviews in (Hinney et al. 2010; Walley et al. 2009; Herrera et al. 2011; Choquet and Meyre 2011)

Nearest Gene	Polymorphism	Chromosome	Phenotype	Odds ratio in adults or adults/children	BMI effect (kg/m^2) per allele or per genotype or WHR effect
FTO	rs9939609	16	BMI, extreme obesity	1.67/1.27	0.33–0.40 (A)
	rs8050136				1.07 (AA)
	rs1421085				0.11 (C)
	rs1558902				0.39 (A)
MC4R	rs17782313/	18	BMI, extreme and childhood obesity	1.12/1.13	0.10–0.22 (C)
	rs12970134				0.36 (AA)
	rs571312				0.24
CTNNBL1	rs6013029	20	BMI	1.42	
TMEM18	rs6548238	2	BMI, childhood obesity	1.19/1.41	0.26 (C)
	rs7561317			1.20	0.70 (GG)
	rs2867125				0.31 (C)
GNPDA2	rs10938397	4	BMI	1.12/1.20	0.18–0.19 (G)
SH2B1	rs7498665	16	BMI	1.11/1.08	0.15 (G)/0.45 (GG)
	rs7359397				0.15 (T)
KCTD15	rs11084753	19	BMI	1.04/0.96	0.06 (G)
	rs29941			1.10	0.06 (C)/0.46 (CC)
MTCH2	rs10838738	11	BMI	1.03	0.07 (G)
	rs3817334				0.06 (T)
NEGR1	rs2815752	1	BMI	1.05	0.10–0.13 (A)
	rs2568958			1.07	0.43 (AA)
NPC1	rs1805081	18	Extreme obesity	0.75/0.75	−0.09 (A)
MAF	rs1424233	16	Extreme obesity	1.39/1.12	0.09 (A)
PTER	rs10598503	10	Extreme obesity	0.68/0.64	

(continued)

Table 24.2 (continued)

Nearest Gene	Polymorphism	Chromosome	Phenotype	Odds ratio in adults or adults/children	BMI effect (kg/m^2) per allele or per genotype or WHR effect
PRL	rs4712652	6	Extreme obesity	0.83	−0.08 (AA)
SEC16B	rs10913469	1	BMI, waist circumference	1.11	0.50 (CC)
	rs543874				0.22 (G)
ETV5	rs7647305	3	BMI	1.11	0.54 (CC)
	rs9816226				0.14 (T)
NCR3/AIF1/BAT2	rs2844479	6	BMI	1.07	
BDNF	rs925946	11	BMI	1.11	0.67 (GG)
	rs6265				0.19 (A)
	rs10767664				
FAIM2/BCDIN3D	rs7138803	12	BMI	1.14	0.12 (A)/0.54 (AA)
SDCCAG8	rs12145833	1	Childhood obesity	1.41–1.52	
TNKS/MSRA	rs17150703	8	Childhood obesity	1.78/1.75	0.040 (G)
	rs7826222		Waist circumference		
TFAP2B	rs987237	6	BMI, waist circumference		0.13 (G)
NRXN3	rs10150332	14	BMI, waist circumference		0.13 (C)
C12orf51	rs2074356	12	WHR		−0.006 (T) (in Asians)
GPRC5B	rs12444979	16	BMI		0.17 (C)
RBJ/ADCY3/POMC	rs713586	2	BMI		0.14 (C)
MAP2K5	rs2241423	15	BMI		0.13 (G)
GIPR	rs2287019	19	BMI		0.15 (C)
FANCL	rs887912	2	BMI		0.10 (T)
TNNI3 K	rs1514175	1	BMI		0.07 (A)
LRRN6C	rs10968576	9	BMI		0.11 (G)
FLJ35779/ HMGCR	rs2112347	5	BMI		0.10 (T)
SLC39A8	rs13107325	4	BMI		0.19 (T)
TMEM160	rs3810291	19	BMI		0.09 (A)

(continued)

Table 24.2 (continued)

Nearest Gene	Polymorphism	Chromosome	Phenotype	Odds ratio in adults or adults/children	BMI effect (kg/m²) per allele or per genotype or WHR effect
CADM2	rs13078807	3	BMI		0.10 (G)
LRP1B	rs2890652	2	BMI		0.09 (C)
PRKD1	rs11847697	14	BMI		0.17 (T)
MTIF3	rs4771122	13	BMI		0.09 (G)
ZNF608	rs48361333	5	BMI		0.07 (A)
PTBP2	rs1555543	1	BMI		0.06 (C)
RPL27A/TUB	rs4929949	11	BMI		0.06 (C)
NUDT3/HMGA1	rs206936	6	BMI		0.06 (G)
RSPO3	rs9491696	6	WHR		0.042 (G) 0.31 (M)/0.50 (W)
VEGFA	rs6905288	6	WHR		0.036 (A) 0.020 (M)/0.052 (W)
TBX15/WARS2	rs984222	1	WHR		0.034 (G)
NFE2L3	rs1055144	7	WHR		0.040 (T)
GRB14	rs10195252	2	WHR		0.033 (T) 0.010 (M)/0.054 (W)
LYPLAL1	rs4846567	1	WHR		0.034 (G) 0.005 (M)/0.059 (W)
DNM3/PIGC	rs1011731	1	WHR		0.028 (G) 0.017 (M)/0.042 (W)
ITPR2/SSPN	rs718314	12	WHR		0.030 (G)
LY86	rs1294421	6	WHR		0.028 (G)
HOXC13	rs1443512	12	WHR		0.031 (G) 0.018 (M)/0.040 (W)

(continued)

Table 24.2 (continued)

Nearest Gene	Polymorphism	Chromosome	Phenotype	Odds ratio in adults or adults/children	BMI effect (kg/m^2) per allele or per genotype or WHR effect
ADAMTS9	rs6795735	3	WHR		0.025 (C) 0.011 (M)/0.038 (W)
ZNRF3/KREMEN1	rs4823006	22	WHR		0.023 (A) 0.015 (M)/0.030 (W)
NISCH/STAB 1	rs6784615	3	WHR		0.043 (T)
CPEB4	rs6861681	5	WHR		0.022 (A)

In a population of 306 women in good health (BMI: 18–53 kg/m^2), the subjects carrying the risk allele had a basal adipocyte lipolysis decreased by 22 % (in vitro experiments) and plasma glycerol levels (in vivo) decreased by 30 % when compared to women homozygotes for the common allele (Wahlen et al. 2008). It has been shown, at least in children, an increased in energy intake but not in energy expenditure (Cecil et al. 2008) and a lower satiety (Wardle et al. 2008) associated with the risk allele.

The *FTO* polymorphisms are the most constantly associated with BMI and obesity, with the most significant results in the GWAS. However, as already mentioned, the relative risk due to these polymorphisms is weak, and moreover, it has been shown to interact with levels of physical activity to confer an increased risk. The at-risk variants have significant effects only in subjects with low activity levels, assessed by either by questionnaire (Andreasen et al. 2008) or by electronic recorder (Rampersaud et al. 2008). This has been replicated many times including in a large cohort (> 20000 subjects), with few exceptions.

Resequencing studies have been performed in obese and control subjects but loss-of-function mutations have been found at a similar frequency in both groups. In a consanguineous family, a rare homozygous missense mutation (changing the sequence of protein) has been described associated with a severe polymalformation syndrome and postnatal lethality, but no evidence of adiposity phenotype (Boissel et al. 2009).

MC4R

The importance of *MC4R* mutations in monogenic obesities has already been developed. In the GWAS, after *FTO*, the *MC4R* locus has been one of the most constantly observed associated with common forms of obesities. Nevertheless, although the patterns of association are compatible with alterations in *MC4R* function, the evidenced polymorphisms are close to but not in the gene. The only gene in the vicinity is involved in apoptosis and likely not a plausible candidate. The SNP rs17782313 located 188 kb downstream of *MC4R* has been associated with BMI variation and obesity risk in adults and children in a very large study, then replicated many times in large populations of different origins (Loos et al. 2008). One copy of the allele at risk is associated with a mean difference of 0.22 kg/m^2 (760 g) and an increase in risk of overweight, obesity, and severe obesity of 8, 12, and 30 %, respectively (Loos et al. 2008). An increase in height has also been observed. It has been associated with eating behavior (Stutzmann et al. 2009). Another polymorphism, rs12970134, has been associated with waist circumference and insulin resistance in Asian Indians, and is also located 150 kb from the MC4R gene (Chambers et al. 2008).

Other Genes

The *FTO* and *MC4R* variants are the most constantly replicated and the most strongly associated polymorphisms in the GWAS. Novel loci were also identified by studies in large groups and meta-analyses (Thorleifsson et al. 2009; Willer et al. 2009; Meyre et al. 2009; Speliotes et al. 2010). More of 30 loci have now been robustly identified, regarding BMI or obesity. When including other phenotypic traits linked to obesity such as waist circumference or waist-to-hips ratio (WHR), the number of loci is over 50 (Table 24.2). Among the BMI-associated loci, three of them show evidence of association with height: *MC4R*, *POMC*, *MCPH2-NDUFS3*. The alleles increasing BMI in *POMC* and *MC4R* have opposite effects on height: decrease for *POMC* and increase for *MC4R*, which is similar to the effects of the severe mutations of these genes.

Several of the likely causal genes are involved in neural signalization and development, or are highly expressed in the central nervous system, including those already involved in monogenic obesities, coding for hypothalamic regulators of energy balance: *POMC*, *BDNF*, *SH2B1*, as well as *MC4R*. This indicates the importance of central regulation of food intake in the susceptibility to obesity and supports the hypothesis of obesity as a neurobehavioral disease. One of the new discovered loci is near the *GIPR*, coding for the receptor to incretin GIP. Nevertheless, for many of these genes, the physiological role explaining their association with BMI variation is still unknown or poorly understood. Other loci are involved in cholesterol metabolism such as *NPC1* (Niemann-Pick C1) and *HMGCR* (HMGcoa reductase). NPC1 is a transmembrane protein involved in the cholesterol transport between different cellular compartments. It plays a role in caveolin expression involved in the transport of fatty acids and storage of triglycerides. Nevertheless, it is not known whether the non-synonymous alleles associated with obesity increase or decrease the expression of NPC1 protein. *HMGCR* codes for a key enzyme in cholesterol synthesis, which is inhibited by statins in order to increase the number of LDL-receptors. However, it is not known whether among the genes located in the same region, *HMGCR* is the one involved in the susceptibility to obesity.

Many genes involved in BMI variation overlap with those associated with extreme obesity, in children as well as in adults. Therefore, most of the time, severe obesity seems to represent an extreme BMI phenotype rather than a distinct condition.

Attempts have been done to assess cumulative effects of polymorphisms and predictive values of genetic scores calculated by counting the number of obesity-susceptibility variants (Speliotes et al. 2010; Li et al. 2010; Sandholt et al. 2010). In the population based ARIC study ($n = 8120$), a genetic score was calculated after genotyping 32 GWAS evidenced SNPs (observed range of this score 16–44) (Speliotes et al. 2010). Each of these variant accounts for a very small part of the BMI variance. Altogether, they explain only 1.45 % of BMI variance, while the main *FTO* SNP alone accounts for 0.34 %. There is a very good linear relationship between the genetic score and BMI. For each additional at-risk allele, BMI increases by 0.17 kg/m^2 (435–551 g gain in body weight for adults of 160–180 cm in height). The difference in BMI between individuals with a high susceptibility score (≥38, 1.5 % of the sample) and those with a low susceptibility score (≤21, 2.2 % of the sample) was 2.75 kg/m^2 (6.99–8.85 kg body weight). The predictive value for BMI and obesity risk of the 32 variants combined is significant but modest. The area under the receiver operating characteristic (ROC) curve for prediction of risk of obesity using a model with age and sex was increased by 6 % when adding the genetic score in the model. In other studies using 12 (Li et al. 2010) or 20 SNPs (Sandholt et al. 2010), the gain in predictive value was smaller, around 2 %.

Epigenetics and obesity

Epigenetics refer to mitotically or meiotically (germline cells) heritable changes in gene expression that do not involve a change in DNA sequence (Jirtle and Skinner 2007). Epigenetic marks include DNA methylations and histone modifications which mediate biological processes such as imprinting. Genomic imprinting determines genomic expression of alleles according to their maternal or paternal origin. Approximately 1 % of autosomal genes are imprinted with expression of only one parental allele. One cluster of genes submitted to imprinting is located in a region at 15q11. A paternal deletion in this region results in the Prader–Willi syndrome. A moderate obesity appears in Albright hereditary osteodystrophy due to disruption of imprinting at the *GNAS* locus (20q13.11). In utero environment, nutrition may influence epigenetic variation affecting permanently metabolism and chronic disease risk such as obesity and diabetes. Transgenerational effects have been observed, due to epigenetic modifications in the germ line leading to inheritance of a phenotype. Abnormal nutrition during gestation has been shown to influence obesity and diabetes susceptibility in children of the 1st generation, but also in the 2nd generation (Herrera et al. 2011; Jirtle and Skinner 2007).

Interactions between environment, epigenetic mechanisms and genetic variation likely play a role in the susceptibility to obesity. High-fat diets modulate the methylation profile of *MC4R* and of *POMC* and *LEP* gene promoters. The PPARγ protein interacts with histone acetyltransferases during adipogenesis. The *FTO* gene, which is a DNA demethylase might play a role by modifying the epigenetic regulations.

Conclusion

These last years, progress in genomic analysis allowed to uncover a large number of loci responsible of the susceptibility to obesity. A large part of them are involved in the neurobehavioral axis, but other pathways are also contributive. The adipose tissue plays a role in this genetic susceptibility through the genetic variability of its differentiation factor PPARγ and adipokines, leptin, and adiponectin.

Genetic studies have paradoxically demonstrated the importance of the environmental influence in the susceptibility to obesity. For many genes, the genetic effect is modulated by food intake and/or the level of physical activity, which is very consistent with the recent epidemic of obesity in the context of an obesogenic environment. Therefore, opposite to the *cliché* that genetic susceptibility means that nothing can be done to cure your illness or to prevent it, the search of environmental susceptibility factors would have efficacy in the treatment or prevention. At cellular and molecular levels, this interaction is the consequence of epigenetic modifications interacting with the genetic variation. Very recently, the gut microbiome has been involved in the development of obesity (Tilg and Kaser 2011). Studies on the intestinal flora have shown in obese individuals that the "obese microbiota" seems to extract more energy from the diet. This can be considered as a new kind of environmental influence. Since the microflora is estimated to contain 150-fold more genes than the human host genome, this interaction with endogenous bacterial environment can be considered as a gene–gene interaction as well.

It might be deceiving that the combined effects of the genetic variants explain only few percent of the inherited variation of obesity risk, which is called the "missing heritability", and have consequently a very small predictive value. Collaborative studies for larger GWAS meta-analyses are currently running. For example the GIANT consortium includes more than 100,000 individuals. The increased statistical power of such studies will allow to detect new variants with smaller effects (to only 50 g of body weight). Till now, the GWAS included SNPs with a minor allele frequency equal or superior to 5 %. Systematic exome sequencing is currently being performed to find rare alleles (missense or nonsense mutations) with larger effects than those previously detected by GWAS. The 1,000 Genomes project has been designed to find rare variants, which will be used in new genotype arrays with wider spectra of allele frequencies (2010). Other kind of variants than SNPs should be studied, in particular CNVs (Copy Number Variations of sequences ≥1 kb). In the *NEGR1* locus, the associated SNPs tag a 45 kb deletion polymorphism, which is a candidate causal variant. The use of CNVs has allowed detecting deletions and duplications in 16p21 responsible for extreme obesity or familial leanness (Jacquemont et al. 2011).

These new approaches will increase the predictive value and the part of the BMI variance explained by genetic variants. Nevertheless, it is difficult to conceive that these methods will allow discovering enough gene variants to explain 95 % of the heritability. By definition, the new variants with small effects will have small effects, and rare variants (which might have larger effects) will be rare! We must keep in mind that heritability of obesity and BMI might have been overestimated. In this case, the discovery of many variants will never have enough predictive value for clinical use.

The main interest in the genetic research concerning obesity is to reveal unknown pathways. To this respect, the unsuspected loci revealed by GWAS open a large field of investigations. It remains to find the true causal genes and functional variants at each locus by fine mapping and molecular biology experiments, then to launch research at the molecular, cellular, and physiological levels to find out the mechanisms and pathways involved in the susceptibility to obesity. These new pathways then should be tested as therapeutic/pharmacologic targets. In this case, the genetic information would be useful in predicting the response to targeted treatment.

References

Allison DB, Faith MS, Nathan JS (1996) Risch's lambda values for human obesity. Int J Obes Relat Metab Disord 20:990–999

Andreasen CH, Stender-Petersen KL, Mogensen MS et al (2008) Low physical activity accentuates the effect of the FTO rs9939609 polymorphism on body fat accumulation. Diabetes 57:95–101

Benzinou M, Walley A, Lobbens S et al (2006) Bardet-Biedl syndrome gene variants are associated with both childhood and adult common obesity in French Caucasians. Diabetes 55:2876–2882

Benzinou M, Creemers JW, Choquet H et al (2008) Common nonsynonymous variants in PCSK1 confer risk of obesity. Nat Genet 40:943–945

Blakemore AI, Froguel P (2010) Investigation of Mendelian forms of obesity holds out the prospect of personalized medicine. Ann N Y Acad Sci 1214:180–189

Bochukova EG, Huang N, Keogh J et al (2010) Large, rare chromosomal deletions associated with severe early-onset obesity. Nature 463:666–670

Boissel S, Reish O, Proulx K et al (2009) Loss-of-function mutation in the dioxygenase-encoding FTO gene causes severe growth retardation and multiple malformations. Am J Hum Genet 85:106–111

Bouatia-Naji N, Meyre D, Lobbens S et al (2006) ACDC/adiponectin polymorphisms are associated with severe childhood and adult obesity. Diabetes 55:545–550

Bouchard C (2010) Defining the genetic architecture of the predisposition to obesity: a challenging but not insurmountable task. Am J Clin Nutr 91:5–6

Bouchard C, Perusse L, Leblanc C et al (1988) Inheritance of the amount and distribution of human body fat. Int J Obes 12:205–215

Bouchard C, Tremblay A, Despres JP et al (1990) The response to long-term overfeeding in identical twins. N Engl J Med 322:1477–1482

Cecil JE, Tavendale R, Watt P et al (2008) An obesity-associated FTO gene variant and increased energy intake in children. N Engl J Med 359:2558–2566

Chagnon YC, Wilmore JH, Borecki IB et al (2000) Associations between the leptin receptor gene and adiposity in middle-aged Caucasian males from the HERITAGE family study. J Clin Endocrinol Metab 85:29–34

Chambers JC, Elliott P, Zabaneh D et al (2008) Common genetic variation near MC4R is associated with waist circumference and insulin resistance. Nat Genet 40:716–718

Church C, Moir L, McMurray F et al (2010) Overexpression of Fto leads to increased food intake and results in obesity. Nat Genet 42:1086–1092

Clément K, Vaisse C, Lahlou N et al (1998) A mutation in the human leptin receptor gene causes obesity and pituitary dysfunction. Nature 392:398–401

Cole SA, Butte NF, Voruganti VS et al (2010) Evidence that multiple genetic variants of MC4R play a functional role in the regulation of energy expenditure and appetite in Hispanic children. Am J Clin Nutr 91:191–199

Cummings DE, Schwartz MW (2003) Genetics and pathophysiology of human obesity. Annu Rev Med 54:453–471
Dina C, Meyre D, Gallina S et al (2007) Variation in FTO contributes to childhood obesity and severe adult obesity. Nat Genet 39:724–726
Dubern B, Clément K, Pelloux V et al (2001) Mutational analysis of melanocortin-4 receptor, agouti-related protein, and alpha-melanocyte-stimulating hormone genes in severely obese children. J Pediatr 139:204–209
Farooqi IS (2011) Genetic, molecular and physiological insights into human obesity. Eur J Clin Invest 41:451–455
Farooqi IS, Jebb SA, Langmack G et al (1999) Effects of recombinant leptin therapy in a child with congenital leptin deficiency. N Engl J Med 341:879–884
Farooqi IS, Yeo GS, Keogh JM et al (2000) Dominant and recessive inheritance of morbid obesity associated with melanocortin 4 receptor deficiency. J Clin Invest 106:271–279
Farooqi IS, Keogh JM, Kamath S et al (2001) Partial leptin deficiency and human adiposity. Nature 414:34–35
Farooqi IS, Keogh JM, Yeo GS et al (2003) Clinical spectrum of obesity and mutations in the melanocortin 4 receptor gene. N Engl J Med 348:1085–1095
Frayling TM, Timpson NJ, Weedon MN et al (2007) A common variant in the FTO gene is associated with body mass index and predisposes to childhood and adult obesity. Science 316:889–894
Fumeron F, Aubert R, Siddiq A et al (2004) Adiponectin gene polymorphisms and adiponectin levels are independently associated with the development of hyperglycemia during a 3-year period: the epidemiologic data on the insulin resistance syndrome prospective study. Diabetes 53:1150–1157
Gu W, Tu Z, Kleyn PW et al (1999) Identification and functional analysis of novel human melanocortin-4 receptor variants. Diabetes 48:635–639
Herrera BM, Keildson S, Lindgren CM (2011) Genetics and epigenetics of obesity. Maturitas 69:41–49
Hinney A, Schmidt A, Nottebom K et al (1999) Several mutations in the melanocortin-4 receptor gene including a nonsense and a frameshift mutation associated with dominantly inherited obesity in humans. J Clin Endocrinol Metab 84:1483–1486
Hinney A, Vogel CI, Hebebrand J (2010) From monogenic to polygenic obesity: recent advances. Eur Child Adolesc Psychiatry 19:297–310
Hoffstedt J, Eriksson P, Mottagui-Tabar S, Arner P (2002) A polymorphism in the leptin promoter region (-2548 G/A) influences gene expression and adipose tissue secretion of leptin. Horm Metab Res 34:355–359
Holder JL Jr, Butte NF, Zinn AR (2000) Profound obesity associated with a balanced translocation that disrupts the SIM1 gene. Hum Mol Genet 9:101–108
Jackson RS, Creemers JW, Ohagi S et al (1997) Obesity and impaired prohormone processing associated with mutations in the human prohormone convertase 1 gene. Nat Genet 16:303–306
Jacquemont S, Reymond A, Zufferey F et al (2011) Mirror extreme BMI phenotypes associated with gene dosage at the chromosome 16p11.2 locus. Nature 478:97–102
Jirtle RL, Skinner MK (2007) Environmental epigenomics and disease susceptibility. Nat Rev Genet 8:253–262
Krude H, Biebermann H, Luck W et al (1998) Severe early-onset obesity, adrenal insufficiency and red hair pigmentation caused by POMC mutations in humans. Nat Genet 19:155–157
Lamri A, Abi Khalil C, Jaziri R, et al. (2011) Dietary fat intake and polymorphisms at the PPARG locus modulate BMI and type 2 diabetes risk in the D.E.S.I.R. prospective study. Int J Obes (Lond) 2011 May 3. [Epub ahead of print]
Le Stunff C, Le Bihan C, Schork NJ, Bougneres P (2000) A common promoter variant of the leptin gene is associated with changes in the relationship between serum leptin and fat mass in obese girls. Diabetes 49:2196–2200
Li WD, Reed DR, Lee JH et al (1999) Sequence variants in the 5′ flanking region of the leptin gene are associated with obesity in women. Ann Hum Genet 63:227–234

Li S, Zhao JH, Luan J et al (2010) Cumulative effects and predictive value of common obesity-susceptibility variants identified by genome-wide association studies. Am J Clin Nutr 91:184–190
Loos RJ, Lindgren CM, Li S et al (2008) Common variants near MC4R are associated with fat mass, weight and risk of obesity. Nat Genet 40:768–775
Mammès O, Betoulle D, Aubert R et al (1998) Novel polymorphisms in the 5′ region of the LEP gene: association with leptin levels and response to low-calorie diet in human obesity. Diabetes 47:487–489
Mammès O, Betoulle D, Aubert R et al (2000) Association of the G-2548A polymorphism in the 5′ region of the LEP gene with overweight. Ann Hum Genet 64:391–394
Mammès O, Aubert R, Betoulle D et al (2001) LEPR gene polymorphisms: associations with overweight, fat mass and response to diet in women. Eur J Clin Invest 31:398–404
Mergen M, Mergen H, Ozata M et al (2001) A novel melanocortin 4 receptor (MC4R) gene mutation associated with morbid obesity. J Clin Endocrinol Metab 86:3448
Meyre D, Delplanque J, Chevre JC et al (2009) Genome-wide association study for early-onset and morbid adult obesity identifies three new risk loci in European populations. Nat Genet 41:157–159
Montague CT, Farooqi IS, Whitehead JP et al (1997) Congenital leptin deficiency is associated with severe early-onset obesity in humans. Nature 387:903–908
O'Rahilly S (2009) Human genetics illuminates the paths to metabolic disease. Nature 462:307–314
Ozata M, Ozdemir IC, Licinio J (1999) Human leptin deficiency caused by a missense mutation: multiple endocrine defects, decreased sympathetic tone, and immune system dysfunction indicate new targets for leptin action, greater central than peripheral resistance to the effects of leptin, and spontaneous correction of leptin-mediated defects. J Clin Endocrinol Metab 84:3686–3695
Ramachandrappa S, Farooqi IS (2011) Genetic approaches to understanding human obesity. J Clin Invest 121:2080–2086
Rampersaud E, Mitchell BD, Pollin TI et al (2008) Physical activity and the association of common FTO gene variants with body mass index and obesity. Arch Intern Med 168:1791–1797
Rankinen T, Zuberi A, Chagnon YC et al (2006) The human obesity gene map: the 2005 update. Obesity (Silver Spring) 14:529–644
Ravussin E, Valencia ME, Esparza J et al (1994) Effects of a traditional lifestyle on obesity in Pima Indians. Diabetes Care 17:1067–1074
Sandholt CH, Sparso T, Grarup N et al (2010) Combined analyses of 20 common obesity susceptibility variants. Diabetes 59:1667–1673
Saunders CL, Chiodini BD, Sham P et al (2007) Meta-analysis of genome-wide linkage studies in BMI and obesity. Obesity (Silver Spring) 15:2263–2275
Sina M, Hinney A, Ziegler A et al (1999) Phenotypes in three pedigrees with autosomal dominant obesity caused by haploinsufficiency mutations in the melanocortin-4 receptor gene. Am J Hum Genet 65:1501–1507
Speliotes EK, Willer CJ, Berndt SI et al (2010) Association analyses of 249,796 individuals reveal 18 new loci associated with body mass index. Nat Genet 42:937–948
Stunkard AJ, Sorensen TI, Hanis C et al (1986) An adoption study of human obesity. N Engl J Med 314:193–198
Stunkard AJ, Harris JR, Pedersen NL, McClearn GE (1990) The body-mass index of twins who have been reared apart. N Engl J Med 322:1483–1487
Stutzmann F, Tan K, Vatin V et al (2008) Prevalence of melanocortin-4 receptor deficiency in Europeans and their age-dependent penetrance in multigenerational pedigrees. Diabetes 57:2511–2518
Stutzmann F, Cauchi S, Durand E et al (2009) Common genetic variation near MC4R is associated with eating behaviour patterns in European populations. Int J Obes (Lond) 33:373–378
Thorleifsson G, Walters GB, Gudbjartsson DF et al (2009) Genome-wide association yields new sequence variants at seven loci that associate with measures of obesity. Nat Genet 41:18–24
Tilg H, Kaser A (2011) Gut microbiome, obesity, and metabolic dysfunction. J Clin Invest 121:2126–2132

Vaisse C, Clément K, Guy-Grand B, Froguel P (1998) A frameshift mutation in human MC4R is associated with a dominant form of obesity. Nat Genet 20:113–114
Vaisse C, Clément K, Durand E et al (2000) Melanocortin-4 receptor mutations are a frequent and heterogeneous cause of morbid obesity. J Clin Invest 106:253–262
Wahlen K, Sjolin E, Hoffstedt J (2008) The common rs9939609 gene variant of the fat mass- and obesity-associated gene FTO is related to fat cell lipolysis. J Lipid Res 49:607–611
Walley AJ, Asher JE, Froguel P (2009) The genetic contribution to non-syndromic human obesity. Nat Rev Genet 10:431–442
Walters RG, Jacquemont S, Valsesia A et al (2010) A new highly penetrant form of obesity due to deletions on chromosome 16p11.2. Nature 463:671–675
Wardle J, Carnell S, Haworth CM et al (2008) Obesity associated genetic variation in FTO is associated with diminished satiety. J Clin Endocrinol Metab 93:3640–3643
Willer CJ, Speliotes EK, Loos RJ et al (2009) Six new loci associated with body mass index highlight a neuronal influence on body weight regulation. Nat Genet 41:25–34
Yeo GS, Farooqi IS, Aminian S et al (1998) A frameshift mutation in MC4R associated with dominantly inherited human obesity. Nat Genet 20:111–112
Zhang Y, Proenca R, Maffei M et al (1994) Positional cloning of the mouse obese gene and its human homologue. Nature 372:425–432
The 1000 Genome Project Consortium (2010) A map of human genome variation from population-scale sequencing. Nature 467:1061–1073
Cho YS, Go MJ, Kim YJ et al (2009) A large-scale genome-wide association study of Asian populations uncovers genetic factors influencing eight quantitative traits. Nat Genet 41:527–534
Choquet H, Meyre D (2011) Molecular basis of obesity: current status and future prospects. Curr Genomics 12:154–168
Frayling TM (2007) Genome-wide association studies provide new insights into type 2 diabetes aetiology. Nat Rev Genet 8:657–662
Heid IM, Jackson AU, Randall JC et al (2010) Meta-analysis identifies 13 new loci associated with waist-hip ratio and reveals sexual dimorphism in the genetic basis of fat distribution. Nat Genet 42:949–960
Lindgren CM, Heid IM, Randall JC et al (2009) Genome-wide association scan meta-analysis identifies three Loci influencing adiposity and fat distribution. PLoS Genet 5:e1000508
Scherag A, Dina C, Hinney A et al (2010) Two new Loci for body-weight regulation identified in a joint analysis of genome-wide association studies for early-onset extreme obesity in French and german study groups. PLoS Genet 6:e1000916

Chapter 25
Genetic and Acquired Lipodystrophic Syndromes

Corinne Vigouroux, Véronique Béréziat, Martine Caron-Debarle and Jacqueline Capeau

Introduction

Increased fat amount as observed in android obesity and metabolic syndrome is clearly associated with metabolic complications. However, conversely, absence or paucity of fat due to genetic or acquired human lipodystrophies leads to even more severe metabolic alterations resulting in premature complications. In the recent years, the genetic etiology of a number of lipodystrophic syndromes has been

C. Vigouroux · V. Béréziat · M. Caron-Debarle · J. Capeau (✉)
Institut des maladies cardiométaboliques et de la nutrition (IHU ICAN), Paris, France
e-mail: Jacqueline.capeau@inserm.fr

C. Vigouroux
e-mail: corinne.vigouroux@inserm.fr

V. Béréziat
e-mail: Veronique.bereziat@inserm.fr

M. Caron-Debarle
e-mail: Martine.debarle@inserm.fr

C. Vigouroux · V. Béréziat · J. Capeau
Faculté de Médecine Pierre et Marie Curie site Saint-Antoine, INSERM, UPMC, UMR_S938, 27 rue Chaligny, 75012 Paris, France

C. Vigouroux · J. Capeau
Service de Biochimie et Hormonologie, Hôpital Tenon, 4 rue de la Chine, 75970 Paris cedex 20, France

M. Caron-Debarle
Faculté de Médecine Pierre et Marie Curie site Saint-Antoine, INSERM, UPMC, UMR_S938, 27 rue Chaligny, 75012 Paris, France

Faculté de Médecine Lyon Sud-BP 12, 165 Chemin du Grand Revoyet, 69921 Oullins Cedex, France

J.-P. Bastard and B. Fève (eds.), *Physiology and Physiopathology of Adipose Tissue*, DOI: 10.1007/978-2-8178-0343-2_25, © Springer-Verlag France 2013 elucidated (Barroso et al. 1999; Cao and Hegele 2000; Magré et al. 2001; Agarwal et al. 2002; George et al. 2004; Kim et al. 2008; Hayashi et al. 2009; Rubio-Cabezas et al. 2009; Gandotra et al. 2011), pointing out the importance of several proteins in adipose tissue (AT) physiology. Otherwise, acquired lipodystrophies could be of iatrogenic origin, linked to treatment with some HIV-antiretrovirals or corticoids, these forms being quite common, or could be associated with endocrine or immune deregulations.

Lipodystrophic syndromes represent a heterogeneous group of diseases all defined by partial or generalized loss of body fat. If partial, lipoatrophy is often associated with fat hypertrophy in other depots, varying according to the type of lipodystrophy. Lipodystrophy is generally associated with severe metabolic alterations including insulin resistance, dyslipidemia and glucose intolerance, stressing for the importance of fat for the correct regulation of whole-body metabolism. Total absence of fat leads to very severe metabolic disturbances. Partial loss of fat can lead to different phenotypes: if AT is reduced in the lower part of the body and often increased in the upper part as in the familial partial lipodystrophies (FPLD), severe metabolic alterations are observed (Garg 2011; Vigouroux et al. 2011). When fat is reduced with a reverse phenotype, as observed in the Barraquer-Simons syndrome, metabolic alterations are generally mild or absent (Misra et al. 2004). AT has the capacity to buffer excess lipid and protect cells against its toxicity (Virtue and Vidal-Puig 2010). In addition, AT releases a number of important factors that control insulin sensitivity, inflammation, and energy metabolism including adipokines and anti-inflammatory cytokines, but also pro-inflammatory cytokines and chemokines, secreted in excess when AT is adversely affected. It now appears that maintaining a healthy fat amount and repartition is an essential requirement for the protection of metabolic homeostasis.

A number of human syndromes have been described, and some remain at present not deciphered with regard to their genetic origin and/or their pathophysiological mechanisms. We will only present in this chapter the main ones. The reader can benefit from a recent complete clinical review from A Garg on the subject (Garg 2011).

Studying human lipodystrophies raises important questions on human fat physiology:

- the markedly different severity of FPLD in men and women further stresses the role of sexual hormones in fat development, repartition, and physiology
- the delayed appearance of FPLD questions the possible role of aging together with hormones
- the association of fat hypertrophy with atrophy in different depots as a result of a single affected protein further questions the different physiology of these depots
- the discovery that mutations in proteins associated with the adipocyte lipid droplet lead to the most severe forms of lipodystrophy emphasizes the under-recognized importance of this cellular organelle.
- the role of mitochondria in adipocyte function needs to be revisited since mitochondrial dysfunction and increased oxidative stress are involved in several forms of lipodystrophies and insulin resistance.

- fat remodeling with fibrosis, classically linked to obesity and resulting in lipotoxicity in other tissues, is probably also an important actor of human lipodystrophies, with lipotoxicity-related alterations also present inside AT.
- endocrine factors have to be considered, as cortisol, growth hormone (GH), and androgens, due to their ability to control adipose tissue distribution.
- the role of immune factors is also emerging, possibly leading to immune-linked fat destruction.

Human Lipodystrophies: Clinical Data

At the clinical level, peripheral lipoatrophy of subcutaneous adipose tissue (SAT) can be easily diagnosed when it affects regions with a natural large fat thickness. Loss of fat into cheeks and temples gives a gaunt face and, in the limbs, makes muscles and veins highly visible. However, lipoatrophy can be difficult to diagnose if mild, in particular at the lower limb level in men, who have physiologically a lower fat amount than women. Whole-body dual-energy X-ray absorptiometry (DEXA) evaluates the total body fat amount, and, importantly, the distribution of adipose tissue, which can help for the diagnosis of partial lipoatrophy. A CT-scan at the thigh level can be useful but requires comparisons with normal subjects. The diagnosis of visceral fat atrophy (or hypertrophy) requires imaging techniques: a CT-scan or an MRI at the lumbar L4 level allows precise evaluation of the SAT and visceral adipose tissue (VAT) areas.

Human lipodystrophies are generally associated with severe insulin resistance. Therefore, clinical signs of insulin resistance can help diagnosis: the presence of skin lesions of *acanthosis nigricans,* a skin brownish lesion present in the axillae, neck, and other body folds is an excellent indication of marked insulin resistance, in particular in normal-weight patients. Long-term insulin resistance can lead to acromegaloid features, striking at the level of face and extremities, frequently observed in congenital forms of lipodystrophies. Insulin resistance can result in increased size of genital organs in prepubertal children and in, ovarian hyperandrogenism leading to hirsutism and sometimes virilization with polycystic ovary syndrome or hyperthecosis in women. Post-receptor insulin resistance is also commonly associated with hepatomegaly and liver steatosis.

At the metabolic level, lipodystrophies are characterized by glucose and lipid alterations, which can be mild or even absent during childhood and increase in severity when patients get older. Lipid alterations associate increased triglyceride (TG) level, which can be raised up to 100 mmol/L, leading to a high risk of acute pancreatitis, while HDL-cholesterol is decreased. Glucose values could remain in the normal range in young patients, if insulin secretion is able to compensate for insulin resistance, but increase progressively leading to glucose intolerance then diabetes, difficult to control.

Regarding complications, the main short-term complication is acute pancreatitis due to very high TG level. Chronic complications are related to long-term diabetic complications, microangiopathy, affecting retina, kidney and nerves, macroangiopathy, leading to early atherosclerosis, and to hepatic complications of steatosis with steatohepatitis, sometimes leading to cirrhosis with portal hypertension and hepatic failure.

The alternative diagnosis with syndromes of insulin resistance due to alterations at the insulin receptor level (type A and type B syndromes) could be difficult. However, in these latter syndromes, lipodystrophy, dyslipidemia, and liver steatosis are absent and very high adiponectin levels have been recently reported (Semple et al. 2008). Sex-hormone binding globulin (SHBG) and insulin-like growth factor binding protein 1 (IGFBP1) levels were also reported to be preserved or even elevated in insulin receptor linked-insulin resistance. In contrast, in lipodystrophies, post-receptor insulin resistance is associated with low levels of adiponectin, SHBG, and IGFBP1 (Semple et al. 2008). Lipomatosis represents a different disease, characterized by localized fat tumors. Lipomas can be multiple, affecting mainly the proximal limbs areas and the neck in the familial lipomatosis. They are sometimes associated with mutations in mitochondrial DNA (*MERRF* mutations in particular). The Launois-Bensaude lipomatosis, of unknown origin, is often associated with peripheral neuropathy and increased alcohol intake.

Human Genetic Generalized Lipodystrophies

Generalized lipodystrophies are rare diseases characterized by complete lipoatrophy, affecting all metabolic fat depots, including in some cases the mechanical fat depots.

In the congenital forms, named Berardinelli-Seip congenital lipodystrophy (BSCL), fat is lost at birth or very early after. Lipoatrophy is associated with muscular hypertrophy and organomegaly, in particular cardiac hypertrophy, and insulin resistance. Growth could be accelerated in the children, but height is generally normal in adults. During adulthood, acromegalic features often occur at the face, hand, and feet levels.

BSCL is generally inherited according to a recessive pattern, most patients exhibiting mutations in one of 4 different genes, encoding seipin (*BSCL2*) (Magré et al. 2001), AGPAT2 (1-acylglycerol-3-phosphate-O-acyltransferase 2) (*BSCL1*) (Agarwal et al. 2002), caveolin-1 (*BSCL3*) (Kim et al. 2008), or cavin-1/PTRF (polymerase I and transcript release factor) (*BSCL4*) (Hayashi et al. 2009), all acting on the pathways of triglyceride synthesis and/or storage in the adipocyte lipid droplet (Fig. 25.1).

The clinical presentation is quite similar whatever the genotype of BSCL, but patients with BSCL2 have the most severe phenotype and metabolic disorders and often present with mild mental retardation. In addition, patients with cavin/PTRF mutations often present with muscular dystrophy associating muscular hypertrophy and weakness, and elevated creatine kinase levels.

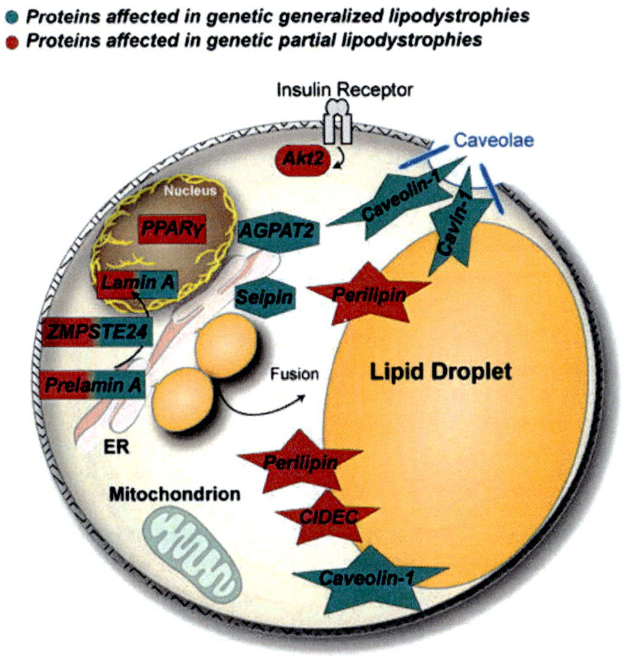

Fig. 25.1 Main proteins affected in genetic lipodystrophies. *AGPAT2* 1-acylglycerol-3-phosphate-O-acyltransferase 2; *PPARγ* Peroxisome proliferator-activated receptor gamma; *CIDEC* Cell death-inducing DFF45-like effector C

AGPAT2

The first BSCL locus, BSCL1, was mapped on chromosome 9q34 and then identified as *AGPAT2* (Agarwal et al. 2002). Thirty-three different mutations have been described among 110 patients: they are null mutations except nine missense mutations. This gene is mutated in about 50 % of patients with typical BSCL. At the clinical level, lipoatrophy implicates all fat depots, except mechanical fat.

AGPAT2 is an enzyme that esterifies the sn-2 position of lysophosphatidic acid (LPA) using acyl-CoA as a substrate to form phosphatidic acid (PA), a key intermediate step in the biosynthesis of glycerophospholipids and TG. AGPAT2 is localized in the endoplasmic reticulum ER (Fig. 25.1), the site of TG synthesis and lipid droplet formation and mainly expressed in adipocytes (Gale et al. 2006).

The mechanism by which AGPAT2 deficiency induces lipodystrophy is likely related to the failure to synthesize TG and to form mature, lipid-loaded adipocytes. Most AGPAT2 mutations underlying BSCL cause near-complete loss of AGPAT activity in vitro. However, in addition to reduced TG storage, AGPAT2 knockdown/deficiency causes an accumulation of several phospholipid species that could be deleterious.

Seipin

Seipin was identified in our laboratory by Jocelyne Magré as the protein mutated in BSCL2 (Magré et al. 2001). Up to now, 31 different mutations have been identified among 137 patients, the majority being "null" mutations that cause dramatic disruption of the protein. Mutations in BSCL2 account for about half of BSCL patients. At the clinical level, patients present the most severe BSCL phenotype with an almost complete lack of AT (Van Maldergem et al. 2002; Simha and Garg 2003), a very low level of circulating leptin but surprisingly detectable adiponectin levels (Antuna-Puente et al. 2010).

BSCL2 encodes an integral protein of the ER (Fig. 25.1). Homo-oligomers of the seipin yeast orthologue were recently shown to form toroids at the junctions between ER and cytosolic lipid droplets (Szymanski et al. 2007; Binns et al. 2010).

The function of seipin in lipid droplets was recently illuminated by data obtained with yeast and human cells. Seipin deficiency results in severe alterations in the morphology of the cellular lipid droplet indicative of a defect in the formation or maturation of this organelle (Szymanski et al. 2007; Boutet et al. 2009), associated with abnormalities in fatty acid composition of phospholipids (Boutet et al. 2009). In lymphoblastoid cell-lines issued from patients with seipin mutations, our lab evidenced a decreased activity of the stearoyl-CoA desaturase-1, the key enzyme of fatty acid monoinsaturation (Boutet et al. 2009). Seipin deficiency is also associated with defective adipogenesis.

Caveolin-1

A homozygous nonsense mutation was identified in *CAV1,* encoding caveolin-1, in one patient by Jocelyne Magré in our laboratory (Kim et al. 2008) inducing a complete loss of caveolin-1 expression in skin fibroblasts. The phenotype of the patient was very similar to that of patients with BSCL1 or BSCL2, although she presented in addition a short stature and a resistance to vitamin D. Additional heterozygous *CAV1* mutations have thereafter been observed but patients presented with atypical partial lipodystrophy and hypertriglyceridemia (Cao et al. 2008).

Caveolin-1 is the major member of the caveolin family (which also comprises caveolin-2 and the muscle-specific caveolin-3), which are integral components of caveolae—specialized microdomains of the plasma membrane enriched in cholesterol and sphingolipids. In adipocytes, caveolin-1 containing caveolae represent 20–30 % of the plasma membrane surface. In adipocytes, caveolae are implicated in the entry of nutrients including free fatty acid (FFA) uptake and their conversion to TG in the lipid droplet. Caveolin-1 is also expressed on the surface of the adipocyte lipid droplet, and probably plays a regulatory role on its expandability (Blouin et al. 2010) (Fig. 25.1). A role for altered autophagy was recently proposed in *CAV1* knockout mice (Le Lay et al. 2010). Possible mechanisms responsible for lipodystrophy could include impairments in the structure and metabolic regulation of the lipid droplet, and/or FFA entry, leading to altered TG synthesis and lipid storage, and to autophagy.

PTRF/Cavin-1

Mutations in *PTRF* encoding the polymerase I and transcript release factor (also known as cavin-1) have been identified in 2009 in Japanese patients with a mixed phenotype including generalized lipodystrophy and muscular dystrophy (Hayashi et al. 2009) and confirmed thereafter. The phenotype of patients with cavin-1 mutations can vary with generally muscular dystrophy with elevated creatine kinase levels and generalized lipodystrophy not always noticed at diagnosis, together with metabolic derangements milder than those reported in other BSCL. Pyloric stenosis and cardiac arrythmia have also been described in some patients.

Cavin-1 was recently shown as an essential factor required for the stabilization of caveolae at the plasma membrane (Hill et al. 2008) that would participate to the last phase of their biogenesis. Loss of cavin-1 causes loss of caveolae and a reduction of expression and mislocalization of the three caveolins (Hayashi et al. 2009) that likely participate to the lipodystrophic and myopathic phenotype. Cavin-1 is also colocalized with caveolin-1 on the adipocyte lipid droplet surface (Fig. 25.1) and could participate to the regulation of triglycerides storage in adipocytes (Blouin et al. 2010).

Other Genes

In addition, mutations in other genes have been associated with generalized lipodystrophy together with signs of premature aging, occurring progressively in the post-natal period. First, heterozygous mutations of *LMNA*, the gene encoding A-type lamins, are responsible for the typical form of Hutchinson-Gilford progeria (De Sandre-Giovannoli et al. 2003; Eriksson et al. 2003) and other progeroid syndromes (Caux et al. 2003; Garg et al. 2009), and homozygous mutations in the gene encoding the zinc metalloproteinase ZMPSTE24, which cleaves prelamin A into mature lamin A, are involved in mandibuloacral dysplasia (MAD) type B (Agarwal et al. 2003).

Other complex syndromes linked to DNA repair defects, including Werner and Bloom syndromes, could present with lipodystrophy and severe insulin resistance, in addition to signs of premature aging (Parker et al. 2011).

Genetic Partial Lipodystrophies

A number of patients present partial forms of lipodystrophies collectively named familial partial lipodystrophies (FPLD) with typically fat loss in the limbs and buttocks and preserved fat or even gain of fat in the upper part of the body, face and neck, and visceral fat.

FPLD are generally dominantly inherited but the lipodystrophic phenotype and the metabolic complications appear progressively after puberty. The severity of the phenotype is generally sex-related, and women have a more severe phenotype than men. In a few cases, the disease is transmitted according to a codominant (metabolic laminopathy linked to *LMNA* p.T655fsX49 mutation (Le Dour et al. 2011)) recessive pattern (FPLD due to mutations in Cell death-inducing DFF45-like effector C, CIDEC) (Rubio-Cabezas et al. 2009), or MAD type A also due to mutations in *LMNA* (Novelli et al. 2002)). The reason for the delayed appearance and the sexual dimorphism is not well understood but suggests the involvement of age-related and possible hormone-related processes. The proteins affected in FPLD are either nuclear proteins as PPARγ and lamin A/C, or proteins involved in insulin signaling as Akt2, or in the adipocyte lipid droplet formation, regulation or maintenance as CIDEC and perilipin (Table 25.1, Fig. 25.1).

Lamin A/C

Familial Partial Lipodystrophy of the Dunnigan type (or FPLD2) is the most typical form of partial lipodystrophic syndrome linked to *LMNA* mutations. The *LMNA* gene encodes the nuclear intermediate filaments A-type lamins, which polymerize with B-type lamins to form the nuclear lamina at the nucleoplasmic side of the inner nuclear membrane in all differentiated cells (Worman 2012) (Fig. 25.1). In addition, A-type lamins are also found in the nucleoplasm, where they interact with chromatin, DNA and transcription factors, and regulate gene positioning, DNA replication, and gene transcription.

The main isoforms of A-type lamins are lamin A and lamin C, which arise from alternative splicing of *LMNA*. Lamin C is produced as a mature protein while a complex post-translational maturation of prelamin A, involving, among several enzymatic processes, the farnesylation of the CAAX C-terminal motif of the protein and ZMPSTE24-mediated proteolysis steps, gives rise to non-farnesylated mature lamin A.

Mutations in *LMNA* cause laminopathies, a group of rare disorders with wide clinical heterogeneity ranging from Emery-Dreifuss muscular dystrophy, limb-girdle muscular dystrophy type 1B, dilated cardiomyopathy with conduction defects, Charcot-Marie-Tooth axonal neuropathy type 2B1, lipodystrophic syndromes and metabolic laminopathies to severe or lethal premature aging syndromes, including MAD-A, Hutchinson-Gilford progeria and progeroid syndromes, and restrictive dermopathy (Worman 2012).

Table 25.1 - Human genetic lipodystrophic syndromes and main affected genes

Gene/protein	Disease	Generalized (G) or partial (P) lipodystrophy	AD or AR transmission
LMNA/lamin A/C	FPLD2	P	AD
LMNA/lamin A/C	Metabolic laminopathies	P	AD
LMNA/lamin A/C	MAD-A	P	AR
LMNA/lamin A/C	Hutchinson-Gilford progeria and progeroid syndromes	G	Heterozygous, de novo mutations
ZMP STE24	MAD-B	G	AR
PPARG/PPARγ	FPLD3	P	AD
PLIN1/perilipin	FPLD4	P	AD
CIDEC	FPLD	P	AR
AGPAT2	BSCL1	G	AR
BSCL2/seipin	BSCL2	G	AR
CAV1/caveolin 1	BSCL3	G	AR
PTRF/cavin 1	BSCL4	G	AR
PSMB8	JMP	G	AR

AD autosomal dominant, *AR* autosomal recessive, *AGPAT2* 1-acylglycerol-3-phosphate O-Acyltransferase 2, *PSMB8* proteasome subunit, beta-type, 8, *FPLD* familial partial lipodystrophy of the Dunnigan type, *MAD* Mandibuloacral dysplasia, *BSCL* Berardinelli-Seip Congenital Lipodystrophy, *JMP* Joint contractures, muscle atrophy, microcytic anemia, and panniculitis-induced lipodystrophy

Most *LMNA* mutations responsible for FPLD2 are heterozygous substitutions of the 482nd codon, in the globular C-terminal domain of the protein, involved in the binding of lamin A to DNA and several transcription factors including SREBP-1c. The FPLD2 phenotype develops after puberty and is characterized by limbs and trunk subcutaneous lipoatrophy, facio-cervical accumulation of fat, insulin resistance, and metabolic alterations with early cardiovascular complications (Cao and Hegele 2000; Shackleton et al. 2000). We have recently shown that the accumulation of cervical fat is not only due to a compensatory storage of triglycerides in these patients, but is a peculiar dystrophic tissue, showing fibrosis, small adipocytes, and remodeling characteristics toward brown-like fat phenotype (Béréziat et al. 2011). Women with FPLD2 are more severely affected, both at the clinical and biological levels than men (Vigouroux et al. 2000). In addition to the typical FPLD2 phenotype, we have described other forms of "metabolic laminopathies" with insulin resistance being the prominent sign, frequently associated with myopathy, and/or cardiac conduction disturbances, in the absence of obvious clinical lipodystrophy (Decaudain et al. 2007; Young et al. 2005). Although typical FPLD2 is an autosomal dominant disease, we have recently described a syndrome linked to a new LMNA mutation, expressed at the heterozygous or the homozygous state, associating partial lipodystrophy, severe insulin resistance, and, in some patients, cardiac conduction defects (Le Dour et al. 2011).

Lipodystrophies due to *LMNA* mutations can also lead to even more complex phenotypes, with progeria-like features (Caux et al. 2003; Garg et al. 2009). In deed, partial or generalized lipoatrophy is part of the phenotype of typical premature aging syndromes due to *LMNA* mutations, MAD and Hutchinson-Gilford progeria. In these diseases, lipoatrophy is associated with post-natal growth retardation, craniofacial dysmorphy, skeletal and skin abnormalities (De Sandre-Giovannoli et al. 2003; Eriksson et al. 2003; Novelli et al. 2002). Interestingly, null mutations in ZMPSTE24, leading to the synthesis of a permanently farnesylated prelamin A, also cause premature aging syndromes with lipodystrophy, in humans and mice (Agarwal et al. 2003).

The molecular mechanisms by which mutations in ubiquitously expressed A-type lamins lead to disorders specifically affecting highly specialized tissues (muscle, heart, AT, bone, cartilage) remain to be elucidated. Three main hypotheses, not mutually exclusive, have been proposed: lamin A/C mutations could result in cellular mechanical stress, in gene expression alterations (leading in particular to altered mesenchymal cell differentiation), and/or in premature aging or lipodystrophic syndromes due to toxicity of farnesylated mutated lamins.

PPARγ

The PPARγ gene encodes the two main isoforms PPARγ-1 and PPARγ-2 that differ by an extra 28-amino acid sequence at the NH2-terminal end of PPARγ-2. PPARγ-2 is mainly expressed in AT while PPARγ-1 is more ubiquitously distributed. The transcriptional activity of PPARγ is induced by endogenous ligands or by synthetic compounds as thiazolidinediones (TZDs). PPARγ transactivation activity is regulated by a number of co-regulators and co-repressors. PPARγ regulates a series of genes involved in adipogenesis (Rosen et al. 1999), in the release, transport, and storage of fatty acids (lipoprotein lipase and the fatty acid transporter CD36). PPARγ also controls insulin-dependent glucose and lipid metabolism, and exerts anti-inflammatory and antioxidant effects in endothelial cells, smooth muscle cells, and macrophages (Duan et al. 2009). PPARγ also regulates bone homeostasis.

Rare dominant-negative and loss-of-function mutations affecting the ligand-binding domain of PPARγ have been identified in patients with FPLD3. The mutations P495L, V318 M (Barroso et al. 1999), and those that lead to truncated proteins (315X and R357X) disrupt the interaction of PPARγ with ligands and coactivators. Several dominant-negative mutations have been described (C114R, C131Y, C162 W) that disrupt the DNA binding domain. Other *PPARG* mutations induce conformational changes in the ligand or DNA binding domains, thus impairing PPARγ transactivation (Jeninga et al. 2009).

The patients are characterized by fat loss in the limbs and gluteal region, hepatic steatosis, dyslipidemia, severe insulin resistance, diabetes and cardiovascular complications (Jeninga et al. 2009). Importantly, the majority of lipodystrophic patients bearing PPARγ mutations present also with severe hypertension, providing evidence for a critical role for PPARγ in blood pressure regulation that may be independent of altered insulin sensitivity.

Akt2

The gene encoding the protein kinase B, *AKT2*, has been involved in one family affected with FPLD. This kinase, playing an important role in the insulin metabolic pathway, was mutated in several members of a family with a dominant transmission of hyperinsulinemia and diabetes (George et al. 2004) (Fig. 25.1). The probant presented with partial lipodystrophy. However, this gene was not found mutated in other lipodystrophic patients and two other missense mutations in *AKT2* do not clearly segregate with insulin resistance in the families and do not alter Akt2 kinase activity (Tan et al. 2007). Interestingly, Akt2 is the major Akt isoform found in adipocytes. Akt has been clearly involved in the inhibition of lipolysis exerted by insulin since it activates the phosphodiesterase PDE3B, which hydrolyses cyclic AMP, therefore decreasing activation of hormone sensitive lipase (HSL).

Recently, activating mutations of Akt2 were found to be responsible for genetic syndromes of hypoglycemia (Hussain et al. 2011).

CIDEC

A homozygous non-sense mutation of CIDEC (p.E186X), leading to the synthesis of a protein truncated in its C-terminal part, has been identified in a young woman presenting with partial lipoatrophy, predominantly affecting limbs and abdomen, muscular hypertrophy, insulin resistant diabetes with *acanthosis nigricans*, hypertriglyceridemia, and liver steatosis (Rubio-Cabezas et al. 2009). The CIDE (cell death-inducing DFF45-like effector) family of proteins comprises three members, CIDEA, CIDEB, and CIDEC, the latter being the human homolog of the murine protein FSP27. CIDE proteins play important roles in fat energy metabolism. FSP27 and CIDEC are predominantly expressed in AT (Fig. 25.1) and in steatotic liver, where they localize to the lipid droplets. CIDEC participates in adipose lipid droplet formation and contributes to the human preadipocyte differentiation process (Li et al. 2010).

Histology of subcutaneous fat from the patient with CIDEC mutation revealed many adipocytes with multiple small lipid droplets and increased mitochondrial density. In accordance, the resting metabolic rate of this patient was increased. In vitro functional studies showed that the mutated CIDEC remained cytosolic, did not localize to lipid droplets and failed to increase their size (Rubio-Cabezas et al. 2009).

Perilipin

A collaborative work between the group of S O'Rahilly and D Savage and our group has recently shown that heterozygous inactivating mutations of the perilipin gene (*PLIN1*) are responsible for a new FPLD syndrome, now called FPLD4 (Gandotra et al. 2011). We have identified two different heterozygous *PLIN1* frameshift mutations (p.L404VfsX158 and p.V398GfsX166) in three non-related women with partial lipoatrophy, insulin resistant diabetes, hypertriglyceridemia, and liver steatosis, leading to the synthesis of aberrant amino acids in the C-terminal region of the protein. These mutations co-segregated with the disease according to an autosomal dominant transmission. Subcutaneous AT from the patients showed an increased fibrosis with macrophage infiltration and adipocytes of decreased size. In accordance, 3T3L1 preadipocytes expressing the mutated forms of perilipin 1 had reduced lipid droplet size and increased basal lipolysis.

Perilipin belongs to the ancient family of the PAT proteins, which bind to intracellular lipid droplet and share a highly-related structure. The unifying nomenclature of "perilipin" proteins (and *PLIN* genes), numbered sequentially from 1 to 5, has been recently adopted to identify the five members of this family. Perilipin (now perilipin 1, encoded by the *PLIN1* gene, chr. 15q26) is specifically localized at the surface of the lipid droplet in adipocytes (Fig. 25.1) and steroidogenic cells.

Perilipin 1 is the most abundant lipid droplet coat protein. It is characterized by the presence of five consensus sequences for serine phosphorylation by c-AMP-dependent protein kinase A (PKA), allowing the regulation of adipocyte lipid storage and lipolysis in response to the cell metabolic status. Perilipin 1 has a dual metabolic function, serving as a dynamic scaffold to coordinate the access of lipases to the lipid droplets (Brasaemle 2007). In the basal state (*i.e.*, in the absence of lipolytic stimuli), perilipin 1 associates with the protein CGI-58 (comparative gene identification-58), a co-activator of ATGL (adipose triglyceride lipase), and restricts the access of cytosolic HSL to stored triglycerides, therefore protecting the lipid droplet by reducing basal lipolysis. Conversely, in response to catecholamines, increased cAMP levels activate PKA, which phosphorylates HSL and perilipin 1. Consequently, CGI-58 dissociates from polyphosphorylated perilipin 1, which recruits HSL at the surface of the lipid droplet, thus activating lipolysis. *PLIN1* mutations evidenced in patients with FPLD4 have been shown to impair the binding of perilipin 1 to CGI-58 in the basal state, thus inducing a sustained constitutive lipolysis (Gandotra et al. 2011a; b).

Human Acquired Lipodystrophies

Acquired Generalized Primary Lipodystrophies

Acquired primary loss of AT is a rare condition. Acquired complete lipoatrophy, named the Lawrence syndrome, can occur in childhood or adulthood. It is associated with the occurrence of severe metabolic disorders and insulin resistance and leads to early cardiovascular, diabetic, and hepatic complications. The presentation is very similar to that of BSCL patients. In addition, in some patients, panniculitis could precede the progressive loss of fat. This form could be associated with signs of auto-immune disorders as hepatitis or anemia. Low complement factor C3, C4, and/or the presence of the C3 nephritic factor is recorded (Misra et al. 2004; Savage et al. 2009). It has been proposed that the progressive disappearance of fat was due to autoantibodies directed against AT (Misra et al. 2004; Savage et al. 2009).

Recently, the group of A Garg described a complex syndrome associating lipodystrophy together with major osteoarticular disorders and a striking dysmorphic appearance. Lipodystrophy resulted from panniculitis-induced fat atrophy and was associated with joint contracture, muscle atrophy, microcytic anemia but with minor metabolic alterations. These patients had an autosomal recessive auto-inflammatory syndrome associated with a homozygous mutation in a gene encoding a catalytic subunit of the immunoproteasome, PSMB8. Patients' lymphoblasts showed reduced chymotrypsin-like proteolytic activity mediated by immunoproteasome probably affecting MHC class I antigen processing and resulting in overall inflammation (Agarwal et al. 2010). A Japanese group has recently confirmed these data and has shown that the protein encoded by PSMB8 also played a role in adipocyte differentiation (Kitamura et al. 2011).

Acquired Partial Primary Lipodystrophies

Regarding acquired partial lipodystrophies, with fat loss in some depots and fat gain in others, the altered body composition could resemble the FPLD phenotype associated with metabolic disorders. The origin of a number of cases, genetic or not, is lacking.

Among the different forms of partial acquired lipodystrophy, one form can be identified since its phenotype is the reverse of that found in FPLD. Patients with the Barraquer-Simmons syndrome, more frequent in women, present a decreased fat amount in the upper part of the body (face, trunk, arms) while fat in the lower part is in excess (buttocks, hips, legs). The majority of the cases are sporadic, and their etiology is unknown, even if auto-immunity has been reported in some cases. A membrano-proliferative glomerulonephritis affects one-third of the patients and more than half of them show signs of activation of the alternate complement pathway: low circulating levels of C3 and presence of C3 nephritic factor. Heterozygous alterations in *LMNB2* encoding lamin B2 have been reported by Hegele et al. (2006) but were not confirmed by the other groups involved in the genetics of lipodystrophies.

Interestingly, while patients with loss of fat in the lower part of the body present with severe metabolic alterations, patients with the reverse phenotype display generally mild or absent alterations, in agreement with the neutral or even beneficial metabolic role of subcutaneous fat from lower limbs (Manolopoulos et al. 2010).

HIV-Related Lipodystrophy

The most frequent form of acquired lipodystrophy is related to HIV-infection and its treatment. In the 1980s, HIV-infection was associated with a devastating burst of morbidity and mortality together with the occurrence of a severe wasting syndrome. The introduction of antiretroviral drugs, at first nucleoside analogue reverse transcriptase inhibitors (NRTI) and, in 1996, HIV-protease inhibitors (PI), allowed a spectacular control of the infection and a marked decrease in AIDS-related mortality and morbidity but resulted in the occurrence of lipodystrophy. In the early 2000s, about half HIV-infected patients (20–80 % according to the studies) were diagnosed with lipodystrophy. A severe lipoatrophy affecting subcutaneous AT was commonly reported in lipodystrophic patients and could be generalized or associated with abdominal fat accumulation, mainly at the intra-visceral level, in some of them. The two phenotypes were often associated. The responsibility of the antiretroviral treatment was rapidly ascertained with a major role for thymidine analogue reverse transcriptase inhibitors (tNRTI), mainly stavudine but also zidovudine, in the occurrence of lipoatrophy. Otherwise, first generation PIs were proposed to act in synergy with tNRTIs and to play a role in visceral and upper body fat hypertrophy. Lipodystrophy was generally associated with metabolic abnormalities, in particular severe dyslipidemia and markedly increased levels of triglycerides and with insulin resistance, altered glucose tolerance, and even diabetes (Caron-Debarle et al. 2010).

At present, lipodystrophy is less prevalent with more recent antiretrovirals used in developed countries but stavudine is largely proposed for treatment initiation in developing countries. Moreover, long-standing lipoatrophy is only partially and slowly reversible. Recently, an increased incidence of abdominal lipohypertrophy has been outlined, associated with the antiretroviral treatment but not clearly associated with any specific class or drug. This lipohypertrophy participates in the increased cardio-metabolic risk observed in these patients. In addition, a state of increased insulin resistance could be linked to immune deficiency in HIV-infected patients, even before the initiation of the antiretroviral treatment (Boufassa et al. 2012). In a large cohort of HIV-infected patients, lipoatrophy and/or abdominal adiposity, as well as the use of some antiretrovirals, were risk factors for incident diabetes (Capeau et al. 2012).

When considering the possible mechanism of antiretroviral toxicity on adipose tissue, stavudine and zidovudine, which cause severe mitochondrial toxicity in part by inhibiting the mtDNA polymerase γ, decrease fat amount and lead to lipoatrophy.

Some HIV-PI are able to inhibit ZMPSTE24 and induce prelamin A accumulation resulting in oxidative stress and inflammation (Caron et al. 2007). The ability of pravastatin or farnesyl transferase inhibitors, which preclude prelamin A farnesylation, to reverse PI-induced-oxidative stress and inflammation has been clearly shown in vitro in fibroblasts and endothelial cells (Caron et al. 2007; Lefèvre et al. 2010).

Taken as a whole, HIV-related lipodystrophy is now considered as a complex multifactorial process. A role for the virus itself is possible but the antiretroviral drugs exert the main effects.

Lipodystrophies Linked to an Excess of Cortisol

Finally, endogenous or exogenous hypercortisolism is associated with the development of a lipodystrophic syndrome with fat loss in the extremities and central fat gain including moon-like facies, buffalo hump (increased subcutaneous fat in the posterior neck and the upper back) and increased visceral AT, leading to the Cushingoid appearance. In addition, these subjects undergo bone loss, hypertension, hyperandrogenism and metabolic alterations with insulin resistance, altered glucose tolerance, diabetes, dyslipidemia, and increased cardiovascular risk (Fardet et al. 2007).

The role of cortisol in adipose tissue is important to consider, since fat is able to convert inactive cortisone into active cortisol due to the presence of the enzyme 11β-hydroxysteroid dehydrogenase type 1. It has been observed that abdominal subcutaneous AT is able to secrete cortisol while the role of visceral fat in that setting has been recently questioned in men. Cortisol increases adipocyte size leading to insulin resistant large adipocytes while it inhibits adipocyte proliferation. However, the mechanisms by which cortisol induces lipodystrophy remain unclear.

The Consequences of Defective Fat on Others Tissues and on Metabolism

Lipodystrophy is always associated with a loss of fat, at least in some depots. Defective AT differentiation, as expected in the case of mutations affecting PPARγ will obviously result in lipodystrophy due to the major role of this factor in adipogenesis, but other tissues are affected since PPARγ is expressed in a number of cell-types where it regulates immunity and inflammation. Otherwise, in most cases of lipodystrophies, it seems that the major concern is the decreased ability to store TG in the adipocyte lipid droplet leading to decreased adipocyte lipid content and therefore lipoatrophy. Indeed, most of the proteins mutated in human lipodystrophies are directly involved in the synthesis and maintenance of the lipid droplet (seipin, perilipin, CIDEC, caveolin 1, and cavin-1) or are acting in the pathways leading to lipid synthesis or storage, at the level of the nucleus, endoplasmic reticulum, or caveolae (lamins A/C through SREBP-1c, AGPAT2, caveolin 1, and cavin-1). Altered lipid storage could result in lipotoxicity inside adipocyte, due to an excess level of intracellular fatty acids or their derivatives. These fatty acids could no longer be stored inside the lipid droplet, leading to mitochondrial dysfunction and increased oxidative stress and therefore to inflammation, remodeling, and lipoatrophy. Dysfunctional AT could result in altered metabolism, insulin resistance, impaired fasting glucose, and dyslipidemia.

An interesting hypothesis has been proposed related to the limitation of fat expandability (Virtue and Vidal-Puig 2010; Huang-Doran et al. 2010). In the general population, it has been observed that some obese individuals, even with very high BMI, are metabolically healthy while, in contrast, less obese subjects could present major metabolic alterations. It is considered that AT has a limited capacity to expand through combination of hyperplasia (limited in adults) and hypertrophy (Virtue and Vidal-Puig 2010). States of positive energy balance are associated with expansion of fat depots to accommodate excess energy intake. But expandability is finite up to a particular set point that varies on an individual basis including genetic factors. Beyond this set point, additional energy excess results in AT failure with oxidative stress, recruitment of macrophages, release of cytokines and of FFA, and decreased adiponectin leading to lipotoxicity. In lipodystrophies, due to limited expansion of fat through defective lipid and lipid droplet formation and maintenance, the set point for fat storage is very low and very rapidly the storage capacity is overwhelm leading to ectopic storage of lipids and severe metabolic deregulation.

The severe insulin resistance and metabolic complications associated with complete lipoatrophy probably result from a high level of lipotoxicity in these patients, that are completely unable to store fat in AT and therefore present with massive ectopic lipid depots in the muscles, heart, liver, and pancreas. In addition, their very low leptin level may contribute to ectopic fat deposition that is, at least partially, reverted by a treatment with recombinant leptin (Chong et al. 2010). The alterations linked to lipotoxicity are probably less severe in patients with partial forms of lipodystrophy, in which lipid storage in some AT depots is still possible.

Therapeutic Options for Human Lipodystrophies

Treatment of metabolic alterations can benefit from diet and exercise recommendations that can ameliorate insulin sensitivity and hypertriglyceridemia. When diabetes is present, it is generally insulin resistant and difficult to control. Insulin sensitizers are used at first, as metformin. A treatment with TZD resulted in favorable effects on glucose control in several patients, even those with mutations in *PPARG,* but is no longer available in France and the use is restricted in other countries. Very high doses of insulin are frequently required. In some cases, medium chain fatty acids supplementation can contribute to lower TG. Otherwise, hypolipidemic drugs such as fibrates are required to avoid major hypertriglyceridemia, which can lead to acute pancreatitis.

Since these patients often present very low leptin levels, replacement of leptin with recombinant human leptin has been evaluated and resulted in markedly improved metabolic values, and regression of liver steatosis, particularly in severe lipoatrophic and hypoleptinemic patients (Oral et al. 2002).

The altered body fat repartition can benefit from plastic surgery. Patients with FPLD2 sometimes undergo successful removal of excess fat at the neck level. In patients with HIV-related facial lipoatrophy, plastic surgery is able to provide amelioration, even if often transitory: the Coleman technique consists in injection into the cheeks of autologous fat. Otherwise, the use of slowly resorbable polylactic acid gels or with non-resorbable fillers such as alkyl-imide allows partial correction. In some patients with severe hypertrophy of fat as buffalo humps, fat removal can be proposed but with a risk of recurrence.

Some medications could possibly improve lipoatrophy: a treatment with troglitazone, a first generation TZD, was initially shown to restore some fat in the limbs in patients with generalized lipodystrophy not related to HIV-infection (Arioglu et al. 2000). In patients with HIV-related lipodystrophy, TZD revealed poor efficacy on fat restoration. This was probably due to the presence of stavudine in the patients' treatment, which impeded fat restitution. When pioglitazone was given to patients not treated by stavudine, a moderate improvement in peripheral fat was reported (Slama et al. 2008).

Conclusion

Lipodystrophies represent a heterogeneous group of severe diseases leading to early diabetic, cardiovascular, and hepatic complications. Alterations in adipose tissue distribution could result from mutations in several genes: the presence of lipodystrophy outlines the importance of these genes in adipose tissue function. The role of the adipocyte lipid droplet as a new organelle playing a leading role in adipocyte functions is shown by the discovery that several genes mutated in lipodystrophies act at that level (Vigouroux et al. 2011). Active researches are looking for mutations in other candidate genes. Even if the pathophysiology of lipodystrophies remains largely unknown, it is obvious that all situations with fat loss, in particular in the lower body fat depots, are associated with severe metabolic disturbances and insulin resistance, while the only lipodystrophic syndrome with the reverse repartition of fat (the Barraquer-Simons syndrome) is generally not associated with severe metabolic alterations. This is reminiscent of the android obesity associated with abnormal metabolic parameters while the gynoid form is largely devoid of them.

Human partial lipodystrophies commonly associate loss of fat in some depots while others are increased. This points to the different physiology of the different fat depots, since the same genetic alteration or drug-induced toxicity results in opposite phenotypes depending on the fat localization.

The presence of mitochondrial dysfunction in lipodystrophies has been revealed in adipose tissue from patients with FPLD2 and other *LMNA* mutations but also in HIV-related lipodystrophies. Interestingly, some forms of lipomatosis result from mutations in mitochondrial DNA. Therefore, the relation between mitochondria and adipose tissue is probably important and complex and could result either in lipoatrophy but also in hypertrophied fat. Mitochondrial dysfunction has been also involved in muscular insulin resistance found during aging and in diabetic patients.

The specific role of lamin A/C in adipose tissue is important to consider. Accumulation of farnesylated prelamin A is involved in several diseases associated with premature aging but also in *LMNA* and HIV-linked lipodystrophies, which also present signs of premature aging. During normal aging, redistribution of fat from the lower to the upper body fat depots is observed. Whether there is a link between type A-lamins and normal aging remains to be demonstrated.

Therefore, studies on human lipodystrophies help to understand the complex physiology and pathophysiology of fat. They point to new genes and new targets, which could lead to the discovery of new therapeutic clues in order to help treatment of patients with lipodystrophies but also, more generally, of patients with common forms of fat redistribution as observed in the metabolic syndrome and type 2 diabetes.

Summary

Human lipodystrophic syndromes are rare conditions in which total or partial fat loss is associated with severe alterations of lipid and glucose parameters together with insulin resistance leading to early diabetes, cardiovascular, and hepatic complications. Lipodystrophies can be classified as genetic or acquired and as generalized or partial. The genetic origin of a number of human lipodystrophic syndromes has been recently unraveled and 10 genes are associated at present with human lipodystrophies, revealing the genetic heterogeneity of these syndromes that are transmitted according to a monogenic recessive or dominant pattern. Acquired forms of lipodystrophy are iatrogenic or linked to immune and endocrine factors, the main ones being secondary to treatments with some human immunodeficiency virus (HIV)-antiretrovirals or to endogenous or exogenous excess of cortisol. Overall, limited fat storage in adipose tissue leads to an altered buffering capacity of excess caloric intake leading to ectopic lipid accumulation in liver, muscles, heart, and pancreas, responsible for impaired insulin sensitivity and cellular dysfunctions. In partial lipodystrophies, increased lipolysis and production of inflammatory mediators by the remaining fat also participates to this process called "lipotoxicity" with increased oxidative stress, insulin resistance, and fat remodeling. Indeed, most of the proteins or functions affected by mutations or some antiretrovirals result in altered adipogenesis, triglyceride storage, and/or formation of the unique adipocyte lipid droplet. Some mutations or antiretrovirals could affect adipogenesis directly or indirectly, through the transcription factors (PPARγ peroxisome proliferator-activated receptor gammaγ, mutated or secondarily altered) or sterol regulatory element binding protein 1c (SREBP1c, secondarily altered by lamin A/C mutations or some HIV protease inhibitors), and insulin signaling through post-receptor alterations (including Akt2 mutations). Other mutated genes are involved in the control of triglyceride synthesis toward the adipocyte lipid droplet (AGPAT2, 1-acylglycerol-3-phosphate-O-acyltransferase 2), or in the formation or maintenance of the lipid droplet (seipin, cell death-inducing DFF45-like effector C (CIDEC), perilipin, caveolin-1, cavin-1). In partial lipodystrophies, remaining fat appears abnormal, with fibrosis and mitochondrial dysfunction. Endocrine factors, such as increased cortisol, decreased growth hormone, and androgens probably worsen the phenotype of upper body and/or visceral fat hypertrophy. In all cases of lipodystrophies, "lipotoxicity" results in adverted liver and heart functions leading to early complications. The management of lipodystrophic patients is difficult: lifestyle modifications with control of caloric intake and increased exercise are required but rarely sufficient, insulin sensitizers and molecules lowering lipids should be used, and high doses of insulin are frequently needed. New therapeutic options as recombinant human leptin (metreleptin) substitution appear helpful in patients with severe metabolic complications.

Acknowledgments Funding sources for most of our studies were from the Institut National de la Santé et de la Recherche Médicale (INSERM), the Agence Nationale de Recherche sur le SIDA et les hépatites (ANRS) and the Société Française de Diabétologie (SFD)

We would like to thank Betty Jacquin for secretarial assistance and Caroline Le Dour for figure drawing.

References

Agarwal AK, Arioglu E, De Almeida S et al (2002) AGPAT2 is mutated in congenital generalized lipodystrophy linked to chromosome 9q34. Nature Genet 31:21–23

Agarwal AK, Fryns JP, Auchus RJ et al (2003) Zinc metalloproteinase, ZMPSTE24, is mutated in mandibuloacral dysplasia. Hum Mol Genet 12:1995–2001

Agarwal AK, Xing C, DeMartino GN et al (2010) PSMB8 encoding the beta5i proteasome subunit is mutated in joint contractures, muscle atrophy, microcytic anemia, and panniculitis-induced lipodystrophy syndrome. Am J Hum Genet 87:866–872

Antuna-Puente B, Boutet E, Vigouroux C et al (2010) Higher adiponectin levels in patients with Berardinelli-Seip congenital lipodystrophy due to seipin as compared with 1-acylglycerol-3-phosphate-o-acyltransferase-2 deficiency. J Clin Endocrinol Metab 95:1463–1468

Arioglu E, Duncan-Morin J, Sebring N et al (2000) Efficacy and safety of troglitazone in the treatment of lipodystrophy syndromes. Ann Intern Med 133:263–274

Barroso I, Gurnell M, Crowley VE et al (1999) Dominant negative mutations in human PPARgamma associated with severe insulin resistance, diabetes mellitus and hypertension. Nature 402:880–883

Béréziat V, Cervera P, Le Dour C et al (2011) LMNA mutations induce a non-inflammatory fibrosis and a brown fat-like dystrophy of enlarged cervical adipose tissue. Am J Pathol 179:2445–2453

Binns DD, Lee S, Hilton CL et al (2010) Seipin is a discrete homooligomer. Biochemistry 49:10747–10755

Blouin CM, Le Lay S, Eberl A et al (2010) Lipid droplet analysis in caveolin-deficient adipocytes: alterations in surface phospholipid composition and maturation defects. J Lipid Res 51:945–956

Boufassa F, Goujard C, Viard JP et al (2012) Immune deficiency could be an early risk factor for altered insulin sensitivity in antiretroviral-naïve HIV-1-infected patients: the ANRS COPANA cohort. Antiviral Ther 17:91–100

Boutet E, El Mourabit H, Prot M et al (2009) Seipin deficiency alters fatty acid Delta9 desaturation and lipid droplet formation in Berardinelli-Seip congenital lipodystrophy. Biochimie 91:796–803

Brasaemle DL (2007) Thematic review series: adipocyte biology. The perilipin family of structural lipid droplet proteins: stabilization of lipid droplets and control of lipolysis. J Lipid Res 48:2547–2559

Cao H, Hegele RA (2000) Nuclear lamin A/C R482Q mutation in canadian kindreds with Dunnigan-type familial partial lipodystrophy. Hum Mol Genet 9:109–112

Cao H, Alston L, Ruschman J et al (2008) Heterozygous CAV1 frameshift mutations (MIM 601047) in patients with atypical partial lipodystrophy and hypertriglyceridemia. Lipids Health Dis 7:3

Capeau J, Bouteloup V, Katlama C et al (2012) Ten-year diabetes incidence in 1,046 HIV-infected patients started on a combination antiretroviral treatment: the ANRS CO8 APROCO-COPILOTE Cohort. AIDS 26:303–314

Caron M, Auclair M, Donadille B et al (2007) Human lipodystrophies linked to mutations in A-type lamins and to HIV protease inhibitor therapy are both associated with prelamin A accumulation, oxidative stress and premature cellular senescence. Cell Death Differ 14:1759–1767

Caron-Debarle M, Lagathu C, Boccara F et al (2010) HIV-associated lipodystrophy: from fat injury to premature aging. Trends Mol Med 16:218–229

Caux F, Dubosclard E, Lascols O et al (2003) A new clinical condition linked to a novel mutation in lamins A and C with generalized lipoatrophy, insulin-resistant diabetes, disseminated leukomelanodermic papules, liver steatosis, and cardiomyopathy. J Clin Endocrinol Metab 88:1006–1113

Chong AY, Lupsa BC, Cochran EK et al (2010) Efficacy of leptin therapy in the different forms of human lipodystrophy. Diabetologia 53:27–35

De Sandre-Giovannoli A, Bernard R, Cau P et al (2003) Lamin A truncation in Hutchinson-Gilford progeria. Science 300:2055

Decaudain A, Vantyghem MC, Guerci B et al (2007) New metabolic phenotypes in laminopathies: LMNA mutations in patients with severe metabolic syndrome. J Clin Endocrinol Metab 92:4835–4844

Duan SZ, Usher MG, Mortensen RM (2009) PPARs: the vasculature, inflammation and hypertension. Curr Opin Nephrol Hypertens 18:128–133

Eriksson M, Brown WT, Gordon LB et al (2003) Recurrent de novo point mutations in lamin A cause Hutchinson-Gilford progeria syndrome. Nature 423:293–298

Fardet L, Cabane J, Lebbe C et al (2007) Incidence and risk factors for corticosteroid-induced lipodystrophy: a prospective study. J Am Acad Dermatol 57:604–609

Gale SE, Frolov A, Han X et al (2006) A regulatory role for 1-acylglycerol-3-phosphate-O-acyltransferase 2 in adipocyte differentiation. J Biol Chem 281:11082–11089

Gandotra S, Le Dour C, Bottomley W et al (2011a) Perilipin deficiency and autosomal dominant partial lipodystrophy. NEJM 364:740–748

Gandotra S, Lim K, Girousse A et al (2011b) Human Frame Shift Mutations Affecting the Carboxyl Terminus of Perilipin Increase Lipolysis by Failing to Sequester the Adipose Triglyceride Lipase (ATGL) Coactivator AB-hydrolase-containing 5 (ABHD5). J Biol Chem 286:34998–35006

Garg A (2011) Lipodystrophies: genetic and acquired body fat disorders. J Clin Endocrinol Metab 96:3313–3325

Garg A, Subramanyam L, Agarwal AK et al (2009) Atypical progeroid syndrome due to heterozygous missense LMNA mutations. J Clin Endocrinol Metab 94:4971–4983

George S, Rochford JJ, Wolfrum C et al (2004) A family with severe insulin resistance and diabetes due to a mutation in AKT2. Science 304:1325–1328

Hayashi YK, Matsuda C, Ogawa M et al (2009) Human PTRF mutations cause secondary deficiency of caveolins resulting in muscular dystrophy with generalized lipodystrophy. J Clin Invest 119:2623–2633

Hegele RA, Cao H, Liu DM et al (2006) Sequencing of the reannotated LMNB2 gene reveals novel mutations in patients with acquired partial lipodystrophy. Am J Hum Genet 79:383–389

Hill MM, Bastiani M, Luetterforst R et al (2008) PTRF-Cavin, a conserved cytoplasmic protein required for caveola formation and function. Cell 132:113–124

Huang-Doran I, Sleigh A, Rochford JJ et al (2010) Lipodystrophy: metabolic insights from a rare disorder. J Endocrinol 207:245–255

Hussain K, Challis B, Rocha N et al (2011) An activating mutation of AKT2 and human hypoglycemia. Science 334:474

Jeninga EH, Gurnell M, Kalkhoven E (2009) Functional implications of genetic variation in human PPARgamma. Trends Endocrinol Metab 20:380–387

Kim C, Delépine M, Boutet E et al (2008) Association of a homozygous nonsense Caveolin-1 mutation with Berardinelli-Seip Congenital Lipodystrophy. J Clin Endocrinol Metab 93:1129–1134

Kitamura A, Maekawa Y, Uehara H et al (2011) A mutation in the immunoproteasome subunit PSMB8 causes autoinflammation and lipodystrophy in humans. J Clin Invest 121:4150–4160

Le Dour C, Schneebeli S, Bakiri F et al (2011) A homozygous mutation of prelamin-A preventing its farnesylation and maturation leads to a severe lipodystrophic phenotype: new insights into the pathogenicity of nonfarnesylated prelamin-A. J Clin Endocrinol Metab 96:E856–E862

Le Lay S, Briand N, Blouin CM et al (2010) The lipoatrophic caveolin-1 deficient mouse model reveals autophagy in mature adipocytes. Autophagy 6:754–763

Lefèvre C, Auclair M, Boccara F et al (2010) Premature senescence of vascular cells is induced by HIV protease inhibitors: implication of prelamin A and reversion by statin. Arterioscler Thromb Vasc Biol 30:2611–2620

Li F, Gu Y, Dong W et al (2010) Cell death-inducing DFF45-like effector, a lipid droplet-associated protein, might be involved in the differentiation of human adipocytes. FEBS J 277:4173–4183

Magré J, Delépine M, Khallouf E et al (2001) Identification of the gene altered in Berardinelli-Seip congenital lipodystrophy on chromosome 11q13. Nature Genet 28:365–370

Manolopoulos KN, Karpe F, Frayn KN (2010) Gluteofemoral body fat as a determinant of metabolic health. Int J Obes 34:949–959

Misra A, Peethambaram A, Garg A (2004) Clinical features and metabolic and autoimmune derangements in acquired partial lipodystrophy: report of 35 cases and review of the literature. Medicine (Baltimore) 83:18–34

Novelli G, Muchir A, Sangiuolo F et al (2002) Mandibuloacral dysplasia is caused by a mutation in LMNA-encoding lamin A/C. Am J Hum Genet 71:426–431

Oral EA, Simha V, Ruiz E et al (2002) Leptin-replacement therapy for lipodystrophy. N Engl J Med 346:570–578

Parker VE, Savage DB, O'Rahilly S et al (2011) Mechanistic insights into insulin resistance in the genetic era. Diabetic Med 28:1476–1486

Rosen ED, Sarraf P, Troy AE et al (1999) PPAR gamma is required for the differentiation of adipose tissue in vivo and in vitro. Mol Cell 4:611–617

Rubio-Cabezas O, Puri V, Murano I et al (2009) Partial lipodystrophy and insulin resistant diabetes in a patient with a homozygous nonsense mutation in CIDEC. EMBO Mol Med 1:280–287

Savage DB, Semple RK, Clatworthy MR et al (2009) Complement abnormalities in acquired lipodystrophy revisited. J Clin Endocrinol Metab 94:10–16

Semple RK, Cochran EK, Soos MA et al (2008) Plasma adiponectin as a marker of insulin receptor dysfunction: clinical utility in severe insulin resistance. Diabetes Care 31:977–979

Shackleton S, Lloyd DJ, Jackson SN et al (2000) LMNA, encoding lamin A/C, is mutated in partial lipodystrophy. Nature Genet 24:153–156

Simha V, Garg A (2003) Phenotypic heterogeneity in body fat distribution in patients with congenital generalized lipodystrophy caused by mutations in the AGPAT2 or Seipin genes. J Clin Endocrinol Metab 88:5433–5437

Slama L, Lanoy E, Valantin MA et al (2008) Effect of pioglitazone on HIV-1-related lipodystrophy: a randomized double-blind placebo-controlled trial (ANRS 113). Antivir Ther 13:67–76

Szymanski KM, Binns D, Bartz R et al (2007) The lipodystrophy protein seipin is found at endoplasmic reticulum lipid droplet junctions and is important for droplet morphology. Proc Natl Acad Sci USA 104:20890–20895

Tan K, Kimber WA, Luan J et al (2007) Analysis of genetic variation in Akt2/PKB-beta in severe insulin resistance, lipodystrophy, type 2 diabetes, and related metabolic phenotypes. Diabetes 56:714–719

Van Maldergem L, Magré J, Gedde-DahlJr T et al (2002) Genotype-phenotype relationships in Berardinelli-Seip congenital lipodystrophy. J Med Genet 39:722–733
Vigouroux C, Magré J, Vantyghem MC et al (2000) Lamin A/C gene: sex-determined expression of mutations in Dunnigan-type familial partial lipodystrophy and absence of coding mutations in congenital and acquired generalized lipoatrophy. Diabetes 49:1958–1962
Vigouroux C, Caron-Debarle M, Le Dour C et al (2011) Molecular mechanisms of human lipodystrophies: from adipocyte lipid droplet to oxidative stress and lipotoxicity. Int J Biochem Cell Biol 43:862–876
Virtue S, Vidal-Puig A (2010) Adipose tissue expandability, lipotoxicity and the Metabolic Syndrome–an allostatic perspective. Biochim Biophys Acta 1801:338–349
Worman HJ (2012) Nuclear lamins and laminopathies. J Pathol 226:316–325
Young J, Morbois-Trabut L, Couzinet B et al (2005) Type A insulin resistance syndrome revealing a novel lamin A mutation. Diabetes 54:1873–1878

Chapter 26
Is the Adipose Tissue a Relevant Target for Obesity Treatment?

Olivier Ziegler and Michel Krempf

Introduction

An excess in adiposity has been clearly associated with numerous comorbidities and pathophysiologic processes, including mechanical and metabolic complications (insulin resistance, altered glucose, or lipid metabolism) influenced by inherited or acquired factors.

In order to define relevant therapeutic targets, it is critical to distinguish obese individuals at high risk for obesity-related metabolic diseases from those who are metabolically healthy (Blüher 2009; Dulloo et al. 2010; Primeau et al. 2010). Sex (male), age, body mass index (BMI), and central adipose tissue (AT) distribution are clearly the main parameters (Després and Lemieux 2006; Jensen 2008).

However, new concepts have emerged during the last 10 years, as limited AT expandability, AT dysfunction, abnormal free fatty acid (FFA) trafficking, specific role for visceral AT (VAT), ectopic fat accumulation, low-grade inflammation, acquired partial lipodystrophy or lipotoxicity (Dulloo et al. 2010; Jensen 2008; Danforth 2000; Frayn 2002; McGarry 2002; Ravussin and Smith 2002; Lewis et al. 2002; Bays et al. 2006; Boden 2008; Heilbronn et al. 2004; Unger 2003; Unger et al. 2010; Virtue and Vidal-Puig 2008, 2010; Mittendorfer 2011; Arsenault et al. 2011; Guri and Bassaganya-Riera 2010).

O. Ziegler (✉)
Service de Diabétologie, Maladies Métaboliques, Nutrition,
CHU de Nancy—Hôpital Brabois Adultes, Rue du Morvan,
Vandoeuvre-Lès-Nancy, 54500 Nancy, France
e-mail: o.ziegler@chu-nancy.fr

M. Krempf
Institut du Thorax, Endocrinologie, Maladies métaboliques et Nutrition
Hôpital Laennec, 44093 Nantes CX, France
e-mail: mkrempf@sante.univ-nantes.fr

J.-P. Bastard and B. Fève (eds.), *Physiology and Physiopathology of Adipose Tissue*,
DOI: 10.1007/978-2-8178-0343-2_26,

In other words, metabolic abnormalities could be viewed as the result of impaired AT function, a phenomenon described in the literature as « adiposopathy » (Bays et al. 2006), « adipocyte dysfunction » (Guri and Bassaganya-Riera 2010; Guilherme et al. 2008) « hazardous fat » *versus* « protective fat » (Dulloo et al. 2010), « sick adipose tissue » (Mittendorfer 2011), « metabolically healthy *versus* unhealthy obese individuals » (Blüher 2009; Primeau et al. 2010) or « lipocentric approach of type 2 diabetes » (T2DM) (McGarry 2002). Of note, this new hypothesis was defended by McGarry (McGarry 2002) during the 2001 Banting lecture: obesity-related insulin resistance could be secondary to ectopic fat accumulation in liver and muscle.

AT is the main lipid storage depot in our body, and is of crucial importance in buffering the daily influx of dietary fat entering the circulation (Frayn 2002). The aim of this paper is to review how impaired AT function contributes to the metabolic syndrome, a pro-inflammatory, atherogenic, and diabetogenic state (Després and Lemieux 2006; Arsenault et al. 2011).

There is a need to revise the clinical approach to the problems of obesity treatment, as further discussed below. Targeting not only caloric imbalance but also AT dysfunction may represent a novel strategy to prevent obesity-related diseases. In short, to the question: is the AT a relevant therapeutic target for obesity management? The answer is yes, but guidelines should help to identify who are the healthy or unhealthy obese patients, and who will benefit the most of treatments including novel strategies targeting AT dysfunction.

A New Paradigm

Normal Adipose Tissue Function

Healthy AT is characterized by a larger number of small fat cells, and a low level of local inflammation. It is well vascularized leading to an adequate supply of oxygen and nutrients for the fat cells and minimally fibrotic. It also displays preserved insulin sensitivity and mitochondrial function.

The main function of adipocytes is to store FFA during energy overload and release them during fasting or starvation. AT has a special place in "buffering" lipid fluxes because it is the single major tissue where triacylglycerol (TG) clearance and FA trapping are up-regulated in the post-prandial period (Jensen 2008; Frayn 2002). It is also the only site of an adapted release of FFA into the circulation, when energy is required. These two functions are the key of AT metabolic flexibility.

AT is also a major endocrine organ with a very active secretory pathway, shown to link adiposity to insulin resistance, T2DM and cardiovascular diseases (CVD). As an example, adiponectin has insulin-sensitizing, anti-inflammatory, anti-apoptotic, and pro-angiogenic properties. The beneficial metabolic effects of this adipokine is to increase the metabolic flexibility of AT.

Impaired Metabolic Flexibility

FFA Flux From Adipose Tissue, a Mass Effect

Since the pool of FFA of adipocytes is released into the circulation in relation to its size, the greater overall fat mass of AT in obese individuals will result in an elevation of FFA flux to non-adipose tissues (Frayn 2002; Boden 2008), even in the absence of a qualitative abnormality in AT metabolism (Lewis et al. 2002). The accumulation of TG in AT leads to increased lipolysis by a mass effect. *De facto*, severe obesity is directly associated with systemic insulin resistance related to enhance FFA flux (Boden 2008).

Role of Insulin Resistance

Insulin resistance of any cause, is associated with metabolic flexibility impairment, both for uptake or release of FFA. Indeed insulin increases lipoprotein lipase action, which is the key enzyme for FFA uptake by adipocytes from TG-rich lipoproteins. Insulin inhibits FFA release (lipolysis) from AT into the circulation via suppressing hormone-sensitive lipase.

As insulin resistance develops, TG accumulation in "fat-buffering" AT is limited (i.e. expandability is limited) (Unger 2003). Moreover insulin-mediated suppression of lipolysis is overwhelmed and increased release of FFA in the circulation ensues leading to fat accumulation into the liver and other non-adipose tissues (McGarry 2002; Lewis et al. 2002; Boden 2008).

Regional Differences in Fat Cell Metabolism

Adipocytes from different fat depots do not have the same metabolic properties, which explain some differences in whole-body lipid metabolism between men and women. Visceral fat cells are more sensitive than subcutaneous (sc) fat cells to the lipolytic effect of catecholamines and less sensitive to the antilipolytic effect of insulin. Adipocytes from abdominal sc AT (SAT) have intermediate properties between visceral fat cells and cells from femoral and gluteal region (Jensen 2008). The response of omental adipocytes to stress via cortisol or NPY local action is particularly high.

However, the respective size of fat depot has to be taken into consideration while explaining FFA flux. Excess release of FFA from upper body sc fat is a major contributor of systemic FFA flux (Jensen 2008; Nielsen et al. 2004). Moreover, the adipogenic ability of visceral adipocytes is lower as compared with sc adipocytes.

The AT distribution is quite different between males and females but also between individuals, as reported by Thomas et al. (2012), even after adjustment for

BMI and age. A total of 477 white volunteers (243 males, 234 females), aged from 17 to 71 years (mean: 37 years) with a wide range of BMI (27.3 ± 4.8 for men; 26 ± 6.8 for women) participated to this large UK-based cohort. The distribution of AT was assessed using anthropometry and whole-body magnetic resonance imaging (MRI) and proton magnetic resonance spectroscopy (MRS) was used to determine intrahepatocellular (IHCL) and intramyocellular (IMCL) lipid content (see results below).

From Dysfunctional Adipose Tissue to Lipotoxicity

Adipose Tissue Dysfunction

The potential abnormal functions of AT include (1) a reduced ability to take up and store circulating FAs; (2) abnormal and excess FFA release (due to increased rate of lipolysis) (Jensen 2008; Frayn 2002). Adipocyte dysfunction is also characterized by insulin resistance and dysregulated adipokine production, e.g., adiponectin is produced at lower levels, whereas the production of pro-inflammatory adipokines is increased (Guilherme et al. 2008; Westerink and Visseren 2011).

The primary signal leading to AT dysfunction could be adipocyte hypertrophy, since enlarged adipocytes are more insulin resistant and produce a pro-inflammatory adipokine pattern (Guri and Bassaganya-Riera 2010; Guilherme et al. 2008; Sun et al. 2011). In the presence of a positive energy balance, AT expands as a result of cellular hypertrophy and hyperplasia. The storage capacity for additional FAs would be dependent on the number and size of fat cells already present and the capacity of AT to activate local pre-adipocytes to undergo the differentiation process in order to form new mature fat cells.

Extensive AT growth has been associated with increased fibrosis and inflammation (Guri and Bassaganya-Riera 2010; Sun et al. 2011; Divoux et al. 2010) and the development of hypoxic conditions (Sun et al. 2011). Adipocytes and macrophages from inflamed fat in obesity release harmful pro-inflammatory cytokines, leading to systemic insulin resistance (Guri and Bassaganya-Riera 2010; Guilherme et al. 2008; Divoux et al. 2010).

Some individuals who gradually gain weight over years in adulthood can initially preserve AT function because they may increase their adipocyte size to a certain threshold before recruiting new fat cells from committed precursor cells or mesenchymal stem (Blüher 2009). A potential dangerous state, from a metabolical point-of-view, is one where existing adipocytes are already highly repleted with lipids, small in number, associated with a low capacity for new adipocyte formation (Danforth 2000; Ravussin and Smith 2002; Hajer et al. 2008). The adaptation of the storage process to the load may also be too slow in case of rapid heavy loads of dietary fat, as can be seen in binge eating disorder, resulting in adipocyte dysfunction.

Interindividual Variations of Fat Storage Capacity

The expandability of the AT stores differs globally among individuals (Virtue and Vidal-Puig 2008; Virtue and Vidal-Puig 2010). Some individuals (BMI > 50 kg/m^2) may be able to undergo almost unlimited expansion of their AT, and hence remain protected from the metabolic syndrome and the toxic effects of FAs. Those with large hip and thigh circumferences (gluteo-femoral phenotype or callipygia) would belong to this category.

The ability to expand AT has a strong sexually dimorphic component. Women, at any BMI, have a greater fat mass and are protected against insulin resistance more than men (Unger et al. 2010; Virtue and Vidal-Puig 2008; Virtue and Vidal-Puig 2010). Gluteo-femoral fat depots with much more expandability are able to limit accumulation of fat in the intraabdominal cavity and the lipotoxicity phenomenon.

Increased FFA Flux and Ectopic Fat Depots

When the storage capacity of AT is exceeded, then a continuous exposure to FAs, not balanced by increased FA oxidation, will lead to an overaccumulation of unoxidized FFA and induces a lipid "spill over"to other tissues. Lipid is deposited in other fat depots (VAT, epicardial AT and peri-vascular tissue) and in non-adipose tissue organs such as liver, skeletal muscle, heart muscle, pancreatic β-cells, and kidney (Boden 2008; Virtue and Vidal-Puig 2010; Mittendorfer 2011; Guri and Bassaganya-Riera 2010).

Fat deposition in liver and muscle is a fundamental pathogenic mechanism leading to the metabolic syndrome (Després and Lemieux 2006; McGarry 2002; Lewis et al. 2002; Arsenault et al. 2011). Increased exposure and uptake of FFA by the liver promote hepatic steatosis which exacerbates the degree of hepatic insulin resistance, and in turn increases hepatic glucose production and very low density lipoprotein (VLDL) secretion, and accelerates its subsequent transition to non-alcoholic steatohepatitis (NASH) and fibrosis. Increased lipid content in the pancreatic islets and lipoapoptosis are the mechanisms, linking obesity and insulin resistance to β cell dysfunction in the pathogenesis of T2DM. Fatty heart or lipotoxic cardiomyopathy is associated with diastolic then systolic dysfunction (Szczepaniak et al. 2007).

Excess Visceral Adiposity: A Marker of Ectopic Fat Depot Syndrome

VAT is significantly linked to increased cardiometabolic risk and perhaps the best candidate among the various fat depots (Després and Lemieux 2006; Arsenault et al. 2011), but it could be a surrogate marker and not a causal factor. The preferential accumulation of fat in the intraabdominal cavity could be viewed as the consequence of the failure of lipid storage in SAT depots. It may be considered as an ectopic deposition, even though it is a physiological fat depot. According to this alternative hypothesis, increased visceral fat mass may represent a marker of the inability of patients with insulin-resistant obesity to store fat in subcutaneous depots (Blüher 2009; Després and Lemieux 2006).

However, VAT may be considered as a major cause of metabolic complications for many reasons. This depot is more active on a per unit weight basis than sc fat (Jensen 2008). Hypertrophic adipocytes and visceral adipose tissue resident macrophages produce more pro-inflammatory cytokines like tumor necrosis factor-alpha (TNF-α) and interleukin-6 (IL-6) and less adiponectin (Jensen 2008). VAT is also more pathogenic because its venous blood drainage is direct via the portal vein to the liver. As a consequence of enlarged visceral adipocytes, the flux of FFA and pro-inflammatory adipokines to the liver increases markedly. This contributes to fat deposition in the liver and to hepatic insulin resistance (Portal hypothesis of insulin resistance).

In the already mentioned study of Thomas et al. (2012), VAT is the fat depot that best predicts ectopic fat accumulation in liver and muscles (see above). Finally, in many studies visceral fat area might be a surrogate parameter for ectopic fat deposition in other sites. As demonstrated by Klöting et al. (Klöting et al. 2010), prediction of variance in glucose infusion rate (GIR) during clamp studies, using visceral fat area ($r^2 = 0.74$) or estimated hepatic steatosis ($r^2 = 0.7$) alone was similar, suggesting that the association between insulin sensitivity and visceral fat area or liver fat content is exchangeable.

FFA, Lipotoxicity, and Insulin Resistance

When lipid overload in non-adipose tissues exceeds the oxidative or storage capacity, resultant cellular dysfunction or cell death is termed lipotoxicity (Unger et al. 2010; Virtue and Vidal-Puig 2010). This cellular stress depends on organelle specific vulnerability (especially endoplasmic reticulum and mitochondria) (Carobbio et al. 2011).

Saturation of TG biosynthesis (a safe form of lipid) and the accumulation of toxic activated lipids (Boden 2008; Mittendorfer 2011) disrupt function such as mitochondrial oxidative phosphorylation and insulin signaling, thus triggering insulin resistance, cellular dysfunction, and finally apoptotic cell death (Unger 2003; Unger et al. 2010; Virtue and Vidal-Puig 2010; Carobbio et al. 2011). Saturated FA such as palmitate, have been shown to induce toxicity by incorporation into ceramide, stimulating the activation of specific signaling kinases that have pro-inflammatory action.

Activated FA may be incorporated into cellular lipids, lysophosphatidic acid (LPA), phosphatidic acid (PA), diacylglycerol (DAG), and finally TG. LPA, PA, and DAG are, as ceramide, lipotoxic species. Harmless TG can be stored in lipid droplets (Guilherme et al. 2008). This response to FFA overload also includes membrane lipid remodeling (Carobbio et al. 2011) that can increase Reactive Oxygen Species (ROS) production (Carobbio et al. 2011).

Another way of getting rid of FFA and their lipotoxic effects is through oxidation. Interestingly, it has been suggested that mitochondrial dysfunction leads to incomplete FA oxidation and mediates lipotoxicity. This could be either inherited (from a genetic or epigenetic cause) or secondary to mitochondrial stress induced by excess of lipid supply, particularly ceramides that also promote the generation of ROS (Virtue and Vidal-Puig 2010; Mittendorfer 2011; Guilherme et al. 2008).

A New Clinical Paradigm: Obesity Subphenotypes

There is a need for new parameters based on pathophysiology (in particular the lipotoxicity hypothesis) in order to improve the classification of obesity phenotypes. During the last 60 years a lot of parameters or techniques have been used for estimating adiposity and fat distribution: (1) anthropometric measures (BMI, waist circumference, waist-to-hip ratio, waist-to-height ratio, skin folds), (2) fat mass, fat depot distribution or ectopic fat deposition assessment using Dual Energy X-ray absorptiometry (DEXA), MRS or MRI; (3) biological parameters to evaluate metabolic complications or inflammation (see below) and finally (4) biomarkers (adipokine production, adipocyte morphology, and transcriptome, …) to evaluate adipocyte function or AT remodeling.

These recent advances in concepts about metabolic risk related to body composition and fat distribution lead us to consider five obesity phenotypes, from a clinical point-of-view. This "redefinition of obesity" could be useful for classification of patients and for risk evaluation, before choosing the appropriate management approach.

Metabolically Healthy but Obese Subjects

The definition of metabolic normality is not easy, because there is a continuous relationship between changes in the previously described parameters and the deterioration of components of metabolic syndrome (Blüher 2009). To date, there is no standardized terminology or method to identify Metabolically Healthy but Obese subjects (MHO) for research protocols or in clinical practice (Primeau et al. 2010; Pataky et al. 2011). The hyperinsulinemic euglycemic clamp, which is the gold standard to determine insulin sensitivity, has been used to evaluate MHO individuals (Klöting et al. 2010; Pataky et al. 2011). But according to Pataky et al. (Pataky et al. 2011), fasting insulinemia and BMI are also useful to characterize the metabolic profile. However, few metabolic differences were observed between MHO subjects identified using metabolic risk factors and insulin-sensitive obese individuals identified using insulin sensitivity indexes (Blüher 2009; Primeau et al. 2010). Of course, the absence of any metabolic disorder including abnormal glucose tolerance, T2DM, dyslipidemia, or hypertension and CVD in an overweight or obese individual is required.

If a consensus definition of the MHO phenotype (Blüher 2009; Primeau et al. 2010; Pataky et al. 2011) does not exist, several recurrent characteristics in MHO subjects have been used, such as plasma levels of TG, apolipoprotein B, high sensitivity C Reactive Protein (hs-CRP), fasting insulin or insulin sensitivity index (e.g. homeostasis model assessment of insulin resistance: HOMA-IR), and TG/HDL-cholesterol ratio. The waist circumference could be excluded as a potential marker because most class 2 obese individuals (IMC $\geq$ 35 kg/m^2) have large waist circumferences. So, this parameter becomes nondiscriminatory in the identification of MHO individuals.

In clinical practice, the main criterion to classify MHO subjects is the absence of metabolic complication (0 or one NCEP ATP III or IDF metabolic syndrome parameters) (Primeau et al. 2010). But if these criteria might thus represent a basis for a consensus definition of the MHO phenotype, the discussion on the cutoff value is still open.

The prevalence of healthy obesity ranges from 10 to 35 % in different studies using different approaches when the obese state is defined using BMI criteria (>30 kg/m^2) (Primeau et al. 2010). Data from Shea et al. (2010) which included 1907 men and women from Canada at the age of 18–85 years, demonstrate that the prevalence of MHO phenotype is higher if obesity state is defined using body fat percentage (determined using DEXA) as compared with BMI criteria (47.7 % *vs.* 34 %, $p < 0.05$).

Mechanisms that could explain the MHO phenotype are poorly understood. However, recent studies (Blüher 2009; Primeau et al. 2010) suggest that lower VAT content, higher gluteal and femoral deposition or lower VAT/abdominal SAT ratio (23), normal AT function (see below, biomarkers), absence of ectopic fat deposition and a normal insulin sensitivity are the major parameters (Blüher 2009; Primeau et al. 2010). With regard to pathological alterations of AT, small adipose cell size, lack of inflammation or fibrosis and gene expression-encoding markers of adipose cell differentiation (adipogenesis) (Basdevant and Clément 2011) could be related to the MHO phenotype. In these conditions, lipid storage in AT is considered as safe and efficient (Fig. 26.1) (Basdevant and Clément 2011).

Unhealthy Obesity: Metabolically Abnormal Obese Patients

Metabolic syndrome related to central obesity may be viewed as « Unhealthy Obesity » and the concerned individuals as metabolically abnormal obese (MAO) patients (Blüher 2009; Primeau et al. 2010) or vulnerable patients (19).

AT dysfunction could be the key link between unhealthy obesity and insulin resistance. Abnormal circulating FFA, visceral and ectopic fat storage that

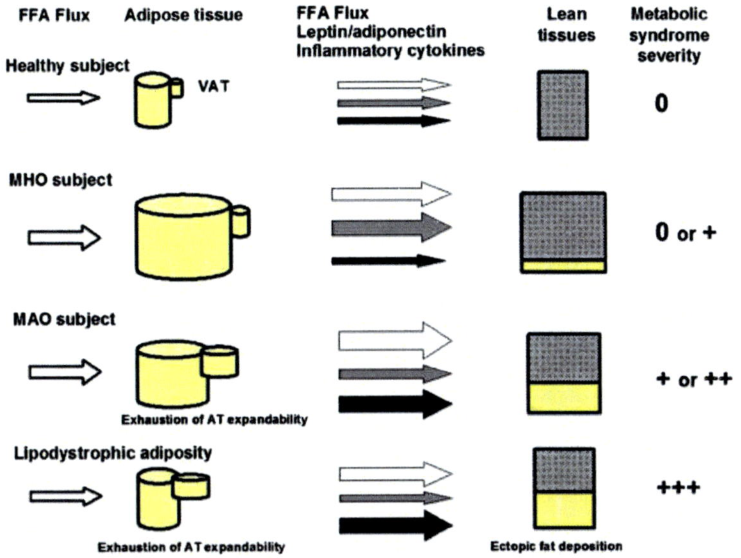

Fig. 26.1 Lipid partitioning between adipose tissue (Visceral AT included) and non-adipose tissues (liver, muscle…), from Unger et al. (Unger 2003) modified. The left arrow indicates excess lipid flux (resulting from increased lipid intake and reduced energy expenditure, i.e. fat oxidation). The three central arrows are respectively (from *up to bottom*) *1* Free fatty acids that are neither stored in AT nor oxidized; *2* Protective anti-lipotoxic adipokines (leptin, adiponectin); *3* Pro-inflammatory cytokines. Ectopic fat deposition appears when AT reaches the limit of its storage capacity; VAT may be considered as an ectopic fat depot. Metabolic syndrome severity depends on obesity phenotype. *AT* adipose tissue; *FFA* free fatty acid; *MAO* metabolically abnormal obese patient; *MHO*: metabolically healthy but obese subject; *VAT* visceral AT

contribute to whole body insulin resistance, may be considered as a consequence of AT dysfunction, when energy balance is positive due to hyperphagia or excess sedentarity (Després and Lemieux 2006; Unger 2003; Unger et al. 2010; Arsenault et al. 2011). A deteriorated cardiometabolic profile or metabolic syndrome including novel components such as NASH and lipotoxic cardiomyopathy (Després and Lemieux 2006; Unger 2003; Unger et al. 2010) may be viewed as a consequence, even if the direction of causality is unclear. It is well documented that metabolic syndrome is a risk factor for T2DM, CVD, and probably for some cancers.

Clinical and biological characteristics of MAO subjects include often a class 1 obesity (BMI < 35 kg/m^2), a predominantly upper body fat distribution, commonly associated with increased visceral fat and the presence of three out five parameters of the metabolic syndrome. Ectopic fat depots are usual (Fig. 26.1). Other relevant plasma parameters to define MAO are HOMA-IR, atherogenic indexes, abnormal liver enzymes (NASH), lower circulating adiponectin, increased inflammatory cytokines levels. As far as AT cellular phenotype is concerned, increased adipocyte diameter, higher macrophage infiltration into fat depot (especially VAT), fibrosis and change in inflammation, and AT remodeling in gene expression (Divoux et al. 2010; Cancello et al. 2005; Capel et al. 2009; Rizkalla et al. 2011).

Lipodystrophic Adiposity

Partial Lipodystrophy

Partial lipodystrophy, either inherited (e.g. Dunnigan syndrome or familial partial lipodystrophy (FPLD) or acquired (HIV-lipodystrophy) and obesity (MAO) are opposite in terms of a deficiency *versus* excess of AT mass (Villarroya et al. 2009; Garg 2011). Yet, these two conditions are accompanied by relatively similar metabolic consequences, including insulin resistance, dysglycemia, dyslipidemia, ectopic fat accumulation (hepatic steatosis and excess VAT), and finally lipotoxicity (Fig. 26.1) (Unger et al. 2010; Virtue and Vidal-Puig 2010). These abnormalities are predictors of an increased risk for diabetes and atherosclerosis in both situations.

The alterations in AT consist of partial loss of peripheral (lipoatrophy) SAT (in the face, limbs, and buttocks) associated with truncal and or/visceral fat accumulation, and lipomatosis, especially in the dorso-cervical area. But clinical presentations vary widely.

Acquired Lipodystrophic Adiposity

The metabolic syndrome is defined in relationship to generalized or regional adiposity, but it has been hypothesized that an acquired type of partial lipodystrophy, could be a cause of insulin resistance for some overweight adults (Ravussin and Smith 2002; Hajer et al. 2008), children (Taksali et al. 2008) or elderly persons (Gavi et al. 2007; Ziegler and Quilliot 2008).

In a preliminary study (600 patients, fat mass distribution measured by DEXA), we described a partial lipodystrophy syndrome, commonly associated with metabolic syndrome that is closely linked with the severity of metabolic complications over a wide range of BMI.

However, the clinical presentation shows less severe loss of peripheral SAT as compared with lipodystrophy associated with HIV-1 infection and anti-retroviral treatment or Dunnigan-type FPLD. When BMI is below 25 kg/m^2 this syndrome may be described as lipodystrophic adiposity and for patients with BMI $\geq$ 30 kg/m^2 as lipodystrophic obesity (Fig. 26.2).

Screening was made easily using two clinical criteria: (1) wiry legs (extremities appear muscular with visible veins due to partial lipoatrophy and (2) tummy or potbelly. A more accurate assessment of central and peripheral adiposity was done using DEXA, the reference method to estimate body fat distribution. The best

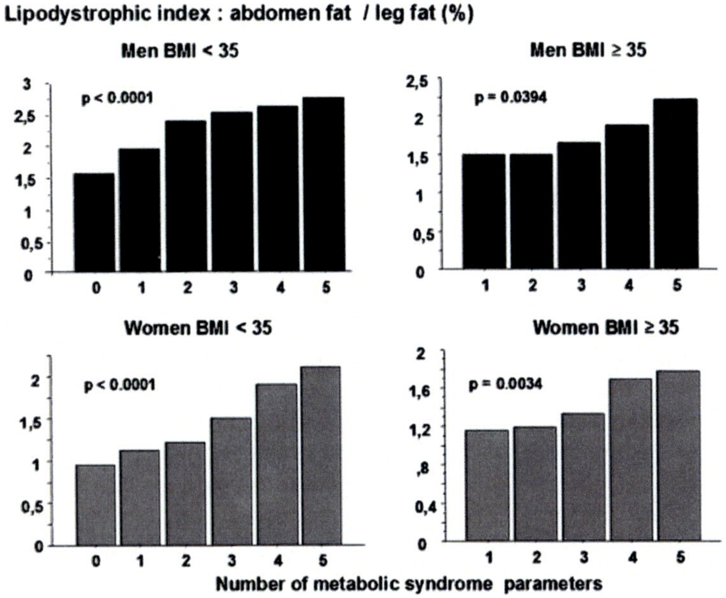

Fig. 26.2 Relationship between the lipodystrophic adiposity phenotype and the severity of the metabolic syndrome. 600 subjects (224 men and 376 women, BMI:18.5–44 kg/m^2, aged from 18 to 77 years, 50 % with metabolic syndrome) were included in this study. Levels of abdominal, leg, and total body fat were measured by DEXA. A lipodystrophic index (abdomen fat to leg fat ratio) was used to assess the degree of partial lipodystrophy (loss of appendicular fat and accumulation of central fat) and the number of metabolic syndrome components (NCEP ATP III definition) its severity. The results showed that the lipodystrophic index increased significantly with the number of metabolic syndrome parameters. All the severely obese subjects (BMI $\geq$ 35 kg/m^2) had an increased waist circumference (no patient with 0 parameter). In this subgroup, the lipodystrophic index was significantly increased only in patients with severe metabolic syndrome (with 4 or 5 components)

lipodystrophic criterion was, in our hands, the leg fat to abdomen fat ratio, by analogy with the V/S ratio (or VAT/SAT) measured with MRI or computerized tomography scan. Our data showed a continuum, as far as AT distribution is concerned, from central adiposity related to common metabolic syndrome to more severe partial lipodystrophy associated with severe metabolic complications (e.g. T2DM with severe insulin resistance or marked hypertriglyceridemia). A few Dunnigan-type FPLD due to mutation in LMNA and 2 FLDP due to mutation in PPARγ were found in this study, confirming the recent data from Dutour et al. (2011).

As described for the genetic type, the results showed that acquired lipodystrophic adiposity is more pronounced in women, but clinical presentation is heterogeneous. Fat loss was often confined to the extremities (legs or forearm) suggesting more a Köbberling type of FPLD (Herbst et al. 2003) (but without a specific edge where the sc fat ends) than a Dunnigan type (Garg 2011). There was a normal or increased distribution of fat on the face, neck, and trunk; both abdominal SAT and VAT (not measured here) seem to be involved in many cases. Buffalo hump (dorso-cervical lipomatosis) was common. Some patients have been evaluated negatively for Cushing's syndrome, because fat accumulation in the face and chin could look like a "moon face" but skin atrophy or muscle weakness were lacking (Dulloo et al. 2010; Unger 2003; Unger et al. 2010).

Many studies have provided evidence that lower amounts of adiposity in the lower extremities is independently associated with metabolic syndrome parameters. In a recent cross-sectional study (Cardia) of 1,579 middle-aged men and women, Shay et al. (Shay et al. 2011) demonstrated that lower amounts of DEXA-assessed lower extremity adiposity are associated with higher insulin resistance in overweight and obese men and women. This association was more pronounced, when examined at a given level of abdominal adiposity. In other words, increased adiposity in the lower extremities may attenuate the metabolic risk observed in overweight and obese individuals. The gluteo-femoral fat depot is viewed as a protective metabolic sink with a good capacity for expansion due to its intrinsic properties (insulin-sensitive adipocytes, low lipolytic rate, no metabolic inflammation, increased secretion of metabolically beneficial adipokines).

Aging may represent another acquired partial lipodystrophy syndrome and this usual phenotype could be, together with sedentarity, one of the major risk factor for the development of insulin resistance in the elderly population (Gavi et al. 2007; Ziegler and Quilliot 2008).

This lipodystrophy phenotype is also characterized by loss of peripheral fat and accumulation of fat in the trunk (Gavi et al. 2007). A significant correlation between age and the lipodystrophic ratio (abdomen fat/leg fat) has been found in our preliminary study. As demonstrated by Thomas et al. (2012), physiological and ectopic fat depots increase with age in both sexes; the same is true for the VAT/SAT ratio, but a reduction of SAT is related to the menopausal status for Caucasian women.

A phenotype reminiscent of partial lipodystrophy has been described in obese children or adolescents. These "obese insulin-resistant" individuals have increased VAT and decreased SAT (see below).

Metabolically Obese Normal Weight Subjects

As described by Ruderman et al. (Ruderman et al. 1998), metabolically obese normal-weight (MONW) individuals exhibit features of the metabolic syndrome but lack the obesity that is usually associated. These subjects have been characterized as having central adiposity, low physical activity, low VO_{2max}, and low insulin sensitivity despite having a normal BMI.

Normal weight obesity is associated with significant cardiometabolic dysregulation, including metabolic syndrome and cardiovascular risk factors (Wildman et al. 2008). Srinivasan et al. (2009) reported the study of 639 normal weight (BMI: 18.5–24.9 kg/m2), black and white adults (75 % white and 36 % men), 20–

44 years old, from the Bogalusa Heart Study. The subjects with central adiposity (waist-to-height ratio ≥0.5) had a significant greater prevalence of hypertension, dyslipidemia (increased TG/HDL-cholesterol ratio), insulin resistance (HOMA-IR index), and elevated hs-CRP (relative risk: 2–3).

Thomas et al. (2012) proposed the "thin-on-the-outside, fat-on-the-inside" (TOFI) as a subphenotype for individuals with proportionally elevated VAT, who are at increased metabolic risk. The definition is based on the VAT/abdominal SAT ratio. Individuals with a BMI between 18.5 and 25 kg/m^2 and increased VAT/SAT ratio, were classified as TOFI; this corresponded to 12 % of women and 14 % of men in their cohort (23). Significantly increased ectopic fat depots (liver and muscle) were observed in TOFI male and female subjects.

A recent study demonstrated that MONW is independently associated with increased risk for cardiovascular mortality in women (Romero-Corral et al. 2009).

It will be of interest to determine if (1) MONW (or TOFI) represents a condition of impaired adipogenesis and expandability of AT; (2) this condition is also associated with a lipodystrophic distribution of AT (see above).

Sarcopenic Obesity

Sarcopenic obesity is an alternate subtype of obesity characterized by loss of muscle and a concomitant increase in fat (Zamboni et al. 2008; Stenholm et al. 2008). Sarcopenic obese patients have more body fat and less lean body mass than non-sarcopenic patients of similar weight. Unfortunately, there is no universal definition for sarcopenic obesity. Moreover, muscle impairment has also to be taken into account, because poor muscle strength is associated with important negative health outcomes. So, sarcopenia is an important cause of frailty, disability, and loss of independence in older obese individuals (Ziegler and Quilliot 2008).

The link between the low-grade inflammatory state associated with obesity and sarcopenia remains to be fully established, but in both cases, inflammatory cytokines could be involved (Stenholm et al. 2008).

Phenotyping of Obese Children and Adolescents

Time Course of Metabolic Phenotype from Childhood to Adulthood

A community-based cohort, the Bogalusa Heart Study (BHS) (1,344 children and adolescents, aged from 4 to 18 years, followed for more than 20 years) has been designed to study the early natural history of T2DM and metabolic syndrome (Srinivasan et al. 2002; Nguyen et al. 2008). Pre-diabetic and diabetic subjects *versus* normoglycemic subjects displayed since childhood, significantly higher levels of BMI and subscapular skinfold (a central adiposity index), and plasma metabolic abnormalities (excess glucose, insulin, HOMA-IR and TG, and low levels of HDL-cholesterol). BMI and insulinemia in childhood are predictors of metabolic syndrome at adulthood (Srinivasan et al. 2002).

BMI and subscapular skinfolds (triceps and subscapular) were consistently higher from childhood to adulthood in offspring of parents with coronary artery disease (CAD) history; higher insulinemia from offspring were associated with positive parental history of CAD after an age of 20 years (Youssef et al. 2002).

Results from BHS have been confirmed by a recent cohort study (Sherar et al. 2011), using DEXA body composition measurements. Young adults with high cardiometabolic risk (26 years old), compared to low, have significantly steeper trajectories of trunk fat mass development, as early as 8 years of age (Sherar et al. 2011).

Ectopic fat depots are already present in obese children and adolescents (Taksali et al. 2008; Liska et al. 2007; Kursawe et al. 2010; Weiss et al. 2005). There seems to be some clear ethnic differences in the fat accumulation in skeletal muscle, liver, and abdominal cavity that suggest the role of genetic, environmental or cultural factors. Obese Hispanic adolescents have a greater amount of lipid accumulated in skeletal muscle than both Caucasians and African Americans, whereas excess visceral fat and liver steatosis were significantly lower in African Americans as compared with both Caucasians and Hispanics (Liska et al. 2007).

Young South Asians had more body fat and more central adiposity than white Caucasians but a lower BMI. Body weight at birth is also lower (Dulloo et al. 2010). South Asians compared to white Caucasians have a higher visceral to superficial AT ratio, more liver fat and a greater adipocyte area (SAT biopsy) (Anand et al. 2011). Ethnic differences in plasma metabolic markers (HDL-cholesterol, insulin, adiponectin) and liver steatosis can be explained by differences in adipocyte size and storage capacity of SAT (Anand et al. 2011).

Obese adolescents with a high proportion of VAT and relatively low abdominal SAT have a phenotype reminiscent of partial lipodystrophy, as reported by Taskali et al. (Taksali et al. 2008). They are at a high risk of having the metabolic syndrome. Hepatic fat was increased, both leptin and total adiponectin were significantly lower, but BMI and fat mass were lower in the lipodystrophic group. As in adults, this phenotype was associated with severe metabolic abnormalities and ectopic fat deposition (liver and VAT) (Taksali et al. 2008). A lower expression of adipogenic and lipogenic genes in abdominal SAT was linked to the phenotype of high visceral and low sc fat depots in obese adolescents (Kursawe et al. 2010).

The MHO phenotype does exist in obese adolescent. Obese insulin-sensitive adolescents are characterized by lower lipid deposition in muscle and VAT and greater levels of adiponectin, as compared with MAO phenotype (Weiss et al. 2005).

Early Programing

The Barker Hypothesis (the association between the prenatal environment, i.e. fetal programing) and disease risk in adult life was an important step in the "developmental origins of health and disease" (DOHaD) field. Low body weight at birth is a significant risk factor for trunk fat accumulation, insulin resistance, metabolic syndrome, and finally CAD in adult life (Barker et al. 2005; Bhargava et al. 2004). In a retrospective longitudinal study of 8760 subjects from Finland, Barker et al. (Barker et al. 2005) reported that boys and girls who had coronary events as adults, had low birth weight and were thin at 2 years of age, after which they tended to increase their BMI rapidly. In the India population-based study of Bhargava et al. (Bhargava et al. 2004), the growth of children in whom impaired glucose tolerance or diabetes later developed was characterized by a low BMI between birth and 2 years of age, a young age at adiposity rebound, and a sustained accelerated gain in BMI until adulthood.

An early development of AT could be protective and contribute to metabolically healthy obesity phenotype, according to an attractive hypothesis (Bouhours-Nouet et al. 2008). Obese 10-year-old children with high birth weight (without gestational diabetes) and postnatal weight gain between 0 and 2 years have better insulin sensitivity as compared with children of the same BMI, but with normal or low birth weight. The high birth weight subjects had less central obesity (lower trunk fat mass/limb fat mass ratio as measured by DEXA), higher insulin sensitivity indexes from an oral glucose tolerance test (lower insulin and FFA concentrations) and higher adiponectin level, than eutrophic and low birth weight subjects. One can speculate that a greater ability of SAT to expand capacity and a better insulin sensitivity (lipogenesis) may protect non-adipose tissues from FFA "spillover", leading to lipotoxicity.

Children who are large-for-gestational-age (LGA) at birth and exposed to an intrauterine environment of diabetes are at increased risk of developing early metabolic syndrome, as shown by the longitudinal study of Boney et al. (2005) who have followed children between 6 and 11 years of age. Analysis of insulin resistance at 11 years in a multivariate logistic regression model revealed that childhood obesity and the combination of LGA status and maternal GDM were associated with insulin resistance, with odds ratios of 4.3 (1.5–11.9) and 10.4 (1.5–74.4), respectively (Boney et al. 2005).

Goals of Obesity Treatment

Definition of Targets

The treatment of obesity has wider objectives than weight loss alone, but treatment targets should be based on the individual's comorbidities and risks that are clearly related to the obesity subphenotypes, and to the natural course of the disease (Fig. 26.3). Reduction in body fat and changes in fat distribution are the primary goals for any obesity management program, but other plasma or tissue biomarkers have to be considered to define appropriate guidelines (Table 26.1).

Phenotypes	Adipose tissue	Adipocytes
MHO Healthy AT 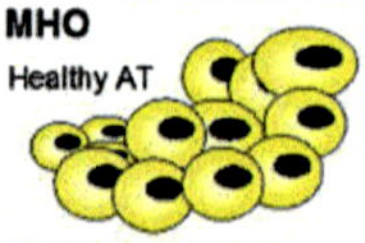	Preserved storage capacity for additional FA in AT Generalized obesity	Hypertrophy: + Hyperplasia: + or +++ (↑ with BMI) Insulin sensitivity ↑ Lipolysis (mass effect)
MAO « Sick » AT AT dysfunction 	Exhaustion of appendicular AT expandability Central Obesity Hypoxia, inflammation ↑ Proinflammatory cytokines	Hypertrophy: ++ or +++ Hyperplasia: 0 or + or ++ (↑ with BMI) Insulin resistance: + or ++ ↑↑ Lipolysis
Lipodystrophic Adiposity « Sick » AT Severe AT dysfunction	Total inability to expand sc fat depots Excess of central AT Loss of peripheral sc AT Hypoxia, inflammation ++ ↑↑ Pro-inflammatory cytokines	Hypertrophy: ++ or +++ (central AT) ↓ Cell size or apoptosis (peripheral sc AT) Insulin resistance: +++ ↑↑↑ Lipolysis

Fig. 26.3 Characteristics of adipose tissue according to obesity phenotypes. MHO: individuals may gain weight due to the almost unlimited expansion capacity of their AT; insulin sensitivity is preserved; they remain protected from the toxic effects of FAs. MAO: the AT reaches the limit of its storage capacity. Metabolic complications are due to ectopic deposition of lipid excess in non-adipose organs (liver, muscle). Lipodystrophic adiposity: The capacity of lipid storage is completely exhausted, due to a partial loss of sc limb fat depots, except in the truncal depots, leading to a severe AT dysfunction. *AT* adipose tissue, *MAO* metabolically abnormal obese patient; *MHO* metabolically healthy but obese subject; *SC* subcutaneous; *VAT* visceral AT

Target According to Obesity Phenotypes

Obesity class but also the identification of the adiposity phenotypes in a clinical setting could have important implications for therapeutic medical decision making.

Severe Obesity (BMI > 35 kg/m^2)

For most patients with BMI >35 kg/m^2 the necessary weight loss for significant improvement of risk factors is larger than previously thought (>15–20 %) from short-term studies. It ranges from 10 to 44 kg according to a recent paper based on 10-year data from the Swedish obese subjects (SOS) study (Sjöström et al. 2011). For example, a 20 kg weight loss yielded a significant decrease in plasma glucose from baseline, whereas the effect of aging is taken into account (Sjöström et al. 2011). The same is true for change in quality of life: Improvements and deteriorations in health-related quality of life were associated with the magnitude of weight loss or regain (Karlsson et al. 2007).

Table 26.1 Treatment goals should be based on the individual comorbidities and risk, i.e., obesity phenotype, rather than on degree of obesity alone

Parameters	Targets according to obesity phenotypes and history	
Weight	Weight loss	
	Obesity class 1 or 2 (BMI < 40 kg/m^2)	5–15 % weight loss
	Obesity class 3 (BMI ≥ 40 kg/m^2)	>30 % weight loss
	weight maintenance	Plateau: no weight fluctuations (yoyo)
Body composition	Excess body fat	Fat mass reduction
	Sarcopenia	Maintenance or accretion of muscle mass
AT distribution	Excess central adiposity	↓ waist circumference
	Excess VAT	↓ VAT
	Partial loss of appendicular AT	↑ limb SAT (?)
Mechanical complications. Clinical symptoms (pain, disability, obstructive sleep apnea syndrome)	Health-related quality of life	HRQOL improvement Large weight loss (larger is better !)
Metabolic syndrome and cardiovascular disease	Insulin resistance	5 % weight loss
	Dysglycemia,dyslipidemia and other metabolic abnormalities Hypertension	30 % VAT reduction ↓ trunk fat mass ↓ body fat mass
Adipocyte function	AT expendability	↑ AT expendability
	Insulin sensitivity	↓ pro-inflammatory cytokines
	Catecholamine sensitivity	↑ adiponectin
	Adipokine secretion	↓ local and systemic insulin resistance
		↓ leptin resistance

AT adipose tissue; *BMI* Body mass index, *HRQOL* Health-related quality of life; *SAT* subcutaneous AT, *VAT* visceral AT

BMI < 35 kg/m^2 Associated with Adipose Tissue Dysfunction: Need for Tailored Weight Loss Intervention

In patients with BMI 25–35 kg/m^2 a modest weight loss (5–10 % of initial weight) is an appropriate goal, as it is achievable and results in improvements in obesity related comorbidities and risks. Three complementary targets can be discussed, beyond weight loss and fat mass reduction.

FFA as Target for Therapy

A sustained reduction in FFA flux from AT to non-adipose tissue would be predicted to result in improvement in the adverse metabolic consequences of lipotoxicity, as discussed throughout this review. It could be crucial when AT buffering is impaired (Boden 2008).

The current risk of metabolic dysfunctions, thus, appears to be determined by the balance between the rate of loading of the body with FA and the rate of eliminating the FA by either TG storage or oxidation. Therefore, the challenges include correction of central adiposity and elevated plasma FFA levels through different approaches: reduction in total daily calorie intake (especially saturated FA intake) and/or an increase in energy expenditure, i.e., FFA oxidation. Paradoxically, agents such as PPARγ activators that overcome insulin resistance of AT by improving adipocyte FFA storage are postulated to effectively reduce the deleterious metabolic effects of AT dysregulation.

Preferential Visceral Adipose Tissue Loss

Visceral adiposity is clearly a significant risk factor for metabolic syndrome and probably CAD, so it is tempting to consider VAT reduction as a critical target. Decrease in VAT/SAT ratio is correlated with improvement in dysmetabolic profile (Després and Lemieux 2006; Arsenault et al. 2011). Some reports suggest that weight loss through diet and exercise decreases visceral fat quicker, and to a greater extent, than sc fat and that exercise promotes visceral fat loss, sometimes without changes in BMI (Bays et al. 2006; Chaston and Dixon 2008). VAT is lost preferentially, relative to abdominal SAT, with modest weight loss, but the effect is attenuated with greater weight loss, according to a recent review (Chaston and Dixon 2008). Very-low-calorie diets (VLCDs) provided early short-term (<4 weeks) preferential VAT loss, but this effect was lost by 12–14 weeks (Chaston and Dixon 2008). However, the MHO, MAO, or lipodystrophy phenotypes have not been taken into consideration in most of these studies.

Improvement in Adipose Tissue Metabolic Dysfunction

According to the AT expandability hypothesis, an obese individual becomes metabolically compromised, when its fat mass expansion reaches a critical limit (Unger et al. 2010; Virtue and Vidal-Puig 2008; Virtue and Vidal-Puig 2010). It indicates that the capacity of an individual to expand its fat mass to store lipid is a more important determinant of metabolic abnormalities than the absolute amount of fat mass. Metabolic syndrome, related to AT dysfunction and ectopic fat deposition appears when the AT became saturated, or reached the limit of its storage capacity.

So, the weight loss program has to be tailored to an individual's threshold of metabolic complications (Virtue and Vidal-Puig 2008; Virtue and Vidal-Puig 2010). However, other studies are needed to demonstrate the validity of this new concept of personalized weight loss program, which could result in better compliance and therapeutic efficacy, if the specific target is well understood and accepted by the patient.

Remodeling of dysfunctional AT into a healthier tissue seems to be another promising new approach to treat MAO; some drugs could have a direct effect on AT inflammation and cytokine secretion (see below). Today, the reduction in fat mass through hypocaloric diet or increased physical activity remains the best solution, mainly by decreasing adipocyte size and by improving AT dysfunction.

Treatment approaches that reduce energy balance by 400–700 kcal/day may produce modest reductions in body weight (5 to 8 %) but a substantial reduction in VAT (15 to 30 %) (Després and Lemieux 2006; Arsenault et al. 2011). This selective mobilization of abdominal/visceral fat has been shown to improve the metabolic profile of abdominally obese patients (Després and Lemieux 2006). As far as lipodystrophy or MONW phenotypes are concerned, a weight loss goal of 2–5 % of initial weight could be appropriate.

Sarcopenic Obesity

Weight loss is not necessary for the development of sarcopenia, but it can accelerate it. Prevention is needed especially for sarcopenic obese patients and for high risk patients (aging, sedentarity, insulin resistance).

It is of critical importance that management strategies focus on maintenance or accretion of muscle mass as well as fat loss, in order to maintain muscle function. Physical activity is absolutely necessary, each of aerobic and resistance training has specific advantages.

Reduction in fat-free mass has been reported to be correlated with change in BMI. Fat mass reduction represents 70–80 % of weight loss in bariatric surgery studies (Ciangura et al. 2009), but interindividual variations are large.

Metabolically Healthy Obese Individuals

The effect of weight loss was shown to be associated with few improvements in cardiometabolic risk factors in MHO individuals (Pataky et al. 2011). However, it should be noted that MHO individuals are not 'without risk' of morbi-mortality (Primeau et al. 2010). They may be exposed to other obesity-associated complications, such as osteoarthritis and obstructive sleep apnea. So, MHO is not a benign condition. In MHO individuals the treatment goals should be focused on health-related quality of life.

Antiobesity Treatment Strategies

The concept that weight loss is the sole indicator of treatment efficacy has been brought into question. Two important points regarding obesity management have been underlined in this review. The first question is "are we aiming at the right target" (Arsenault et al. 2011) ? In other words, we have to characterize the profile of overweight patients who are vulnerable or who are metabolically healthy. As already seen, treatments goals should be determined according to obesity phenotypes. There is a need for a new approach based on criteria to assess the excess of visceral/ectopic fat and for interventions affecting AT function. Further plasma or cellular biomarkers will be available in the next future (Rizkalla et al. 2011; Marquez-Quinones et al. 2010; Gogebakan et al. 2011).

The second question is related to treatment strategy, because different approaches are needed for the initial phase of weight loss and for the subsequent phase of weight maintenance. During the initial phase, the weight loss is more driven by the energy deficiency than by the composition of the diet in itself. However, when the weight losing phase is over, the dietary composition with regard to macronutrient may be crucial.

Dietary Approaches

Weight Loss Period

To induce weight loss or to decrease body fat, a temporary negative energy balance has to be created. The results depend only on energy deficit level and patient compliance. Substantially decreased adherence after a few months is typical in weight loss trials. Experience has shown that most patients are unable to continue losing weight for longer than 5–7 months, and weight loss reaches a plateau (mean weight loss: 6 kg with a large range: 4–12 kg). All diets are equally successful in promoting weight loss (Sacks et al. 2009). For example, both low carbohydrate and mediterranean style eating patterns have been shown to promote weight loss with similar results after 1–2 years (Giugliano and Esposito 2008; Nordmann et al. 2006).

Weight Maintenance Period

An optimal macronutrient distribution and dietary pattern of weight loss diet has not been definitely established. Diets can be tailored to individual patients on the basis of their personal and cultural preference, in order to have the best chance of long-term success (Sacks et al. 2009; Shai et al. 2008).

Mediterranean diets might enhance weight loss by providing a sustainable dietary pattern that offers a variety of healthy palatable foods. Moreover, its effect on long-term weight control (2 years) has been convincingly demonstrated. Energy-restricted high-protein diet with a low glycemic index and soluble fiber has been shown to be effective on weight maintenance (Larsen et al. 2010).

Phenotype Tailored Approach

Dietary Management of Metabolic Syndrome

Management of metabolic comorbidities may require specific dietary features. However, the optimal mix of macronutrients for meal plans of patients with metabolic syndrome or diabetes has not been identified. As shown by a meta-analysis, at least at 6 months, low-carbohydrate diets were associated with greater improvement in plasma TG and HDL-cholesterol concentrations than low-fat diets, but LDL-cholesterol level was significantly higher on the low-carbohydrate diets (Nordmann et al. 2006).

A recent systematic review and meta-analysis of 50 research studies (both observational prospective studies and randomised controlled trials) evaluated the effect of the mediterranean diet on the metabolic syndrome and its individual components (Kastorini et al. 2011). Adherence to this diet reduced significantly the risk of developing the metabolic syndrome by 31 % but the effect on each individual components is low (e.g.: −0.42 cm (95 % CI: −0.82 to −0.02) for waist circumference or +1.17 mg/dl (0.38–1.96) for HDL-cholesterol).

Interventions Affecting Adipocyte Tissue Dysfunction

Diet-induced weight loss can modify systemic markers of AT dysfunction, but results may differ according to the metabolic profile of the selected population and the intervention itself (Bays et al. 2006; Capel et al. 2009; Fisher et al. 2010; Klimcakova et al. 2010). Intentional weight loss improves insulin resistance or endothelial dysfunction, and reduces plasma levels of hs-CRP, IL-6, and the soluble TNF-α receptors, without an effect on plasma levels of TNF-α (Bays et al.

2006; Westerink and Visseren 2011). A dose–effect relationship between the degree of weight loss and the improvement in these plasma makers has been demonstrated (Klimcakova et al. 2010; Madsen et al. 2008). But at least 10 % weight loss seems necessary for long-term combined improvement of adiponectin, hs-CRP, and fibrinogen levels (Madsen et al. 2008). As reported by Poitou et al. (Poitou et al. 2011), a reduction of at least 5 % of fat mass was sufficient to observe a significant decrease of plasma $CD14^{dim}CD16^{+}$ monocytes, whose number is increased in obese individuals.

The mediterranean diet ensures adequate intake of micronutrients with antioxidants, and anti-inflammatory effects which could protect from diseases that are related to low-grade chronic inflammation and AT dysfunction (Giugliano and Esposito 2008). Low-glycemic-index diets have been associated with reduced levels of hs-CRP or plasma cytokines (Gogebakan et al. 2011).

Direct or indirect protective effects of n-3 FA against AT dysfunction and lipotoxicity have been suggested, the opposite could be true for saturated FA or *trans* FA (Virtue and Vidal-Puig 2010).

Reducing fructose intake may also have beneficial effects on adipose function and on VAT accumulation (Bays et al. 2006; Westerink and Visseren 2011). As demonstrated by Stanhope et al. (Stanhope et al. 2009) dietary fructose specifically increases visceral adiposity, promotes post-prandial lipemia and increases de novo lipogenesis, in overweight/obese adults, as compared with glucose (glucose- or fructose-sweetened beverages providing 25 % of energy requirements for 10 weeks).

Effect on Cellular Obesity Phenotype: Adipocyte Diameter and Adipose Tissue Remodeling

AT transcriptome changes during weight loss or weight maintenance periods have been analyzed in a few studies. As recently shown by Riskalla et al. (2011), an energy-restricted high-protein diet with a low glycemic index and soluble fiber (LC-P-LGI) would be more effective than a low-calorie conventional diet (LC-CONV) on weight loss and related metabolic risk factors. Moreover, it was associated with a greater reduction in adipocyte size (abdominal SAT biopsy). Diet-induced changes in gene expression were correlated with decreased adipocyte size, according to a coordinated program related to weight loss. No major differences were seen between the two dietary arms (LC-P-LGI diet versus LC-CONV diet). Profiles of genes involved in inhibiting adipogenesis, cell migration, adhesion, and angiogenesis but increasing apoptosis were correlated with decreased adipocyte size. The same was true for the decreased expression of a subset of genes implicated in inflammation.

An AT transcriptome study (Capel et al. 2009) reveals that molecular adaptations in AT and their relation to insulin sensitivity vary strikingly between different dietary periods. When energy restriction and weight stabilization were compared, an opposite pattern of regulation (up- or down-regulation) was observed between adipocyte metabolism genes and genes belonging to pathways chiefly operating in macrophages. During energy restriction (i.e. weight loss period), a salient feature was the down-regulation of genes involved in adipocyte metabolism, whereas during the weight stabilization phase (isocaloric diet) there was a down-regulation of macrophage gene expression (Capel et al. 2009).

In the Diogenes program (Marquez-Quinones et al. 2010), there were differences in the SAT transcriptome of subjects who continue to lose weight as compared with subjects who regained weight, independently of the diet-macronutrient composition. Mitochondrial oxidative phosphorylation was the major pattern associated with continued weight loss (Marquez-Quinones et al. 2010).

Whether the up- or down-regulation of the SAT transcriptome according to the type of macronutrient reflects also variations between obesity phenotypes remains to be elucidated.

In summary, dietary approach results in improvement in plasma and cellular markers related to obesity metabolic complications, suggesting that the metabolically unhealthy obese phenotype can be partly reversed, at least in selected patients.

Physical and Metabolic Fitness

Physical exercise has been shown to improve weight maintenance after weight reduction, but also glucose or lipid metabolism in the insulin-resistant state (Bays et al. 2006; Westerink and Visseren 2011). Reduced physical activity and low cardio-respiratory fitness may contribute to unhealthy obesity. The changes in metabolic parameters may be independent of exercise-related changes in body mass (Magkos 2010).

The beneficial effects of being ‘fat and fit’ may be mediated by lower visceral fat mass and lower liver fat content, i.e., by an improvement in adipocyte dysfunction and ectopic fat deposition. Both aerobic and resistance training may exert beneficial metabolic effects, so exercise might be a key treatment option for the reversal of unhealthy obesity. However, exercise training does not have a further independent effect on markers of inflammation (adipokines and hs-CRP) in patients who participate to a hypocaloric weight loss intervention (Fisher et al. 2010).

Pharmacological Treatment of Obesity

Long-term weight maintenance after weight reduction is the most important challenge for anti-obesity drugs (Ziegler 2011). According to the double target concept (drugs simultaneously act on weight control and on metabolic abnormalities), the weight loss therapeutic agents may even affect metabolic parameters and cardiometabolic risk independently of weight loss alone.

Orlistat or sibutramine administration decreases both visceral and subcutaneous fat, with possibly more reduction of visceral fat, but data are scarce (Bays et al. 2006; Ziegler 2011).

Cannabinoid-1 receptor (CB1R) blockers such as rimonabant have been demonstrated to reduce weight, abdominal adiposity (SAT and VAT), and the metabolic abnormalities associated with adipocyte dysfunction. The results of the RIO-Lipids study (Després et al. 2005) provide evidence for a weight-loss-independent effect of rimonabant on adiponectin levels (about 50 % of the increase in adiponectin levels could not be attributed to weight loss).

GLP1 analogs are new drugs that can be considered as useful for the treatment of both obesity and T2DM. Liraglutide has been shown to provide sustained weight loss over 2 years (Astrup et al. 2011).

Bariatric Surgery

Bariatric surgery results in sustainable long-term weight loss and improvement or remission of metabolic diseases associated with obesity, suggesting that the unhealthy obese phenotype can be reversed, even in severely obese patients. According to a recent expert consensus, a marked decrease in fat mass and normalization of AT function may explain the beneficial effects of bariatric surgery on obesity-related disorders (Bays et al. 2009).

Ciangura et al. (Ciangura et al. 2009) studied the dynamics of change in body composition in obese women using sequential DEXA measurements during the first year of follow-up after gastric bypass.

Two main results have to be underlined. Lean body mass loss plateaued after 6 months, but fat mass showed a continuous decrease over time. There was no evidence of a decrease in total and appendicular lean body mass as compared with non-surgical control subjects, matched for age and body fatness (Ciangura et al. 2009). Fat mass loss appeared more important on the trunk, as compared to the extremities (decreased trunk to appendicular fat mass ratio) (Ciangura et al. 2009).

Finally, surgery-induced weight reduction improves the systemic and AT inflammatory states associated with obesity, suggesting a beneficial effect on AT remodeling (Cancello et al. 2005; Clément 2011). In particular, weight loss was associated with decreased number of SAT macrophages and with a switch toward a less pro-inflammatory state (Cancello et al. 2005).

Other Interventions Affecting Adipose Tissue Function

As already underlined, increasing physical activity and diet-induced weight reduction are two important lifestyle change measures to reduce insulin resistance and visceral adiposity, mainly by decreasing the size of existing adipocytes. These two interventions are central and effective in modifying AT dysfunction. However, other more or less specific interventions affecting AT function could be useful, whereas weight reduction resulting from lifestyle intervention is often modest. Three approaches for remodeling dysfunctional AT into a healthier tissue are briefly discussed below.

Adipose Tissue Expandability and Lipotoxicity

Paradoxically with regard to obese individuals, a capacity to recruit, proliferate, and differentiate adipocytes may allow AT to maintain normal metabolic functions during times of positive caloric balance.

Thiazolidinediones

Thiazolidinediones (TZD or glitazones) have been developed to improve insulin resistance. These agents are PPAR-γ agonists that are able to modulate fat storage through multiple mechanisms (Heilbronn et al. 2004; Westerink and Visseren 2011): (1) they promote adipogenesis and SAT development, so fat storage may be increased in a physiological depot; (2) in some studies, TZD have been shown to decrease the VAT/SAT ratio, with either a decrease or no change in visceral fat; (3) they may decrease ectopic fat deposition in liver and muscle; (4) they improve adipocyte endocrine functions, decreasing production of inflammatory adipokines and increasing production of adiponectin and leptin.

As shown by Gastaldelli et al. (Gastaldelli et al. 2009), pioglitazone is a good candidate for the treatment of NASH, at least in individuals with prediabetes or diabetes. Patients with NASH have severe AT insulin resistance (Adipo-IR) independently of the degree of obesity. Pioglitazone treatment (45 mg/day or placebo for 6 months) resulted in a 47 % Adipo-IR decrease that was significantly correlated with the reduction of hepatic fat and with the reduction in hepatic necroinflammation (Gastaldelli et al. 2009).

TZDs have been used tot treat either genetic or acquired partial lipodystrophies. The benefits on AT distribution were often small, but a good efficacy on insulin resistance, blood glucose control, and dyslipidemia has been reported. In our experience, therapy with pioglitazone resulted in marked and sustained improvement in metabolic control, as far as diabetic patients with lipodystrophic adiposity phenotype are concerned. However, to the best of our knowledge, no controlled intervention study to verify this interesting effect has been published.

Remodeling of Adipose Tissue by Virus

Human adenovirus type 36 increases adiposity, but improves metabolic profile and insulin sensitivity in experimentally infected animals (Basdevant and Clément 2011; Rogers et al. 2008). This virus is able to facilitate (1) the differentiation of preadipocytes and adipogenesis; (2) lipogenesis; (3) metabolically favorable remodeling of AT (reduction of inflammatory cytokines production). The ability of adenovirus to induce these changes has been successfully investigated in human primary AT explants (Rogers et al. 2008). The potential of viral proteins, as a therapeutic target for treating insulin resistance by expanding AT, remains to be established.

Lipid Partitioning and Oxidation of Excess Lipids

Deficiency of Leptin and Leptin Resistance

According to the Unger hypothesis (Unger 2003; Unger et al. 2010), leptin resistance is a major cause of lipotoxicity. The anti-lipotoxic effect of leptin may be mediated by an increase in lipid oxidation in non-adipose tissues. Relative hypoleptinemia has been reported in visceral obesity (Unger 2003), but the majority of obese individuals exhibit leptin resistance. It is tempting to use a pharmacological approach to increase leptin sensitivity. Endoplasmic reticulum stress could be resolved by the administration of chemical chaperones including 4-phenyl butyric acid and tauroursodeoxycholic acid that cause hypophagia and weight loss in mice (Ziegler 2011).

Further, amylin works synergistically with leptin to decrease body weight in rodents, an effect which was attributed to amylin receptor stimulation reversing leptin resistance (Ziegler 2011). Results of a 20 weeks proof-of-concept study have been encouraging, patients treated with the combination of pramlintide, a synthetic amylin analog, and a recombinant human leptin (metreleptin) experienced a mean weight loss of 12 % (Ziegler 2011).

Early clinical trials of sc leptin therapy (recombinant leptin: R-metHuleptin and metreleptin) for the treatment of obesity failed, except for leptin-deficient states, i.e., morbidly obese patients with hypoleptinemia due to a very rare leptin gene mutation or polymorphism. Leptin replacement therapy has also been shown very effective in patients with familial generalized or partial lipodystrophy (Unger 2003; Unger et al. 2010). As recently reported by Simha et al. (Simha et al. 2011), metreleptin replacement therapy is equally effective in familial partial lipodystrophy patients of the Dunnigan variety (24 women treated for 6 months) with both severe and moderate hypoleptinemia in reducing body weight, fat mass, serum and hepatic TG levels.

Adiponectin

The anti-lipotoxic action of TZD and rimonabant may be mediated by an increase in adiponectin production, an insulin-sensitizing hormone with multiples metabolic and anti-atherosclerotic actions. The AT expandability hypothesis has been supported in a transgenic rodent model. A mouse overexpressing adiponectin in AT on an obese *ob/ob* background becomes very obese without metabolic complications and with no ectopic fat deposition in liver (Virtue and Vidal-Puig 2010).

Sirtuin Activators

Sirtuin 1 (SIRT1) is NAD-dependent deacetylase that is up-regulated by caloric restriction in rodents (Guarente 2006). SIRT1 enhance fat mobilization and FA mitochondrial oxidation by interacting with PPAR γ and PGC1 α. A SIRT 1 activator, resveratrol, has proven to be effective in rodent models of diet-induced obesity). Specific SIRT1 activators have been investigated in phase I or II trials for the treatment of T2DM and obesity (Ziegler 2011).

Other Approaches to Increase Energy Expenditure

Four new potential therapeutic approaches could be envisioned, according to Kahn et al. (Tseng et al. 2010): (1) increasing brown fat differentiation from progenitor cells of skeletal muscle and white AT, using cytokines or growth factors; (2) activating brown fat thermogenesis, using $\beta 3$ adrenergic receptor agonists; recently a new pathway to promote energy expenditure was observed in response to bile acids that bind to a novel G-protein-coupled receptor TGR5; a TGR5 agonist, INT-777, has already shown efficacy, reducing adiposity in mice with diet-induced obesity; (3) promoting skeletal muscle thermogenesis by different pathways including SIRT1 and AMP-activated protein kinase (AMPK), an energy sensor widely considered to be involved in the pharmacologic action of metformin (see below); (4) increasing general mitochondrial uncoupling that leads to energy inefficiency; new molecules may be attractive, if more selective and less toxic than dinitrophenol.

Effects on Adipose Tissue Function and Remodeling

Direct Actions of Pharmacologic Agents

Several classes of drugs have unintended influences on AT function, but the clinical relevance of these actions remains to be determined (Bays et al. 2006; Westerink and Visseren 2011). As already reported, TZD have pleiotropic effects (e.g.: AT inflammation, insulin resistance). Metformin is thought to increase FA oxidation by stimulating AMPK. Another possible role of metformin on AMPK-dependent lipolysis in adipocytes has been suggested, that could decrease FFA plasma levels. Moreover metformin may inhibit PAI-1 production by human

SAT. So, direct metformin effects on AT may have beneficial effects, beyond its modest effect through weight reduction (2 kg weight loss).

Statins are capable of reducing systemic inflammation (Westerink and Visseren 2011). Due to differences in lipophilicity, statins may have different effects on adiponectin production. By a PPARα agonistic mechanism, fibrates increase the uptake and oxidation of FA in the liver, muscle, and heart. They may also modulate the production of some cytokines, such as adiponectin and TNFα.

Salicylates have direct anti-inflammatory effects but some data suggests a possible role of PPAR-γ agonism. High dose acetylsalicylic acid may improve insulin resistance and increase adiponectin (Westerink and Visseren 2011).

Aldosterone antagonists that inhibit AT mineralcorticosteroid receptor activation may improve AT dysfunction (Westerink and Visseren 2011). In particular, a decrease in the number of hypertrophic adipocytes and infiltrating macrophages has been reported. Angiotensin Converting Enzyme Inhibitors (ACE-i) are known to slightly improve insulin resistance. Angiotensin II could have a direct pro-inflammatory effect on AT. Angiotensin II type 1 Receptor Blockers (ARB) exert PPARγ partial agonistic activity (Westerink and Visseren 2011). Some ARB have been shown to increase adiponectin plasma levels and to decrease VAT volume.

Adipose Tissue Fibrosis

Excessive deposition of collagen in AT is involved in the pathophysiology of AT dysregulation. AT fibrosis inhibition may improve insulin resistance (Basdevant and Clément 2011; Clément 2011). Whether a therapeutic approach is able to control the fibrosis amount and to modify AT remodeling remains to be established. It is tempting to speculate that changes in the fibroinflammation status may contribute to the metabolic amelioration after diet-induced weight loss.

Peripheral Cannabinoid-1 Receptor Antagonist

Targeting peripheral CB1R has therapeutic potential for improving cardiometabolic abnormalities in obese patients. Treatment of obese mice with a peripherally restricted CB1R antagonist improves glycemic control and dyslipidemia and reverses hepatic steatosis (Tam et al. 2010; Patti 2010). Selective targeting of peripheral CB1R may result in an improved hormonal-metabolic profile in obesity without the untoward neuropsychiatric effects (depression and suicidal tendencies) observed following treatment with brain-penetrant CB1R antagonists (rimonabant or taranabant). It could be a novel exciting therapeutic option for NASH (Ziegler 2011).

Discussion

Benefit–Risk Ratio Analysis According to Obesity Phenotype

As already underlined, a better understanding of obesity phenotypes is needed to educate healthcare professionals, physicians, and even researchers. Most intervention studies do not take into account the heterogeneous clinical presentation of this complex disease. For example, the identification of the MHO individual in a clinical setting could have important implications for therapeutic medical decision making, in order to avoid useless and not efficient weight loss treatments.

So, the AT expansion limit hypothesis, leads to the concept of a "set point" for capacity of AT to expand beyond which an overweight individual becomes metabolically unhealthy. A weight management program has to be personalized to an individual's threshold of metabolic complications. Early tailored approach can prevent the ectopic fat accumulation that causes insulin resistance (Virtue and Vidal-Puig 2008).

The Natural History of Unhealthy Obesity

Obesity is a chronic progressive disease that evolves over time (dynamic weight gain phase, followed by a static phase when obesity reaches a plateau, and a weight loss resistance phase) (Clément 2011). It is obvious from a clinical point-of-view that obesity subphenotypes change according to the natural history of the disease, but also with aging (lipodystrophy and sarcopenia). AT dysfunction appears gradually, so that metabolic syndrome most often appears in late middle age, indicating that many years of relatively good health can precede the onset of comorbidities. An obese individual may be considered as MHO during many years, before to become MAO (Unger et al. 2010). No prospective studies have demonstrated that MHO subjects can maintain insulin sensitivity during the entire life (Blüher 2009).

That is why the duration of obesity is a crucial parameter to be considered, when explaining the development of AT dysfunction (Blüher 2009). A healthy obese phenotype may reverse to obesity-related insulin resistance, especially when energy balance remains positive for a long time. About 20 years are needed to develop diabetes or coronary heart disease but genetic and environmental factors and early (epigenetic) programing are also crucial contributors.

The resistance to weight loss could be related to the accumulation of fibrosis that may hamper fat mass loss induced by caloric restriction (Clément 2011). The propensity to regain weight after diet-induced weight loss is well known (Yoyo syndrome), so that dietary treatments have limited efficacy at this period of natural history. Bariatric surgery or a pharmacologic treatment (to be defined) could be a useful option for patients with biological alterations of AT, to be discussed according to the benefit–risk ratio. However, the pathophysiological and clinical relevance of fibrosis in the AT is not clear: early event for some patients or a late complication of AT dysfunction? This question is still open.

Contribution of Other Organs to the Pathogenesis of Metabolic Alterations

AT is not the only therapeutic target. The dialog between AT and other central (brain) or peripheral organs (liver, gut, muscle…) provided a conceptual framework to analyze the pathogenesis of insulin resistance and other metabolic complications of obesity including systemic inflammation. Two examples are presented here.

Gut

The gut is considered as a central player in development of systemic inflammation and the metabolic syndrome (Clément 2011; Lam et al. 2011; Tilg and Kaser 2011). Hyperphagia, pro-inflammatory dietary components (e.g., saturated fat), and genetic predisposition may combine to alter the gut microbiota.

Impairment or leakiness of the gut barrier and local inflammation will result in mesenteric AT dysfunction (adipocyte hypertrophy, macrophage infiltration). The adipose-derived cytokines and increased FFA flux from the expanded fat mass will impact on the liver via direct portal access, resulting in liver steatosis, insulin resistance and inflammation. Increased intestinal permeability causing elevated systemic levels of lipopolysaccharides (LPS), and intestinal microbiota as supported by many works, might have a profound role in the development of metabolic syndrome and systemic inflammation (Lam et al. 2011). The relevance of the probiotic approaches to modulate gut microbiota in the management of obesity and metabolic syndrome in humans needs further studies.

Liver

Increased fat accumulation in the liver is a marker of hepatic insulin resistance that predicts the metabolic syndrome. It is probably an early event in the development of systemic insulin resistance that promotes the development of AT dysfunction. In adipocytes, insulin resistance from any cause is associated with FFA elevation, due to decreased anti-lipolytic effects of insulin. This results in an increased flux of FFA to the liver, together with pro-inflammatory factors from the portal circulation. Liver steatosis is probably an early marker of ectopic fat deposition, at least in some individuals. Early diagnosis and management is necessary to delay the progression to NASH and fibrosis but also other complications of the metabolic syndrome (T2DM and CVD). The same strategies are needed to approach NASH and AT dysfunction by behavioral programs to reduce excess nutrition and increase exercise.

Conclusion

AT is a master regulatory tissue in controlling whole-body lipid flux, thereby modulating both glucose and lipid homeostasis. A lot of data support the hypothesis that AT dysfunction, especially adipocyte hypertrophy, inflammation and decreased adiponectin secretion, is an important contributor to insulin resistance in obesity. This could be the primary defects in obesity-related insulin resistance.

It is important to distinguish early the ‘unhealthy’ from the ‘healthy’ obese phenotype, if possible in childhood, using pertinent biomarkers. Anti-obesity treatment strategies even with moderate reduction of fat mass, may reverse unhealthy into healthy obesity. These interventions include calorie-restricted diets, adapted to the weight loss or the weight maintenance phase, increased physical activity (especially during the maintenance phase) as well as anti-obesity drugs or bariatric surgery, if necessary. Indications have to be discussed according to the obese phenotype and the severity of adiposity excess.

References

Anand SS, Tarnopolsky MA, Rashid S et al (2011) Adipocyte hypertrophy, fatty liver and metabolic risk factors in South Asians: the Molecular Study of Health and Risk in Ethnic Groups (mol-SHARE). PLoS One 6:e22112

Arsenault BJ, Beaumont EP, Després JP, Larose E (2011) Mapping body fat distribution: A key step towards the identification of the vulnerable patient? Ann Med. doi: 10.3109/07853890.2011.605387

Astrup A, Carraro R, Finer N, et al (2011) Safety, tolerability and sustained weight loss over 2 years with the once-daily human GLP-1 analog, liraglutide. Int J Obes (Lond) Aug 16. doi: 10.1038/ijo.2011.158

Barker DJ, Osmond C, Forsen TJ, Kajantie E, Eriksson JG (2005) Trajectories of growth among children who have coronary events as adults. N Engl J Med 353:1802–1809

Basdevant A, Clément K (2011) Histoire naturelle et origine des obésités. In: Basdevant A, Bouillot JL, Clément K, Oppert JM, Tounian P (eds) Traité de Médecine et chirurgie de l’obésité Lavoisier. Médecine Sciences publications, Paris, pp 10–20

Bays H, Blonde L, Rosenson R (2006) Adiposopathy: how do diet, exercise and weight loss drug therapies improve metabolic disease in overweight patients? Expert Rev Cardiovasc Ther 4:871–895

Bays HE, Laferrere B, Dixon J et al (2009) Adiposopathy and bariatric surgery: is ‘sick fat’ a surgical disease? Int J Clin Pract 63:1285–1300

Bhargava SK, Sachdev HS, Fall CH et al (2004) Relation of serial changes in childhood body-mass index to impaired glucose tolerance in young adulthood. N Engl J Med 350:865–875

Blüher M (2009) The distinction of metabolically ‘healthy’ from ‘unhealthy’ obese individuals. Curr Opin Lipidol 21:38–43

Boden G (2008) Obesity and free fatty acids. Endocrinol Metab Clin North Am 37: 635–646, viii-ix
Boney CM, Verma A, Tucker R, Vohr BR (2005) Metabolic syndrome in childhood: association with birth weight, maternal obesity, and gestational diabetes mellitus. Pediatrics 115:e290–e296
Bouhours-Nouet N, Dufresne S, de Casson FB et al (2008) High birth weight and early postnatal weight gain protect obese children and adolescents from truncal adiposity and insulin resistance: metabolically healthy but obese subjects? Diabetes Care 31:1031–1036
Cancello R, Henegar C, Viguerie N et al (2005) Reduction of macrophage infiltration and chemoattractant gene expression changes in white adipose tissue of morbidly obese subjects after surgery-induced weight loss. Diabetes 54:2277–2286
Capel F, Klimcakova E, Viguerie N et al (2009) Macrophages and adipocytes in human obesity: adipose tissue gene expression and insulin sensitivity during calorie restriction and weight stabilization. Diabetes 58:1558–1567
Carobbio S, Rodriguez-Cuenca S, Vidal-Puig A (2011) Origins of metabolic complications in obesity: ectopic fat accumulation. The importance of the qualitative aspect of lipotoxicity. Curr Opin Clin Nutr Metab Care 14:520–526
Chaston TB, Dixon JB (2008) Factors associated with percent change in visceral versus subcutaneous abdominal fat during weight loss: findings from a systematic review. Int J Obes (Lond) 32:619–628
Ciangura C, Bouillot JL, Lloret-Linares C et al (2009) Dynamics of change in total and regional body composition after gastric bypass in obese patients. Obesity (Silver Spring) 18:760–765
Clément K (2011) Bariatric surgery, adipose tissue and gut microbiota. Int J Obes (Lond) 35(Suppl 3):S7–S15
Danforth E Jr (2000) Failure of adipocyte differentiation causes type II diabetes mellitus? Nat Genet 26:13
Després JP, Lemieux I (2006) Abdominal obesity and metabolic syndrome. Nature 444:881–887
Després JP, Golay A, Sjöström L (2005) Effects of rimonabant on metabolic risk factors in overweight patients with dyslipidemia. N Engl J Med 353:2121–2134
Divoux A, Tordjman J, Lacasa D et al (2010) Fibrosis in human adipose tissue: composition, distribution, and link with lipid metabolism and fat mass loss. Diabetes 59:2817–2825
Dulloo AG, Jacquet J, Solinas G, Montani JP, Schutz Y (2010) Body composition phenotypes in pathways to obesity and the metabolic syndrome. Int J Obes (Lond) 34(Suppl 2):S4–S17
Dutour A, Roll P, Gaborit B et al (2011) High prevalence of laminopathies among patients with metabolic syndrome. Hum Mol Genet 20:3779–3786
Fisher G, Hyatt TC, Hunter GR, Oster RA, Desmond RA, Gower BA (2010) Effect of diet with and without exercise training on markers of inflammation and fat distribution in overweight women. Obesity (Silver Spring) 19:1131–1136
Frayn KN (2002) Adipose tissue as a buffer for daily lipid flux. Diabetologia 45:1201–1210
Garg A (2011) Clinical review: Lipodystrophies: genetic and acquired body fat disorders. J Clin Endocrinol Metab 96:3313–3325
Gastaldelli A, Harrison SA, Belfort-Aguilar R et al (2009) Importance of changes in adipose tissue insulin resistance to histological response during thiazolidinedione treatment of patients with nonalcoholic steatohepatitis. Hepatology 50:1087–1093
Gavi S, Feiner JJ, Melendez MM, Mynarcik DC, Gelato MC, McNurlan MA (2007) Limb fat to trunk fat ratio in elderly persons is a strong determinant of insulin resistance and adiponectin levels. J Gerontol A Biol Sci Med Sci 62:997–1001
Giugliano D, Esposito K (2008) Mediterranean diet and metabolic diseases. Curr Opin Lipidol 19:63–68
Gogebakan O, Kohl A, Osterhoff MA et al (2011) Effects of weight loss and long-term weight maintenance with diets varying in protein and glycemic index on cardiovascular risk factors: the diet, obesity, and genes (DiOGenes) study: a randomized, controlled trial. Circulation 124:2829–2838
Guarente L (2006) Sirtuins as potential targets for metabolic syndrome. Nature 444:868–874
Guilherme A, Virbasius JV, Puri V, Czech MP (2008) Adipocyte dysfunctions linking obesity to insulin resistance and type 2 diabetes. Nat Rev Mol Cell Biol 9:367–377

Guri AJ, Bassaganya-Riera J (2010) Systemic effects of white adipose tissue dysregulation and obesity-related inflammation. Obesity (Silver Spring) 19:689–700

Hajer GR, van Haeften TW, Visseren FL (2008) Adipose tissue dysfunction in obesity, diabetes, and vascular diseases. Eur Heart J 29:2959–2971

Heilbronn L, Smith SR, Ravussin E (2004) Failure of fat cell proliferation, mitochondrial function and fat oxidation results in ectopic fat storage, insulin resistance and type II diabetes mellitus. Int J Obes Relat Metab Disord 28(Suppl 4):S12–S21

Herbst KL, Tannock LR, Deeb SS, Purnell JQ, Brunzell JD, Chait A (2003) Kobberling type of familial partial lipodystrophy: an underrecognized syndrome. Diabetes Care 26:1819–1824

Jensen MD (2008) Role of body fat distribution and the metabolic complications of obesity. J Clin Endocrinol Metab 93:S57–S63

Karlsson J, Taft C, Ryden A, Sjöström L, Sullivan M (2007) Ten-year trends in health-related quality of life after surgical and conventional treatment for severe obesity: the SOS intervention study. Int J Obes (Lond) 31:1248–1261

Kastorini CM, Milionis HJ, Esposito K, Giugliano D, Goudevenos JA, Panagiotakos DB (2011) The effect of Mediterranean diet on metabolic syndrome and its components: a meta-analysis of 50 studies and 534,906 individuals. J Am Coll Cardiol 57:1299–1313

Klimcakova E, Kovacikova M, Stich V, Langin D (2010) Adipokines and dietary interventions in human obesity. Obes Rev 11:446–456

Klöting N, Fasshauer M, Dietrich A et al (2010) Insulin-sensitive obesity. Am J Physiol Endocrinol Metab 299:E506–E515

Kursawe R, Eszlinger M, Narayan D et al (2010) Cellularity and adipogenic profile of the abdominal subcutaneous adipose tissue from obese adolescents: association with insulin resistance and hepatic steatosis. Diabetes 59:2288–2296

Lam YY, Mitchell AJ, Holmes AJ et al (2011) Role of the gut in visceral fat inflammation and metabolic disorders. Obesity (Silver Spring) 19:2113–2120

Larsen TM, Dalskov SM, van Baak M et al (2010) Diets with high or low protein content and glycemic index for weight-loss maintenance. N Engl J Med 363:2102–2113

Lewis GF, Carpentier A, Adeli K, Giacca A (2002) Disordered fat storage and mobilization in the pathogenesis of insulin resistance and type 2 diabetes. Endocr Rev 23:201–229

Liska D, Dufour S, Zern TL et al (2007) Interethnic differences in muscle, liver and abdominal fat partitioning in obese adolescents. PLoS One 2:e569

Madsen EL, Rissanen A, Bruun JM et al (2008) Weight loss larger than 10 % is needed for general improvement of levels of circulating adiponectin and markers of inflammation in obese subjects: a 3-year weight loss study. Eur J Endocrinol 158:179–187

Magkos F (2010) Exercise and fat accumulation in the human liver. Curr Opin Lipidol 21: 507–517

Marquez-Quinones A, Mutch DM, Debard C et al (2010) Adipose tissue transcriptome reflects variations between subjects with continued weight loss and subjects regaining weight 6 mo after caloric restriction independent of energy intake. Am J Clin Nutr 92:975–984

McGarry JD (2002) Banting lecture 2001: dysregulation of fatty acid metabolism in the etiology of type 2 diabetes. Diabetes 51:7–18

Mittendorfer B (2011) Origins of metabolic complications in obesity: adipose tissue and free fatty acid trafficking. Curr Opin Clin Nutr Metab Care 14:535–541

Nguyen QM, Srinivasan SR, Xu JH, Chen W, Berenson GS (2008) Changes in risk variables of metabolic syndrome since childhood in pre-diabetic and type 2 diabetic subjects: the Bogalusa Heart Study. Diabetes Care 31:2044–2049

Nielsen S, Guo Z, Johnson CM, Hensrud DD, Jensen MD (2004) Splanchnic lipolysis in human obesity. J Clin Invest 113:1582–1588

Nordmann AJ, Nordmann A, Briel M et al (2006) Effects of low-carbohydrate vs low-fat diets on weight loss and cardiovascular risk factors: a meta-analysis of randomized controlled trials. Arch Intern Med 166:285–293

Pataky Z, Makoundou V, Nilsson P et al (2011) Metabolic normality in overweight and obese subjects. Which parameters? Which risks? Int J Obes (Lond) 35:1208–1215

Patti ME (2010) Rehashing endocannabinoid antagonists: can we selectively target the periphery to safely treat obesity and type 2 diabetes? J Clin Invest 120:2646–2648

Poitou C, Dalmas E, Renovato M et al (2011) CD14dimCD16 + and CD14 + CD16 + monocytes in obesity and during weight loss: relationships with fat mass and subclinical atherosclerosis. Arterioscler Thromb Vasc Biol 31:2322–2330

Primeau V, Coderre L, Karelis AD et al (2010) Characterizing the profile of obese patients who are metabolically healthy. Int J Obes (Lond) 35:971–981

Ravussin E, Smith SR (2002) Increased fat intake, impaired fat oxidation, and failure of fat cell proliferation result in ectopic fat storage, insulin resistance, and type 2 diabetes mellitus. Ann N Y Acad Sci 967:363–378

Rizkalla SW, Prifti E, Cotillard A et al (2011) Differential effects of macronutrient content in 2 energy-restricted diets on cardiovascular risk factors and adipose tissue cell size in moderately obese individuals: a randomized controlled trial. Am J Clin Nutr 95:49–63

Rogers PM, Mashtalir N, Rathod MA et al (2008) Metabolically favorable remodeling of human adipose tissue by human adenovirus type 36. Diabetes 57:2321–2331

Romero-Corral A, Somers VK, Sierra-Johnson J et al (2009) Normal weight obesity: a risk factor for cardiometabolic dysregulation and cardiovascular mortality. Eur Heart J 31:737–746

Ruderman N, Chisholm D, Pi-Sunyer X, Schneider S (1998) The metabolically obese, normal-weight individual revisited. Diabetes 47:699–713

Sacks FM, Bray GA, Carey VJ et al (2009) Comparison of weight-loss diets with different compositions of fat, protein, and carbohydrates. N Engl J Med 360:859–873

Shai I, Schwarzfuchs D, Henkin Y et al (2008) Weight loss with a low-carbohydrate, Mediterranean, or low-fat diet. N Engl J Med 359:229–241

Shay CM, Carnethon MR, Church TR et al (2011) Lower extremity fat mass is associated with insulin resistance in overweight and obese individuals: the CARDIA study. Obesity (Silver Spring) 19:2248–2253

Shea JL, Randell EW, Sun G (2010) The prevalence of metabolically healthy obese subjects defined by BMI and dual-energy X-ray absorptiometry. Obesity (Silver Spring) 19:624–630

Sherar LB, Eisenmann JC, Chilibeck PD et al (2011) Relationship between trajectories of trunk fat mass development in adolescence and cardiometabolic risk in young adulthood. Obesity (Silver Spring) 19:1699–1706

Simha V, Subramanyam L, Szczepaniak L, et al. (2011) Comparison of efficacy and safety of Leptin replacement therapy in moderately and severely hypoleptinemic patients with familial partial Lipodystrophy of the Dunnigan Variety. J Clin Endocrinol Metab Dec 14 [Epub ahead of print]

Sjöström CD, Lystig T, Lindroos AK (2011) Impact of weight change, secular trends and ageing on cardiovascular risk factors: 10-year experiences from the SOS study. Int J Obes (Lond) 35:1413–1420

Srinivasan SR, Myers L, Berenson GS (2002) Predictability of childhood adiposity and insulin for developing insulin resistance syndrome (syndrome X) in young adulthood: the Bogalusa Heart Study. Diabetes 51:204–209

Srinivasan SR, Wang R, Chen W, Wei CY, Xu J, Berenson GS (2009) Utility of waist-to-height ratio in detecting central obesity and related adverse cardiovascular risk profile among normal weight younger adults (from the Bogalusa Heart Study). Am J Cardiol 104:721–724

Stanhope KL, Schwarz JM, Keim NL et al (2009) Consuming fructose-sweetened, not glucose-sweetened, beverages increases visceral adiposity and lipids and decreases insulin sensitivity in overweight/obese humans. J Clin Invest 119:1322–1334

Stenholm S, Harris TB, Rantanen T, Visser M, Kritchevsky SB, Ferrucci L (2008) Sarcopenic obesity: definition, cause and consequences. Curr Opin Clin Nutr Metab Care 11:693–700

Sun K, Kusminski CM, Scherer PE (2011) Adipose tissue remodeling and obesity. J Clin Invest 121:2094–2101

Szczepaniak LS, Victor RG, Orci L, Unger RH (2007) Forgotten but not gone: the rediscovery of fatty heart, the most common unrecognized disease in America. Circ Res 101:759–767

Taksali SE, Caprio S, Dziura J et al (2008) High visceral and low abdominal subcutaneous fat stores in the obese adolescent: a determinant of an adverse metabolic phenotype. Diabetes 57:367–371
Tam J, Vemuri VK, Liu J et al (2010) Peripheral CB1 cannabinoid receptor blockade improves cardiometabolic risk in mouse models of obesity. J Clin Invest 120:2953–2966
Thomas EL, Parkinson JR, Frost GS, et al (2012) The missing risk: MRI and MRS phenotyping of abdominal adiposity and ectopic fat. Obesity (Silver Spring) 20:76–87
Tilg H, Kaser A (2011) Gut microbiome, obesity, and metabolic dysfunction. J Clin Invest 121:2126–2132
Tseng YH, Cypess AM, Kahn CR (2010) Cellular bioenergetics as a target for obesity therapy. Nat Rev Drug Discov 9:465–482
Unger RH (2003) Lipid overload and overflow: metabolic trauma and the metabolic syndrome. Trends Endocrinol Metab 14:398–403
Unger RH, Clark GO, Scherer PE, Orci L (2010) Lipid homeostasis, lipotoxicity and the metabolic syndrome. Biochim Biophys Acta 1801:209–214
Villarroya F, Domingo P, Giralt M (2009) Drug-induced lipotoxicity: lipodystrophy associated with HIV-1 infection and antiretroviral treatment. Biochim Biophys Acta 1801:392–399
Virtue S, Vidal-Puig A (2008) It's not how fat you are, it's what you do with it that counts. PLoS Biol 6:e237
Virtue S, Vidal-Puig A (2010) Adipose tissue expandability, lipotoxicity and the Metabolic Syndrome: an allostatic perspective. Biochim Biophys Acta 1801:338–349
Weiss R, Taksali SE, Dufour S et al (2005) The "obese insulin-sensitive" adolescent: importance of adiponectin and lipid partitioning. J Clin Endocrinol Metab 90:3731–3737
Westerink J, Visseren FL (2011) Pharmacological and non-pharmacological interventions to influence adipose tissue function. Cardiovasc Diabetol 10:13
Wildman RP, Muntner P, Reynolds K et al (2008) The obese without cardiometabolic risk factor clustering and the normal weight with cardiometabolic risk factor clustering: prevalence and correlates of 2 phenotypes among the US population (NHANES 1999–2004). Arch Intern Med 168:1617–1624
Youssef AA, Valdez R, Elkasabany A, Srinivasan SR, Berenson GS (2002) Time-course of adiposity and fasting insulin from childhood to young adulthood in offspring of parents with coronary artery disease: the Bogalusa Heart Study. Ann Epidemiol 12:553–559
Zamboni M, Mazzali G, Fantin F, Rossi A, Di Francesco V (2008) Sarcopenic obesity: a new category of obesity in the elderly. Nutr Metab Cardiovasc Dis 18:388–395
Ziegler O (2011) Traitements médicamenteux de l'obésité. In: Basdevant A, Bouillot JL, Clément K, Oppert JM, Tounian P (eds) Traité de Médecine et chirurgie de l'obésité Lavoisier. Médecine Sciences publications, Paris, pp 450–461
Ziegler O, Quilliot D (2008) Obésité de la personne âgée : épidémiologie et conséquences. In Hébuterne X, Alix E, Raynaud-Simon A, Vellas B (eds) Traité de nutrition de la personne âgée. Springer, Paris pp 111–121

Index

J.-P. Bastard and B. Fève (eds.), *Physiology and Physiopathology of Adipose Tissue*,
DOI: 10.1007/978-2-8178-0343-2, © Springer-Verlag France 2013